Study Guide

Charles M. Seiger

Damian Hill

Fundamentals of
Anatomy &
Physiology

NINTH EDITION

Frederic H. Martini
Judi L. Nath
Edwin F. Bartholomew

Benjamin Cummings

Boston Columbus Indianapolis New York San Francisco Upper Saddle River
Amsterdam Cape Town Dubai London Madrid Milan Munich Paris Montréal Toronto
Delhi Mexico City São Paulo Sydney Hong Kong Seoul Singapore Taipei Tokyo

Executive Editor: Leslie Berriman
Project Editor: Robin Pille
Editorial Assistant: Nicole McFadden
Managing Editor: Deborah Cogan
Production Project Manager: Dorothy Cox
Senior Manufacturing Buyer: Stacey Weinberger
Project Management and Composition: S4Carlisle Publishing Services
Cover Photograph: Mike Powell/Getty Images
Main Text Cover Designer: tani hasegawa
Marketing Manager: Derek Perrigo
Supplement Cover Design: 17th Street

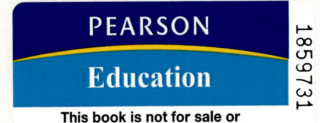

PEARSON

Education

1859731

This book is not for sale or distribution in the U.S.A. or Canada

Benjamin Cummings
is an imprint of

www.pearsonhighered.com

ISBN-10: 0-321-74167-6
ISBN-13: 978-0-321-74167-7
1 2 3 4 5 6 7 8 9 10—SCI—15 14 13 12 11
www.pearsonhighered.com

Contents

Preface

This Study Guide has been written to accompany Martini/Nath/Bartholomew *Fundamentals of Anatomy & Physiology*, Ninth Edition, and serve as a resource to help students master and reinforce difficult concepts.

In the Ninth Edition of this Study Guide, the questions and exercises have been reviewed, refined, and, when needed, rewritten in an effort to correspond closely with the concepts and organization in the new edition of the text. The three-level learning system in the Study Guide is aligned with the three levels that students encounter in the text's end-of-chapter material. Level 1 questions are keyed directly to the numbered chapter Learning Outcomes and offer a review of facts and terms. Level 2 questions encourage students to review concepts. Level 3 questions promote critical thinking and clinical applications.

I would like to acknowledge the students and colleagues who have made an impact on this work through their thoughtful feedback. I am grateful to Damian Hill for his help with preparing the revision of the Study Guide for this edition. My thanks to Angela Edwards of Trident Technical College for her careful accuracy review of the Answer Key. Thanks also to the editorial and production teams at Pearson Science and S4Carlisle for their work in organizing and producing this revision. Finally and most importantly, I thank my family for their continuing support.

Of course, any errors or omissions found by the reader are attributable to the author, rather than to the reviewers. Readers with comments, suggestions, or corrections should contact me at the address below.

Charles Seiger
c/o Pearson Science
1301 Sansome Street
San Francisco,
CA 94111

Format and Features

The three-level review system utilized in the Martini/Nath/Bartholomew text and all of its supplements affords each student a logical framework for the progressive development of skill as he or she advances through material of increasing levels of complexity. The three levels of organization are designed to help students (1) learn basic facts, ideas, and principles; (2) increase their capacity for abstraction and concept integration; and (3) think critically in applying what they have learned to specific situations.

Level 1, *Reviewing Facts and Terms,* consists of exercises that test the student's mastery of vocabulary and recall of information through objective-based multiple-choice, completion, matching, and illustration labeling questions.

Level 2 focuses on *Reviewing Concepts,* a process that actively involves the student in combining, integrating, and relating the basic facts and concepts mastered in Level 1. In addition to multiple-choice, completion, and short-answer/essay questions, this level incorporates a feature that is unique to this study guide: the development and completion of concept maps. The examples provided give students a framework of information that encourages them to complete the map. Once completed, these maps help the user to understand relationships between ideas and to differentiate related concepts. In addition, they have been designed to serve as models that will encourage students to develop new maps of their own.

Another feature of Level 2 is the use of chapter overviews designed for the student to complete and study important concepts by integrating and relating the basic facts in the chapter. The chapter overviews are written to demonstrate the interrelationships of concepts and facts in an understandable, dynamic way.

Level 3 activities promote *Critical Thinking and Clinical Applications* through the use of life experiences, plausible clinical situations, and common diagnostic problems. These techniques draw on the students' analytical and organizational powers, enabling them to develop associations and relationships that ultimately may benefit their decision-making ability in clinical situations and everyday life experiences.

An answer key for each chapter has been provided at the end of the Study Guide.

An Introduction to Anatomy and Physiology

OVERVIEW

Are you interested in knowing something about your body? Have you ever wondered what makes your heart beat or why and how muscles contract to produce movement? If you are curious to understand the what, why, and how of the human body, then the study of anatomy and physiology is essential for you. The term *anatomy* is derived from a Greek word that means to cut up or dissect representative animals or human cadavers, which historically has served as the basis for understanding the structure of the human body. *Physiology* is the science that attempts to explain the physical and chemical processes occurring in the body. Anatomy and physiology provide the foundation for personal health and clinical applications.

Chapter 1 is an introduction to anatomy and physiology *citing* some of the basic functions of living organisms, *defining* various specialties of anatomy and physiology, *identifying* levels of organization in living things, *explaining* homeostasis and regulation, and *introducing* some basic anatomical terminology. The information in this chapter will provide the framework for a better understanding of anatomy and physiology, and includes basic concepts and principles necessary to get you started on a successful and worthwhile trek through the human body.

LEVEL 1: REVIEWING FACTS AND TERMS

Review of Learning Outcomes

After completing this chapter, you should be able to do the following:

OUTCOME 1-1 Explain the importance of studying anatomy and physiology.

OUTCOME 1-2 Identify basic study skill strategies to use in this course.

OUTCOME 1-3 Define anatomy and physiology, describe the origins of anatomical and physiological terms, and explain the significance of *Terminologia Anatomica (International Anatomical Terminology)*.

OUTCOME 1-4 Explain the relationship between anatomy and physiology, and describe various specialties of each discipline.

OUTCOME 1-5 Identify the major levels of organization in organisms, from the simplest to the most complex, and identify major components of each organ system.

OUTCOME 1-6   Explain the concept of homeostasis.

OUTCOME 1-7   Describe how negative feedback and positive feedback are involved in homeostatic regulation, and explain the significance of homeostasis.

OUTCOME 1-8   Use anatomical terms to describe body sections, body regions, and relative positions.

OUTCOME 1-9   Identify the major body cavities and their subdivisions, and describe the functions of each.

Multiple Choice

Place the letter corresponding to the best answer in the space provided.

OUTCOME 1-1   _____   1. The study of anatomy and physiology provides

    a. the foundation for understanding all other basic life sciences.

    b. an understanding of how your body works under normal and abnormal conditions.

    c. a basis for making commonsense decisions about your own life.

    d. all of the above are correct.

OUTCOME 1-2   _____   2. Reading the textbook and memorizing important facts and concepts will assure you of success in the A&P course.

    a. True

    b. False

OUTCOME 1-3   _____   3. The study of internal and external body structures is

    a. physiology.

    b. histology.

    c. anatomy.

    d. cytology.

OUTCOME 1-3   _____   4. Physiology is the study of

    a. how living organisms perform functions.

    b. internal and external body structures.

    c. maintenance and repair of body tissues.

    d. all of the above are correct.

OUTCOME 1-3   _____   5. The four basic building blocks of medical terms are

    a. cells, tissues, organs, and systems.

    b. the brain, the heart, the lungs, and the kidneys.

    c. word roots, prefixes, suffixes, and combining forms.

    d. all of the above are correct.

OUTCOME 1-3   _____   6. The worldwide official standard of anatomical vocabulary is

    a. Martini's A&P textbook.

    b. the International Association of Anatomists.

    c. eponyms and related historical details.

    d. *Terminologia Anatomica.*

OUTCOME 1-4   _____   7. Anatomy and physiology are closely related because

    a. anatomical information provides clues about functions.

    b. all specific functions are performed by specific structures.

    c. physiological mechanisms can be explained only in terms of underlying anatomy.

    d. all of the above are correct.

OUTCOME 1-4 _____ 8. The study of general form and superficial anatomical markings is called
   a. developmental anatomy.
   b. surface anatomy.
   c. comparative anatomy.
   d. systemic anatomy.

OUTCOME 1-4 _____ 9. The anatomical specialty that provides a bridge between the realms of macroscopic anatomy and microscopic anatomy is
   a. gross anatomy.
   b. regional anatomy.
   c. developmental anatomy.
   d. surgical anatomy.

OUTCOME 1-4 _____ 10. The specialized study that analyzes the structure of individual cells is
   a. histology.
   b. microbiology.
   c. cytology.
   d. pathology.

OUTCOME 1-4 _____ 11. The scientist who studies the effects of *diseases* on organ or system functions would be classified as a
   a. histophysiologist.
   b. cell physiologist.
   c. system physiologist.
   d. pathological physiologist.

OUTCOME 1-5 _____ 12. The smallest *living* units in the body are
   a. elements.
   b. subatomic particles.
   c. cells.
   d. molecules.

OUTCOME 1-5 _____ 13. The level of organization that reflects the interactions between organ systems is the
   a. cellular level.
   b. tissue level.
   c. molecular level.
   d. organism.

OUTCOME 1-6 _____ 14. The two regulatory systems in the human body are the
   a. nervous and endocrine.
   b. digestive and reproductive.
   c. muscular and skeletal.
   d. cardiovascular and lymphoid.

OUTCOME 1-6 _____ 15. *Homeostasis* refers to
   a. the chemical operations under way in the body.
   b. individual cells becoming specialized to perform particular functions.
   c. changes in an organism's immediate environment.
   d. the maintenance of a stable internal environment.

OUTCOME 1-7   _____   16. When a variation outside of normal limits triggers an automatic response that corrects the situation, the mechanism is called
    a. positive feedback.
    b. crisis management.
    c. negative feedback.
    d. homeostasis.

OUTCOME 1-7   _____   17. When an initial disturbance produces a response that exaggerates the disturbance, the mechanism is called
    a. autoregulation.
    b. negative feedback.
    c. extrinsic regulation.
    d. positive feedback.

OUTCOME 1-8   _____   18. An erect body, with the feet together, eyes directed forward, and the arms at the side of the body with the palms of the hands turned forward, represents the
    a. supine position.
    b. prone position.
    c. anatomical position.
    d. proximal position.

OUTCOME 1-8   _____   19. Moving from the wrist toward the elbow is an example of moving in a _____ direction.
    a. proximal
    b. distal
    c. medial
    d. lateral

OUTCOME 1-8   _____   20. RLQ is an abbreviation used as a reference to designate a specific
    a. section of the vertebral column.
    b. area of the cranial vault.
    c. region of the pelvic girdle.
    d. abdominopelvic quadrant.

OUTCOME 1-8   _____   21. Making a sagittal section results in the separation of
    a. anterior and posterior portions of the body.
    b. superior and inferior portions of the body.
    c. dorsal and ventral portions of the body.
    d. right and left portions of the body.

OUTCOME 1-9   _____   22. The two major subdivisions of the ventral body cavity are the
    a. pleural and pericardial cavity.
    b. thoracic and abdominopelvic cavity.
    c. pelvic and abdominal cavity.
    d. cranial and spinal cavity.

OUTCOME 1-9   _____   23. The heart and the lungs are located in the _____ cavity.
    a. pericardial
    b. thoracic
    c. pleural
    d. abdominal

OUTCOME 1-9   _____   24. The ventral body cavity is divided by a flat muscular sheet called the
a. mediastinum.
b. pericardium.
c. diaphragm.
d. peritoneum.

Completion

Using the terms below, complete the following statements. Use each term only once.

autoregulation	histologist	physiology	medial
organs	mediastinum	distal	peritoneal
regulation	digestive	equilibrium	molecules
tissues	liver	embryology	urinary
integumentary	homeostasis	procrastination	body structures
transverse	pericardial	positive feedback	eponym
endocrine			

OUTCOME 1-1   1. The study of anatomy and physiology provides an explanation of how the body responds to normal and abnormal conditions and maintains _____.

OUTCOME 1-2   2. One of the best strategies for success in an A&P course is to avoid _____.

OUTCOME 1-3   3. Anatomy is the study of internal and external _____.

OUTCOME 1-3   4. A comparative name for a structure that was originally named for a real or mythical person is an _____.

OUTCOME 1-4   5. A person who specializes in the study of tissue is called a _____.

OUTCOME 1-4   6. The study of early developmental processes is called _____.

OUTCOME 1-4   7. The study of the *functions* of the living cell is called cell _____.

OUTCOME 1-5   8. In complex organisms such as the human being, cells unite to form _____.

OUTCOME 1-5   9. At the chemical level of organization, chemicals interact to form complex _____.

OUTCOME 1-5   10. An organ system is made up of structural units called _____.

OUTCOME 1-5   11. The kidneys, bladder, and ureters are organs that belong to the _____ system.

OUTCOME 1-5   12. The esophagus, large intestine, and stomach are organs that belong to the _____ system.

OUTCOME 1-5   13. The organ system to which the skin belongs is the _____ system.

OUTCOME 1-6   14. The term that refers to the adjustments in physiological systems is _____.

OUTCOME 1-6   15. When opposing processes or forces are in balance, it can be said that they have reached a state of _____.

OUTCOME 1-6   16. When the activities of a cell, tissue, organ, or system change automatically due to environmental variation, the homeostatic mechanism that operates is called _____.

OUTCOME 1-7   17. A response that is important in accelerating processes that must proceed to completion rapidly is called _____.

OUTCOME 1-7   18. The two systems often controlled by negative feedback mechanisms are the nervous and _____ systems.

OUTCOME 1-8   19. Tenderness in the right upper quadrant (RUQ) might indicate problems with the
_____.

OUTCOME 1-8   20. A term that means "close to the long axis of the body" is _____.

OUTCOME 1-8   21. A term that means "away from an attached base" is _____.

OUTCOME 1-8   22. A horizontal or cross-section view of the human body at a right angle to the long
axis of the body is a _____ view.

OUTCOME 1-9   23. The subdivision of the thoracic cavity that houses the heart is the _____
cavity.

OUTCOME 1-9   24. The region that lies between and separates the two pleural cavities is the
_____.

OUTCOME 1-9   25. The abdominopelvic cavity is also known as the _____ cavity.

Matching

Match the terms in column B with the terms in column A. Use letters for answers in the spaces provided.
Use each term only once.

Part I

		Column A	**Column B**
OUTCOME 1-1	_____	1. study of A&P	A. success strategy
OUTCOME 1-2	_____	2. diligent study	B. *Terminologia Anatomica*
OUTCOME 1-3	_____	3. Latin origins	C. understand basic life sciences
OUTCOME 1-4	_____	4. cytology	D. disease
OUTCOME 1-4	_____	5. histology	E. organelles
OUTCOME 1-4	_____	6. gross anatomy	F. endocrine
OUTCOME 1-4	_____	7. pathology	G. study of tissues
OUTCOME 1-5	_____	8. internal cell structures	H. cardiovascular
OUTCOME 1-5	_____	9. heart	I. study of cells
OUTCOME 1-5	_____	10. pituitary	J. macroscopic

Part II

		Column A	**Column B**
OUTCOME 1-6	_____	11. homeostasis	K. skull
OUTCOME 1-6	_____	12. automatic system change	L. rapid response
OUTCOME 1-7	_____	13. receptor	M. steady state
OUTCOME 1-7	_____	14. nervous system	N. ventral body cavity
OUTCOME 1-8	_____	15. cranial	O. stimulus
OUTCOME 1-8	_____	16. prone	P. abdominopelvic
OUTCOME 1-9	_____	17. peritoneal	Q. autoregulation
OUTCOME 1-9	_____	18. coelom	R. face down

Drawing/Illustration Labeling

Identify each numbered plane. Place your answers in the spaces provided.

OUTCOME 1-8 **FIGURE 1-1** Planes of the Body

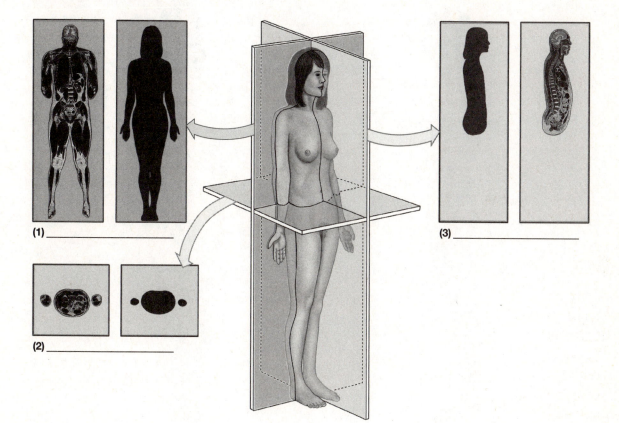

(1) _____

(2) _____

(3) _____

Identify each numbered relative direction. Place your answers in the spaces provided.

OUTCOME 1-8   **FIGURE 1-2**   Human Body Orientation and Direction

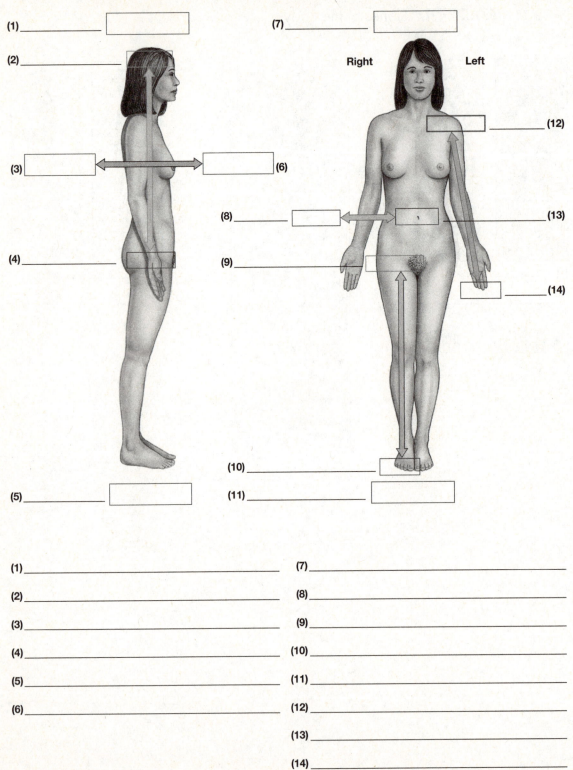

(1) _____

(2) _____

(3) _____

(4) _____

(5) _____

(6) _____

(7) _____

(8) _____

(9) _____

(10) _____

(11) _____

(12) _____

(13) _____

(14) _____

Right   Left

(1) _____

(2) _____

(3) _____

(4) _____

(5) _____

(6) _____

(7) _____

(8) _____

(9) _____

(10) _____

(11) _____

(12) _____

(13) _____

(14) _____

Label each numbered anatomical landmark with the appropriate anatomical term. Place your answers in the spaces provided.

OUTCOME 1-8   **FIGURE 1-3**   Anatomical Landmarks

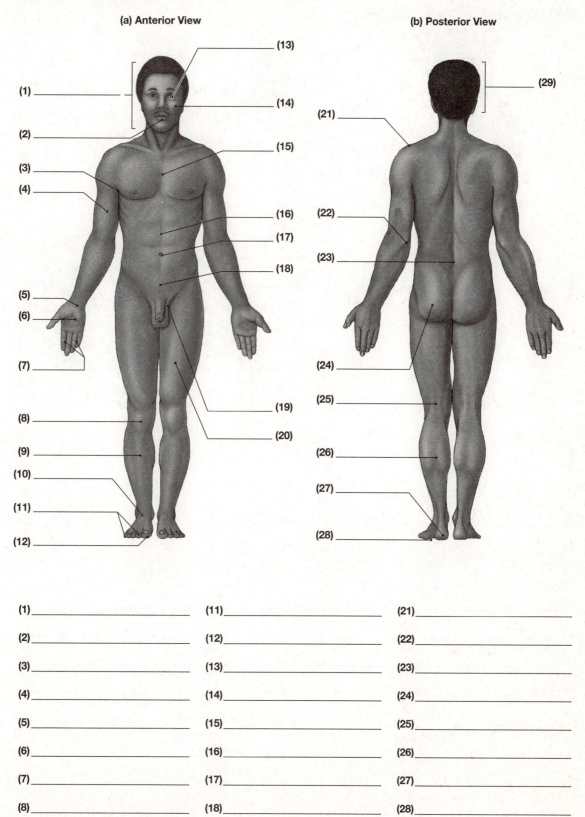

(a) Anterior View

(b) Posterior View

(1) _____	(11) _____	(21) _____
(2) _____	(12) _____	(22) _____
(3) _____	(13) _____	(23) _____
(4) _____	(14) _____	(24) _____
(5) _____	(15) _____	(25) _____
(6) _____	(16) _____	(26) _____
(7) _____	(17) _____	(27) _____
(8) _____	(18) _____	(28) _____
(9) _____	(19) _____	(29) _____
(10) _____	(20) _____	

Label each numbered body cavity. Place your answers in the spaces provided.

OUTCOME 1-9 **FIGURE 1-4** Body Cavities

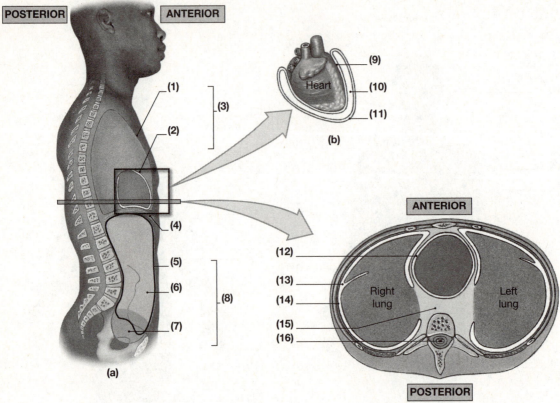

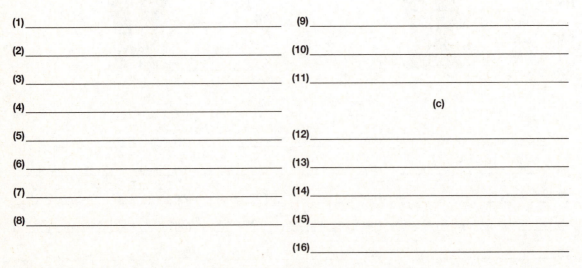

(a)	(b)
(1)_____	(9)_____
(2)_____	(10)_____
(3)_____	(11)_____
(4)_____	(c)
(5)_____	(12)_____
(6)_____	(13)_____
(7)_____	(14)_____
(8)_____	(15)_____
	(16)_____

LEVEL 2: REVIEWING CONCEPTS

Chapter Overview

Levels of Organization

Using the terms below, fill in the blanks to consider events and structures at several independent levels of organization. Use each term only once.

tissues	systems	organism
atoms	cells	organs
organelles	molecules	

At the chemical (or molecular) level, the smallest stable units of matter are the (1) _____. These units, invisible to the naked eye, can combine to form (2) _____ with complex shapes. The combinations of complex shapes interact to form the structural and functional components of cells, called (3) _____. These microscopic components, along with a living substance, comprise the makeup of the individual units of structure and function in all living things called (4) _____. Many of these individual units may combine with one another to form four kinds of (5) _____ that work together to perform one or more specific functions. These multi-unit, more complex structural types that work in combination to perform several functions are called (6) _____, which, when performing in an interactive functional capacity in humans, make up the 11 body (7) _____. At the highest level of organization, this complex complete living being is referred to as an (8) _____.

Concept Map I

Using the following terms, fill in the circled, numbered, blank spaces to correctly complete the concept map. Use each term only once.

surgical anatomy   structure of organ systems   cytology
embryology   regional anatomy   macroscopic anatomy
tissues

```
                              ┌──────────┐
                              │ Anatomy  │
                              └──────────┘
                          Anatomical Specialties

        ┌──────────────────────┼──────────────────────┐
        ▼                      ▼                        ▼
  ┌───────────┐        ┌──────────────┐         ┌──────────────┐
  │Gross anatomy│      │Developmental │         │ Microscopic  │
  └───────────┘        │   anatomy    │         │   anatomy    │
                       └──────────────┘         └──────────────┘
  Sometimes called        Study of                 Called
        ▼                      ▼                        ▼
  ┌───────────┐        ┌──────────────┐         ┌──────────────┐
  │① _____ │        │⑤ _____   │         │ Fine anatomy │
  └───────────┘        └──────────────┘         └──────────────┘
  Strategies to study                          Includes study of
```

```
 1      2    3    4    5    6
 ▼      ▼         ▼    ▼    ▼
```

Surface anatomy	Systemic anatomy	Radiographic anatomy

Study of

Early developmental processes

⑥ _____
Study of
Cells

Histology
Study of
⑦ _____

Refers to — General form, superficial markings

② _____
Refers to — Features in specific areas of body

Medical anatomy
Study of — Anatomical damages during illness

④ _____
Studies — Landmarks for surgical procedure

③ _____

Anatomical structures by X-ray, MRI, CT scan

Concept Map II

Using the following terms, fill in the circled, numbered, blank spaces to correctly complete the concept map. Use each term only once.

effects of diseases pathological physiology respiratory
cell functions organ physiology specific organ systems
cardiac physiology chemical interactions

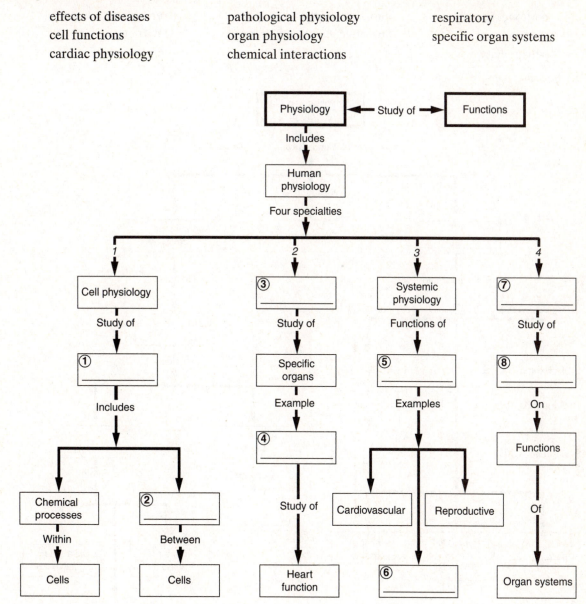

Concept Map III

Using the following terms, fill in the circled, numbered, blank spaces to correctly complete the concept map. Use each term only once.

peritoneal cavity   thoracic cavity   reproductive organs
mediastinum   digestive glands and organs   heart
pelvic cavity   trachea, esophagus   left lung

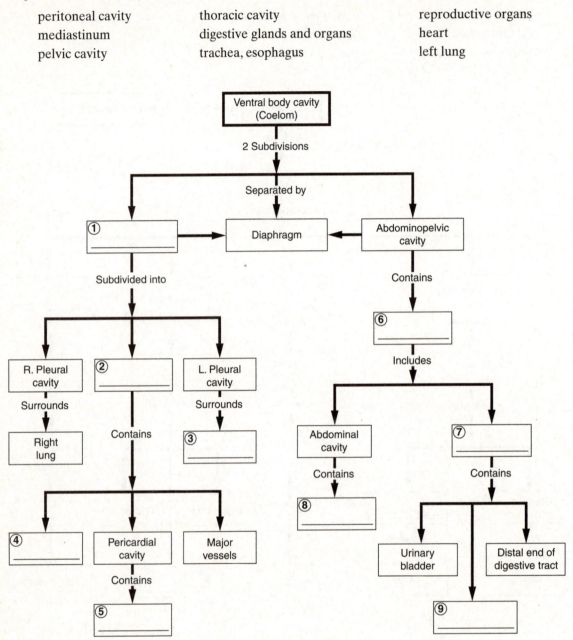

Concept Map IV

Using the following terms, fill in the circled, numbered, blank spaces to correctly complete the concept map. Use each term only once.

imaging technique
internal structures
computerized (axial)
 tomography

magnetic field and radio
 waves
x-rays

MRI
PET scan

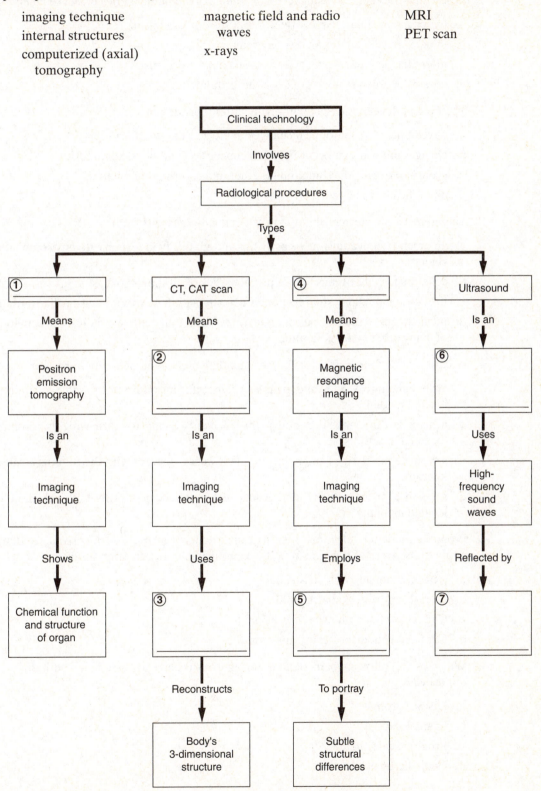

Multiple Choice

Place the letter corresponding to the best answer in the space provided.

_____ 1. Proceeding through increasing levels of complexity in humans, the correct sequence is

    a. molecular, cellular, tissue, organ, organ system, organism.

    b. molecular, cellular, organ, tissue, organism, organ system.

    c. molecular, cellular, organ system, tissue, organ, organism.

    d. organism, organ system, organ, tissue, cellular, molecular.

_____ 2. The field of developmental anatomy includes the study of

    a. development of anatomical changes that occur due to injury.

    b. changes in form that occur between conception and physical maturity.

    c. groups of organs that function together in a coordinated manner.

    d. all of the above are correct.

_____ 3. The homeostatic mechanism of extrinsic regulation results from

    a. an organ system adjusting its activities automatically in response to environmental changes.

    b. cells, tissues, and organs that adjust activities of all the organ systems.

    c. the cardiovascular system responding to changes in heart rate and cardiac output.

    d. the activities of the nervous or endocrine system that control or adjust the activities of many other systems simultaneously.

_____ 4. Anatomical position refers to a person standing erect, feet facing forward, and

    a. arms hanging to sides and palms of hands facing anteriorally and the thumbs located medially.

    b. arms in a raised position and palms of hands facing forward with the thumbs located laterally.

    c. arms hanging to sides and palms of hands facing forward with the thumbs located laterally.

    d. arms in a raised position and palms of hands facing dorsally with the thumbs located medially.

_____ 5. From the following selections, identify the directional terms in _correct sequence_ that apply to the areas of the human body. (ventral, posterior, superior, inferior)

    a. Anterior, dorsal, cephalic, caudal

    b. Dorsal, anterior, caudal, cephalic

    c. Caudal, cephalic, anterior, posterior

    d. Cephalic, caudal, posterior, anterior

_____ 6. Which of the following pairs of anatomical terms is correctly associated with its opposite?

    a. Distal, coronal

    b. Cranial, caudal

    c. Proximal, lateral

    d. Cephalic, posterior

_____ 7. The correct sequence in which a homeostatic regulatory pathway occurs is

    a. an effector, a control center, and a receptor.

    b. a control center, a receptor, and an effector.

    c. a receptor, a control center, and an effector.

    d. a control center, an effector, and a receptor.

_____ 8. From the organ systems listed below, select the organs in correct sequence that are found in each of the systems. (cardiovascular, digestive, endocrine, urinary, integumentary)

    a. Blood vessels, pancreas, kidneys, lungs, nails

    b. Heart, stomach, lungs, kidneys, hair

    c. Heart, liver, pituitary gland, kidneys, skin

    d. Lungs, gall bladder, ovaries, bladder, sebaceous glands

_____ 9. In a negative feedback system, the process that triggers a response that corrects the situation is

    a. the presence of a receptor area and an effector area.

    b. an exaggeration of the stimulus.

    c. temporary repair to the damaged area.

    d. a variation outside of normal limits.

_____ 10. Suppose an individual's body temperature is 37.3°C. This variation from the "normal" value may represent

    a. an illness that has not been identified.

    b. individual variation rather than a homeostatic malfunction.

    c. the need to see a physician immediately.

    d. a variability that is abnormal.

_____ 11. If the temperature of the body climbs above 99°F, negative feedback regulation could trigger

    a. increased heat conservation by restricted blood flow to the skin.

    b. shivering in the individual.

    c. activation of the positive feedback mechanism.

    d. an increased heat loss through enhanced blood flow to the skin and sweating.

_____ 12. The term _medial surface_ refers to an area

    a. close to the long axis of the body.

    b. further away from the long axis of the body.

    c. toward an attached base.

    d. away from an attached base.

_____ 13. The sectional plane that divides the body so the face remains intact is the

    a. sagittal plane.

    b. coronal plane.

    c. midsagittal plane.

    d. transverse plane.

_____ 14. Negative feedback systems

        a. counteract the effects of a stimulus.

        b. provide short-term control over the body's internal conditions.

        c. tend to interfere with homeostasis.

        d. all of the above are correct.

Completion

Using the terms below, complete the following statements. Use each term only once.

appendicitis	histologist	negative feedback
sternum	knee	elbow
nervous	lymphatic	extrinsic regulation
cardiovascular		

1. The specialist who investigates structures at the tissue level of organization is the _____.

2. The system responsible for internal transport of cells and dissolved materials, including nutrients, wastes, and gases, is the _____ system.

3. The system responsible for defense against infection and disease is the _____.

4. Activities of the nervous and endocrine systems to control or adjust the activities of many different systems simultaneously are _____.

5. The system that performs crisis management by directing rapid, short-term, and very specific responses is the _____ system.

6. The popliteal artery can be found near the _____.

7. Tenderness in the right lower quadrant of the abdomen may indicate _____.

8. Moving proximally from the wrist brings you to the _____.

9. To oppose any departure from the norm, physiological systems are typically regulated by _____.

10. If a surgeon makes a midsagittal incision in the inferior region of the thorax, the incision would be made through the _____.

Short Essay

Briefly answer the following questions in the spaces provided below.

1. a. Describe a person who is in the *anatomical* position.

   b. Describe the *supine* position.

   c. Describe the *prone* position.

2. Show your understanding of the *levels of organization* in a human, listing them in correct sequence from the simplest level to the most complex level.

3. What is homeostatic regulation and how is it physiologically important?

4. What is the major difference between negative feedback and positive feedback?

5. References are made by the anatomist to the *front, back, head,* and *tail* of the human body. What directional reference terms are used to describe each one of these directions? Your answer should be in the order previously listed.

6. What is the difference between a *sagittal* section and a *transverse* section?

7. Identify the body cavity in which each of the following organs or organ systems are found:
   a. reproductive organs, urinary bladder
   b. cardiovascular, digestive, and urinary system
   c. heart, lungs
   d. stomach, intestines

8. What are the two essential functions of body cavities?

LEVEL 3: CRITICAL THINKING AND CLINICAL APPLICATIONS

Using principles and concepts learned in Chapter 1, answer the following questions. Write your answers on a separate sheet of paper.

1. Unlike the abdominal viscera, the thoracic viscera is separated into two cavities by an area called the mediastinum. What is the clinical importance of this compartmental arrangement?

2. The process of blood clotting is associated with the process of positive feedback. Describe the events that confirm this statement.

3. A radioactive tracer is induced into the heart to trace the possibility of a blockage in or around the uterus. Give the sequence of *body cavities* that would be included as the tracer travels in the blood from the heart through the aorta and the uterine artery.

4. Suppose autoregulation fails to maintain homeostasis in the body. What is the process called that takes over by initiating activity of both the nervous and endocrine systems? What are the results of this comparable homeostatic mechanism?

5. When there is an increased level of calcium ions in the blood, calcitonin is released from the thyroid gland. If this hormone is controlled by negative feedback, what effect would calcitonin have on blood calcium levels?

6. How do the following systems serve to maintain homeostatic regulation of body fluid volume? (a) urinary (b) digestive (c) integumentary (d) cardiovascular

7. Body temperature is regulated by a control center in the brain that functions like a thermostat. Assuming a normal range of 98°–99°F, identify from the graph below what would happen if there was an increase or decrease in body temperature beyond the normal limits. Use the following selections to explain what would happen at no. 1 and no. 2 on the graph.

- body cools
- shivering
- increased sweating
- temperature declines

- body heat is conserved
- ↑ blood flow to skin
- ↓ blood flow to skin
- temperature rises

CHAPTER 2

The Chemical Level of Organization

OVERVIEW

Technology within the last 40 to 50 years has allowed humans to "see and understand" the unseen within the human body. Today, instead of an "organ system view" of the body, we are able to look through the eyes of scientists using sophisticated technological tools and see the "ultramicro world" within us. Our technology has permitted us to progress from the "macro-view" to the "micro-view" and to understand that the human body is made up of atoms, and that the interactions of these atoms control the physiological processes within the body.

The study of anatomy and physiology begins at the most fundamental level of organization, namely, individual atoms and molecules. The concepts in Chapter 2 provide a framework for understanding how simple components combine to make up the more complex forms that comprise the human body. Information is provided that shows how the chemical processes need to continue throughout life in an orderly and timely sequence if homeostasis is to be maintained.

The student activities in this chapter stress many of the important principles that make up the science of chemistry, both inorganic and organic. The exercises are set up to measure your knowledge of chemical principles and to evaluate your ability to apply these principles to the structure and functions of organ systems within the human body.

LEVEL 1: REVIEWING FACTS AND TERMS

Review of Learning Outcomes

After completing this chapter, you should be able to do the following:

OUTCOME 2-1   Describe an atom and how atomic structure affects interactions between atoms.

OUTCOME 2-2   Compare the ways in which atoms combine to form molecules and compounds.

OUTCOME 2-3   Distinguish among the major types of chemical reactions that are important for studying physiology.

OUTCOME 2-4   Describe the crucial role of enzymes in metabolism.

OUTCOME 2-5   Distinguish between organic and inorganic compounds.

OUTCOME 2-6   Explain how the chemical properties of water make life possible.

OUTCOME 2-7   Discuss the importance of pH and the role of buffers in body fluids.

OUTCOME 2-8   Describe the physiological roles of inorganic compounds.

OUTCOME 2-9   Discuss the structures and functions of carbohydrates.

OUTCOME 2-10   Discuss the structures and functions of lipids.

OUTCOME 2-11   Discuss the structures and functions of proteins.

OUTCOME 2-12   Discuss the structures and functions of nucleic acids.

OUTCOME 2-13   Discuss the structures and functions of high-energy compounds.

OUTCOME 2-14   Explain the relationship between chemicals and cells.

Multiple Choice

Place the letter corresponding to the best answer in the space provided.

OUTCOME 2-1   _____   1. The smallest stable units of matter are
　　　　　　　　　　　　　　a. electrons.
　　　　　　　　　　　　　　b. mesons.
　　　　　　　　　　　　　　c. protons.
　　　　　　　　　　　　　　d. atoms.

OUTCOME 2-1   _____   2. The three subatomic particles that are constituents of atoms are
　　　　　　　　　　　　　　a. carbon, hydrogen, and oxygen.
　　　　　　　　　　　　　　b. protons, neutrons, and electrons.
　　　　　　　　　　　　　　c. atoms, molecules, and compounds.
　　　　　　　　　　　　　　d. cells, tissues, and organs.

OUTCOME 2-1   _____   3. The protons in an atom are found only
　　　　　　　　　　　　　　a. outside the nucleus.
　　　　　　　　　　　　　　b. in the nucleus or outside the nucleus.
　　　　　　　　　　　　　　c. in the nucleus.
　　　　　　　　　　　　　　d. in orbitals.

OUTCOME 2-2   _____   4. The unequal sharing of electrons in a molecule of water is an example of a(n)
　　　　　　　　　　　　　　a. ionic bond.
　　　　　　　　　　　　　　b. polar covalent bond.
　　　　　　　　　　　　　　c. double covalent bond.
　　　　　　　　　　　　　　d. strong covalent bond.

OUTCOME 2-2   _____   5. Chemical bonds created by electrical attraction between anions and cations are
　　　　　　　　　　　　　　a. ionic bonds.
　　　　　　　　　　　　　　b. polar covalent bonds.
　　　　　　　　　　　　　　c. nonpolar covalent bonds.
　　　　　　　　　　　　　　d. double covalent bonds.

OUTCOME 2-3   _____   6. From the following choices, select the one that diagrams a typical *decomposition* reaction.
　　　　　　　　　　　　　　a. $A + B \rightleftharpoons AB$
　　　　　　　　　　　　　　b. $AB + CD \rightarrow AD + CB$
　　　　　　　　　　　　　　c. $AB \rightarrow A + B$
　　　　　　　　　　　　　　d. $C + D \rightarrow CD$

OUTCOME 2-3 _____ 7. A + B $\rightleftharpoons$ AB is an example of a(n) _____ reaction.
    a. reversible
    b. synthesis
    c. decomposition
    d. a and b are correct

OUTCOME 2-3 _____ 8. AB + CD $\rightarrow$ AD + CB is an example of a(n) _____ reaction.
    a. reversible
    b. synthesis
    c. exchange
    d. a, b, and c are correct

OUTCOME 2-4 _____ 9. The presence of an appropriate enzyme affects only the
    a. rate of a reaction.
    b. direction of a reaction.
    c. products that will be formed from a reaction.
    d. a, b, and c are correct.

OUTCOME 2-4 _____ 10. Organic catalysts made by a living cell to promote a specific reaction are called
    a. nucleic acids.
    b. buffers.
    c. enzymes.
    d. metabolites.

OUTCOME 2-5 _____ 11. The major difference between inorganic and organic compounds is that *organic* compounds are usually
    a. small molecules held together partially or completely by ionic bonds.
    b. made up of carbon and hydrogen.
    c. large molecules that are soluble in water.
    d. easily destroyed by heat.

OUTCOME 2-5 _____ 12. The four major classes of organic compounds are
    a. water, acids, bases, and salts.
    b. carbohydrates, fats, proteins, and water.
    c. nucleic acids, salts, bases, and water.
    d. carbohydrates, lipids, proteins, and nucleic acids.

OUTCOME 2-6 _____ 13. The ability of water to maintain a relatively constant temperature and then prevent rapid changes in body temperature is due to its
    a. solvent capacities.
    b. small size.
    c. boiling and freezing points.
    d. large capacity to absorb heat.

OUTCOME 2-7 _____ 14. To maintain homeostasis in the human body, the pH of the blood must remain in which of the following ranges?
    a. 6.80 to 7.20
    b. 7.35 to 7.45
    c. 6.99 to 7.01
    d. 6.80 to 7.80

OUTCOME 2-7   _____ 15. The human body generates significant quantities of acids that may promote a disruptive
   a. increase in pH.
   b. pH of about 7.40.
   c. decrease in pH.
   d. sustained muscular contraction.

OUTCOME 2-8   _____ 16. A solute that dissociates to release hydrogen ions and causes a decrease in pH is
   a. a base.
   b. a salt.
   c. an acid.
   d. water.

OUTCOME 2-8   _____ 17. Of the following ionic compounds, the one that is *least* likely to affect the concentrations of H^+ and OH^- ions is a(n)
   a. base.
   b. neutral salt.
   c. acid.
   d. colloid.

OUTCOME 2-9   _____ 18. A carbohydrate molecule is made up of
   a. carbon, hydrogen, and oxygen.
   b. monosaccharides, disaccharides, and polysaccharides.
   c. glucose, fructose, and galactose.
   d. carbon, hydrogen, and nitrogen.

OUTCOME 2-9   _____ 19. A complex carbohydrate resulting from repeated dehydration synthesis reactions is best described as a
   a. lipid.
   b. monosaccharide.
   c. polysaccharide.
   d. disaccharide.

OUTCOME 2-10   _____ 20. The lipids derived from arachidonic acid that function as chemical messengers coordinating local cellular activities are the
   a. eicosanoids.
   b. glycerides.
   c. fatty acids.
   d. steroids.

OUTCOME 2-10   _____ 21. Which of the following lipids can be either saturated or unsaturated?
   a. Phospholipids
   b. Glycerides
   c. Fatty acids
   d. Steroids

OUTCOME 2-11   _____ 22. The building blocks of proteins are small molecules called
   a. peptide bonds.
   b. amino acids.
   c. R groups.
   d. amino groups.

OUTCOME 2-11 _____ 23. Special proteins that speed up metabolic reactions are called

    a. transport proteins.

    b. contractile proteins.

    c. structural proteins.

    d. enzymes.

OUTCOME 2-12 _____ 24. The three basic components of a *single nucleotide* of a nucleic acid are

    a. purines, pyrimidines, and sugar.

    b. sugar, phosphate group, and nitrogen base.

    c. guanine, cytosine, and thymine.

    d. pentose, ribose, and deoxyribose.

OUTCOME 2-13 _____ 25. The most important high-energy compound found in human cells is

    a. DNA.

    b. UTP.

    c. ATP.

    d. GTP.

OUTCOME 2-13 _____ 26. The three components required to create the high-energy compound ATP are

    a. adenosine, phosphate groups, and appropriate enzymes.

    b. deoxyribose, phosphates, and appropriate enzymes.

    c. carbohydrates, lipids, and proteins.

    d. fatty acids, triglycerides, and steroids.

OUTCOME 2-14 _____ 27. Biochemical building blocks form functional units called

    a. ATP.

    b. DNA.

    c. atoms.

    d. cells.

Completion

Using the terms below, complete the following statements. Use each term only once.

glucose	protons	mass number	solvent
H_2	acidic	carbonic acid	exergonic
decomposition	ionic bond	organic	ATP
salt	covalent bonds	cation	DNA
catalysts	solute	water	metabolic turnover
inorganic	endergonic	lipids	
proteins	synthesis	buffers	

OUTCOME 2-1   1. The atomic number of an atom is determined by the number of _____.

OUTCOME 2-1   2. The total number of protons and neutrons in the nucleus is the _____.

OUTCOME 2-2   3. Atoms that complete their outer electron shells by sharing electrons with other atoms result in molecules held together by _____.

OUTCOME 2-2   4. When one atom loses an electron and another accepts that electron, the result is the formation of a(n) _____.

OUTCOME 2-2   5. The chemical notation that would indicate "one molecule of hydrogen composed of two hydrogen atoms" is _____.

OUTCOME 2-2   6. A chemical with a charge of +1 due to the loss of one electron refers to a
_____.

OUTCOME 2-3   7. Hydrolysis is a type of _____ reaction involving the splitting of water;
a condensation reaction is a type of _____ reaction that results in the
formation water.

OUTCOME 2-4   8. Reactions that release energy are said to be _____, and reactions that
absorb heat are called _____ reactions.

OUTCOME 2-4   9. Compounds that accelerate chemical reactions without themselves being
permanently changed are called _____.

OUTCOME 2-5   10. Compounds that contain the elements carbon and hydrogen, and usually oxygen,
are _____ compounds.

OUTCOME 2-5   11. Compounds whose primary structural ingredients do not contain carbon
and hydrogen atoms are called _____ compounds.

OUTCOME 2-6   12. The fluid medium of a solution is called the _____, and the dissolved
substance is called the _____.

OUTCOME 2-6   13. The most important constituent of the body, accounting for up to two-thirds of the
total body weight, is _____.

OUTCOME 2-7   14. A solution with a pH of 6.0 is _____.

OUTCOME 2-7   15. Compounds in body fluids that help stabilize pH within normal limits are
_____.

OUTCOME 2-8   16. The interaction of an acid and a base in which the hydrogen ions of the acid are
replaced by the positive ions of the base results in the formation of a(n)
_____.

OUTCOME 2-8   17. An example of a weak acid that serves as an effective buffer in the human body is
_____.

OUTCOME 2-9   18. The most important metabolic "fuel" in the body is _____.

OUTCOME 2-10   19. The biochemical building blocks that are components forming the cell membrane
of cells are the _____.

OUTCOME 2-11   20. The most abundant organic components of the human body, accounting for about
20 percent of the total body weight, are the _____.

OUTCOME 2-12   21. The nucleic acid in human cells that determines the body's inherited characteristics
is _____.

OUTCOME 2-13   22. The most abundant high-energy compound used by cells to perform their vital
functions is _____.

OUTCOME 2-14   23. The continuous removal and replacement of organic molecules in cells is part of
a process called _____.

Matching

Match the terms in column B with the terms in column A. Use letters for answers in the spaces provided. Use each term only once.

Part I

		Column A	Column B
OUTCOME 2-1	_____	1. electron	A. two products; two reactants
OUTCOME 2-2	_____	2. N_2	B. sodium chloride
OUTCOME 2-2	_____	3. polar covalent bond	C. catalyst
OUTCOME 2-2	_____	4. NaCl	D. inorganic compound
OUTCOME 2-3	_____	5. exchange reaction	E. triple covalent bond
OUTCOME 2-4	_____	6. enzyme	F. inorganic base
OUTCOME 2-5	_____	7. HCl	G. dissociation
OUTCOME 2-5	_____	8. NaOH	H. "water loving"
OUTCOME 2-6	_____	9. ionization	I. negative electric charge
OUTCOME 2-6	_____	10. hydrophilic	J. unequal sharing of electrons

Part II

		Column A	Column B
OUTCOME 2-7	_____	11. OH^-	K. sugars and starches
OUTCOME 2-7	_____	12. H^+	L. bicarbonate ion
OUTCOME 2-8	_____	13. HCO_3^-	M. chemical messengers
OUTCOME 2-9	_____	14. carbohydrates	N. adenine and guanine
OUTCOME 2-10	_____	15. hormones	O. AMP
OUTCOME 2-11	_____	16. proteins	P. hydrogen ion
OUTCOME 2-12	_____	17. purines	Q. turnover time
OUTCOME 2-13	_____	18. example of substrate	R. amino acids
OUTCOME 2-14	_____	19. synthesis-breakdown	S. hydroxide ion

Drawing/Illustration Labeling

Identify each numbered structure. Place your answers in the spaces provided.

OUTCOME 2-1   **FIGURE 2-1**   Diagram of an Atom

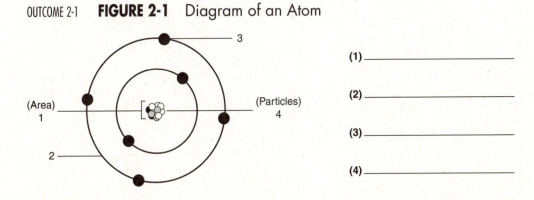

(1) _____

(2) _____

(3) _____

(4) _____

OUTCOME 2-2   **FIGURE 2-2**   Identification of Types of Bonds

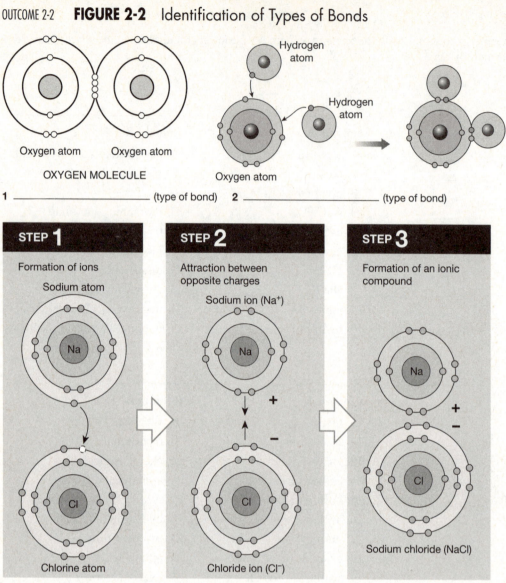

Oxygen atom   Oxygen atom

OXYGEN MOLECULE

Hydrogen atom

Hydrogen atom

Oxygen atom

1 _____ (type of bond)   2 _____ (type of bond)

STEP 1

Formation of ions

Sodium atom

Na

Chlorine atom

STEP 2

Attraction between opposite charges

Sodium ion (Na⁺)

Na

$+$

$-$

Cl

Chloride ion (Cl⁻)

STEP 3

Formation of an ionic compound

Na

$+$

$-$

Cl

Sodium chloride (NaCl)

3 _____ (type of bond)

OUTCOMES 2-9–2-12 **FIGURE 2-3** Identification of Organic Molecules

Select from the following terms to identify each molecule. Use each term only once.

polysaccharide   cholesterol   monosaccharide
amino acid   DNA   saturated fatty acid
disaccharide   polyunsaturated fatty acid

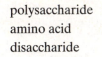

1 _____

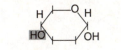

2 _____

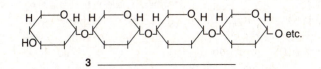

3 _____

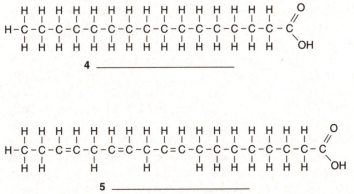

4 _____

5 _____

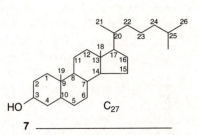

6 _____

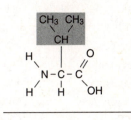

7 _____

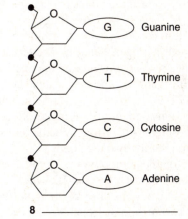

8 _____

LEVEL 2: REVIEWING CONCEPTS

Chapter Overview

Using the terms below, fill in the blanks to complete the chemical organization in the human body. Unless otherwise noted, use each term only once.

cells	carbohydrates	salts	isotope
lipids	double covalent	eight	positively
molecule	bond	DNA	deuterium
electrons	nucleic acids	water	negatively
proteins	electron clouds	six	neutrons
oxygen gas	polar covalent	oxygen (use twice)	protons
ATP	RNA	carbon dioxide	nucleus

Atoms consist of negative particles called (1) _____ that whirl around in (2) _____ at high speed. These particles are circling stationary particles in a central region called the (3) _____. This central region contains positive charges called (4) _____. With the exception of hydrogen in its natural state, all the central areas contain particles that are neutral, the (5) _____. There are some particles that look exactly like hydrogen; however, if a hydrogen has one neutron, it is a "heavier" hydrogen called (6) _____, chemically identified as a(n) (7) _____, commonly used in research laboratories as radioactive tracers. An example of an atom that contains eight protons, eight neutrons, and eight electrons is (8) _____. Atoms of this kind appear to be chemically active because the outermost cloud only has (9) _____ negative particles, and a full complement of (10) _____ is necessary if the atom is to maintain chemical stability. In the atmosphere, these "active atoms" are constantly in "search" of atoms of their own kind. When two of these atoms contact one another, each one shares two electrons with the other, forming a (11) _____, resulting in one (12) _____ of (13) _____. When a single oxygen atom comes in contact with two hydrogen atoms, a (14) _____ bond is formed due to an unequal sharing of electrons. The result is a region around the oxygen atom that is (15) _____ charged and a region around the hydrogen atoms that is (16) _____ charged.

Inorganic compounds usually lack carbon. The most important inorganic compounds in the body are (17) _____, a by-product of cell metabolism; (18) _____, an atmospheric gas required in important metabolic reactions; (19) _____, which accounts for most of our body weight; and inorganic acids, bases, and (20) _____, compounds held together partially or completely by ionic bonds.

Organic compounds always contain carbon. Organic molecules that contains carbon, hydrogen, and oxygen in a ratio near 1:2:1 are (21) _____, which are most important as energy sources that are catabolized rather than stored. Other organic molecules, called (22) _____, also have carbon and hydrogen in roughly a 1:2 ratio, but have much less oxygen and provide twice as much energy. The most abundant organic components of the human body formed from amino acids are the (23) _____, which provide a variety of essential functions and determine cell shapes and properties of tissues. The large organic molecules that store and process information at the molecular level inside cells are the (24) _____. The two classes of these large molecules are (25) _____, which determines inherited characteristics, and (26) _____, necessary to manufacture specific proteins in cells.

To perform vital functions, cells must use energy obtained by breaking down organic substances. The most abundant high-energy compound used to perform the body's vital functions is (27) _____, a nucleotide-based, high-energy compound that transfers energy in specific enzymatic reactions. All of these biochemical building blocks, both organic and inorganic compounds, are found in the functional units called (28) _____.

Concept Map I

Using the following terms, fill in the circled, numbered, blank spaces to correctly complete the concept map. Use each term only once.

monosaccharides   sucrose   complex carbohydrates
glucose   glycogen   plants
disaccharides

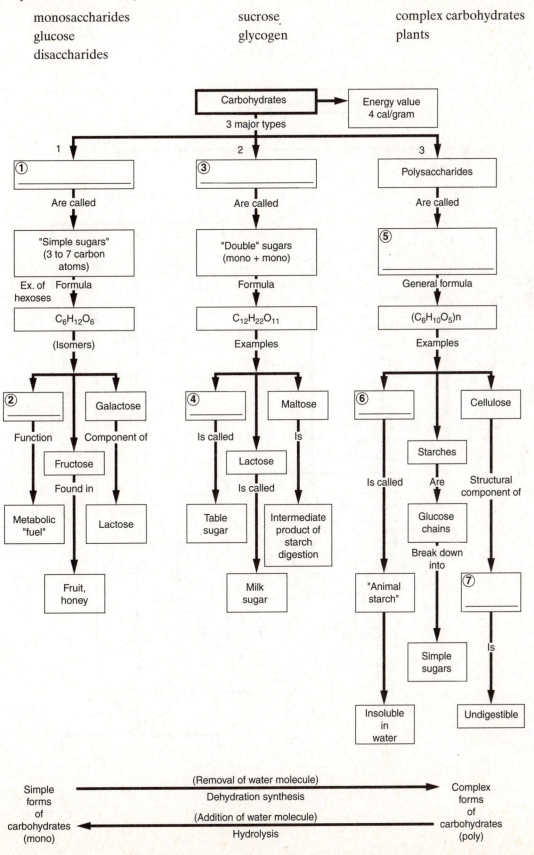

Simple forms of carbohydrates (mono) ← (Addition of water molecule) Hydrolysis — Dehydration synthesis (Removal of water molecule) → Complex forms of carbohydrates (poly)

Concept Map II

Using the following terms, fill in the circled, numbered, blank spaces to correctly complete the concept map. Use each term only once.

local hormones	Di-	saturated
glycerides	phospholipid	carbohydrates + diglyceride
steroids	glycerol + 3 fatty acids	

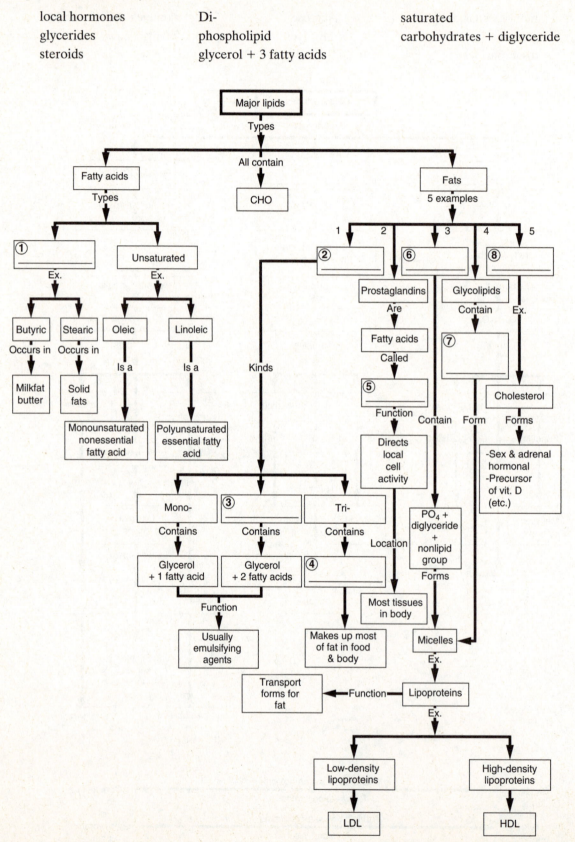

Concept Map III

Using the following terms, fill in the circled, numbered, blank spaces to correctly complete the concept map. Use each term only once.

variable group
enzymes
structural proteins
amino group

$-COOH$
keratin
amino acids
primary

globular proteins
quaternary
alpha helix

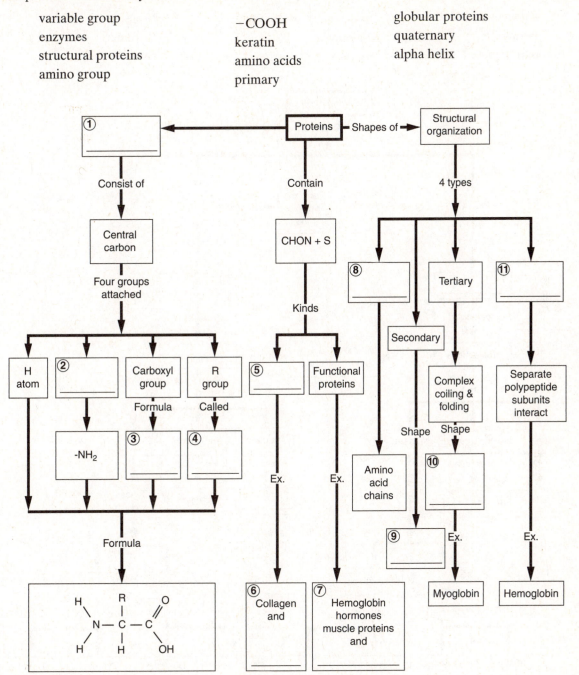

Concept Map IV

Using the following terms, fill in the circled, numbered, blank spaces to correctly complete the concept map. Use each term only once.

ribonucleic acid   deoxyribose nucleic acid   N bases
pyrimidines   adenine   thymine
deoxyribose   purines   ribose

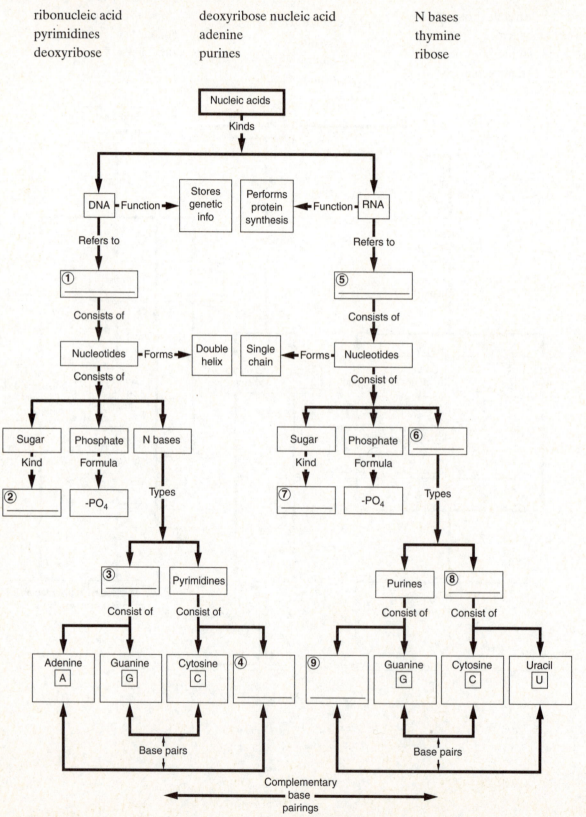

Multiple Choice

Place the letter corresponding to the best answer in the space provided.

_____ 1. The chemical properties of every element are determined mostly by the

  a. number and arrangement of electrons in the outer energy level.
  b. number of protons in the nucleus.
  c. number of protons and neutrons in the nucleus.
  d. atomic weight of the atom.

_____ 2. Whether an atom will react with another atom will be determined primarily by the

  a. number of protons present in the atom.
  b. number of electrons in the outermost energy level.
  c. atomic weight of the atom.
  d. number of subatomic particles present in the atom.

_____ 3. In the formation of *nonpolar* covalent bonds, there is

  a. equal sharing of protons and electrons.
  b. donation of electrons.
  c. unequal sharing of electrons.
  d. equal sharing of electrons.

_____ 4. The symbol $2H_2O$ means that two identical molecules of water are each composed of

  a. 4 hydrogen atoms and 2 oxygen atoms.
  b. 2 hydrogen atoms and 2 oxygen atoms.
  c. 4 hydrogen atoms and 1 oxygen atom.
  d. 2 hydrogen atoms and 1 oxygen atom.

_____ 5. Water is a particularly effective solvent because

  a. it has a dipolar charge distribution.
  b. cations and anions are produced by hydration.
  c. hydrophobic molecules have many polar covalent bonds.
  d. it has a high heat capacity, which dissolves molecules.

_____ 6. The formation of a complex molecule by the removal of water is called

  a. dehydration synthesis.
  b. hydrolysis.
  c. activation energy.
  d. reversible reaction.

_____ 7. A *salt* may best be described as an

  a. organic molecule created by chemically altering an acid or base.
  b. inorganic molecule that buffers solutions.
  c. inorganic molecule created by the reaction of an acid and a base.
  d. organic molecule used to flavor food.

_____ 8. The chemical makeup of a lipid molecule is different from a carbohydrate in that the lipid molecule

    a. contains much less oxygen than a carbohydrate having the same number of carbon atoms.

    b. contains twice as much oxygen as the carbohydrate.

    c. contains equal amounts of carbon and oxygen in its molecular structure.

    d. the chemical makeup is the same.

_____ 9. Lipid deposits are important as _energy reserves_ because

    a. they appear as fat deposits on the body.

    b. they are readily broken down to release energy.

    c. the energy released from lipids is metabolized quickly.

    d. lipids provide twice as much energy as carbohydrates.

_____ 10. Proteins differ from carbohydrates in that they

    a. are not energy nutrients.

    b. do not contain carbon, hydrogen, and oxygen.

    c. always contain nitrogen.

    d. are inorganic compounds.

_____ 11. Compared to the other major organic compounds, nucleic acids are unique in that they

    a. contain nitrogen.

    b. store and process information at the molecular level.

    c. are found only in the nuclei of cells.

    d. control the metabolic activities of the cell.

_____ 12. Compounds that stabilize the pH of a solution by binding or releasing hydrogen ions are called

    a. suspensions.

    b. colloids.

    c. hydrophilic.

    d. buffers.

_____ 13. From the selections that follow, choose the one that represents the symbols for each of the following elements in the correct order. (carbon, sodium, phosphorus, iron, oxygen, nitrogen, sulfur)

    a. C, So, Ph, I, O, Ni, S

    b. C, Na, P, I, O, N, S

    c. C, Na, P, Fe, O, N, S

    d. C, Na, P, Fe, O_2, N_2, S

_____ 14. If an atom has an atomic number of 92 and its atomic weight is 238, how many protons does the atom have?

    a. 238

    b. 92

    c. 146

    d. 54

_____ 15. If the second energy level of an atom has one electron, how many more does it need to fill it to its maximum capacity?

    a. 1

    b. 2

    c. 5

    d. 7

_____ 16. The atomic structure of hydrogen looks like which one of the following?

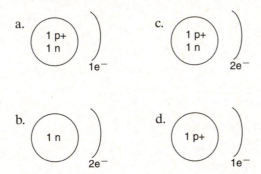

_____ 17. The number of neutrons in $^8O^{17}$ is

    a. 8.

    b. 9.

    c. 7.

    d. 17.

_____ 18. Which one of the following selections represents the pH of the *weakest* acid?

    a. 7.0

    b. 1.3

    c. 3.2

    d. 6.7

_____ 19. The type of bond that has the most important effects on the properties of water and the shapes of complex molecules is the

    a. hydrogen bond.

    b. ionic bond.

    c. covalent bond.

    d. polar covalent bond.

_____ 20. If oxygen has an atomic weight of 16, what is the molecular weight of an oxygen molecule?

    a. 16

    b. 8

    c. 32

    d. 2

_____ 21. If the concentration of hydrogen ions is (0.000001), what is the pH?

    a. 1

    b. 5

    c. 6

    d. 7

_____ 22. Two simple sugars joined together form a disaccharide. The reaction necessitates the

     a. removal of water to create a more complex molecule.

     b. addition of water to create a more complex molecule.

     c. presence of cations and anions to initiate electrical attraction.

     d. disassembling of molecules through hydrolysis.

_____ 23. The presence of a *carboxylic acid group* at the end of a carbon chain demonstrates a characteristic common to all

     a. amino acids.

     b. inorganic acids.

     c. nucleic acids.

     d. organic acids.

_____ 24. The three important functions of triglycerides are

     a. solubility, reactivity, and lubrication.

     b. energy, insulation, and protection.

     c. support, movement, and transport.

     d. buffering, metabolic regulation, and defense.

Completion

Using the terms below, complete the following statements. Use each term only once.

nucleic acids	isomers	molecular weight
peptide bond	ions	hydrolysis
ionic bond	hydrophobic	salts
mole	saturated	dehydration synthesis
molecule	alkaline	inert

1. Elements that do not readily participate in chemical processes are said to be _____.

2. For every element, a quantity that has a mass in grams equal to the atomic weight will contain the same number of atoms. The name given to this quantity is a(n) _____.

3. A chemical structure containing more than one atom is a(n) _____.

4. Atoms or molecules that have positive or negative charges are called _____.

5. Electrical attraction between opposite charges produces a(n) _____.

6. The sum of the atomic weights of its component atoms is a molecule's _____.

7. _____ are ionic compounds containing any cation (except H^+) and any anion (except OH^-).

8. Molecules that have few if any polar covalent bonds and do not dissolve in water are _____.

9. If the pH is above 7, the solution is _____.

10. Of the four major classes of organic compounds, the one responsible for storing genetic information is the _____.

11. Molecules that have the same molecular formula but different structural formulas are _____.

12. The process that breaks a complex molecule into smaller fragments by the addition of a water molecule is _____.

13. Glycogen, a branched polysaccharide composed of interconnected glucose molecules, is formed by the process of _____.

14. Butter, fatty meat, and ice cream are examples of sources of fatty acids that are said to be
    _____.

15. The attachment of a carboxylic acid group of one amino acid to the amino group of another forms a connection called a(n) _____.

Short Essay

Briefly answer the following questions in the spaces provided below.

1. Suppose an atom has eight protons, eight neutrons, and eight electrons. Construct a diagram of the atom and identify the subatomic particles by placing them in their proper locations. Identify the element.

2. From the following simulated reaction, identify which is the hydrolysis reaction and which is the dehydration synthesis reaction.
   a. $A–B + H_2O \rightarrow A–H + HO–B$
   b. $A–H + HO–B \rightarrow A – B + H_2O$

3. Why are the elements helium, argon, and neon called *inert gases*?

4. In a water (H_2O) molecule, the unequal sharing of electrons creates a *polar covalent bond*. Why?

5. Calculate the molecular weight (MW) of one molecule of glucose ($C_6H_{12}O_6$). [Note: atomic weights C = 12; H = 1; O = 16]

6. List four chemical properties of water that make life possible.

7. List the four major classes of organic compounds found in the human body and give an example for each one.

8. Differentiate between a saturated and an unsaturated fatty acid.

9. Using the four kinds of nucleotides that make up a DNA molecule, construct a model that shows the arrangement of the components that make up each nucleotide and the pairing of nucleotides. *Name each nucleotide.*

10. What are the three components that make up one nucleotide of ATP?

LEVEL 3: CRITICAL THINKING AND CLINICAL APPLICATIONS

Using principles and concepts learned in Chapter 2, answer the following questions. Write your answers on a separate sheet of paper.

1. Using the letters AB and CD, show how each would react in an exchange reaction.

2. Why might sodium bicarbonate ("baking soda") be used to relieve excessive stomach acid?

3. Using the glucose molecule ($C_6H_{12}O_6$), demonstrate your understanding of dehydration synthesis by writing an equation to show the formation of a molecule of sucrose ($C_{12}H_{22}O_{11}$). (Make sure the equation is balanced.)

4. Even though the recommended dietary intake for carbohydrates is 55 to 60 percent of the daily caloric intake, why do the carbohydrates account for less than 3 percent of our total body weight?

5. An important buffer system in the human body involves carbon dioxide (CO_2) and bicarbonate ion (HCO^-) in the reversible reaction $CO_2 + H_2O \rightleftharpoons H_2CO_3 \rightleftharpoons H^+ + HCO_3^-$. If a person becomes excited and exhales large amounts of CO_2, how will the pH of the person's body be affected?

6. How does excessive boiling or heating of proteins affect the ability of the protein molecule to perform its normal biological functions?

7. You are interested in losing weight, so you decide to eliminate your intake of fats completely. You opt for a fat substitute such as *Olestra*, which contains compounds that cannot be used by the body. Why might this decision be detrimental to you?

8. In a reaction pathway that consists of four steps, how would decreasing the amount of enzyme that catalyzes the second step affect the amount of product produced at the end of the pathway?

CHAPTER 3

The Cellular Level of Organization

OVERVIEW

Cells are the basic units of structure and function in all living things. Although they are typically microscopic in size, they are complex, highly organized units. In the human body all cells originate from a single fertilized egg and additional cells are produced by the division of the preexisting cells. These cells become specialized in a process called differentiation, forming tissues and organs that perform specific functions. The "roots" of the study of anatomy and physiology are established by understanding the basic concepts of cell biology. This chapter provides basic principles related to the structure of cell organelles and the vital physiological functions that each organelle performs.

The following exercises relating to cell biology will help you to learn the subject matter and to synthesize and apply the information in meaningful and beneficial ways.

LEVEL 1: REVIEWING FACTS AND TERMS

Review of Learning Outcomes

After completing this chapter, you should be able to do the following:

OUTCOME 3-1   List the functions of the plasma membrane and the structural features that enable it to perform those functions.

OUTCOME 3-2   Describe the organelles of a typical cell, and indicate the specific functions of each.

OUTCOME 3-3   Explain the functions of the cell nucleus and discuss the nature and importance of the genetic code.

OUTCOME 3-4   Summarize the role of DNA in protein synthesis, cell structure, and cell function.

OUTCOME 3-5   Describe the processes of cellular diffusion and osmosis, and explain their role in physiological systems.

OUTCOME 3-6   Describe carrier-mediated transport and vesicular transport mechanisms used by cells to facilitate the absorption or removal of specific substances.

OUTCOME 3-7   Explain the origin and significance of the transmembrane potential.

OUTCOME 3-8   Describe the stages of the cell life cycle, including mitosis, interphase, and cytokinesis, and explain their significance.

OUTCOME 3-9   Discuss the regulation of the cell life cycle.

OUTCOME 3-10   Discuss the relationship between cell division and cancer.

OUTCOME 3-11   Define differentiation, and explain its importance.

Multiple Choice

Place the letter corresponding to the best answer in the space provided.

OUTCOME 3-1 _____ 1. The major components of the cell membrane are

    a. carbohydrates, fats, proteins, and water.

    b. carbohydrates, lipids, ions, and vitamins.

    c. amino acids, fatty acids, carbohydrates, and cholesterol.

    d. phospholipids, proteins, glycolipids, and cholesterol.

OUTCOME 3-1 _____ 2. Most of the communication between the interior and exterior of the cell occurs by way of

    a. receptor sites.

    b. peripheral proteins.

    c. integral protein channels.

    d. the phospholipid bilayer.

OUTCOME 3-1 _____ 3. Because of its chemical–structural composition, the cell membrane is called a

    a. phospholipid bilayer.

    b. glycocalyx.

    c. microfilament.

    d. hydrophobic barrier.

OUTCOME 3-2 _____ 4. The primary components of the cytoskeleton, which gives the cell strength and rigidity and anchors the position of major organelles, are

    a. microvilli.

    b. microtubules.

    c. microfilaments.

    d. thick filaments.

OUTCOME 3-2 _____ 5. Which of the following selections contains only *membranous* organelles?

    a. Cytoskeleton, microvilli, centrioles, cilia, ribosomes

    b. Centrioles, lysosomes, nucleus, endoplasmic reticulum, cilia, ribosomes

    c. Mitochondria, nucleus, endoplasmic reticulum, Golgi apparatus, lysosomes, peroxisomes

    d. Nucleus, mitochondria, lysosomes, centrioles, microvilli

OUTCOME 3-2 _____ 6. Approximately 95 percent of the energy needed to keep a cell alive is generated by the activity of the

    a. mitochondria.

    b. ribosomes.

    c. nucleus.

    d. microtubules.

OUTCOME 3-2 _____ 7. *Nucleoli* are nuclear organelles that

    a. contain the chromosomes.

    b. are responsible for producing DNA.

    c. control nuclear operations.

    d. synthesize the components of ribosomes.

OUTCOME 3-2   _____   8. The three major functions of the *endoplasmic reticulum* are
        a. hydrolysis, diffusion, and osmosis.
        b. detoxification, packaging, and modification.
        c. synthesis, storage, transport.
        d. pinocytosis, phagocytosis, and storage.

OUTCOME 3-2   _____   9. The functions of the *Golgi apparatus* include
        a. synthesis, storage, alteration, and packaging.
        b. isolation, protection, sensitivity, and organization.
        c. strength, movement, control, and secretion.
        d. neutralization, absorption, assimilation, and secretion.

OUTCOME 3-2   _____   10. Peroxisomes, which are smaller than lysosomes, are primarily responsible for:
        a. the control of lysosomal activities.
        b. absorption and neutralization of toxins.
        c. synthesis and packaging of secretions.
        d. renewal and modification of the cell membrane.

OUTCOME 3-3   _____   11. The *major* factor that allows the nucleus to control cellular operations is through its
        a. ability to communicate chemically through nuclear pores.
        b. location within the cell.
        c. regulation of protein synthesis.
        d. "brainlike" sensory devices that monitor cell activity.

OUTCOME 3-3   _____   12. Ribosomal proteins and RNA are produced primarily in the
        a. nucleus.
        b. nucleolus.
        c. cytoplasm.
        d. mitochondria.

OUTCOME 3-3   _____   13. Along the length of the DNA strand, information is stored in the sequence of
        a. the sugar-phosphate linkages.
        b. nitrogen bases.
        c. the hydrogen bonds.
        d. the amino acid chain.

OUTCOME 3-3   _____   14. A sequence of three nitrogen bases can specify the identity of a
        a. specific gene.
        b. single DNA molecule.
        c. single amino acid.
        d. specific peptide chain.

OUTCOME 3-4   _____   15. The process in which RNA polymerase uses the genetic information to assemble a strand of mRNA is
        a. translation.
        b. transcription.
        c. initiation.
        d. elongation.

OUTCOME 3-4   _____ 16. If the DNA triplet is TAG, the corresponding codon on the mRNA strand will be
    a. AUC.
    b. AGC.
    c. ATC.
    d. ACT.

OUTCOME 3-4   _____ 17. If the mRNA has the codons (GGG) – (GCC) – (AAU), it will bind to the tRNAs with anticodons
    a. (CCC) – (CGG) – (TTA).
    b. (CCC) – (CGG) – (UUA).
    c. (CCC) – (AUC) – (UUA).
    d. (CCC) – (UAC) – (TUA).

OUTCOME 3-5   _____ 18. The transport process that requires the presence of specialized integral membrane protein is
    a. diffusion.
    b. carrier-mediated transport.
    c. reticular transport.
    d. osmosis.

OUTCOME 3-5   _____ 19. Ions and other small water-soluble materials cross the cell membrane only by passing through
    a. ligands.
    b. membrane anchors.
    c. a channel protein.
    d. a receptor protein.

OUTCOME 3-6   _____ 20. All transport through the cell membrane can be classified as either
    a. active or passive.
    b. diffusion or osmosis.
    c. pinocytosis or phagocytosis.
    d. permeable or impermeable.

OUTCOME 3-6   _____ 21. The mechanism by which glucose can enter the cytoplasm without expending ATP is via
    a. secondary active transport.
    b. diffusion.
    c. an ion pump.
    d. a recognition protein.

OUTCOME 3-6   _____ 22. The major difference between diffusion and bulk flow is that when molecules move by *bulk flow,* they move
    a. at random.
    b. as a unit in one direction.
    c. as individual molecules.
    d. slowly over long distances.

OUTCOME 3-6 _____ 23. In osmosis, water flows across a membrane toward the solution that has the
   a. higher concentration of water.
   b. lower concentration of solutes.
   c. concentration of solutes at equilibrium.
   d. higher concentration of solutes.

OUTCOME 3-7 _____ 24. The transmembrane potential results from the
   a. unequal distribution of ions across the membrane.
   b. functional reverse of endocytosis.
   c. formation of endosomes filled with extracellular fluid.
   d. equal distribution of ions across the membrane.

OUTCOME 3-7 _____ 25. The factor(s) that interact(s) to create and maintain the transmembrane potential is (are) the
   a. membrane permeability for sodium.
   b. membrane permeability for potassium.
   c. presence of the Na-K exchange pump.
   d. a, b, and c are correct.

OUTCOME 3-8 _____ 26. The correct sequence of the cell cycle beginning with interphase is
   a. G_1, G_0, G_2, S_1, M.
   b. G_0, G_1, S, G_2, M.
   c. G_1, G_0, S_1, M, G_2.
   d. G_0, S_1, G_1, M, G_2.

OUTCOME 3-8 _____ 27. The process of mitosis takes place in the
   a. S phase.
   b. M phase.
   c. G_1 phase.
   d. G_0 phase.

OUTCOME 3-8 _____ 28. The four stages of mitosis in correct sequence are
   a. prophase, anaphase, metaphase, telophase.
   b. prophase, metaphase, telophase, anaphase.
   c. prophase, metaphase, anaphase, telophase.
   d. prophase, anaphase, telophase, metaphase.

OUTCOME 3-9 _____ 29. As the cell life cycle proceeds, cyclin levels climb, causing the maturation promoting factor (MPF) to appear in the cytoplasm, initiating the process of
   a. metabolic turnover.
   b. cell differentiation.
   c. repair and development.
   d. mitosis.

OUTCOME 3-9 _____ 30. At the chromosome level, the number of cell divisions performed by a cell and its descendants is regulated by structures called

    a. neoplasms.

    b. telomeres.

    c. chalones.

    d. oncogenes.

OUTCOME 3-10 _____ 31. The spreading process of a primary tumor is called _____, and the dispersion of malignant cells to establish a secondary tumor is called _____.

    a. invasion; metastasis

    b. penetration; circulation

    c. metastasis; invasion

    d. circulation; penetration

OUTCOME 3-11 _____ 32. The process of differentiation is cell specialization as a result of

    a. irreversible alteration in protein structure.

    b. the presence of an oncogene.

    c. gene activation or repression.

    d. the process of fertilization.

Completion

Using the terms below, complete the following statements. Use each term only once.

transmembrane	differentiation	phospholipids	interphase
translation	MPF	cytoskeleton	anaphase
endocytosis	gene	cancer	nucleolus
ribosomes	nuclear pores	fixed ribosomes	osmosis

OUTCOME 3-1   1. Most of the surface area of the cell membrane consists of _____.

OUTCOME 3-1   2. Intracellular membrane proteins are bound to a network of supporting filaments called the _____.

OUTCOME 3-2   3. The areas of the endoplasmic reticulum that are called the rough ER contain _____.

OUTCOME 3-3   4. Chemical communication between the nucleus and cytosol occurs through the _____.

OUTCOME 3-3   5. The transient nuclear organelle that synthesizes ribosomal RNA is the _____.

OUTCOME 3-3   6. A _____ contains all the DNA triplet codes needed to produce specific peptides.

OUTCOME 3-4   7. The construction of a functional polypeptide using the information provided by an mRNA strand is called _____.

OUTCOME 3-4   8. The cell organelles responsible for the synthesis of proteins using information provided by nuclear DNA are _____.

OUTCOME 3-5   9. The special name given to the net diffusion of water across a membrane is _____.

OUTCOME 3-6   10. The packaging of extracellular materials in a vesicle at the cell surface for importation into the cell is called _____.

OUTCOME 3-7   11. The separation of positive and negative ions by the cell membrane produces a potential difference called the _____ potential.

OUTCOME 3-8   12. Somatic cells spend the majority of their functional lives in a state known as _____.

OUTCOME 3-8   13. The phase of mitosis in which the chromatid pairs separate and the daughter chromosomes move toward opposite ends of the cells is _____.

OUTCOME 3-9   14. One of the internal stimuli that causes many cells to set their own pace for mitosis and cell division is the _____.

OUTCOME 3-10   15. In adults, telomerase activation is a key step in the development of _____.

OUTCOME 3-11   16. The specialization process that causes a cell's functional abilities to become more restricted is called _____.

Matching

Match the terms in column B with the terms in column A. Use letters for answers in the spaces provided. Use each term only once.

Part I

		Column A	Column B
OUTCOME 3-1	_____	1. recognition protein	A. membranous
OUTCOME 3-1	_____	2. cell membrane	B. chromosomes
OUTCOME 3-2	_____	3. ribosomes	C. DNA–nitrogen base
OUTCOME 3-2	_____	4. Golgi apparatus	D. RNA–nitrogen base
OUTCOME 3-3	_____	5. nucleus	E. elongation
OUTCOME 3-3	_____	6. DNA strands	F. transcription
OUTCOME 3-4	_____	7. thymine	G. cell control center
OUTCOME 3-4	_____	8. uracil	H. phospholipid bilayer
OUTCOME 3-4	_____	9. mRNA formation	I. glycoprotein
OUTCOME 3-4	_____	10. produces peptide chains	J. protein factories

Part II

		Column A	Column B
OUTCOME 3-5	_____	11. plasma membrane	K. slightly negatively charged
OUTCOME 3-5	_____	12. osmosis	L. DNA replication
OUTCOME 3-6	_____	13. active transport	M. transmembrane potentials
OUTCOME 3-7	_____	14. inside cell membrane	N. selectively permeable
OUTCOME 3-7	_____	15. resting potential	O. cell specialization
OUTCOME 3-7	_____	16. potential difference	P. visible chromosomes
OUTCOME 3-8	_____	17. S phase	Q. shorten with each division
OUTCOME 3-8	_____	18. mitosis	R. net movement of water
OUTCOME 3-9	_____	19. telomere	S. undisturbed cell
OUTCOME 3-10	_____	20. metastasis	T. carrier-mediated
OUTCOME 3-11	_____	21. differentiation	U. malignant cell division

Drawing/Illustration Labeling

Identify each numbered structure. Place your answers in the spaces provided.

OUTCOME 3-2 **FIGURE 3-1** Anatomy of a Model Cell

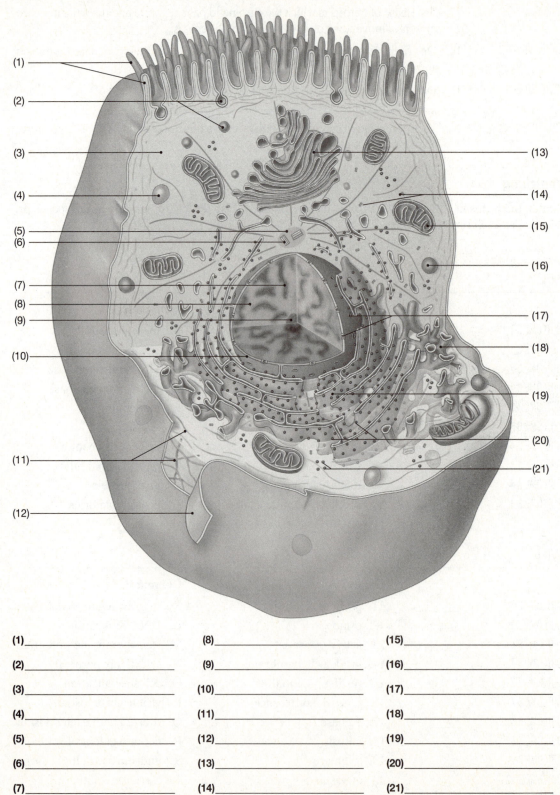

(1)_____	(8)_____	(15)_____
(2)_____	(9)_____	(16)_____
(3)_____	(10)_____	(17)_____
(4)_____	(11)_____	(18)_____
(5)_____	(12)_____	(19)_____
(6)_____	(13)_____	(20)_____
(7)_____	(14)_____	(21)_____

OUTCOME 3-1   **FIGURE 3-2** The Plasma Membrane

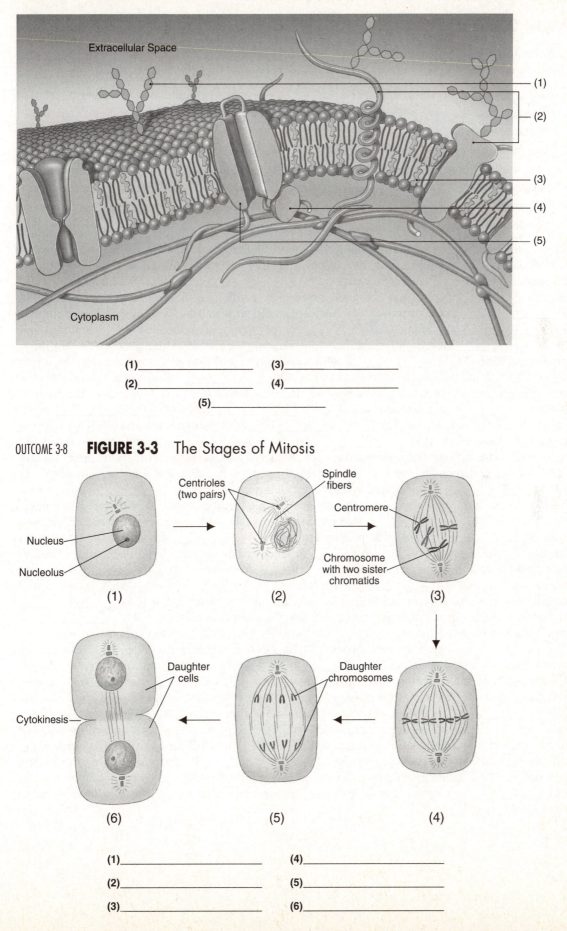

(1)_____   (3)_____

(2)_____   (4)_____

(5)_____

OUTCOME 3-8   **FIGURE 3-3** The Stages of Mitosis

(1)_____   (4)_____

(2)_____   (5)_____

(3)_____   (6)_____

LEVEL 2: REVIEWING CONCEPTS

Chapter Overview

Using the terms below, fill in the blanks to complete the concepts in the cellular level of organization. Use each term only once.

autolysis	proteins	protein synthesis	cilia
waste	ribosomes	peripheral	cell membrane
nuclear pores	saccules	ATP	cristae
nucleoli	organelles	cytosol	integral
extracellular fluid	lysosomes	Golgi apparatus	phospholipids
cytoskeleton	nucleus	metabolic enzymes	mitochondria
digestive enzymes	ions	channels	chromosomes
nuclear envelope	nucleoplasm	matrix	

Cells are the building blocks of all plants and animals. They are the smallest units that perform all vital physiological functions. The study of cells begins by looking at the anatomy of a model cell that is surrounded by a watery medium known as the (1) _____. The outer boundary of the cells, the (2) _____, may contain extensions containing microtubule doublets called (3) _____, used to move material over the cell surface. The membrane structure consists of a (4) _____ bilayer with (5) _____ proteins embedded in the membrane and (6) _____ proteins attached to the inner membrane surface. These proteins line the inner surface and form (7) _____ to allow materials to pass into and out of the cell. The inside of the cell contains the intracellular fluid called (8) _____. Resistance to movement is greater inside the cell than outside because of the presence of dissolved nutrients in the form of (9) _____, soluble and insoluble (10) _____, and (11) _____ products. A protein framework, the (12)_____, gives the cytosol strength and flexibility. Formed functional structures, the (13) _____, are in abundance, some attached and others floating freely in the cytosol. Some of these structures appear to lack membranes and others are membranous. The ribosomes, some of which are attached and some free-floating, are involved in (14) _____. Small "cashew-shaped" structures, the (15) _____, float freely within the fluid intracellular medium. This unusual double membranous formation contains numerous folds called (16) _____ that serve to increase the surface area exposed to the fluid contents, the (17) _____. The presence of (18) _____ attached to the folds would indicate that this is where (19) _____ is generated by the mitochondria. Floating in the cytosol close to or in the center of the model cell is the "control center," the (20) _____ enclosed by a (21) _____ containing (22) _____ that allow small particles/materials to pass through. Smaller "nucleus-like" structures, the (23) _____, engage in activities indicating that they synthesize the components of the (24) _____, since RNA ribosomal proteins are in abundance. The fluid content of the nucleus, the (25) _____, contains ions, enzymes, RNA, DNA, and their nucleotides. The DNA strands form complex structures called (26) _____, which contain information to synthesize thousands of different proteins and controls the synthesis of RNA. Some of the synthesized protein molecules will be stored, while others via transport vesicles will be delivered to the (27) _____, a system of flattened membrane discs called (28) _____. These discs look like a stack of dinner plates in which synthesis and packaging of secretions along with cell membrane renewal and modifications are taking place. The breakdown and recycling processes in cells are performed by special vesicles called (29) _____. Produced at the Golgi apparatus, they provide an isolated environment for potentially dangerous chemical reactions. They contain (30) _____, which can destroy the cell's proteins and organelles in a process called (31) _____.

Concept Map I

Using the following terms, fill in the circled, numbered, blank spaces to correctly complete the concept map. Use each term only once.

ribosomes   nucleolus   membranous   centrioles

lysosomes   lipid bilayer   proteins   organelles

fluid component

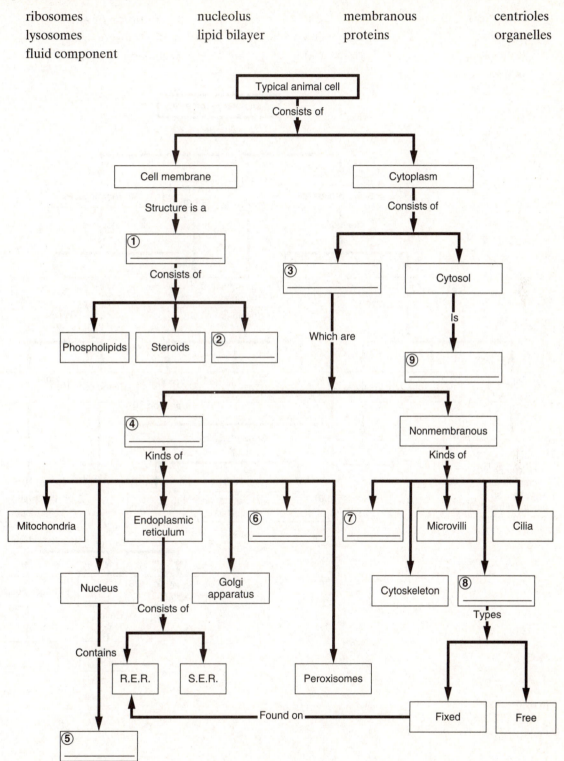

Concept Map II

Using the following terms, fill in the circled, numbered, blank spaces to correctly complete the concept map. Use each term only once.

exocytosis

pinocytosis

net diffusion of water

diffusion

channel mediated

active transport

"cell eating"

vesicular transport

specificity

molecular size

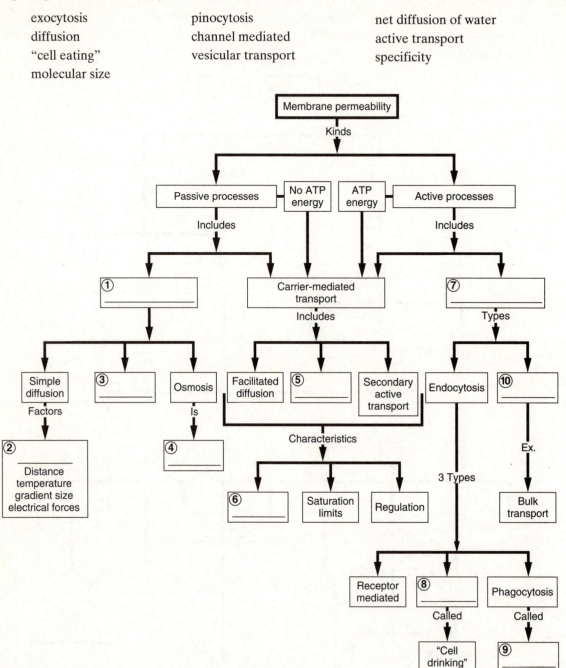

Concept Map III

Using the following terms, fill in the circled, numbered, blank spaces to correctly complete the concept map. Follow the numbers to comply with the organization of the concept map. Use each term only once.

metaphase   somatic cells   telophase   cytokinesis
DNA replication G_2 phase   mitosis G_1 phase

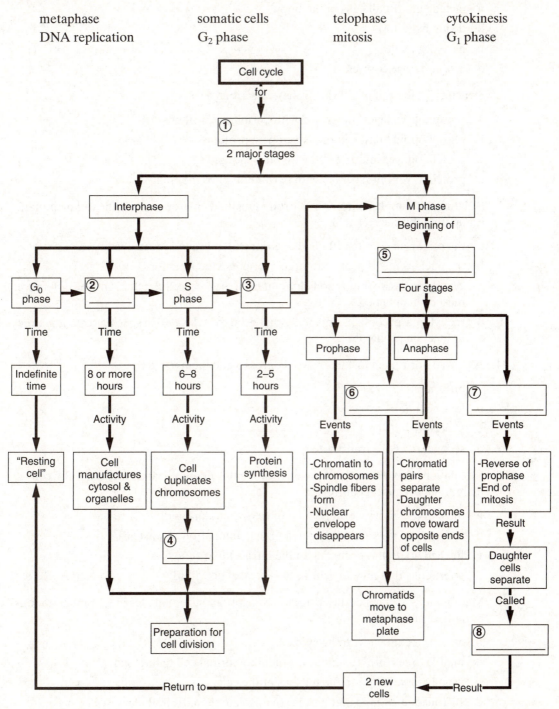

Multiple Choice

Place the letter corresponding to the best answer in the space provided.

_____ 1. Somatic cells

    a. include almost all cells in the body.

    b. divide by mitosis.

    c. have organelles.

    d. a, b, and c are correct.

_____ 2. Structurally, the cell membrane is best described as a

    a. phospholipid layer integrated with peripheral proteins.

    b. phospholipid bilayer interspersed with proteins.

    c. protein bilayer interspersed with phospholipids.

    d. protein layer interspersed with peripheral phospholipids.

_____ 3. Isolating the cytoplasm from the surrounding fluid environment by the cell membrane is important because

    a. the cell organelles lose their shape if the membrane is destroyed.

    b. the nucleus needs protection to perform its vital functions.

    c. cytoplasm has a composition different from the extracellular fluid and the differences must be maintained.

    d. the cytoplasm contains organelles that need to be located in specific regions in order to function properly.

_____ 4. A steroid hormone most likely would cross a cell membrane by

    a. diffusion.

    b. leak channel.

    c. exocytosis.

    d. secondary active transport.

_____ 5. Regulation of exchange with the environment is an important function of the cell membrane because it controls

    a. most of the activity occurring in the extracellular fluid.

    b. the entry of ions and nutrients, and the elimination of wastes.

    c. the activity that is occurring in the intracellular fluid.

    d. alterations that might occur in the extracellular fluid.

_____ 6. Membranous organelles differ from nonmembranous organelles in that membranous organelles are

    a. always in contact with the cytosol.

    b. unable to perform functions essential to normal cell maintenance.

    c. usually found close to the nucleus of the cell.

    d. surrounded by lipid membranes that isolate them from the cytosol.

_____ 7. The major functional difference between flagella and cilia is that flagella

    a. move fluid past a stationary cell.

    b. move fluids or secretions across the cell surface.

    c. move a cell through the surrounding fluid.

    d. move DNA molecules during cell division.

_____   8. The smooth ER (SER) has a variety of functions that center around the synthesis of

    a. lipids and carbohydrates.

    b. proteins and lipids.

    c. glycogen and proteins.

    d. carbohydrates and proteins.

_____   9. The reason lysosomes are sometimes called "cellular suicide packets" is

    a. the lysosome fuses with the membrane of another organelle.

    b. lysosomes fuse with endocytic vesicles with solid materials.

    c. the breakdown of lysosomal membranes can destroy a cell.

    d. lysosomes have structures that penetrate other cells.

_____   10. The energy-producing process in the mitochondria involves a series of reactions in which _____ is consumed and _____ is generated.

    a. carbon dioxide; oxygen

    b. water; oxygen

    c. carbon dioxide; water

    d. oxygen; carbon dioxide

_____   11. The most notable characteristic of the G_0 phase of an interphase cell is that

    a. the cell is manufacturing cell organelles.

    b. the cell carries on normal functions.

    c. the cell duplicates its chromosomes.

    d. DNA polymerase binds to nitrogen bases.

_____   12. The replication of DNA occurs primarily during the

    a. S phase.

    b. G_1 phase.

    c. G_2 phase.

    d. M phase.

_____   13. The process of *cytokinesis* refers to the

    a. constriction of chromosomes along the metaphase plate.

    b. formation of spindle fibers between the centriole pairs.

    c. reorganization of the nuclear contents.

    d. physical separation of the daughter cells.

_____   14. Which of the following most accurately describes a factor contributing to the transmembrane potential in a human cell?

    a. Cations are concentrated inside the cell.

    b. Phospholipids are positively charged, so negatively charged particles stay inside the cell.

    c. K^+ and Na^+ diffuse equally across the cell membrane and the interior of the cell develops a negative charge.

    d. Proteins inside the cell have a net negative charge relative to the extracellular fluid.

_____   15. The sodium-potassium exchange pump

    a. moves sodium and potassium ions along their concentration gradients.

    b. is composed of a carrier protein located in the cell membrane.

    c. is not necessary for the maintenance of homeostasis.

    d. does not require ATP cellular energy.

_____ 16. The reason that *water-soluble* ions and molecules cannot easily enter certain regions of the cell membrane is because of the presence of

    a. hydrophilic ends exposed to the solution.

    b. channels too small for ions and molecules to enter.

    c. hydrophobic tails on the interior of the membrane.

    d. gated channels that close when ions are present.

_____ 17. The effect of diffusion in body fluids is that it tends to

    a. increase the concentration gradient of the fluid.

    b. scatter the molecules and inactivate them.

    c. repel like charges and attract unlike charges.

    d. equilibrate local concentration gradients.

_____ 18. During *osmosis,* water will always flow across a membrane toward the solution that has the

    a. highest concentration of solvents.

    b. highest concentration of solutes.

    c. equal concentrations of solutes.

    d. equal concentrations of solvents.

_____ 19. A solution that is hypotonic to a cell's cytoplasm causes

    a. water to move into the cell.

    b. water to move out of the cell.

    c. crenation.

    d. no effect on water movement.

_____ 20. Red blood cells undergo crenation when the cells are placed in contact with a(n)

    a. hypotonic solution.

    b. hypertonic solution.

    c. isotonic solution.

    d. salt solution.

_____ 21. Injecting a concentrated salt solution into the circulatory system would result in

    a. little or no effect on the red blood cells.

    b. hemolysis of the red blood cells.

    c. a slight increase in the osmotic pressure inside the red blood cell.

    d. increased osmotic pressure of the ECF outside the red blood cells.

_____ 22. *Facilitated diffusion* differs from ordinary diffusion in that

    a. ATP is expended during facilitated diffusion.

    b. molecules move against a concentration gradient.

    c. carrier proteins are involved.

    d. it is an active process utilizing carriers.

_____ 23. One of the great advantages of moving materials by *active transport* is

    a. carrier proteins are not necessary.

    b. the process is not dependent on a concentration gradient.

    c. the process has no energy cost.

    d. receptor sites are not necessary for the process to occur.

_____ 24. In the human body, the process of *phagocytosis* is illustrated by

     a. air expelled from the lungs.

     b. a specific volume of blood expelled from the left ventricle.

     c. vacuolar digestion of a solvent.

     d. a white blood cell engulfing a bacterium.

_____ 25. Epsom salts exert a laxative effect due to the process of

     a. osmosis.

     b. diffusion.

     c. diarrhea.

     d. phagocytosis.

_____ 26. The formation of a malignant tumor indicates that

     a. the cells are remaining within a connective tissue capsule.

     b. the tumor cells resemble normal cells, but they are dividing faster.

     c. mitotic rates of cells are no longer responding to normal control mechanisms.

     d. metastasis is necessary and easy to control.

Completion

Using the terms below, complete the following statements. Use each term only once.

cilia	hydrophobic	peroxisomes	microvilli
exocytosis	rough ER	phagocytosis	glycocalyx
diffusion	endocytosis	isotonic	cytokinesis
autolysis	permeability	mitosis	

1. Ions and water-soluble compounds cannot cross the lipid portion of a cell membrane because the lipid tails of the phospholipid molecules are highly _____.

2. The viscous layer that provides protection and lubrication for the cell membrane is known as the _____.

3. Cells that are actively engaged in *absorbing* materials from the extracellular fluid, such as the cells of the digestive tract and the kidneys, contain _____.

4. In the respiratory tract, sticky mucus and trapped dust particles are moved toward the throat and away from the delicate respiratory surface because of the presence of _____.

5. Pancreatic cells that manufacture digestive *enzymes* contain an extensive _____.

6. Glycoproteins that cover most cell surfaces are usually released by the process of _____.

7. Cells may remove bacteria, fluids, and organic debris from their surroundings in vesicles at the cell surface by the process of _____.

8. Toxins such as alcohol or hydrogen peroxide that are absorbed from the extracellular fluid or generated by chemical reactions in the cytoplasm are absorbed and neutralized by _____.

9. The property that determines the cell membrane's effectiveness as a barrier is its _____.

10. A drop of ink spreading to color an entire glass of water demonstrates the process of _____.

11. If a solution has a solute concentration that will not cause a net movement in or out of the cells, the solution is said to be _____.

12. When lysosomal enzymes destroy the cell's proteins and organelles, the process is called _____.

13. Pseudopodia are cytoplasmic extensions that function in the process of _____.

14. The division of somatic cells followed by the formation of two daughter cells is a result of the process of _____.

15. The *end* of a cell division is marked by the completion of _____.

Short Essay

Briefly answer the following questions in the spaces provided below.

1. Confirm your understanding of cell specialization by citing five systems in the human body and naming a specialized cell found in each system.

2. List four general functions of the cell membrane.

3. Sequentially list the phases of the interphase stage of the cell life cycle, and briefly describe what occurs in each.

4. What organelles would be necessary to construct a functional "typical" cell? (Assume the presence of cytosol and a cell membrane.)

5. What are the *functional* differences among centrioles, cilia, and flagella?

6. What five major factors influence diffusion rates?

LEVEL 3: CRITICAL THINKING AND CLINICAL APPLICATIONS

Using principles and concepts learned in Chapter 3, answer the following questions. Write your answers on a separate sheet of paper.

1. How does cytokinesis play an important role in the cell cycle?

2. An instructor at the fitness center tells you that bodybuilders have the potential for increased supplies of energy and improved muscular performance because of increased numbers of mitochondria in their muscle cells. Why is this?

3. All cells are enclosed by a cell membrane that is necessary for cell functions and protection. What benefits would cellular organelles have if enclosed by a membrane similar to the cell membrane?

4. Using the principles of *tonicity*, explain the difference between drowning in the ocean versus an inland freshwater lake.

5. Many coaches and personal fitness trainers recommend drinking Gatorade after strenuous exercise because "it gives your body what it thirsts for." Using the principles of tonicity, explain the benefits of ingesting such a drink during and after exercising.

6. A patient recovering from surgery was mistakenly given a transfusion of 5 percent salt solution instead of a 0.9 percent physiological saline solution. The patient immediately went into shock and soon after died. What caused the patient to go into shock?

7. Let's say you were preparing vegetables to be used in a tossed salad several hours later and you wanted those vegetables to remain crisp, so you placed them in a bowl of cold tap water. Osmotically speaking, explain why the veggies remain crisp.

CHAPTER 4

The Tissue Level of Organization

OVERVIEW

Have you ever thought what it would be like to live on an island all alone? Your survival would depend on your ability to perform all of the activities necessary to remain healthy and alive. In today's society, surviving alone would be extremely difficult because of the important interrelationships and interdependence we have with others to support us in all aspects of life inherent in everyday living. So it is with individual cells in the multicellular body.

Individual cells of similar structure and function join together to form groups called *tissues*, which are identified on the basis of their origin, location, shape, and function. Many of the tissues are named according to the organ system in which they are found or the function they perform, such as neural tissue in the nervous system, muscle tissue in the muscular system, or connective tissue that is involved with the structural framework of the body and supporting, surrounding, and interconnecting other tissue types.

The activities in Chapter 4 introduce the discipline of *histology*, the study of tissues, with emphasis on the four primary types: epithelial, connective, muscle, and neural tissue. The questions and exercises are designed to help you organize and conceptualize the interrelationships and interdependence of individual cells that extend to the tissue level of cellular organization.

LEVEL 1: REVIEWING FACTS AND TERMS

Review of Learning Outcomes

After completing this chapter, you should be able to do the following:

OUTCOME 4-1 Identify the four major types of tissues in the body and describe their roles.

OUTCOME 4-2 Discuss the types and functions of epithelial tissue.

OUTCOME 4-3 Describe the relationship between form and function for each type of epithelium.

OUTCOME 4-4 Compare the structures and functions of the various types of connective tissue.

OUTCOME 4-5 Describe how cartilage and bone function as a supporting connective tissue.

OUTCOME 4-6 Explain how epithelial and connective tissues combine to form four types of membranes, and specify the functions of each.

OUTCOME 4-7   Describe how connective tissue establishes the framework of the body.

OUTCOME 4-8   Describe the three types of muscle tissue and the special structural features of each type.

OUTCOME 4-9   Discuss the basic structure and role of neural tissue.

OUTCOME 4-10   Describe how injuries affect the tissues of the body.

OUTCOME 4-11   Describe how aging affects the tissues of the body.

Multiple Choice

Place the letter corresponding to the best answer in the space provided.

OUTCOME 4-1   _____   1. The four primary tissue types found in the human body are
                a. squamous, cuboidal, columnar, and glandular.
                b. adipose, elastic, reticular, and cartilage.
                c. skeletal, cardiac, smooth, and muscle.
                d. epithelial, connective, muscle, and neural.

OUTCOME 4-2   _____   2. The type of tissue that covers exposed surfaces and lines internal passageways and body cavities is
                a. muscle.
                b. neural.
                c. epithelial.
                d. connective.

OUTCOME 4-3   _____   3. The two types of *layering* recognized in epithelial tissues are
                a. cuboidal and columnar.
                b. squamous and cuboidal.
                c. columnar and stratified.
                d. simple and stratified.

OUTCOME 4-3   _____   4. The types of cells that form glandular epithelium that secrete enzymes and buffers in the pancreas and salivary glands are
                a. simple squamous epithelium.
                b. simple cuboidal epithelium.
                c. stratified cuboidal epithelium.
                d. transitional epithelium.

OUTCOME 4-3   _____   5. The type of epithelial tissue found along the ducts that drain sweat glands is
                a. transitional epithelium.
                b. simple squamous epithelium.
                c. stratified cuboidal epithelium.
                d. pseudostratified columnar epithelium.

OUTCOME 4-3   _____   6. A single layer of epithelial cells covering a basal lamina is termed
                a. simple epithelium.
                b. stratified epithelium.
                c. squamous epithelium.
                d. cuboidal epithelium.

OUTCOME 4-3 _____ 7. Simple epithelial cells are characteristic of regions where

    a. mechanical or chemical stresses occur.

    b. support and flexibility are necessary.

    c. padding and elasticity are necessary.

    d. secretion and absorption occur.

OUTCOME 4-3 _____ 8. From a surface view, cells that look like fried eggs laid side by side are

    a. squamous epithelium.

    b. simple epithelium.

    c. cuboidal epithelium.

    d. columnar epithelium.

OUTCOME 4-3 _____ 9. Stratified epithelium has several cell layers above the basal lamina and is usually found in areas where

    a. secretion and absorption occur.

    b. mechanical or chemical stresses occur.

    c. padding and elasticity are necessary.

    d. storage and secretion occur.

OUTCOME 4-3 _____ 10. Cells that form a neat row with nuclei near the center of each cell and that appear square in typical sectional views are

    a. stratified epithelium.

    b. squamous epithelium.

    c. cuboidal epithelium.

    d. columnar epithelium.

OUTCOME 4-3 _____ 11. The major structural difference between columnar epithelia and cuboidal epithelia is that the *columnar epithelia*

    a. are hexagonal and the nuclei are near the center of each cell.

    b. consist of several layers of cells above the basal lamina.

    c. are thin and flat and occupy the thickest portion of the membrane.

    d. are taller and slender and the nuclei are crowded into a narrow band close to the basement membrane.

OUTCOME 4-3 _____ 12. Simple squamous epithelium would be found in the following area(s) of the body

    a. urinary tract and inner surface of circulatory system.

    b. respiratory surface of lungs.

    c. lining of body cavities.

    d. a, b, and c are correct.

OUTCOME 4-3 _____ 13. Stratified columnar epithelia provide protection along portions of which of the following systems?

    a. Skeletal, muscular, endocrine, and integumentary

    b. Lymphoid, cardiovascular, urinary, and reproductive

    c. Nervous, skeletal, muscular, and endocrine

    d. Reproductive, digestive, respiratory, and urinary

OUTCOME 4-3   _____   14. Glandular epithelia contain cells that produce

    a. exocrine secretions only.

    b. exocrine or endocrine secretions.

    c. endocrine secretions only.

    d. secretions released from goblet cells only.

OUTCOME 4-4   _____   15. The three basic components of all connective tissues are

    a. free exposed surface, exocrine secretions, and endocrine secretions.

    b. fluid matrix, cartilage, and osteocytes.

    c. specialized cells, extracellular protein fibers, and ground substance.

    d. satellite cells, cardiocytes, and osteocytes.

OUTCOME 4-4   _____   16. The three classes of connective tissue based on structure and function are

    a. fluid, supporting, and connective tissue proper.

    b. cartilage, bone, and blood.

    c. collagenic, reticular, and elastic.

    d. adipose, reticular, and ground.

OUTCOME 4-4   _____   17. The two *most abundant* cell populations found in connective tissue proper are

    a. fibroblasts and fibrocytes.

    b. mast cells and lymphocytes.

    c. melanocytes and mesenchymal cells.

    d. fixed cells and wandering cells.

OUTCOME 4-4   _____   18. Most of the volume in loose connective tissue is made up of

    a. elastic fibers.

    b. ground substance.

    c. reticular fibers.

    d. collagen fibers.

OUTCOME 4-4   _____   19. The major purposes of adipose tissue in the body are

    a. strength, flexibility, and elasticity.

    b. support, connection, and conduction.

    c. energy storage, cushioning, and insulating.

    d. absorption, compression, and lubrication.

OUTCOME 4-4   _____   20. Reticular tissue forms the basic framework and organization for several organs that have

    a. a complex three-dimensional structure.

    b. tightly packed collagen and elastic fibers.

    c. adipocytes that are metabolically active.

    d. relative proportions of cells, fibers, and ground substance.

OUTCOME 4-4   _____   21. Tendons are cords of dense regular connective tissue that

    a. cover the surface of a muscle.

    b. connect one bone to another.

    c. attach skeletal muscles to bones.

    d. surround organs such as skeletal muscle tissue.

OUTCOME 4-4 _____ 22. Ligaments are bundles of elastic and collagen fibers that
a. connect one bone to another bone.
b. attach skeletal muscle to bones.
c. connect one muscle to another muscle.
d. cover the surface of a muscle.

OUTCOME 4-4 _____ 23. The three major subdivisions of the extracellular fluid in the body are
a. blood, water, and saliva.
b. plasma, interstitial fluid, and lymph.
c. blood, urine, and saliva.
d. spinal fluid, cytosol, and blood.

OUTCOME 4-5 _____ 24. Bone functions as a supporting connective tissue because
a. the collagen fibers act like steel reinforcing rods.
b. the mineralized matrix acts like concrete.
c. the minerals surrounding the collagen fibers produce a strong, flexible combination that is highly resistant to shattering.
d. a, b, and c are correct.

OUTCOME 4-5 _____ 25. The two types of supporting connective tissue are
a. skeletal and smooth.
b. cutaneous and serous.
c. cartilage and bone.
d. columnar and cuboidal.

OUTCOME 4-5 _____ 26. The type of tissue that fills internal spaces and provides structural support and a framework for communication within the body is
a. connective.
b. epithelial.
c. muscle.
d. neural.

OUTCOME 4-5 _____ 27. The three major types of cartilage found in the body are
a. collagen, reticular, and elastic cartilage.
b. regular, irregular, and dense cartilage.
c. hyaline, elastic, and fibrocartilage.
d. interstitial, appositional, and calcified.

OUTCOME 4-5 _____ 28. The pads that lie between the vertebrae of the spinal column contain
a. elastic fibers.
b. fibrocartilage.
c. hyaline cartilage.
d. dense, regular connective tissue.

OUTCOME 4-6 _____ 29. The mucous membranes that are lined by simple epithelia perform the functions of
a. digestion and circulation.
b. respiration and excretion.
c. absorption and secretion.
d. a, b, and c are correct.

OUTCOME 4-6 _____ 30. The mesothelium of serous membranes is very thin, a structural characteristic that makes them
    a. subject to friction.
    b. relatively waterproof and usually dry.
    c. resistant to abrasion and bacterial attack.
    d. extremely permeable.

OUTCOME 4-6 _____ 31. Bone cells found in the lacunae within the matrix are called
    a. osteocytes.
    b. chondrocytes.
    c. adipocytes.
    d. stroma.

OUTCOME 4-7 _____ 32. The layers of connective tissue that create the internal framework of the body are responsible for
    a. providing strength and stability.
    b. maintaining the relative positions of internal organs.
    c. providing a route for the distribution of blood vessels.
    d. a, b, and c.

OUTCOME 4-8 _____ 33. Muscle tissue has the ability to
    a. provide a framework for communication within the body.
    b. carry impulses from one part of the body to another.
    c. cover exposed surfaces of the body.
    d. contract and produce active movement.

OUTCOME 4-8 _____ 34. The three types of muscle tissue found in the body are
    a. elastic, hyaline, and fibrous.
    b. striated, nonstriated, and fibrous.
    c. voluntary, involuntary, and nonstriated.
    d. skeletal, cardiac, and smooth.

OUTCOME 4-8 _____ 35. Skeletal muscle fibers are very unusual because they may be
    a. a foot or more in length, and each cell contains hundreds of nuclei.
    b. subject to the activity of pacemaker cells, which establish contraction rate.
    c. devoid of striations, spindle-shaped, with a single nucleus.
    d. unlike smooth muscle cells capable of division.

OUTCOME 4-9 _____ 36. Neural tissue is specialized to
    a. contract and produce movement.
    b. carry electrical impulses from one part of the body to another.
    c. provide structural support and fill internal spaces.
    d. line internal passageways and body cavities.

OUTCOME 4-9 _____ 37. The major function of *neurons* in neural tissue is to
    a. provide a supporting framework for neural tissue.
    b. regulate the composition of the interstitial fluid.
    c. act as phagocytes that defend neural tissue.
    d. transmit signals that take the form of changes in the transmembrane potential.

OUTCOME 4-9 _____ 38. Structurally, neurons are unique because they are the only cells in the body that have

    a. lacunae and canaliculi.

    b. axons and dendrites.

    c. satellite cells and neuroglia.

    d. soma and stroma.

OUTCOME 4-10 _____ 39. The restoration of homeostasis after an injury involves two related processes, which are

    a. necrosis and fibrous.

    b. infection and immunization.

    c. inflammation and regeneration.

    d. isolation and reconstruction.

OUTCOME 4-10 _____ 40. The release of histamine by mast cells at an injury site produces the following responses

    a. redness, warmth, and swelling.

    b. bleeding, clotting, and healing.

    c. necrosis, fibrosis, and scarring.

    d. hematoma, shivering, and retraction.

OUTCOME 4-11 _____ 41. With advancing age, the

    a. hormonal activity remains unchanged.

    b. energy consumption in general declines.

    c. effectiveness of tissue repair increases.

    d. all of the above are correct.

Completion

Using the terms below, complete the following statements. Use each term only once.

polarity	lamina propria	collagen	neuroglia
necrosis	exocytosis	connective	mesothelium
endothelium	skeletal	reticular	stratified squamous
neural	abscess	increase	superficial fascia
ligaments	declines	aponeuroses	chondrocytes

OUTCOME 4-1   1. The four primary tissue types found in the body are connective, muscle, epithelial, and _____.

OUTCOME 4-2   2. _____ refers to the structural and functional differences between apical and basal surfaces of epithelia.

OUTCOME 4-3   3. The epithelium that lines the body cavity is the _____.

OUTCOME 4-3   4. The lining of the heart and blood vessels is called a(n) _____.

OUTCOME 4-3   5. In merocrine secretion, the product is released through _____.

OUTCOME 4-4   6. Of the four primary types, the tissue that stores energy in bulk quantities is _____.

OUTCOME 4-4   7. The most common fibers in connective tissue proper are _____ fibers.

OUTCOME 4-4   8. Connective tissue fibers forming a branching, interwoven framework that is tough but flexible describes _____ fibers.

OUTCOME 4-4   9. The collagen fibers that connect one bone to another are the _____.

OUTCOME 4-5   10. Cartilage cells, or _____, divide and contribute additional matrix during interstitial growth.

OUTCOME 4-6   11. The loose connective tissue of a mucous membrane is called the _____.

OUTCOME 4-7   12. A layer of areolar tissue and fat, called _____, separates the skin from underlying tissues and organs.

OUTCOME 4-8   13. The only type of muscle tissue that is under voluntary control is _____.

OUTCOME 4-9   14. Neural tissue contains several different kinds of supporting cells called _____.

OUTCOME 4-10   15. The death of cells or tissues from disease or injury is referred to as _____.

OUTCOME 4-10   16. An accumulation of pus in an enclosed tissue space is called an _____.

OUTCOME 4-11   17. With advancing age, tissue repair _____ and cancer rates _____.

Matching

Match the terms in column B with the terms in column A. Use letters for answers in the spaces provided. Use each term only once.

Part I

		Column A	Column B
OUTCOME 4-1	_____	1. contraction	A. pseudostratified columnar
OUTCOME 4-2	_____	2. covering epithelia	B. secrete into ducts
OUTCOME 4-2	_____	3. glandular epithelia	C. secrete hyaluronan and proteins
OUTCOME 4-3	_____	4. exocrine	D. incomplete cellular layer
OUTCOME 4-3	_____	5. cilia	E. muscle tissue
OUTCOME 4-4	_____	6. fibroblasts	F. filled with histamine
OUTCOME 4-4	_____	7. mast cells	G. specialized secretions
OUTCOME 4-5	_____	8. bone cells	H. epidermis
OUTCOME 4-6	_____	9. synovial membrane	I. osteocytes

Part II

		Column A	Column B
OUTCOME 4-7	_____	10. deep fascia	J. thinning epithelia
OUTCOME 4-8	_____	11. muscle tissue	K. supporting cells
OUTCOME 4-8	_____	12. skeletal muscle tissue	L. movement
OUTCOME 4-9	_____	13. dendrites	M. tissue destruction
OUTCOME 4-9	_____	14. neuroglia	N. encapsulates organs
OUTCOME 4-10	_____	15. histamine	O. dilation
OUTCOME 4-10	_____	16. regeneration	P. not all tissues can
OUTCOME 4-10	_____	17. necrosis	Q. myosatellite cells, voluntary
OUTCOME 4-11	_____	18. age-related change	R. receive information

Drawing/Illustration Labeling

Identify each numbered structure. Place your answers in the spaces provided.

OUTCOME 4-3   **FIGURE 4-1**   Types of Epithelial Tissue

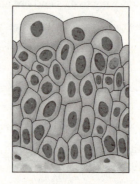

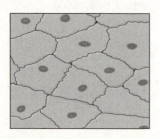

(1)_____

(2)_____

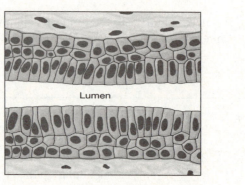

Lumen

(3)_____

(4)_____

(5)_____

(6)_____

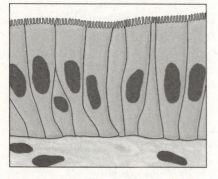

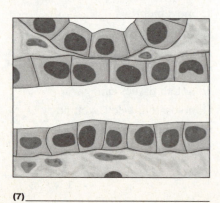

(7)_____

OUTCOME 4-4   **FIGURE 4-2** Types of Connective Tissue

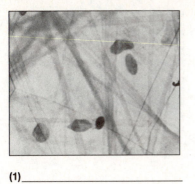

(1)_____

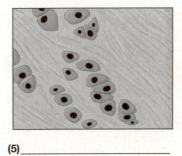

(2)_____

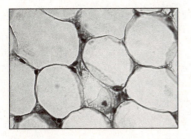

(3)_____

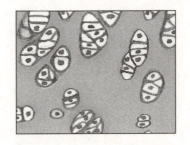

(4)_____

(5)_____

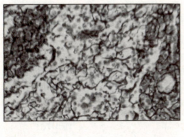

(6)_____

(7)_____

(8)_____

(9)_____

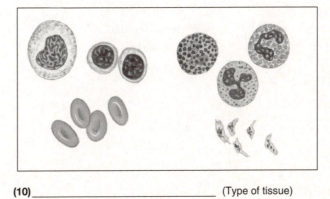

(10)_____ (Type of tissue)

OUTCOME 4-8 **FIGURE 4-3** Types of Muscle Tissue

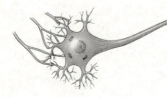

(1)_____ (2)_____ (3)_____

OUTCOME 4-9 **FIGURE 4-4** Identify the Type of Tissue

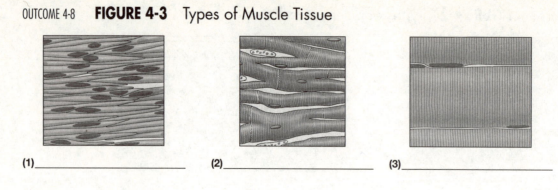

(1)_____

LEVEL 2: REVIEWING CONCEPTS

Chapter Overview

Using the terms below, correctly complete the chart of tissue types by filling in the numbered spaces. Use the columns "Body Location" and "Description/Appearance" to identify each tissue type.

Epithelial (1–8)	**Connective (9–17)**	**Muscle (18–20)**	**Neural (21)**
stratified columnar	areolar	smooth	neural
simple columnar	fibrocartilage	skeletal	
simple squamous	adipose	cardiac	
simple cuboidal	bone or osseous		
stratified squamous	dense irregular		
transitional	elastic cartilage		
pseudostratified columnar	dense regular		
stratified cuboidal	hyaline cartilage		
	blood		

IDENTIFICATION OF TISSUES

Body Location	Tissue Type	Description/Appearance
EPITHELIAL		
Mucous membrane lining of mouth and esophagus	(1) _____	Multiple layers of flattened cells
Mucosa of stomach and large intestine	(2) _____	Single rows of column-shaped cells
Trachea mucosa	(3) _____	One cell layer—cells rest on basal lamina (ciliated)
Respiratory surface of lungs	(4) _____	Single layer of flattened cells (Excessive number of cells and abnormal chromosomes observed)
Sweat glands	(5) _____	Layers of hexagonal or cube-like cells
Collecting tubules of kidney	(6) _____	Hexagonal shape; neat row of single cells
Mucous membrane lining of urinary bladder	(7) _____	Cells with ability to slide over one another; layered appearance
Male urethra	(8) _____	Multiple layers; basal layers shorter than superficial layers
CONNECTIVE		
Subcutaneous tissue; around kidneys, buttocks, breasts	(9) _____	Closely packed fat cells
Widely distributed packages organs; forms basal lamina of epithelia	(10) _____	Loose organization; many possible cells and fibers
Tendons; ligaments	(11) _____	Fibroblasts in matrix; parallel collagenic and elastic fibers
Dermis; capsules of joints; fascia of muscles	(12) _____	Fibroblast in matrix; irregularly arranged fibers
Ends of long bones; costal cartilages of ribs; support nose, trachea, larynx	(13) _____	Chondrocytes in lacunae; groups 2–4 cells
Intervertebral disks; disks of knee joints	(14) _____	Chondrocytes in lacunae
External ear; epiglottis	(15) _____	Chondrocytes in lacunae
Skeleton	(16) _____	Osteocytes in lacunae, vascularized
Cardiovascular system	(17) _____	Liquid—plasma RBC, WBC, platelets
MUSCLE		
Attached to bones	(18) _____	Long; cylindrical; multinucleate
Heart	(19) _____	Cardiocytes; intercalated disks
Walls of hollow organs; blood vessels	(20) _____	Nonstriated, uninucleated cells
NEURAL		
Brain; spinal cord; peripheral nervous system	(21) _____	Neurons (axons, dendrites), neuroglia (support cells)

Concept Map I

Using the following terms, fill in the circled, numbered, blank spaces to correctly complete the concept map. Use each term only once.

connective   muscle   neural   epithelial

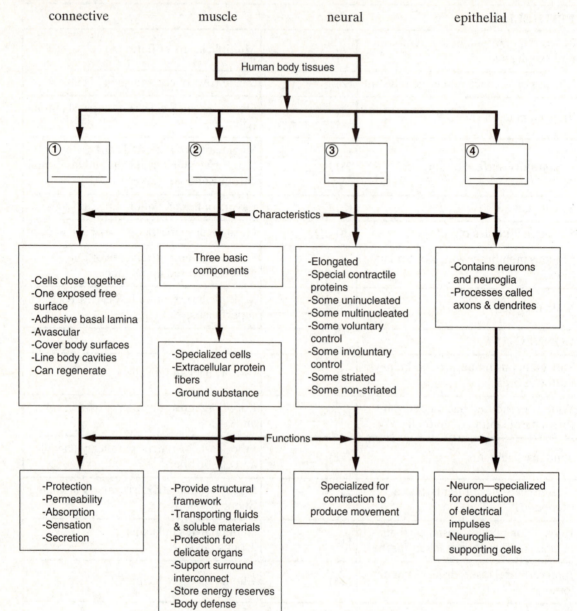

Human body tissues

①_____ ②_____ ③_____ ④_____

Characteristics

①
-Cells close together
-One exposed free
 surface
-Adhesive basal lamina
-Avascular
-Cover body surfaces
-Line body cavities
-Can regenerate

②
Three basic
components

-Specialized cells
-Extracellular protein
 fibers
-Ground substance

③
-Elongated
-Special contractile
 proteins
-Some uninucleated
-Some multinucleated
-Some voluntary
 control
-Some involuntary
 control
-Some striated
-Some non-striated

④
-Contains neurons
 and neuroglia
-Processes called
 axons & dendrites

Functions

-Protection
-Permeability
-Absorption
-Sensation
-Secretion

-Provide structural
 framework
-Transporting fluids
 & soluble materials
-Protection for
 delicate organs
-Support surround
 interconnect
-Store energy reserves
-Body defense

Specialized for
contraction to
produce movement

-Neuron—specialized
 for conduction
 of electrical
 impulses
-Neuroglia—
 supporting cells

Concept Map II

Using the following terms, fill in the circled, numbered, blank spaces to correctly complete the concept map. Use each term only once.

transitional

stratified cuboidal

lining

male urethra mucosa

exocrine glands

endocrine glands

stratified squamous

pseudostratified columnar

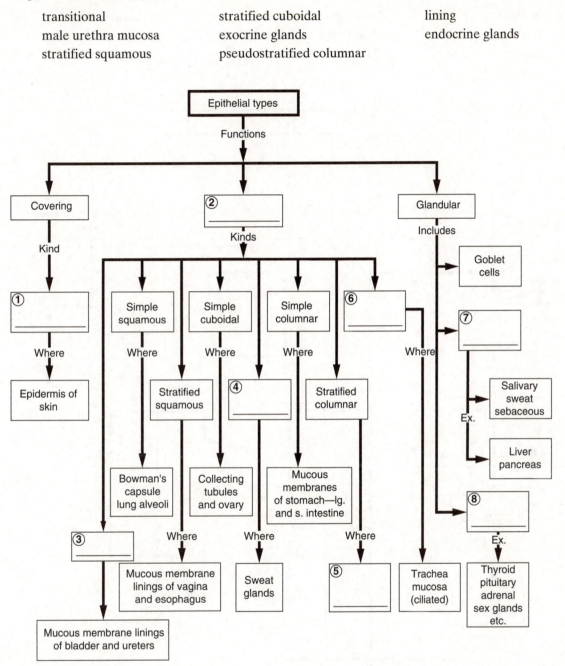

Concept Map III

Using the following terms, fill in the circled, numbered, blank spaces to correctly complete the concept map. Use each term only once.

ligaments	loose connective tissue	regular
bone	chondrocytes in lacunae	tendons
adipose	fluid connective tissue	blood
hyaline		

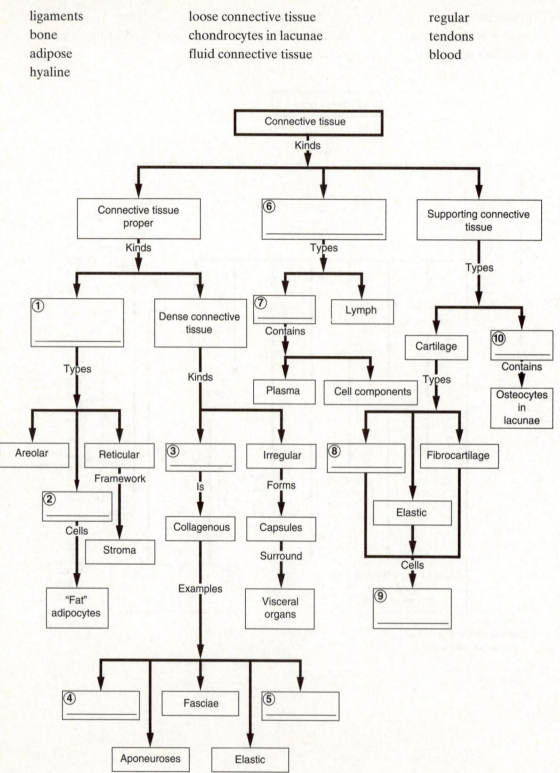

Concept Map IV

Using the following terms, fill in the circled, numbered, blank spaces to correctly complete the concept map. Use each term only once.

ground substance   adipose cells   branching interwoven framework
mast cells   phagocytosis   collagen

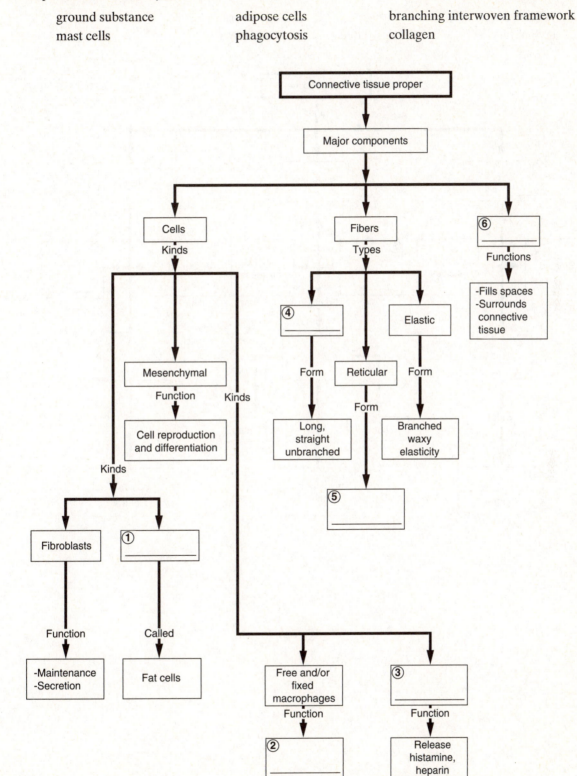

Concept Map V

Using the following terms, fill in the circled, numbered, blank spaces to correctly complete the concept map. Use each term only once.

multinucleate   cardiac   viscera

involuntary   nonstriated voluntary

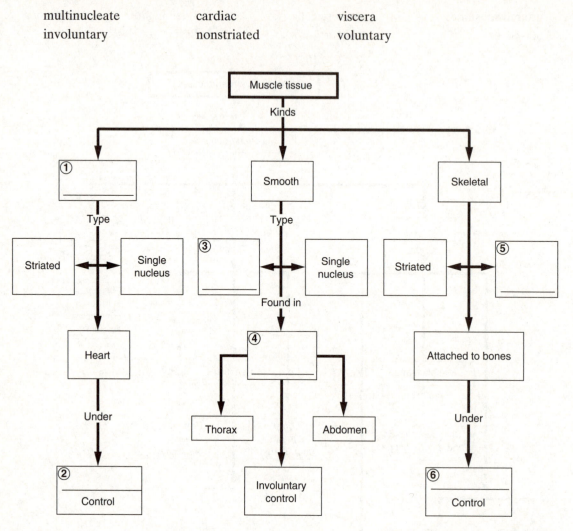

Concept Map VI

Using the following terms, fill in the circled, numbered, blank spaces to correctly complete the concept map. Use each term only once.

soma   neuroglia   dendrites

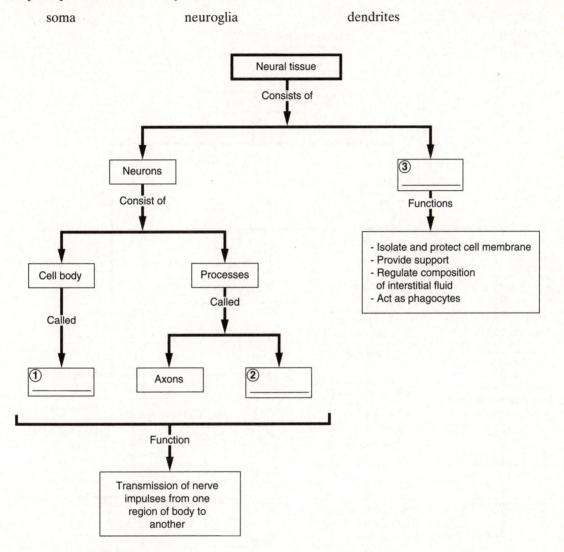

Concept Map VII

Using the following terms, fill in the circled, numbered, blank spaces to correctly complete the concept map. Use each term only once.

thick, waterproof, dry	mucous	goblet cells
pericardium	no basal lamina	skin
synovial	fluid formed on membrane surface	phagocytosis

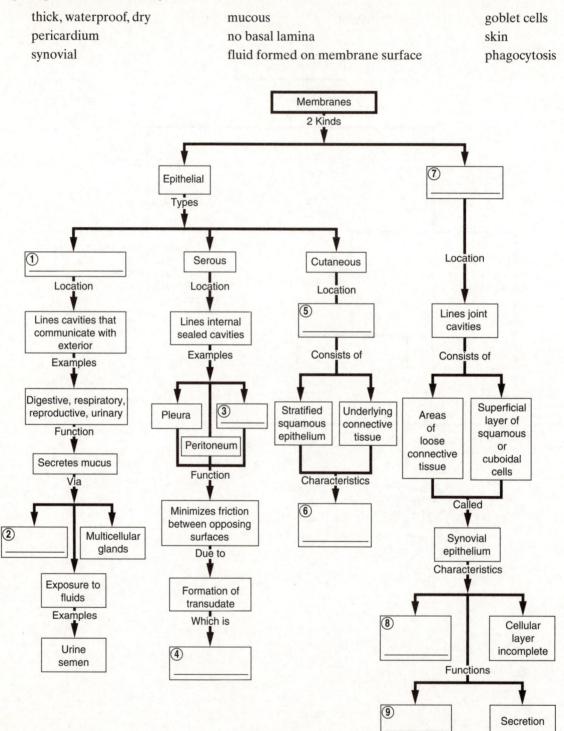

Multiple Choice

Place the letter corresponding to the best answer in the space provided.

_____ 1. If epithelial cells are classified according to their cell *shape*, the classes would include

    a. simple, stratified, and pseudostratified.

    b. squamous, cuboidal, and columnar.

    c. simple, squamous, and stratified.

    d. pseudostratified, stratified, and columnar.

_____ 2. If epithelial cells are classified according to their *function*, the classes would include those involved with

    a. support, transport, and storage.

    b. defense, support, and storage.

    c. lining, covering, and secreting.

    d. protection, defense, and transport.

_____ 3. Epithelial tissues are differentiated from the other tissue types because they

    a. always have a free surface exposed to the environment or to some internal chamber or passageway.

    b. have little space between adjacent epithelial cells.

    c. do not contain blood vessels.

    d. a, b, and c are correct.

_____ 4. Certain epithelial cells are called *pseudostratified* columnar epithelium because

    a. they have a layered appearance but all the cells contact the basal lamina.

    b. they are stratified and all the cells do not contact the basal lamina.

    c. their nuclei are all located the same distance from the cell surface.

    d. they are a mixture of cell types.

_____ 5. Three modes of secretion by glandular epithelial cells are

    a. serous, mucous, and mixed secretions.

    b. alveolar, acinar, and tubuloacinar secretions.

    c. merocrine, apocrine, and holocrine secretions.

    d. simple, compound, and tubular secretions.

_____ 6. Milk production in the breasts occurs through

    a. holocrine and apocrine secretions.

    b. apocrine and merocrine secretions.

    c. merocrine and tubular secretions.

    d. none of the above.

_____ 7. Holocrine secretions differ from other methods of secretion because

    a. some cytoplasm is lost as well as the secretory product.

    b. the secretory product is released through exocytosis.

    c. the product is released but the cell is destroyed.

    d. the secretions leave the cell intact.

_____   8. Examples of exocrine glands that secrete onto some internal or external surface are

      a. pituitary and thyroid.

      b. thymus and salivary.

      c. pancreas and pituitary.

      d. sweat and salivary.

_____   9. The two _fluid connective tissues_ found in the human body are

      a. mucus and matrix.

      b. blood and lymph.

      c. ground substance and hyaluronic acid.

      d. collagen and plasma.

_____   10. _Supporting connective tissues_ found in the body are

      a. muscle and bone.

      b. mast cells and adipocytes.

      c. cartilage and bone.

      d. collagen and reticular fibers.

_____   11. The common factor shared by the three connective tissue fiber types is that all three types are

      a. composed mostly of matrix.

      b. abundant in all major organs in the body.

      c. resistant to stretching due to the presence of ground substance.

      d. springy, resilient structures capable of extensive stretching.

_____   12. During a weight loss program when nutrients are scarce, adipocytes

      a. differentiate into mesenchymal cells.

      b. are normally destroyed and disappear.

      c. tend to enlarge and eventually divide.

      d. deflate like collapsing balloons.

_____   13. Hyaline cartilage serves to

      a. support the pinna of the outer ear.

      b. connect the ribs to the sternum.

      c. support the epiglottis.

      d. support the vocal cords.

_____   14. Summarizing the structural and functional properties of skeletal muscle tissue, it can be considered

      a. nonstriated involuntary muscle.

      b. nonstriated voluntary muscle.

      c. striated voluntary muscle.

      d. striated involuntary muscle.

_____   15. The major identifying feature characteristic of _mucous membranes_ is that

      a. they line cavities that communicate with the exterior.

      b. they line the sealed, internal cavities of the body.

      c. they minimize friction between adjacent organs.

      d. enclosed organs of the body are in close contact at all times.

_____ 16. Mucous membranes would be found primarily in which of the following systems?

    a. Skeletal, muscular, endocrine, and circulatory

    b. Integumentary, lymphoid, nervous, and endocrine

    c. Digestive, respiratory, reproductive, and urinary

    d. Skeletal, lymphoid, circulatory, and muscular

_____ 17. The pleura, peritoneum, and pericardium are examples of

    a. mucous membranes.

    b. abdominal structures.

    c. visceral organs.

    d. serous membranes.

_____ 18. The primary function of a _serous_ membrane is to

    a. provide nourishment and support to the body lining.

    b. reduce friction between the parietal and visceral surfaces.

    c. establish boundaries between internal organs.

    d. line cavities that communicate with the exterior.

_____ 19. In contrast to serous or mucous membranes, the _cutaneous_ membrane is

    a. thin, permeable to water, and usually moist.

    b. lubricated by goblet cells found in the epithelium.

    c. thick, relatively waterproof, and usually dry.

    d. covered with a specialized connective tissue, the lamina propria.

_____ 20. Which of the following are examples of intercellular connections found in epithelia?

    a. Cell adhesion molecules

    b. Gap junctions

    c. Desmosomes

    d. a, b, and c are correct

_____ 21. A component that synovial fluid and ground substance have in common is the presence of

    a. hyaluronic acid.

    b. phagocytes.

    c. ascites.

    d. satellite cells.

_____ 22. The capsules that surround most organs such as the kidneys and organs in the thoracic and peritoneal cavities are bound to the

    a. superficial fascia.

    b. deep fascia.

    c. subserous fascia.

    d. subcutaneous layer.

_____ 23. The only avascular type of connective tissue is

    a. areolar tissue.

    b. blood.

    c. cartilage.

    d. bone.

_____ 24. Tissue destruction that occurs after cells have been injured or destroyed is called

     a. dysplasia.

     b. necrosis.

     c. anaplasia.

     d. metaplasia.

_____ 25. Which of the following best defines *inflammation*?

     a. The secretion of histamine to increase blood flow to the injured area

     b. A coordinated defense that produces redness, swelling, heat, and pain

     c. A restoration process to heal the injured area

     d. The stimulation of macrophages to defend injured tissue

Completion

Using the terms below, complete the following statements. Use each term only once.

stroma	dense regular connective tissue	lacunae
fibroblasts	platelets	ground substance
tendons	lamina propria	plasma
axon	goblet cells	serous
transudate	subserous fascia	synovial fluid

1. The loose connective tissue component of a mucous membrane is called the _____.

2. The fluid formed on the surfaces of a serous membrane is called a(n) _____.

3. The connective tissue fibers that connect skeletal muscles to bone are the _____.

4. The layer of loose connective tissue that lies between the deep fascia and the serous membranes that line body cavities is the _____.

5. The fluid found in connective tissues is known as the _____.

6. The only cells that are *always* present in connective tissue proper are the _____.

7. The formed element in blood that consists of tiny membrane-enclosed packets of cytoplasm is called _____.

8. Chondrocytes in the cartilage matrix occupy small chambers known as _____.

9. Membranes that line the sealed, internal subdivisions of the ventral body cavity (cavities) not open to the exterior are _____ membranes.

10. The joint cavity that contains the ends of articulating bones is filled with _____.

11. The only example of unicellular exocrine glands in the body is that of _____.

12. The matrix substance in blood is _____.

13. The basic framework of reticular tissue found in the liver, spleen, lymph nodes, and bone marrow is the _____.

14. The branching process of a neuron that conducts information to other cells is the _____.

15. Tendons, aponeuroses, fascia, elastic tissue, and ligaments are all examples of _____.

Short Essay

Briefly answer the following questions in the spaces provided below.

1. What are the four primary tissue types in the body?

2. Summarize four essential functions of epithelial tissue.

3. What is the functional difference between microvilli and cilia on the exposed surfaces of epithelial cells?

4. How do the processes of merocrine, apocrine, and holocrine secretions differ?

5. List and describe the types of secretions that can be used to classify an exocrine gland.

6. List five important characteristics of epithelial tissue.

7. What three basic components are shared by all connective tissues?

8. What three types of fibers are found in connective tissue?

9. What four kinds of membranes consisting of epithelial and connective tissues that cover and protect other structures and tissues are found in the body?

10. What are the three types of muscle tissue?

11. What two types of cell populations make up neural tissue, and what is the primary function of each type?

LEVEL 3: CRITICAL THINKING AND CLINICAL APPLICATIONS

Using principles and concepts learned in Chapter 4, answer the following questions. Write your answers on a separate sheet of paper.

1. Skeletal muscle cells in adults are incapable of dividing. How is new skeletal muscle formed?

2. Why is it important that the pharynx, esophagus, anus, and vagina have the same epithelial organization?

3. What effect does a deficiency of vitamin C have on the development of connective tissue?

4. How are connective tissues associated with body immunity?

5. Why is the ability of cardiac muscle to repair itself after damage or injury limited?

6. Why does pinching the skin of a body region not usually distort or damage the underlying muscle?

7. Joints such as the elbow, shoulder, and knee contain considerable amounts of cartilage and fibrous connective tissue. How does this relate to the fact that joint injuries are often very slow to heal?

8. What chemicals do mast cells release and what effect do they produce?

9. Explain the relationship between the functions of epithelial tissue and the control of permeability.

10. Based on the varied cell population in connective tissue proper, what response might you expect at the sites of injury or damaged tissues?

The Integumentary System

OVERVIEW

The integumentary system consists of the skin and associated structures, including hair, nails, and a variety of glands. The four primary tissue types making up the skin comprise what is considered to be the largest structurally integrated organ system in the human body.

Because the skin and its associated structures are readily seen by others, a lot of time is spent caring for the skin, to enhance its appearance and prevent skin disorders that may alter desirable structural features on and below the skin surface. The integument manifests many of the functions of living matter, including protection, excretion, secretion, absorption, synthesis, storage, sensitivity, and temperature regulation. Studying the important structural and functional relationships in the integument provides numerous examples that demonstrate patterns that apply to tissue interactions in other organ systems.

LEVEL 1: REVIEWING FACTS AND TERMS

Review of Learning Outcomes

After completing this chapter, you should be able to do the following:

OUTCOME 5-1	Describe the main structural features of the epidermis, and explain the functional significance of each.
OUTCOME 5-2	Explain what accounts for individual differences in skin color, and discuss the response of melanocytes to sunlight exposure.
OUTCOME 5-3	Describe the interaction between sunlight and vitamin D_3 production.
OUTCOME 5-4	Describe the roles of epidermal growth factor.
OUTCOME 5-5	Describe the structure and functions of the dermis.
OUTCOME 5-6	Describe the structure and functions of the hypodermis.
OUTCOME 5-7	Describe the mechanisms that produce hair, and explain the structural basis for hair texture and color.
OUTCOME 5-8	Discuss the various kinds of glands in the skin, and list the secretions of those glands.
OUTCOME 5-9	Describe the anatomical structure of nails, and explain how they are formed.
OUTCOME 5-10	Explain how the skin responds to injury and repairs itself.
OUTCOME 5-11	Summarize the effects of aging on the skin.

Multiple Choice

Place the letter corresponding to the best answer in the space provided.

OUTCOME 5-1 _____ 1. The layers of the epidermis, beginning with the deepest layer and proceeding outwardly, include the stratum
   a. corneum, granulosum, spinosum, and basale.
   b. granulosum, spinosum, basale, and corneum.
   c. spinosum, basale, corneum, and granulosum.
   d. basale, spinosum, granulosum, and corneum.

OUTCOME 5-1 _____ 2. The layers of the epidermis where mitotic divisions occur are the
   a. basale and spinosum.
   b. corneum and basale.
   c. spinosum and corneum.
   d. mitosis occurs in all the layers.

OUTCOME 5-1 _____ 3. The epidermis consists of a
   a. stratified squamous epithelium.
   b. papillary layer.
   c. reticular layer.
   d. a, b, and c.

OUTCOME 5-2 _____ 4. Differences in skin color between individuals reflect distinct
   a. numbers of melanocytes.
   b. melanocyte distribution patterns.
   c. levels of melanin synthesis.
   d. UV responses and nuclear activity.

OUTCOME 5-2 _____ 5. The basic factors interacting to produce skin color are
   a. sunlight and ultraviolet radiation.
   b. the presence of carotene and melanin.
   c. melanocyte production and oxygen supply.
   d. circulatory supply and pigment concentration and composition.

OUTCOME 5-2, 5-10 _____ 6. Excessive exposure of the skin to UV radiation may cause redness, edema, blisters, and pain. The presence of blisters classifies the burn as
   a. first degree.
   b. second degree.
   c. third degree.
   d. none of these.

OUTCOME 5-3 _____ 7. When exposed to ultraviolet radiation, epidermal cells in the stratum basale and stratum spinosum convert a cholesterol-related steroid into
   a. calcitriol.
   b. MSH.
   c. vitamin D_3.
   d. ACTH.

OUTCOME 5-3 _____ 8. The hormone essential for the normal absorption of calcium and phosphorus by the small intestine is
   a. vitamin D.
   b. epidermal growth factor.
   c. adrenocorticotropic hormone.
   d. calcitriol.

OUTCOME 5-4 _____ 9. Epidermal growth factor EF has widespread effects on epithelia tissue in that it
   a. promotes the divisions of germinative cells.
   b. accelerates the production of keratin.
   c. stimulates epidermal development and repair.
   d. all of the above.

OUTCOME 5-5 _____ 10. The two major components of the dermis are
   a. capillaries and nerves.
   b. dermal papillae and a subcutaneous layer.
   c. sensory receptors and accessory structures.
   d. a papillary layer and a reticular layer.

OUTCOME 5-5 _____ 11. From the following selections, choose the one that identifies what the dermis contains to communicate with other organ systems.
   a. Blood vessels
   b. Lymphatic vessels
   c. Nerve fibers
   d. a, b, and c

OUTCOME 5-6 _____ 12. The primary tissues comprising the hypodermis (subcutaneous layer) are
   a. epithelial and neural.
   b. transitional and glandular.
   c. hyaline and cuboidal.
   d. areolar and adipose.

OUTCOME 5-6 _____ 13. The reason the hypodermis is a good location for subcutaneous injection by hypodermic needle is that it has a
   a. large number of sensory receptors.
   b. large number of lamellated corpuscles.
   c. limited number of capillaries and no vital organs.
   d. all of the above.

OUTCOME 5-6 _____ 14. An important function of the subcutaneous layer is to
   a. stabilize the position of the skin in relation to underlying tissues.
   b. provide sensation of pain and temperature.
   c. adjust gland secretion rates.
   d. monitor sensory receptors.

OUTCOME 5-7 _____ 15. Special smooth muscles in the dermis that, when contracted, produce "goose bumps" are called
    a. tissue papillae.
    b. arrector pili.
    c. root sheaths.
    d. cuticular papillae.

OUTCOME 5-7 _____ 16. Hair production occurs in the
    a. hair follicle, deep in the dermis.
    b. root hair plexus, in the reticular layer.
    c. hair follicle, in the stratum lucidum.
    d. hair follicle, in the hair matrix.

OUTCOME 5-7 _____ 17. The natural factor responsible for varying shades of hair color is the
    a. number of melanocytes.
    b. amount of carotene production.
    c. type of pigment present.
    d. a, b, and c.

OUTCOME 5-8 _____ 18. Accessory structures of the skin include the
    a. dermis, epidermis, and hypodermis.
    b. cutaneous and subcutaneous layers.
    c. hair follicles, sebaceous glands, and sweat glands.
    d. blood vessels, macrophages, and neurons.

OUTCOME 5-8 _____ 19. Sensible perspiration released by the eccrine sweat glands serves to
    a. cool the surface of the skin.
    b. reduce body temperature.
    c. dilute harmful chemicals.
    d. a, b, and c.

OUTCOME 5-8 _____ 20. When the body temperature becomes abnormally high, thermoregulatory homeostasis is maintained by a(n)
    a. increase in sweat gland activity and blood flow to the skin.
    b. decrease in blood flow to the skin and sweat gland activity.
    c. increase in blood flow to the skin and a decrease in sweat gland activity.
    d. increase in sweat gland activity and a decrease in blood flow to the skin.

OUTCOME 5-9 _____ 21. Not visible from the surface, nail production occurs at an epithelial fold called the
    a. eponychium.
    b. cuticle.
    c. nail root.
    d. lunula.

OUTCOME 5-10 _____ 22. The immediate response by the skin to an injury is
    a. bleeding and an inflammatory response triggered by damaged mast cells.
    b. the epidermal cells are immediately replaced.
    c. fibroblasts in the dermis create scar tissue.
    d. the formation of a scab.

OUTCOME 5-10   _____ 23. The practical limit to the healing process of the skin is the formation of inflexible, fibrous, noncellular

    a. scabs.

    b. skin grafts.

    c. ground substance.

    d. scar tissue.

OUTCOME 5-11   _____ 24. Hair turns gray or white due to

    a. a decline in glandular activity.

    b. a decrease in the number of Langerhans cells.

    c. decreased melanocyte activity.

    d. decreased blood supply to the dermis.

OUTCOME 5-11   _____ 25. Sagging and wrinkling of the integument occurs from

    a. the decline of germinative cell activity in the epidermis.

    b. a decrease in the elastic fiber network of the dermis.

    c. a decrease in vitamin D production.

    d. deactivation of sweat glands.

Completion

Using the terms below, complete the following statements. Use each term only once.

apocrine	stratum corneum	epidermis
keloid	stratum lucidum	glandular
connective	melanocytes	eponychium
eccrine glands	follicle	vellus
reticular layer	EGF	granulation tissue
Langerhans cells	sebum	melanin
decrease	MSH	cholecalciferol
blisters	arrector pili	

OUTCOME 5-1   1. The first line of defense against an often hostile environment is the _____.

OUTCOME 5-1   2. In areas where the epidermis is thick, such as the palms of the hands and the soles of the feet, the cells are flattened, densely packed, and filled with keratin. This layer is called the _____.

OUTCOME 5-1   3. Keratinized cells would be found primarily in the _____.

OUTCOME 5-1   4. Mobile macrophages that are a part of the immune system and found scattered among the deeper cells of the epidermis are called _____.

OUTCOME 5-2   5. The peptide secreted by the pituitary gland, which darkens the skin, is _____.

OUTCOME 5-2   6. The pigment that absorbs ultraviolet radiation before it can damage mitochondrial DNA is _____.

OUTCOME 5-3   7. Bone development is abnormal and bone maintenance is inadequate if there is a dietary deficiency or a lack of skin production of _____.

OUTCOME 5-4   8. Stimulating synthetic activity and secretion by epithelial cells is among the roles of _____.

OUTCOME 5-5   9. The type of tissue that comprises most of the dermis is _____.

OUTCOME 5-6   10. The subcutaneous layer, or hypodermis, is extensively interwoven with the connective tissue fibers of the _____.

OUTCOME 5-7   11. The fine "peach fuzz" hairs found over much of the body surface are called _____.

OUTCOME 5-7   12. A bundle of smooth muscle cells, which extends from the papillary layer of the dermis to the connective tissue sheath surrounding the hair follicle, is the _____.

OUTCOME 5-7   13. Hair develops from a group of epidermal cells at the base of a tube-like depression called a(n) _____.

OUTCOME 5-7   14. Variations in hair color reflect differences in structure and variations in the pigment produced by _____.

OUTCOME 5-8   15. The secretion that lubricates and inhibits the growth of bacteria on the skin is called _____.

OUTCOME 5-8   16. The glands in the skin that become active when the body temperature rises above normal are the _____.

OUTCOME 5-8   17. The sweat glands that communicate with hair follicles are called _____.

OUTCOME 5-8   18. If the body temperature drops below normal, heat is conserved by a(n) _____ in the diameter of dermal blood vessels.

OUTCOME 5-9   19. The stratum corneum that covers the exposed nail closest to the root is the _____.

OUTCOME 5-10   20. A thick, raised area of scar tissue covered by a shiny, smooth epidermal surface is a _____.

OUTCOME 5-10   21. A second degree burn is readily identified by the appearance of _____.

OUTCOME 5-10   22. During injury repair, the combination of blood clot, fibroblasts, and an extensive capillary network is called _____.

OUTCOME 5-11   23. In older adults, dry and scaly skin is usually a result of a decrease in _____ activity.

Matching

Match the terms in column B with the terms in column A. Use letters for answers in the spaces provided.

		Column A	**Column B**
OUTCOME 5-1	_____	1. epidermis	A. vitamin D
OUTCOME 5-2	_____	2. carotene	B. thermoregulation
OUTCOME 5-3	_____	3. cholecalciferol	C. secreted lipid product
OUTCOME 5-4	_____	4. EGF	D. cutaneous plexus
OUTCOME 5-5	_____	5. dermis	E. decreased melanocyte activity
OUTCOME 5-6	_____	6. hypodermis	F. embryonic hair
OUTCOME 5-7	_____	7. lanugo	G. blood clot
OUTCOME 5-8	_____	8. sebum	H. keratinocytes
OUTCOME 5-8	_____	9. nervous system	I. subcutaneous layer
OUTCOME 5-9	_____	10. cuticle	J. eponychium
OUTCOME 5-10	_____	11. scab	K. produced by salivary glands
OUTCOME 5-11	_____	12. effects of aging	L. converted to vitamin A

Drawing/Illustration Labeling

Identify each numbered structure. Place your answers in the spaces provided.

OUTCOME 5-1
OUTCOME 5-5
OUTCOME 5-6
OUTCOME 5-7
OUTCOME 5-8

FIGURE 5-1 Components of the Integumentary System

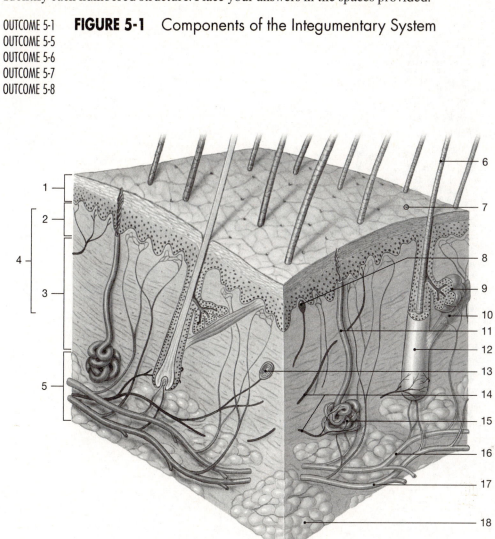

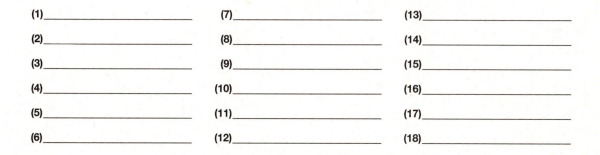

(1)_____

(2)_____

(3)_____

(4)_____

(5)_____

(6)_____

(7)_____

(8)_____

(9)_____

(10)_____

(11)_____

(12)_____

(13)_____

(14)_____

(15)_____

(16)_____

(17)_____

(18)_____

OUTCOME 5-7   **FIGURE 5-2**   Hair Follicles and Hairs

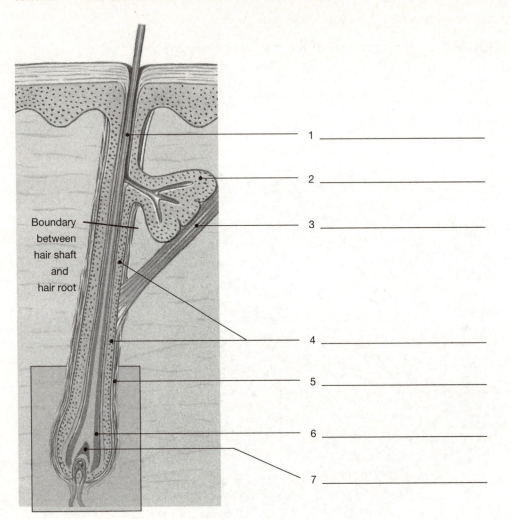

Boundary
between
hair shaft
and
hair root

1 _____

2 _____

3 _____

4 _____

5 _____

6 _____

7 _____

OUTCOME 5-9　**FIGURE 5-3**　Nail Structure (a) Superficial View (b) Longitudinal Section

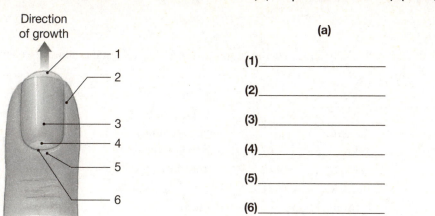

(a)

(1)_____

(2)_____

(3)_____

(4)_____

(5)_____

(6)_____

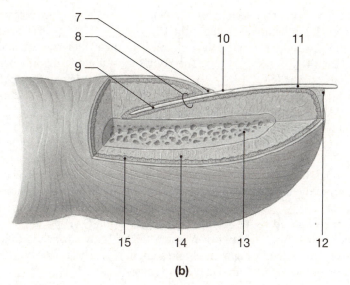

(b)

(b)

(7)_____

(8)_____

(9)_____

(10)_____

(11)_____

(12)_____

(13)_____

(14)_____

(15)_____

LEVEL 2: REVIEWING CONCEPTS

Chapter Overview

Using the terms below, fill in the blanks to complete the chapter overview of the integument. Use each term only once.

granulosum	sweat	mammary
vellus hairs	lunula	hypodermis
cutaneous	collagen	eponychium
environmental	skin	papillary layer
corneum	sebaceous	keratinocytes
hair	basale	ceruminous
hyponychium	terminal hairs	elastic
ear wax	accessory	reticular layer
hair and nails		

The two major components of the integumentary system are the (1) _____ membrane and the (2) _____ structures. The integument includes the (3) _____. The structures located primarily in the dermis and protruding through the epidermis to the skin surface include the multicellular exocrine glands and the (4) _____. The body's most abundant epithelial cells, which are dominant in the epidermis, are the (5) _____. The layers or strata in the epidermis in order from the clear layer toward the free surface are the stratum (6) _____, the stratum spinosum, the stratum (7) _____, the stratum lucidum, and the stratum (8) _____. The two major components of the dermis are the superficial (9) _____ and a deeper (10) _____. The two types of fibers that provide dermal strength and elasticity are very strong, stretch-resistant (11) _____ fibers and the (12) _____ fibers that permit stretching and recoiling to their original length. They serve to prevent damage to the tissue. The tissue beneath the dermis that is not a part of the integument, but is important in stabilizing the position of the skin to underlying tissues, is the (13) _____. The non-living accessory structure composed of dead keratinized cells that have been pushed to the surface of the skin is the (14) _____. In the adult integument, the fine "peach fuzz" located over much of the body surface is called the (15) _____, and the heavy, more deeply pigmented, and sometimes curly accessory structures are the (16) _____. The exocrine glands found in the skin consist of the (17) _____ glands that discharge an oily secretion into hair follicles, and (18) _____ glands that produce and secrete apocrine and merocrine products into hair follicles and onto the surface of the skin. Other integumentary glands include the (19) _____ glands of the breasts, and the (20) _____ glands, which form a mixture called cerumen, or (21) _____. The nails, made of keratinized epidermal cells, protect the tips of the fingers and toes. The free edge of the nail extends over the (22) _____, an area of thickened stratum corneum. The stratum corneum of the nail root extends over the exposed nail, forming the (23) _____, or cuticle. Near the root, the moon-like pole crescent is known as the (24) _____. In perspective, the integumentary system provides mechanical protection against (25) _____ hazards for all the other systems.

Concept Map I

Using the following terms, fill in the circled, numbered, blank spaces to correctly complete the concept map. Use each term only once.

sensory reception   vitamin D synthesis   produce secretions

lubrication   dermis   exocrine glands

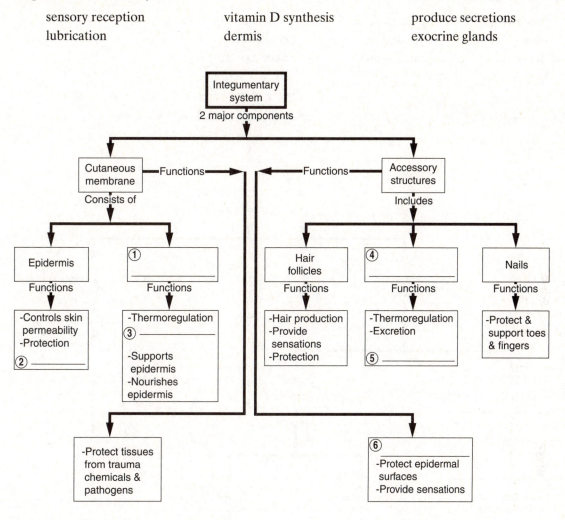

Concept Map II

Using the following terms, fill in the circled, numbered, blank spaces to correctly complete the concept map. Use each term only once.

nerves	epidermis	collagen
skin	hypodermis	connective
fat	granulosum	papillary layer

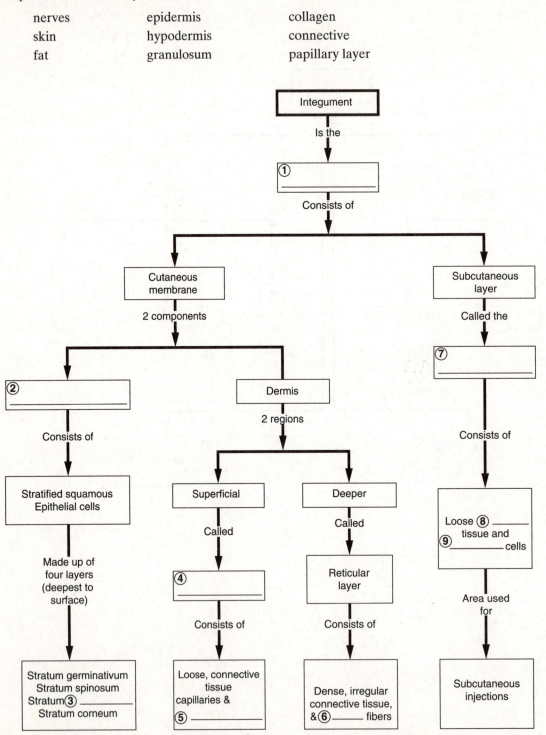

Concept Map III

Using the following terms, fill in the circled, numbered, blank spaces to correctly complete the concept map. Use each term only once.

merocrine or "eccrine"   "peach fuzz"   glands

thickened stratum corneum   arms and legs   lunula

cerumen ("ear wax")   terminal   cuticle

odors   sebaceous

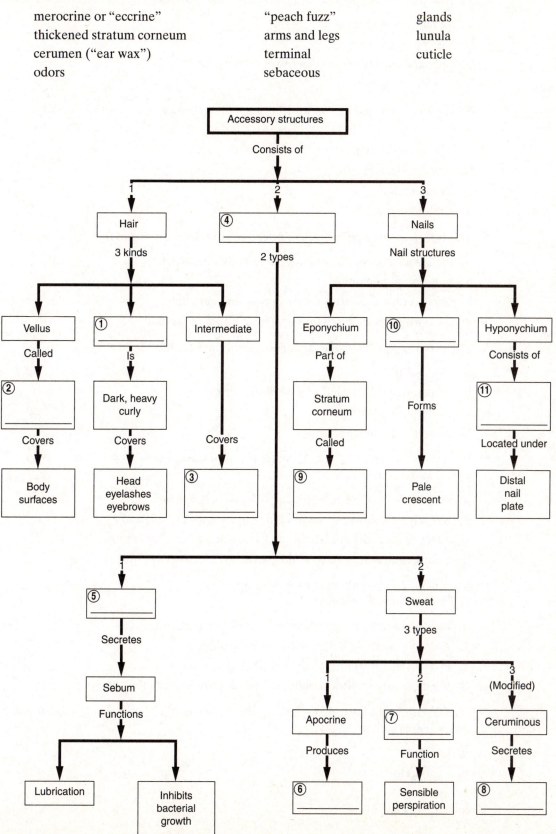

Multiple Choice

Place the letter corresponding to the best answer in the space provided.

_____ 1. The two major components of the integument are the

    a. dermis and epidermis.

    b. hair and skin.

    c. cutaneous membrane and accessory structures.

    d. eleidin and keratin.

_____ 2. Psoriasis is a skin disorder in which there is abnormal increased mitotic activity in the

    a. stratum spinosum.

    b. stratum lucidum.

    c. stratum basale.

    d. stratum corneum.

_____ 3. Third-degree burns differ from first- and second-degree burns in that

    a. the epidermis, dermis, and hypodermis are destroyed.

    b. they are more painful.

    c. fluid accumulates between the dermis and epidermis.

    d. the burn is restricted to the superficial layers of the skin.

_____ 4. The two components of the cutaneous membrane are the

    a. hair and skin.

    b. subcutaneous layer and accessory structures.

    c. epidermis and dermis.

    d. melanin and melanocytes.

_____ 5. Because freshwater is hypotonic to body fluids, sitting in a freshwater bath causes

    a. water to leave the epidermis and dehydrate the tissue.

    b. water from the interstitial fluid to penetrate the surface and evaporate.

    c. water to enter the epidermis and cause the epithelial cells to swell.

    d. complete cleansing because the bacteria on the surface drown.

_____ 6. Malignant melanomas are extremely dangerous and life-threatening because

    a. they develop in the germinative layer of the epidermis.

    b. they form tumors that interfere with circulation.

    c. metastasis is restricted to the dermis and epidermis.

    d. the melanocytes grow rapidly and metastasize through the lymphatic system.

_____ 7. The subcutaneous layer that separates the integument from the deep fascia around other organs is the

    a. dermis.

    b. hypodermis.

    c. epidermis.

    d. reticular layer.

_____ 8. Ceruminous glands are modified sweat glands located in the

    a. reticular layer of the dermis.

    b. stratum spinosum of the epidermis.

    c. nasal passageways.

    d. external auditory canal.

_____ 9. The regulation of salts, water, and organic wastes by the integumentary glands is accomplished by

    a. excretion.

    b. digestion.

    c. respiration.

    d. all of the above.

_____ 10. Scar tissue is best described as

    a. an accumulation of pus in an enclosed tissue space.

    b. a widespread inflammation of the dermis caused by bacterial infection.

    c. a necrosis occurring because of inadequate circulation.

    d. fibrous, noncellular accumulation of tissue at the site of injury repair.

_____ 11. The protein that permits stretch and recoil of the skin is

    a. collagen.

    b. elastin.

    c. keratin.

    d. melanin.

_____ 12. The primary pigments contained in the epidermis are

    a. melanin and chlorophyll.

    b. xanthophylls and melanin.

    c. carotene and melanin.

    d. xanthophylls and carotene.

Completion

Using the terms below, complete the following statements. Use each term only once.

keloid	Langerhans cells
Merkel cells	dermatitis
hypodermis	vitiligo
cyanosis	seborrheic dermatitis
carotene	papillary

1. Skin surfaces that lack hair contain specialized epithelial cells sensitive to touch called _____.

2. An orange-yellow pigment that normally accumulates in epidermal cells is _____.

3. During a sustained reduction in blood supply, the oxygen levels in the tissues decline, causing the skin to take on a bluish coloration called _____.

4. When individuals lose their melanocytes, the condition is known as _____.

5. The layer of the dermis that contains the capillaries, lymphatic vessels, and sensory neurons that supply the surface of the skin is the _____ layer.

6. An inflammation of the skin that primarily involves the papillary layer is called _____.

7. The tissue beneath the dermis important in stabilizing the position of the skin in relation to underlying tissues while permitting independent movement is the _____.

8. "Cradle cap" in infants and "dandruff" in adults result from an inflammation around abnormally active sebaceous glands called _____.

9. A thick, flattened mass of scar tissue that begins at the injury site and grows into the surrounding tissue is called a _____.

10. Due to the onset of aging, increased damage and infection are apparent because of a decrease in the number of _____.

Short Essay

Briefly answer the following questions in the spaces provided below.

1. A friend says to you, "Don't worry about what you say to her; she is thick-skinned." Anatomically speaking, what areas of the body would your friend be referring to? Why are these areas thicker?

2. Two females are discussing their dates. One of the girls says, "I liked everything about him except he had body odor." What is the cause of body odor?

3. A hypodermic needle is used to introduce drugs into the loose connective tissue of the hypodermis. Beginning on the surface of the skin in the region of the thigh, list, in order, the layers of tissue the needle would penetrate to reach the hypodermis.

4. The general public associates a tan with good health. What is wrong with this assessment?

5. Many shampoo advertisements list the ingredients, such as honey, kelp extracts, beer, vitamins, and other nutrients, as being beneficial to the hair. Why could this be considered false advertisement?

6. What benefit does the application of creams and oils have to the natural functioning of the skin?

7. You are a nurse in charge of a patient who has decubitus ulcers, or "bedsores." What causes bedsores, and what should have been done to prevent the skin tissue degeneration or necrosis that has occurred?

8. After a first date, a young man confides to a friend, "Every time I see her, I get 'goose bumps.'" What is happening in this young man's skin?

LEVEL 3: CRITICAL THINKING AND CLINICAL APPLICATIONS

Using principles and concepts learned about in Chapter 5, answer the following questions. Write your answers on a separate sheet of paper.

1. Even though the stratum corneum is water resistant, it is not waterproof. When the skin is immersed in water, osmotic forces may move water in or out of the epithelium. Long-term exposure to seawater endangers survivors of a shipwreck by accelerating dehydration. How and why does this occur?

2. A young Caucasian girl is frightened during a violent thunderstorm, during which lightning strikes nearby. Her parents notice that she is pale, in fact, she has "turned white." Why has her skin color changed to this "whitish" appearance?

3. Tretinoin (Retin-A) has been called the anti-aging cream. Since it is applied topically, how does it affect the skin?

4. A third-degree burn does more damage to the skin but is less painful than a second-degree burn. Explain.

5. Someone asks you, "Is hair really important to the human body?" What responses would you give to show the functional necessities of hair?

6. a. How does exposure to optimum amounts of sunlight promote proper bone maintenance and growth in children?

   b. If exposure to sunlight is rare due to geographical location or environmental conditions, what can be done to minimize impaired maintenance and growth?

7. Individuals who participate in endurance sports must continually provide the body with fluids. Explain why this is necessary.

8. Why do calluses form on the palms of the hands when doing manual labor?

9. Bacterial invasion of the superficial layers of the skin is quite common. Why is it difficult to reach the underlying connective tissues? (Cite at least six features of the skin that help protect the body from invasion by bacteria.)

CHAPTER 6

Osseous Tissue and Bone Structure

OVERVIEW

Osteology is a specialized science that is the study of bone (osseous) tissue and skeletal structure and function. The skeletal system consists of bones and related connective tissues, which include cartilage and ligaments. Chapter 6 addresses the topics of bone development and growth; histological organization; bone classification; the effects of nutrition, hormones, and exercise on the skeletal system; and the mechanisms that maintain skeletal structure and function throughout the lifetime of an individual.

The exercises in this chapter are written to show that although bones have common microscopic characteristics and the same basic dynamic nature, each bone has a characteristic pattern of ossification and growth, a characteristic shape, and identifiable surface features that reflect its functional relationship to other bones and other systems throughout the body. It is important for the student to understand that even though bone tissue is structurally stable, it is living tissue and is functionally dynamic.

LEVEL 1: REVIEWING FACTS AND TERMS

Review of Learning Outcomes

After completing this chapter, you should be able to do the following:

OUTCOME 6-1   Describe the primary functions of the skeletal system.

OUTCOME 6-2   Classify bones according to shape and internal organization, giving examples of each type, and explain the functional significance of each of the major types of bone markings.

OUTCOME 6-3   Identify the cell types in bone, and list their major functions.

OUTCOME 6-4   Compare the structures and functions of compact bone and spongy bone.

OUTCOME 6-5   Compare the mechanisms of endochondral ossification and intramembranous ossification.

OUTCOME 6-6   Describe the remodeling and homeostatic mechanisms of the skeletal system.

OUTCOME 6-7   Discuss the effects of exercise, hormones, and nutrition on bone development and on the skeletal system.

OUTCOME 6-8   Explain the role of calcium as it relates to the skeletal system.

OUTCOME 6-9   Describe the types of fractures, and explain how fractures heal.

OUTCOME 6-10   Summarize the effects of the aging process on the skeletal system.

Multiple Choice

Place the letter corresponding to the best answer in the space provided.

OUTCOME 6-1 _____ 1. The function(s) of the skeletal system is (are)

    a. it is a storage area for calcium and lipids.

    b. it is involved in blood cell formation.

    c. it provides structural support for the entire body.

    d. a, b, and c are correct.

OUTCOME 6-1 _____ 2. Storage of lipids that represent an important energy reserve in bone occur in areas of

    a. red marrow.

    b. yellow marrow.

    c. bone matrix.

    d. ground substance.

OUTCOME 6-2 _____ 3. Which of the following selections is correctly identified as a long bone?

    a. Rib

    b. Sternum

    c. Humerus

    d. Patella

OUTCOME 6-2 _____ 4. Bones forming the roof of the skull and the scapula are referred to as

    a. irregular bones.

    b. flat bones.

    c. short bones.

    d. sesamoid bones.

OUTCOME 6-2 _____ 5. Depressions, grooves, and tunnels in bone indicate

    a. attachment of ligaments and tendons.

    b. where bones articulate.

    c. where blood vessels or nerves lie alongside or penetrate bones.

    d. all of the above.

OUTCOME 6-2 _____ 6. An anatomical term used to describe a rounded passageway for blood vessels or nerves is

    a. sulcus.

    b. fossa.

    c. fissure.

    d. foramen.

OUTCOME 6-3 _____ 7. Mature bone cells found in lacunae are called

    a. osteoblasts.

    b. osteocytes.

    c. osteoclasts.

    d. osteoprogenitors.

OUTCOME 6-3 _____ 8. Giant multinucleated cells involved in the process of osteolysis are

    a. osteocytes.

    b. osteoblasts.

    c. osteoclasts.

    d. osteoprogenitor cells.

OUTCOME 6-4 _____ 9. One of the basic histological differences between compact and spongy bone is that in *compact bone,*

    a. the basic functional unit is the osteon.

    b. there is a lamellar arrangement.

    c. there are plates or struts called trabeculae.

    d. osteons are not present.

OUTCOME 6-4 _____ 10. Spongy (or cancellous) bone, unlike compact bone, resembles a network of bony struts separated by spaces that are normally filled with

    a. osteocytes.

    b. lacunae.

    c. bone marrow.

    d. lamella.

OUTCOME 6-4 _____ 11. Spongy bone is found primarily at the _____ of long bones.

    a. bone surfaces, except inside joint capsules

    b. expanded ends of long bones, where they articulate with other skeletal elements

    c. axis of the diaphysis

    d. exterior region of the bone shaft to withstand forces applied at either end

OUTCOME 6-4 _____ 12. Compact bone is usually thickest where

    a. bones are not heavily stressed.

    b. stresses arrive from many directions.

    c. trabeculae are aligned with extensive cross-bracing.

    d. stresses arrive from a limited range of directions.

OUTCOME 6-5 _____ 13. During intramembranous ossification, the developing bone grows outward from the ossification center in small struts called

    a. spicules.

    b. lacunae.

    c. the osteogenic layer.

    d. dermal bones.

OUTCOME 6-5 _____ 14. When osteoblasts begin to differentiate within a mesenchymal or fibrous connective tissue, the process is called

    a. endochondral ossification.

    b. osteoprogenation.

    c. osteolysis.

    d. intramembranous ossification.

OUTCOME 6-5   _____ 15. The process during which bones begin development as cartilage models and the cartilage is later replaced by bone is called

    a. intramembranous ossification.

    b. endochondral ossification.

    c. articular ossification.

    d. secondary ossification.

OUTCOME 6-5   _____ 16. Within the metaphysis region is the epiphyseal cartilage, where

    a. secondary ossification centers are located.

    b. the epiphysis begins to calcify at birth.

    c. cartilage is being replaced by bone.

    d. collagen fibers become cemented into lamella by osteoblasts.

OUTCOME 6-6   _____ 17. The process of removing and replacing bone's organic and mineral components is called

    a. calcification.

    b. ossification.

    c. remodeling.

    d. osteoprogenesis.

OUTCOME 6-7   _____ 18. The major effect exercise has on bones is that it

    a. provides oxygen for bone development.

    b. enhances the process of calcification.

    c. serves to maintain and increase bone mass.

    d. accelerates the healing process when a fracture occurs.

OUTCOME 6-7   _____ 19. Growth hormone from the pituitary gland and thyroxine from the thyroid gland maintain normal bone growth activity at the

    a. epiphyseal plates.

    b. diaphysis.

    c. periosteum.

    d. endosteum.

OUTCOME 6-8   _____ 20. Which of the following selections describes a homeostatic mechanism of the skeleton?

    a. As one osteon forms through the activity of osteoblasts, another is destroyed by osteoclasts

    b. Adjusting the rate of calcium excretion by the kidneys

    c. Vitamin D stimulating the absorption and transport of calcium and phosphate ions

    d. a, b, and c are correct

OUTCOME 6-8   _____ 21. When large numbers of calcium ions are mobilized into the body fluids, the bones

    a. become weaker.

    b. become denser.

    c. become stronger.

    d. are not affected.

OUTCOME 6-9   _____ 22. In a *greenstick* fracture,

  a. only one side of the shaft is broken and the other is bent.

  b. the shaft bone is broken across its long axis.

  c. the bone protrudes through the skin.

  d. the bone is shattered into small fragments.

OUTCOME 6-9   _____ 23. A Pott's fracture, which occurs at the ankle, is identified primarily by

  a. a transverse break in the bone.

  b. its specific location.

  c. a spiral break in the bone.

  d. nondisplacement.

OUTCOME 6-10   _____ 24. A normal part of the aging process results in bones becoming

  a. thicker and stronger.

  b. massive and fractured.

  c. thinner and weaker.

  d. longer and more slender.

OUTCOME 6-10   _____ 25. The condition that produces a reduction in *bone mass* sufficient to compromise normal function is

  a. osteopenia.

  b. osteitis deformans.

  c. osteomyelitis.

  d. osteoporosis.

OUTCOME 6-10   _____ 26. Osteoporosis is more frequent in older women than older men due to

  a. decline in circulating estrogen.

  b. increased muscle mass in men.

  c. decreased muscle mass in females.

  d. a, b, and c are correct.

Completion

Using the terms below, complete the following statements. Use each term only once.

osteoclasts	osteocytes	osteoporosis
compound	remodeling	osteon
matrix	support	calcitriol
ossification	irregular	epiphysis
calcium	Wormian	comminuted
intramembranous	yellow marrow	endochondral
osteoblasts	condyle	cancers
bone markings	calcitonin	

OUTCOME 6-1   1. The storage of lipids in bones occurs in the _____.

OUTCOME 6-1   2. Of the five major functions of the skeleton, the two that depend on the dynamic nature of bone are mineral storage and _____.

OUTCOME 6-2   3. The surface features of the skeletal system that yield an abundance of anatomical information are referred to as _____.

OUTCOME 6-2   4. A smooth, rounded articular process that articulates with an adjacent bone is a _____.

OUTCOME 6-2   5. Bones with complex shapes and short, flat, notched, or ridged surfaces are termed _____.

OUTCOME 6-2   6. Sutural bones, which are small, flat, odd-shaped bones found between the flat bones of the skull, are also referred to as _____ bones.

OUTCOME 6-3   7. Cuboidal cells that synthesize the organic components of the bone matrix are _____.

OUTCOME 6-3   8. In adults, the cells responsible for maintaining the matrix in osseous tissue are the _____.

OUTCOME 6-4   9. The basic functional unit of compact bone is the _____.

OUTCOME 6-4   10. The expanded region of a long bone consisting of spongy bone is called the _____.

OUTCOME 6-5   11. When osteoblasts differentiate within a mesenchymal or fibrous connective tissue, the process is called _____ ossification.

OUTCOME 6-5   12. The type of ossification that begins with the formation of a hyaline cartilage model is _____.

OUTCOME 6-5   13. The process that refers specifically to the formation of bone is _____.

OUTCOME 6-6   14. The organic and mineral components of the bone matrix are continually being recycled and renewed through the process of _____.

OUTCOME 6-6   15. During bone renewal, as one osteon forms through the activity of osteoblasts, another is destroyed by _____.

OUTCOME 6-7   16. The ability of bone to adapt to new stresses results from the turnover and recycling of _____.

OUTCOME 6-7   17. The hormone synthesized in the kidneys that is essential for normal calcium and phosphate ion absorption in the digestive tract is _____.

OUTCOME 6-8   18. The major mineral associated with the development and mineralization of bone is _____.

OUTCOME 6-8   19. If the calcium ion concentration of the blood rises above normal, C cells in the thyroid gland secrete _____.

OUTCOME 6-9   20. Fractures that shatter the affected area into a multitude of bony fragments are called _____ fractures.

OUTCOME 6-9   21. Fractures that project through the skin are called _____ fractures.

OUTCOME 6-10   22. The reduction of bone mass that occurs with aging is called _____.

OUTCOME 6-10   23. Osteoclast-activating factor is released by tissues affected by many _____.

Matching

Match the terms in column B with the terms in column A. Use letters for answers in the spaces provided.

Part I

		Column A	Column B
OUTCOME 6-1	_____	1. blood cell formation	A. synthesize osteoid
OUTCOME 6-2	_____	2. bone marking	B. spicules
OUTCOME 6-2	_____	3. sulcus	C. Haversian system
OUTCOME 6-3	_____	4. osteoprogenitor cells	D. interstitial and appositional growth
OUTCOME 6-3	_____	5. osteoblasts	E. tuberosity
OUTCOME 6-4	_____	6. spongy bones	F. cancellous bone
OUTCOME 6-4	_____	7. osteon	G. red bone marrow
OUTCOME 6-5	_____	8. intramembranous ossification	H. crack repair
OUTCOME 6-5	_____	9. endochondral ossification	I. narrow groove

Part II

		Column A	Column B
OUTCOME 6-5	_____	10. apositional growth	J. vitamin C deficiency
OUTCOME 6-6	_____	11. bone maintenance	K. stimulates osteoblast activity
OUTCOME 6-7	_____	12. thyroxine	L. increased calcium concentration
OUTCOME 6-7	_____	13. vitamin A	M. bony fragments
OUTCOME 6-7	_____	14. scurvy	N. stimulates bone growth
OUTCOME 6-8	_____	15. rickets	O. remodeling
OUTCOME 6-8	_____	16. parathyroid hormone	P. reduction in bone mass
OUTCOME 6-9	_____	17. comminuted fracture	Q. inadequate ossification
OUTCOME 6-10	_____	18. osteopenia	R. increase in diameter
OUTCOME 6-10	_____	19. osteoporosis	S. vitamin D deficiency

Drawing/Illustration Labeling

Identify each numbered structure. Place your answers in the spaces provided.

OUTCOME 6-4   **FIGURE 6-1**   Structure of Compact Bone

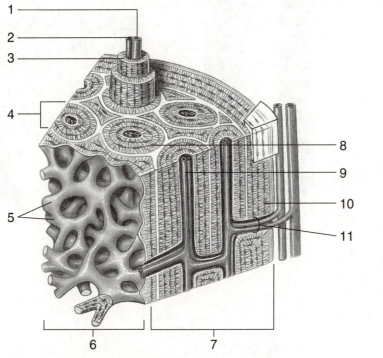

(1)_____

(2)_____

(3)_____

(4)_____

(5)_____

(6)_____

(7)_____

(8)_____

(9)_____

(10)_____

(11)_____

OUTCOME 6-3   **FIGURE 6-2**   Structure of a Long Bone—L.S.

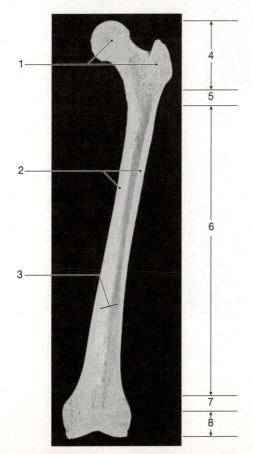

(1)_____

(2)_____

(3)_____

(4)_____

(5)_____

(6)_____

(7)_____

(8)_____

Identify the type of fracture in each of the following images.

OUTCOME 6-9 **FIGURE 6-3** Major Types of Fractures

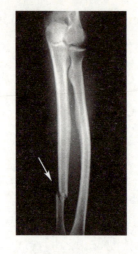

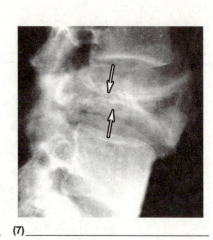

(1)_____ (4)_____ (7)_____

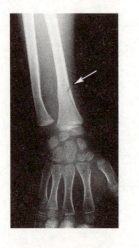

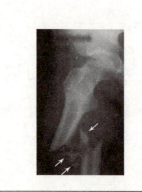

(2)_____ (5)_____

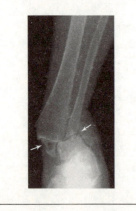

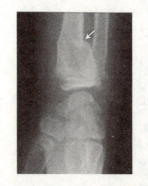

(3)_____ (6)_____

LEVEL 2: REVIEWING CONCEPTS

Chapter Overview

Using the terms below, fill in the blanks to complete the overview of osseous tissue and bone structure. Use each term only once.

sutural bones	flat bones	short bones
periosteum	red bone marrow	diaphysis
lacunae	osteopenia	trabeculae
remodeling	epiphysis	endochondral ossification
sesamoid bones	long bones	canaliculi
ossification	osteocytes	irregular bones
metaphysis	intramembranous	yellow bone marrow
osteoporosis	spongy bone	marrow cavity
osteon		

OSSEOUS TISSUE AND BONE STRUCTURE

The adult skeleton contains 206 major bones, which are divided into six broad categories according to their individual shapes. (1) _____ are generally small, flat bones, such as the patella or kneecap. (2) _____ are relatively long and slender, such as the femur. (3) _____ are small and boxy, such as the wrist bones (carpals). (4) _____ have thin, roughly parallel surfaces, such as the sternum and ribs. (5) _____ have complex shapes with short, flat-notched, or ridged surfaces, such as the spinal vertebrae. (6) _____, or Wormian bones, are small, flat, irregularly shaped bones between the flat bones of the skull. A representative long bone such as the tibia has an extended tubular shaft, or (7) _____, and an expanded area at each end known as the (8) _____. The narrow zone connecting the tubular shaft to the expanded area at each end is the (9) _____. The compact bone of the tubular shaft forms a protective layer that surrounds a central space called the (10) _____. The expanded areas at each end of the tubular shaft consist largely of spongy bone, or (11) _____. The matrix of bone contains specialized cells called (12) _____ within pockets called (13) _____. The branching network for the exchange of nutrients, waste products, and gases is the (14) _____, narrow passageways through the matrix. Except at joints, the (15) _____ covers the outer surfaces of bones. The basic functional unit of mature compact bone is the (16) _____, or Haversian system. In spongy bone, the matrix forms struts and plates called (17) _____. Spongy bone contains (18) _____, which is responsible for blood cell formation. The (19) _____ in spongy bone serves as an important energy reserve. The process of replacing other tissues with bone is called (20) _____. When bone replaces existing cartilage, it is called (21) _____, while in (22) _____, bone develops directly from mesenchyme or fibrous connective tissue. As a part of normal bone mainte-nance, the bone matrix is recycled and renewed through the process of (23) _____. As a normal part of the aging process, the bones become thinner and weaker, causing a reduction in bone mass called (24) _____. When the reduction in bone mass is sufficient to compromise normal functions, the condition is known as (25) _____.

Concept Map I

Using the following terms, fill in the circled, numbered, blank spaces to correctly complete the concept map. Use each term only once.

osteocytes   collagen   intramembranous ossification
periosteum   hyaline cartilage

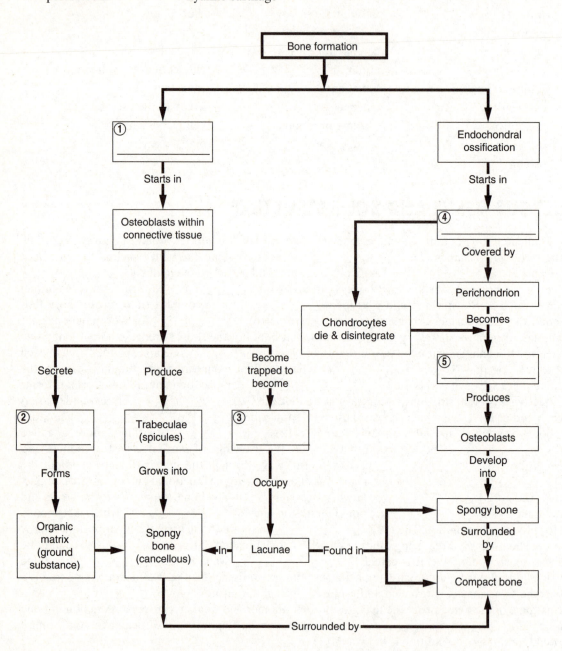

Concept Map II

Using the following terms, fill in the circle, numbered, blank spaces to correctly complete the concept map. Use each term only once.

↓ calcium level
homeostasis
releases stored Ca from bone

↓ Ca concentration
in body fluids
parathyroid glands

calcitonin
↑ Ca concentration
in body fluids

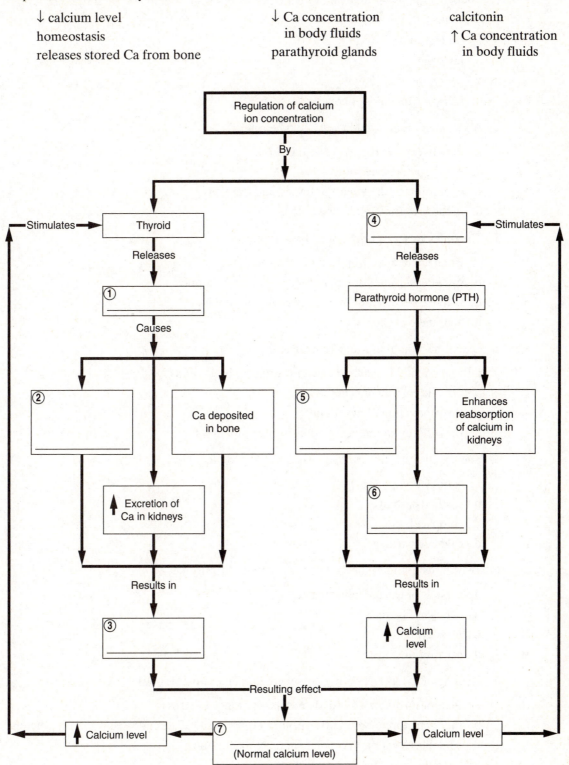

Multiple Choice

Place the letter corresponding to the best answer in the space provided.

_____ 1. Changing the magnitude and direction of forces generated by skeletal muscles is an illustration of the skeletal function of

    a. protection.

    b. leverage.

    c. energy reserve in bones.

    d. storage capability of bones.

_____ 2. A sesamoid bone would most often be found

    a. in between the flat bones of the skull.

    b. in the spinal vertebrae.

    c. near joints at the knee, the hands, and the feet.

    d. in the forearm and the lower leg.

_____ 3. The outer surface of the bone, the periosteum,

    a. isolates the bone from surrounding tissues.

    b. provides a route for circulatory and nervous supply.

    c. actively participates in bone growth and repair.

    d. a, b, and c are correct.

_____ 4. Osteolysis is an important process in the regulation of

    a. calcium and phosphate concentrations in body fluids.

    b. organic components in the bone matrix.

    c. the production of new bone.

    d. the differentiation of osteoblasts into osteocytes.

_____ 5. The calcification of cartilage results in the production of

    a. spongy bone.

    b. ossified cartilage.

    c. compact bone.

    d. calcified cartilage.

_____ 6. Two-thirds of the weight of bone is accounted for by

    a. crystals of calcium phosphate.

    b. collagen fibers.

    c. osteocytes.

    d. calcium carbonate.

_____ 7. Healing of a fracture, even after severe damage, depends on the survival of

    a. the bone's mineral strength and its resistance to stress.

    b. the circulatory supply and the cellular components of the endosteum and periosteum.

    c. osteoblasts within the spongy and compact bone.

    d. the external callus that protects the fractured area.

_____ 8. Three target sites for regulating the calcium ion concentration in body fluids are the

    a. heart, liver, and lungs.

    b. liver, kidneys, and stomach.

    c. pancreas, heart, and lungs.

    d. bones, intestinal tract, and kidneys.

_____ 9. When the calcium ion concentration of the blood rises above normal, secretion of the hormone *calcitonin*

    a. promotes osteoclast activity.

    b. increases the rate of intestinal absorption.

    c. increases the rate of calcium ion excretion.

    d. activates the parathyroid gland to release parathyroid hormone.

_____ 10. When cartilage is produced at the epiphyseal side of the metaphysis at the same rate as bone is deposited on the opposite side, bones

    a. grow wider.

    b. become shorter.

    c. grow longer.

    d. become thicker.

_____ 11. The major advantage(s) for bones to undergo continual remodeling is (are)

    a. it may change the shape of a bone.

    b. it may change the internal structure of a bone.

    c. it may change the total amount of minerals deposited in the bones.

    d. a, b, and c are correct.

_____ 12. The fibers of *tendons* intermingle with those of the periosteum, attaching

    a. skeletal muscles to bones.

    b. the end of one bone to another bone.

    c. the trabecular framework to the periosteum.

    d. articulations with the trabeculae.

_____ 13. Giant cells, called *osteoclasts*, with 50 or more nuclei serve to

    a. synthesize the organic components of the bone matrix.

    b. form the trabecular framework that protects cells of the bone marrow.

    c. line the inner surfaces of the central canals.

    d. secrete acids that dissolve the bony matrix and release the stored minerals.

_____ 14. After fertilization occurs, the skeleton begins to form in about

    a. two weeks.

    b. four weeks.

    c. six weeks.

    d. three months.

_____ 15. The circulating hormones that stimulate bone growth are

    a. calcitonin and parathyroid hormone.

    b. growth hormone and thyroxine.

    c. oxytocin and secretin.

    d. epinepherine and relaxin.

_____ 16. Appositional bone growth at the outer surface results in a(n)

        a. increase in the diameter of a growing bone.

        b. increase in the overall length of a bone.

        c. thickening of the cartilages that support the bones.

        d. increased hardening of the periosteum.

_____ 17. The vitamins that are specifically required for normal bone growth are

        a. vitamins B_1, B_2, and B_3.

        b. vitamins C and B complex.

        c. vitamins A, D, E, and K.

        d. vitamins A, C, and D.

_____ 18. The hormones that coordinate the storage, absorption, and excretion of calcium ions are

        a. growth hormone and thyroxine.

        b. calcitonin and parathyroid hormone.

        c. estrogens and androgens.

        d. calcitriol and cholecalciferol.

_____ 19. A bone is covered by a(n)

        a. endosteum and lined with a periosteum.

        b. periosteum and endosteum.

        c. periosteum and lined with an endosteum.

        d. lamella and lined with a matrix.

_____ 20. A fracture that projects through the skin is a

        a. compound fracture.

        b. comminuted fracture.

        c. Colles' fracture.

        d. greenstick fracture.

Completion

Using the terms below, complete the following statements. Use each term only once.

osteomalacia	endochondral	rickets
diaphysis	intramembranous	Pott's
osteoblasts	epiphyseal plates	canaliculi
osteopenia		

1. The communication pathways from the lacunae, which connect the osteocytes with one another and the blood vessels of the Haversian canal, are _____.

2. The condition in which an individual develops a bow-legged appearance as the leg bones bend under the weight of the body is _____.

3. The type of cells responsible for the production of new bone are _____.

4. Dermal bones, such as several bones of the skull, the lower jaw, and the collarbone, are a result of _____ ossification.

5. Limb bone development is a good example of the process of _____ ossification.

6. Long bone growth during childhood and adolescence continues until the _____ close.

7. A condition in which the bones appear normal but are weak and flexible due to poor mineralization is called _____.

8. Fragile limbs, a reduction in height, and the loss of teeth are a part of the aging process referred to as _____.

9. A common type of ankle fracture that affects both bones of the leg is a _____ fracture.

10. The location of compact bone in the long bone of an adult is the _____.

Short Essay

Briefly answer the following questions in the spaces provided below.

1. What five major functions is the skeletal system responsible for in the human body?

2. What are the primary histological *differences* between compact and spongy bone?

3. What are the differences between the *periosteum* and the *endosteum*?

4. How does the process of *calcification* differ from *ossification*?

5. Differentiate between the *beginning stages* of intramembranous and endochondral ossification.

6. The conditions of *gigantism* and *pituitary dwarfism* are extreme opposites. What hormonal disturbance causes each condition?

7. What six shape classifications are used to divide the bones of the body into categories? Give an example of each classification.

8. What are the fundamental relationships between the skeletal system and other body systems?

LEVEL 3: CRITICAL THINKING AND CLINICAL APPLICATIONS

Using principles and concepts learned in Chapter 6, answer the following questions. Write your answers on a separate sheet of paper.

1. Why is an individual who experiences premature puberty not as tall as expected at age 18?

2. How might a cancerous condition in the body be related to the development of a severe reduction in bone mass and excessive bone fragility?

3. During a conversation between two grandchildren, one says to the other, "Grandpa seems to be getting shorter as he gets older." Why is this a plausible observation?

4. Good nutrition and exercise are extremely important in bone development, growth, and maintenance. If you were an astronaut, what vitamin supplements and what type of exercise would you need to be sure that the skeletal system retained its integrity while in a weightless environment in space?

5. While skateboarding, 9-year-old Christopher falls and breaks his right leg. His parents are informed by a physician that the proximal end of the tibia where the epiphysis meets the diaphysis is fractured. At the age of 18, during a routine physical, Christopher learns that his right leg is 2 cm shorter than his left, probably resulting from his skateboard accident. How might this difference be explained?

6. Thorough examination of the skeletons of ancient peoples provides cultural conclusions about the lifestyles of these ancient cultures. What sorts of clues might bones provide describing the lifestyles of these individuals?

7. The skeletal remains of a middle-aged male have been found buried in a shallow grave. As the pathologist assigned to the case, what type of anatomical information could you provide by examining the bones to determine the identity of the individual?

7

The Axial Skeleton

OVERVIEW

The skeletal system in the human body is composed of 206 bones, 80 of which are found in the axial division, and 126 of which make up the appendicular division. Chapter 7 includes a study of the bones and associated parts of the axial skeleton, located along the body's longitudinal axis and center of gravity. They include 22 skull bones, 6 auditory ossicles, 1 hyoid bone, 26 vertebrae, 24 ribs, and 1 sternum. This bony framework protects and supports the vital organs in the dorsal and ventral body cavities. In addition, the bones serve as areas for muscle attachment, articulate at joints for stability and movement, assist in respiratory movements, and stabilize and position elements of the appendicular skeleton.

Your study and review for this chapter includes the identification and location of bones, the identification and location of bone markings, the functional anatomy of bones, and the articulations that comprise the axial skeleton.

LEVEL 1: REVIEWING FACTS AND TERMS

Review of Learning Outcomes

After completing this chapter, you should be able to do the following:

OUTCOME 7-1   Identify the bones of the axial skeleton, and specify their functions.

OUTCOME 7-2   Identify the bones of the cranium and face, and explain the significance of the markings on the individual bones.

OUTCOME 7-3   Identify the foramina and fissures of the skull, and cite the major structures using the passageways.

OUTCOME 7-4   Describe the structure and functions of the orbital complex, nasal complex, and paranasal sinuses.

OUTCOME 7-5   Describe the key structural differences among the skulls of infants, children, and adults.

OUTCOME 7-6   Identify and describe the curvatures of the spinal column, and indicate the function of each.

OUTCOME 7-7   Identify the vertebral regions, and describe the distinctive structural and functional characteristics of vertebrae in each region.

OUTCOME 7-8   Explain the significance of the articulations between the thoracic vertebrae and the ribs, and between the ribs and sternum.

Multiple Choice

Place the letter corresponding to the best answer in the space provided.

OUTCOME 7-1 _____ 1. The axial skeleton can be recognized because it
    a. includes the bones of the arms and legs.
    b. forms the longitudinal axis of the body.
    c. includes the bones of the pectoral and pelvic girdles.
    d. a, b, and c are correct.

OUTCOME 7-1 _____ 2. The axial skeleton provides an extensive surface area for the attachment of muscles that
    a. adjust the positions of the head, the neck, and the trunk.
    b. perform respiratory movements.
    c. stabilize or position parts of the appendicular skeleton.
    d. all of the above.

OUTCOME 7-1 _____ 3. Which of the following selections includes bones found exclusively in the axial skeleton?
    a. Ear ossicles, scapula, clavicle, sternum, hyoid
    b. Vertebrae, ischium, ilium, skull, ribs
    c. Skull, vertebrae, ribs, sternum, hyoid
    d. Sacrum, ear ossicles, skull, scapula, ilium

OUTCOME 7-1 _____ 4. The axial skeleton creates a framework that supports and protects organ systems in the
    a. ventral body cavities.
    b. pleural cavity.
    c. abdominal cavity.
    d. pericardial cavity.

OUTCOME 7-2 _____ 5. The bones of the *cranium* that exclusively represent *single*, unpaired bones are the
    a. occipital, parietal, frontal, and temporal.
    b. occipital, frontal, sphenoid, and ethmoid.
    c. frontal, temporal, parietal, and sphenoid.
    d. ethmoid, frontal, parietal, and temporal.

OUTCOME 7-2 _____ 6. The *paired* bones of the cranium are the
    a. ethmoid and sphenoid.
    b. frontal and occipital.
    c. occipital and parietal.
    d. parietal and temporal.

OUTCOME 7-2 _____ 7. The *associated* bones of the skull include the
    a. mandible and maxilla.
    b. nasal and lacrimal.
    c. hyoid and auditory ossicles.
    d. vomer and palatine.

OUTCOME 7-2 _____ 8. The *single, unpaired* bones that make up the skeletal part of the face are the

    a. mandible and vomer.

    b. nasal and lacrimal.

    c. mandible and maxilla.

    d. nasal and palatine.

OUTCOME 7-2 _____ 9. The lines, tubercles, crests, ridges, and other processes on the bones represent areas that are used primarily for

    a. attachment of muscles to bones.

    b. attachment of bone to bone.

    c. joint articulation.

    d. increasing the surface area of the bone.

OUTCOME 7-2 _____ 10. The *sutures* that articulate the bones of the skull are the

    a. parietal, occipital, frontal, and temporal.

    b. calvaria, foramen, condyloid, and lacerum.

    c. posterior, anterior, lateral, and dorsal.

    d. lambdoid, sagittal, coronal, and squamous.

OUTCOME 7-3 _____ 11. *Foramina*, located on the bones of the skull, serve primarily as passageways for

    a. airways and ducts for secretions.

    b. sound and sight.

    c. nerves and blood vessels.

    d. muscle fibers and nerve tissue.

OUTCOME 7-4 _____ 12. The bones that make up the lateral wall and rim of the eye orbit are the

    a. lacrimal, zygomatic, and maxilla.

    b. ethmoid, temporal, and zygomatic.

    c. maxilla, zygomatic, and sphenoid.

    d. temporal, frontal, and sphenoid.

OUTCOME 7-4 _____ 13. The *sinuses,* or internal chambers in the skull, are found in the

    a. sphenoid, ethmoid, vomer, and lacrimal bones.

    b. sphenoid, frontal, ethmoid, and maxillary bones.

    c. ethmoid, frontal, lacrimal, and maxillary bones.

    d. lacrimal, vomer, ethmoid, and frontal bones.

OUTCOME 7-4 _____ 14. The nasal complex consists of the

    a. frontal, sphenoid, and ethmoid bones.

    b. maxilla, lacrimal, and ethmoidal concha.

    c. inferior concha.

    d. a, b, and c are correct.

OUTCOME 7-4 _____ 15. The air-filled chambers that communicate with the nasal cavities are the
    a. condylar processes.
    b. paranasal sinuses.
    c. maxillary foramina.
    d. mandibular foramina.

OUTCOME 7-4 _____ 16. The primary function(s) of the paranasal sinus mucus epithelium is (are) to
    a. lighten the skull bones.
    b. humidify and warm incoming air.
    c. trap foreign particulate matter such as dust or microorganisms.
    d. all of the above.

OUTCOME 7-5 _____ 17. Areas of the head that are involved in the formation of the skull are called
    a. fontanelles.
    b. craniocephalic centers.
    c. craniulums.
    d. ossification centers.

OUTCOME 7-5 _____ 18. The reason the skull can be distorted without damage during birth is
    a. fusion of the ossification centers is completed.
    b. the brain is large enough to support the skull.
    c. fibrous connective tissue connects the cranial bones.
    d. the shape and structure of the cranial elements are elastic.

OUTCOME 7-5 _____ 19. The anatomical structures that allow for distortion of the infant skull during the birthing process are called
    a. cranial foramina.
    b. fontanelles.
    c. alveolar process.
    d. cranial ligaments.

OUTCOME 7-5 _____ 20. The most significant growth in the skull occurs before age 5 because
    a. the brain stops growing and cranial sutures develop.
    b. brain development is incomplete until maturity.
    c. the cranium of a child is larger than that of an adult.
    d. the ossification and articulation process is completed.

OUTCOME 7-6 _____ 21. The primary spinal curves that appear late in fetal development
    a. help shift the trunk weight over the legs.
    b. accommodate the lumbar and cervical regions.
    c. become accentuated as the toddler learns to walk.
    d. accommodate the thoracic and abdominopelvic viscera.

OUTCOME 7-6 _____ 22. An abnormal lateral curvature of the spine is called
    a. kyphosis.
    b. lordosis.
    c. scoliosis.
    d. amphiarthrosis.

OUTCOME 7-7   _____ 23. The vertebrae that indirectly effect changes in the volume of the rib cage are the

    a. cervical vertebrae.

    b. thoracic vertebrae.

    c. lumbar vertebrae.

    d. sacral vertebrae.

OUTCOME 7-7   _____ 24. The most massive and least mobile of the vertebrae are the

    a. thoracic.

    b. cervical.

    c. lumbar.

    d. sacral.

OUTCOME 7-7   _____ 25. Which of the following selections correctly identifies the sequence of the vertebra from superior to inferior?

    a. Thoracic, cervical, lumbar, coccyx, sacrum

    b. Cervical, lumbar, thoracic, sacrum, coccyx

    c. Cervical, thoracic, lumbar, sacrum, coccyx

    d. Cervical, thoracic, sacrum, lumbar, coccyx

OUTCOME 7-7   _____ 26. When identifying the vertebra, a numerical shorthand is used, such as C_3. The C refers to the

    a. region of the vertebrae.

    b. position of the vertebrae in a specific region.

    c. numerical order of the vertebrae.

    d. articulating surface of the vertebrae.

OUTCOME 7-7   _____ 27. C_1 and C_2 have specific names, which are the

    a. sacrum and coccyx.

    b. atlas and axis.

    c. cervical and costal.

    d. atlas and coccyx.

OUTCOME 7-7   _____ 28. The sacrum consists of five fused elements that afford protection for

    a. reproductive, digestive, and excretory organs.

    b. respiratory, reproductive, and endocrine organs.

    c. urinary, respiratory, and digestive organs.

    d. endocrine, respiratory, and urinary organs.

OUTCOME 7-7   _____ 29. The primary purpose of the coccyx is to provide

    a. protection for the urinary organs.

    b. protection for the anal opening.

    c. an attachment site for leg muscles.

    d. an attachment site for a muscle that closes the anal opening.

OUTCOME 7-8   _____   30. The first seven pairs of ribs are called true ribs, while the lower five pairs are called *false* ribs because
        a. the fused cartilages merge with the costal cartilage.
        b. they do not attach directly to the sternum.
        c. the last two pairs have no connection with the sternum.
        d. they differ in shape from the true ribs.

OUTCOME 7-8   _____   31. The skeleton of the chest, or thoracic cage, consists of
        a. cervical vertebrae, ribs, and sternum.
        b. cervical vertebrae, ribs, and thoracic vertebrae.
        c. cervical vertebrae, ribs, and pectoral girdle.
        d. thoracic vertebrae, ribs, and sternum.

OUTCOME 7-8   _____   32. The three components of the adult sternum are the
        a. pneumothorax, hemothorax, and tuberculum.
        b. manubrium, body, and xiphoid process.
        c. head, capitulum, and tuberculum.
        d. angle, body, and shaft.

Completion

Using the terms below, complete the following statements. Use each term only once.

vertebral body	costal	floating
axial	cranium	capitulum
fontanelles	cervical	"soft spot"
mucus	paranasal	foramen magnum
compensation	xiphoid process	inferior conchae
muscles		

OUTCOME 7-1   1. The part of the skeletal system that forms the longitudinal axis of the body is the _____ division.

OUTCOME 7-1   2. The bones of the skeleton provide an extensive surface area for the attachment of _____.

OUTCOME 7-2   3. The part of the skull that provides protection for the brain is the _____.

OUTCOME 7-2, 7-3   4. The opening that connects the cranial cavity with the canal enclosed by the spinal column is the _____.

OUTCOME 7-3, 7-4   5. The paired scroll-like bones located on each side of the nasal septum are the _____.

OUTCOME 7-4   6. The airspaces connected to the nasal cavities are the _____ sinuses.

OUTCOME 7-4   7. Irritants are flushed off the walls of the nasal cavities because of the production of _____.

OUTCOME 7-5   8. At birth, the cranial bones are connected by areas of fibrous connective tissues called _____.

OUTCOME 7-5   9. The anterior fontanelle that is easily seen by new parents is referred to as the _____.

OUTCOME 7-6   10. The spinal curves that become accentuated when a child learns to walk and run are called _____ curves.

OUTCOME 7-6   11. The anatomical structure of vertebrae that transfers weight along the axis of the vertebral column is called the _____.

OUTCOME 7-7   12. The vertebrae that stabilize relative positions of the brain and spinal cord are the _____ vertebrae.

OUTCOME 7-8   13. The cartilaginous extensions that connect the ribs to the sternum are the _____ cartilages.

OUTCOME 7-8   14. A typical rib articulates with the vertebral column at the area of the rib called the head, or _____.

OUTCOME 7-8   15. The last two pairs of ribs that do not articulate with the sternum are called _____ ribs.

OUTCOME 7-8   16. The smallest part of the sternum that serves as an area of attachment for the muscular diaphragm and rectus abdominis muscles is the _____.

Matching

Match the terms in column B with the terms in column A. Use letters for answers in the spaces provided.

Part I

		Column A	Column B
OUTCOME 7-1, 7-8	_____	1. respiratory movement	A. calvaria
OUTCOME 7-2	_____	2. hyoid bone	B. premature closure of fontanelle
OUTCOME 7-2	_____	3. skullcap	C. infant skull
OUTCOME 7-2	_____	4. sphenoid	D. olfactory foramina
OUTCOME 7-3	_____	5. ethmoid	E. paranasal sinuses
OUTCOME 7-4	_____	6. air-filled chambers	F. sella turcica
OUTCOME 7-5	_____	7. fontanelle	G. elevation of rib cage
OUTCOME 7-5	_____	8. craniostenosis	H. stylohyoid ligaments

Part II

		Column A	Column B
OUTCOME 7-6	_____	9. primary curves	I. jugular notch
OUTCOME 7-7	_____	10. cervical vertebrae	J. ribs 8–10
OUTCOME 7-7	_____	11. lumbar vertebrae	K. C_1
OUTCOME 7-7	_____	12. atlas	L. C_2
OUTCOME 7-7	_____	13. axis	M. ribs 1–7
OUTCOME 7-8	_____	14. vertebrosternal ribs	N. accommodation
OUTCOME 7-8	_____	15. vertebrochondral ribs	O. neck
OUTCOME 7-8	_____	16. manubrium	P. lower back

Drawing/Illustration Labeling

Identify each numbered structure. Place your answers in the spaces provided.

OUTCOME 7-1   **FIGURE 7-1**   Bones of the Axial Skeleton

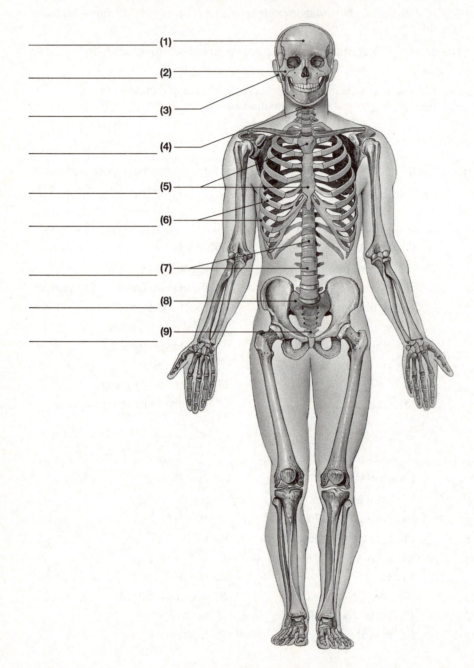

_____ (1)

_____ (2)

_____ (3)

_____ (4)

_____ (5)

_____ (6)

_____ (7)

_____ (8)

_____ (9)

OUTCOME 7-2, 7-3, 7-4 **FIGURE 7-2** Anterior View of the Adult Skull

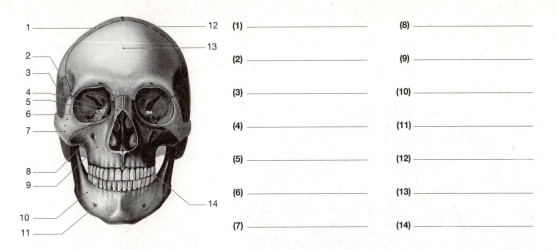

(1) _____	(8) _____
(2) _____	(9) _____
(3) _____	(10) _____
(4) _____	(11) _____
(5) _____	(12) _____
(6) _____	(13) _____
(7) _____	(14) _____

OUTCOME 7-2, 7-3, 7-4 **FIGURE 7-3** Lateral View of the Adult Skull

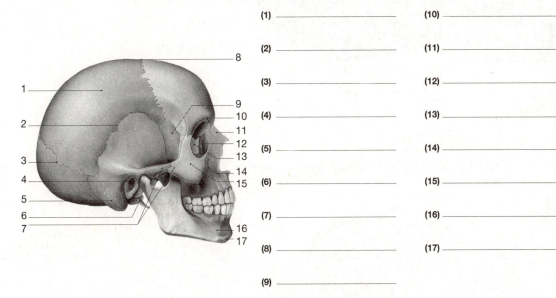

(1) _____	(10) _____
(2) _____	(11) _____
(3) _____	(12) _____
(4) _____	(13) _____
(5) _____	(14) _____
(6) _____	(15) _____
(7) _____	(16) _____
(8) _____	(17) _____
(9) _____	

OUTCOME 7-2, 7-3, 7-4 **FIGURE 7-4** Inferior View of the Adult Skull

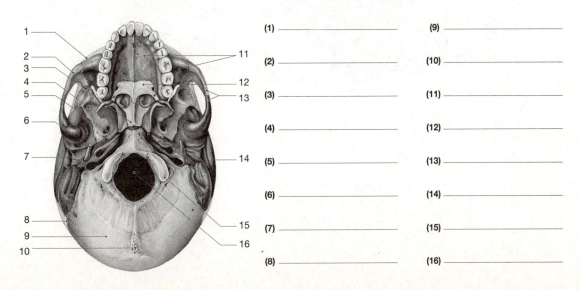

(1) _____	(9) _____
(2) _____	(10) _____
(3) _____	(11) _____
(4) _____	(12) _____
(5) _____	(13) _____
(6) _____	(14) _____
(7) _____	(15) _____
(8) _____	(16) _____

OUTCOME 7-5   **FIGURE 7-5**   Fetal Skull—Lateral View

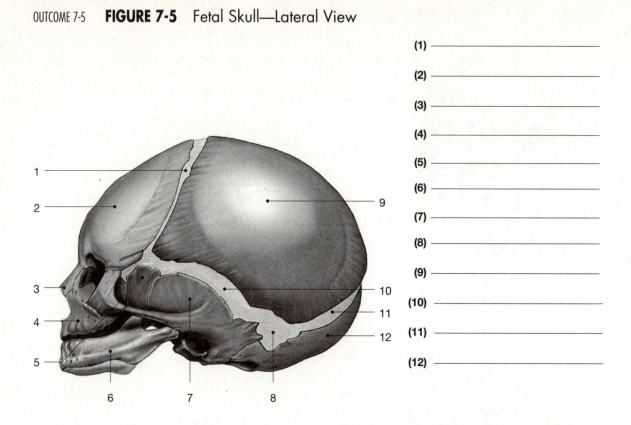

(1) _____

(2) _____

(3) _____

(4) _____

(5) _____

(6) _____

(7) _____

(8) _____

(9) _____

(10) _____

(11) _____

(12) _____

OUTCOME 7-5   **FIGURE 7-6**   Fetal Skull—Superior View

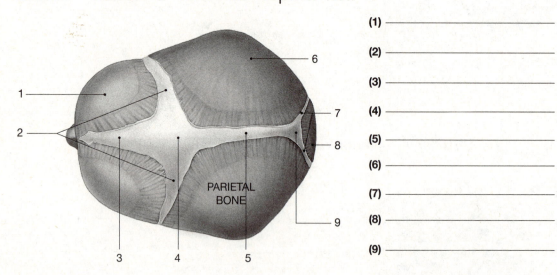

PARIETAL BONE

(1) _____

(2) _____

(3) _____

(4) _____

(5) _____

(6) _____

(7) _____

(8) _____

(9) _____

OUTCOME 7-6, 7-7   **FIGURE 7-7**   The Vertebral Column

SPINAL CURVES   VERTEBRAL REGIONS

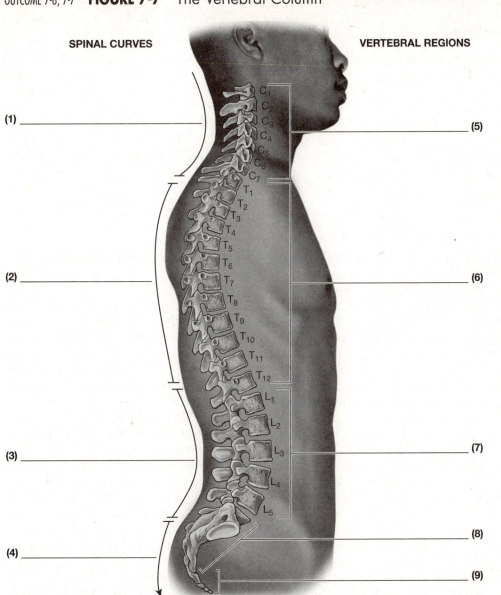

(1) _____

(2) _____

(3) _____

(4) _____

(5) _____

(6) _____

(7) _____

(8) _____

(9) _____

OUTCOME 7-8   **FIGURE 7-8**   The Ribs

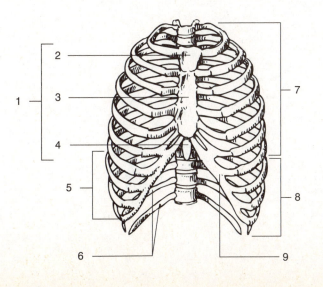
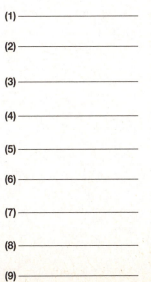

(1) _____

(2) _____

(3) _____

(4) _____

(5) _____

(6) _____

(7) _____

(8) _____

(9) _____

LEVEL 2: REVIEWING CONCEPTS

Chapter Overview

Using the terms below, fill in the blanks to complete the chapter overview of the axial skeleton.

coccyx	skull	hyoid
axial	lumbar	sagittal
lacrimal	manubrium	zygomatic
thoracic	mandible	false
sphenoid	nasal	Wormian
sternum	xiphoid process	occipital
cranium	true	ribs
sacrum	vertebrae	cervical
parietal	floating	vomer

The bones of the (1) _____ division of the skeleton form the longitudinal axis of the body. The most superior part, the (2) _____, consists of the (3) _____, the face, and associated bones. The face consists of six paired bones, which include the (4) _____, or cheek bone; the maxilla; the (5) _____ bones; the palatine; the (6) _____; and the nasal concha. The lower jaw bone, the (7) _____, and the (8) _____ bone make up the single bones of the face. The cranial part of the skull includes four unpaired bones, the (9) _____, frontal, (10) _____, and the ethmoid. The paired bones of the cranium are the (11) _____ and the temporal bones. The bones that form the sutures of the skull are called the (12) _____ bones. These suture bones form immovable joints named the lambdoid, coronal, squamous, and (13) _____. The associated skull bones include the ear ossicles, and the (14) _____, the only bone in the body not articulated with another bone. Just below the neck and looking anteriorly at the skeleton, the 12 pairs of (15) _____ and the mid-line (16) _____ form the anterior part of the thorax. The mid-line, broad-surfaced bone with three individual articulated bones includes the superiorly located (17) _____, a broad middle body, and the (18) _____ located at the inferior end. The first seven pairs of ribs are the (19) _____ ribs, and 8 through 12 are the (20) _____ ribs. The last two pairs of ribs do not attach to the cartilage above, identifying them as (21) _____ ribs. Looking at a posterior view of the skeleton, the 26 uniquely articulated (22) _____ are exposed. The first seven make up the (23) _____ area, the next 12 the (24) _____ region, the five that follow the (25) _____, and the last two fused areas, the (26) _____ consisting of five fused bones. The last, the tail bone, consists of three to five fused bones, called the (27) _____. The axial skeleton consists of 80 bones, roughly 40 percent of the bones in the body.

Concept Map I

Using the following terms, fill in the circled, numbered, blank spaces to correctly complete the concept map. Use each term only once.

floating ribs   temporal   Wormian bones

hyoid   sacral   vertebral column

lacrimal   xiphoid process   occipital

sternum   skull   mandible

thoracic   lambdoid

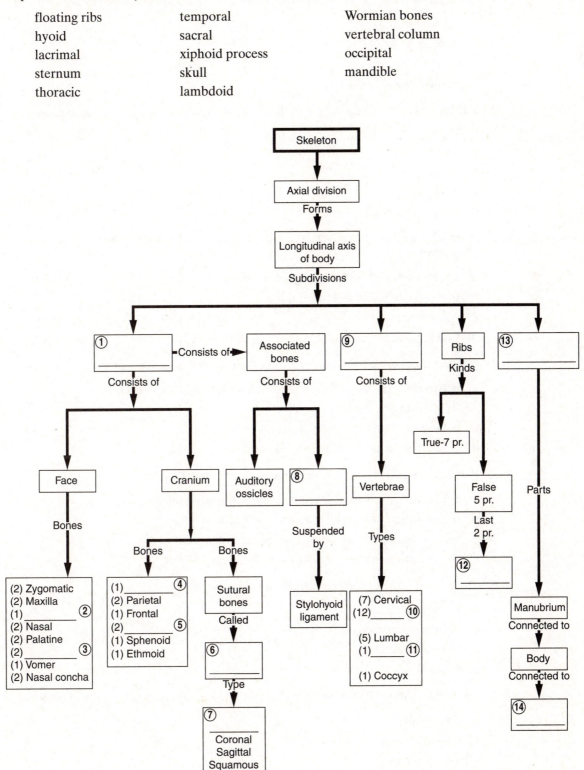

Multiple Choice

Place the letter corresponding to the best answer in the space provided.

_____ 1. Brain growth, skull growth, and completed cranial suture development occur

    a. before birth.

    b. right after birth.

    c. by age 5.

    d. between age 5 and 10.

_____ 2. The area of the greatest degree of flexibility along the vertebral column is found from

    a. C_3–C_7.

    b. T_1–T_6.

    c. T_7–T_{12}.

    d. L_1–L_5.

_____ 3. After a hard fall, compression fractures or compression-dislocation fractures most often involve the

    a. last thoracic and first two lumbar vertebrae.

    b. first two cervical vertebrae.

    c. sacrum and coccyx.

    d. fifth lumbar and the sacrum.

_____ 4. Intervertebral discs are found in between all the vertebrae except

    a. between C_1 and T_1, and T_{12} and L_1; the sacrum and the coccyx.

    b. the sacrum and the coccyx.

    c. between L_5 and the sacrum.

    d. between C_1 and C_2 and the sacrum and coccyx.

_____ 5. Part of the loss in *height* that accompanies aging results from

    a. degeneration of osseous tissue in the diaphysis of long bones.

    b. degeneration of skeletal muscles attached to bones.

    c. the decreasing size and resiliency of the intervertebral discs.

    d. the reduction in the number of vertebrae due to aging.

_____ 6. The skull articulates with the vertebral column at the

    a. foramen magnum.

    b. occipital condyles.

    c. lambdoid sutures.

    d. C_1 and C_2.

_____ 7. The saddle-shaped feature of the sphenoid that houses the pituitary gland is the

    a. crista galli.

    b. styloid process.

    c. sella turcica.

    d. frontal squama.

_____ 8. The growth of the cranium is usually associated with the

      a. expansion of the brain.

      b. development of the fontanelles.

      c. closure of the sutures.

      d. time of birth.

_____ 9. Beginning at the superior end of the vertebral canal and proceeding inferiorally, the

      a. diameter of the cord and the size of the vertebral arch increase.

      b. diameter of the cord increases and the size of the vertebral arch decreases.

      c. diameter of the cord decreases and the size of the vertebral arch increases.

      d. diameter of the cord and the size of the vertebral arch decrease.

_____ 10. The vertebrae that are directly articulated with the ribs are the

      a. cervical and thoracic.

      b. thoracic only.

      c. cervical only.

      d. thoracic and lumbar.

_____ 11. During CPR, proper positioning of the hands is important so that an excessive pressure will not break the

      a. manubrium.

      b. xiphoid process.

      c. body of the sternum.

      d. costal cartilages.

Completion

Using the terms below, complete the following statements. Use each term only once.

kyphosis	mental foramina	auditory tube
auditory ossicles	lordosis	alveolar processes
metopic	lacrimal	scoliosis
compensation		

1. The air-filled passageway that connects the pharynx to the tympanic cavity is the _____.

2. At birth, the two frontal bones that have not completely fused are connected at the _____ suture.

3. The smallest facial bones, which house the tear ducts and drain into the nasal cavity, are the _____ bones.

4. The associated skull bones of the middle ear that conduct sound vibrations from the tympanum to the inner ear are the _____.

5. The oral margins of the maxillae that provide the sockets for the teeth are the _____.

6. Small openings that serve as nerve passageways on each side of the body of the mandible are the _____.

7. The lumbar and cervical curves that appear several months after birth and help to position the body weight over the legs are known as _____ curves.

8. A normal thoracic curvature that becomes exaggerated, producing a "roundback" appearance, is a _____.

9. An exaggerated lumbar curvature or "swayback" appearance is a _____.

10. An abnormal lateral curvature that usually appears in adolescence during periods of rapid growth is _____.

Short Essay

Briefly answer the following questions in the spaces provided below.

1. What are the four primary functions of the axial skeleton?

2. a. List the *paired* bones of the *cranium*.
   b. List the *unpaired* (single) bones of the *cranium*.

3. a. List the *paired* bones of the *face*.
   b. List the *unpaired* (single) bones of the *face*.

4. Why are the auditory ossicles and hyoid bone referred to as associated bones of the skull?

5. What is *craniostenosis* and what are the results of this condition?

6. What is the difference between a *primary curve* and a *secondary curve* of the spinal column?

7. Distinguish among the abnormal spinal curvature distortions of kyphosis, lordosis, and scoliosis.

8. What is the difference between the *true* ribs and the *false* ribs?

LEVEL 3: CRITICAL THINKING AND CLINICAL APPLICATIONS

Using principles and concepts learned in Chapter 7, answer the following questions. Write your answers on a separate sheet of paper.

1. What senses are provided by the sense organs housed by cranial and facial bones?

2. A friend of yours tells you that she has been diagnosed as having TMJ. What bones are involved with this condition, how are they articulated, and what symptoms are apparent?

3. Your nose has been crooked for years and you have had a severe sinus condition to accompany the disturbing appearance of the nose. You suspect there is a relationship between the crooked nose and the chronic sinus condition. How do you explain your suspicion?

4. During a car accident you become a victim of *whiplash*. You experience pains in the neck and across the upper part of the back. Why?

5. During a child's routine physical examination, it is discovered that one leg is shorter than the other. Relative to the spinal column, what might be a plausible explanation for this condition?

6. An archaeologist at an excavation site finds several small skull bones. From his observations, he concludes that the skulls are those of children less than 1 year old. How can he tell their ages from an examination of their bones?

CHAPTER 8

The Appendicular Skeleton

OVERVIEW

How many things can you think of that require the use of your hands? Your arms? Your legs? Your feet? If you are an active person, the list would probably be endless. There are few daily activities, if any, that do not require the use of the arms and/or legs performing in a dynamic, coordinated way to allow you to be an active, mobile individual. This intricately articulated framework of 126 bones makes up the appendicular division of the skeleton. The bones consist of the pectoral girdle and the upper limbs, and the pelvic girdle and the lower limbs. The bones in this division provide support, are important as sites for muscle attachment, articulate at joints, are involved with movement and mobility, and are utilized in numerous ways to control the environment that surrounds you every second of your life.

Your study and review for Chapter 8 includes the identification and location of bones and bone markings, and the functional anatomy of the bones and articulations that comprise the appendicular skeleton.

LEVEL 1: REVIEWING FACTS AND TERMS

Review of Learning Outcomes

After completing this chapter, you should be able to do the following:

OUTCOME 8-1   Identify the bones that form the pectoral girdle, their functions, and their superficial features.

OUTCOME 8-2   Identify the bones of the upper limbs, their functions, and their superficial features.

OUTCOME 8-3   Identify the bones that form the pelvic girdle, their functions, and their superficial features.

OUTCOME 8-4   Identify the bones of the lower limbs, their functions, and their superficial features.

OUTCOME 8-5   Summarize sex differences and age-related changes in the human skeleton.

Multiple Choice

Place the letter corresponding to the best answer in the space provided.

OUTCOME 8-1   _____   1. The clavicles articulate with a bone of the sternum called the

     a. xiphoid process.

     b. scapula.

     c. manubrium.

     d. deltoideus.

OUTCOME 8-1   _____   2. The surfaces of the scapulae and clavicles are extremely important as sites for

     a. muscle attachment.

     b. positions of nerves and blood vessels.

     c. nourishment of muscles and bones.

     d. support and flexibility.

OUTCOME 8-1   _____   3. The conoid tubercle and costal tuberosity are processes located on the

     a. scapulae.

     b. clavicle.

     c. sternum.

     d. manubrium.

OUTCOME 8-1   _____   4. The bones of the *pectoral* girdle include the

     a. clavicle and scapula.

     b. ilium and ischium.

     c. humerus and femur.

     d. ulna and radius.

OUTCOME 8-1   _____   5. The large posterior process on the scapula is the

     a. coracoid process.

     b. acromion.

     c. olecranon fossa.

     d. styloid process.

OUTCOME 8-1   _____   6. The primary function of the *pectoral* girdle is to

     a. protect the organs of the thorax.

     b. provide areas for articulation with the vertebral column.

     c. position the shoulder joint and provide a base for arm movement.

     d. support and maintain the position of the skull.

OUTCOME 8-2   _____   7. The parallel bones that support the forearm are the

     a. humerus and femur.

     b. ulna and radius.

     c. tibia and fibula.

     d. scapula and clavicle.

OUTCOME 8-2   _____   8. The large rough elevation on the lateral surface of the shaft of the humerus is the
   a. greater tubercle.
   b. radial fossa.
   c. deltoid tuberosity.
   d. lesser tubercle.

OUTCOME 8-3   _____   9. The bones of the *pelvic* girdle include the
   a. tibia and fibula.
   b. ilium, pubis, and ischium.
   c. ilium, ischium, and acetabulum.
   d. coxa, patella, and acetabulum.

OUTCOME 8-3   _____   10. The joint that limits movements between the two pubic bones is the
   a. pubic symphysis.
   b. obturator foramen.
   c. greater sciatic notch.
   d. pubic tubercle.

OUTCOME 8-4   _____   11. The large medial bone of the lower leg is the
   a. femur.
   b. fibula.
   c. tibia.
   d. humerus.

OUTCOME 8-4   _____   12. A prominent elevation that runs along the center of the posterior surface of the femur, which serves as an attachment site for muscles that abduct the femur, is the
   a. greater trochanter.
   b. trochanteric crest.
   c. trochanteric lines.
   d. linea aspera.

OUTCOME 8-5, 8-3   _____   13. The general appearance of the pelvis of the female compared to the male is that the female pelvis is
   a. heart-shaped.
   b. robust, heavy, and rough.
   c. relatively deep.
   d. broad, light, and smooth.

OUTCOME 8-5   _____   14. The shape of the pelvic inlet in the female is
   a. heart-shaped.
   b. triangular.
   c. circular.
   d. somewhat rectangular.

OUTCOME 8-5   _____ 15. Which of the following selections would be used to estimate muscular development and muscle mass?

     a. Sex and age

     b. Degenerative changes in the normal skeletal system

     c. Normal timing of skeletal development

     d. Appearance of various ridges and general bone mass

OUTCOME 8-5   _____ 16. Skeletal remains can provide evidence of healed fractures that could help build an understanding of that person's medical history.

     a. True

     b. False

OUTCOME 8-5   _____ 17. The specific *areas* of the skeleton that are generally used to identify significant differences between a male and female are the

     a. arms and legs.

     b. ribs and vertebral column.

     c. skull and pelvis.

     d. a, b, and c are correct.

OUTCOME 8-5   _____ 18. The two important *skeletal elements* that are generally used to determine sex and age are the

     a. teeth and healed fractures.

     b. presence of muscular and fatty tissue.

     c. condition of the teeth and muscular mass.

     d. bone weight and bone markings.

OUTCOME 8-5   _____ 19. In determining the age of a skeleton, which of the following selections would be used?

     a. The size of the bones

     b. The size and roughness of bone markings

     c. The mineral content of the bones

     d. a, b, and c are correct

Completion

Using the terms below, complete the following statements. Use each term only once.

patella	clavicle	childbearing
pubic symphysis	age	pectoral girdle
femur	humerus	pelvis
coxae	glenoid cavity	teeth
styloid	acetabulum	acromion

OUTCOME 8-1   1. The clavicle articulates with a process of the scapulae called the _____.

OUTCOME 8-1   2. The only direct connection between the pectoral girdle and the axial skeleton is an articulation between the sternum and the _____.

OUTCOME 8-1   3. The shoulder area and its component bones comprise a region referred to as the _____.

OUTCOME 8-1   4. The scapula articulates with the proximal end of the humerus at the _____.

OUTCOME 8-2   5. The ulna and the radius both have long shafts that contain pointed processes called _____ processes.

OUTCOME 8-2   6. The bone in the upper arm, or brachium, that extends from the scapula to the elbow is the _____.

OUTCOME 8-3   7. The pelvic girdle consists of six bones collectively referred to as the _____.

OUTCOME 8-3   8. Ventrally, the coxae are connected by a pad of fibrocartilage at the _____.

OUTCOME 8-4   9. The longest and heaviest bone in the body, which articulates with the tibia at the knee joint, is the _____.

OUTCOME 8-4   10. At the hip joint to either side, the head of the femur articulates with the _____.

OUTCOME 8-4   11. In its normal motion, the _____ glides in a superior-inferior direction.

OUTCOME 8-5, 8-3   12. An enlarged pelvic outlet in the female is an adaptation for _____.

OUTCOME 8-5   13. A major means of determining the medical history of a person is to examine the condition of the individual's _____.

OUTCOME 8-5   14. Skeletal differences between males and females are usually identified axially on the skull and/or the appendicular aspects of the _____.

OUTCOME 8-5   15. Reduction in mineral content of the bony matrix is an example of a skeletal change related to _____.

Matching

Match the terms in column B with the terms in column A. Use letters for answers in the spaces provided.

		Column A	**Column B**
OUTCOME 8-1	_____	1. scapular process	A. lower arm bones
OUTCOME 8-1	_____	2. pectoral girdle	B. determine approximate age
OUTCOME 8-1	_____	3. clavicle	C. lower leg bones
OUTCOME 8-2	_____	4. radius, ulna	D. hip bones
OUTCOME 8-3	_____	5. pelvic girdle	E. oval to round
OUTCOME 8-4	_____	6. patella	F. acromion process
OUTCOME 8-4	_____	7. tibia, fibula	G. sternal end
OUTCOME 8-5	_____	8. female pelvic inlet	H. heavier bone weight
OUTCOME 8-5	_____	9. male pelvic inlet	I. smooth bone markings
OUTCOME 8-5	_____	10. epiphyseal cartilages	J. shoulder girdle
OUTCOME 8-5	_____	11. male characteristic	K. heart-shaped
OUTCOME 8-5	_____	12. process of aging	L. kneecap

Drawing/Illustration Labeling

Identify each numbered structure. Place your answers in the spaces provided.

OUTCOME 8-1, 8-2, 8-3, 8-4   **FIGURE 8-1**   Appendicular Skeleton

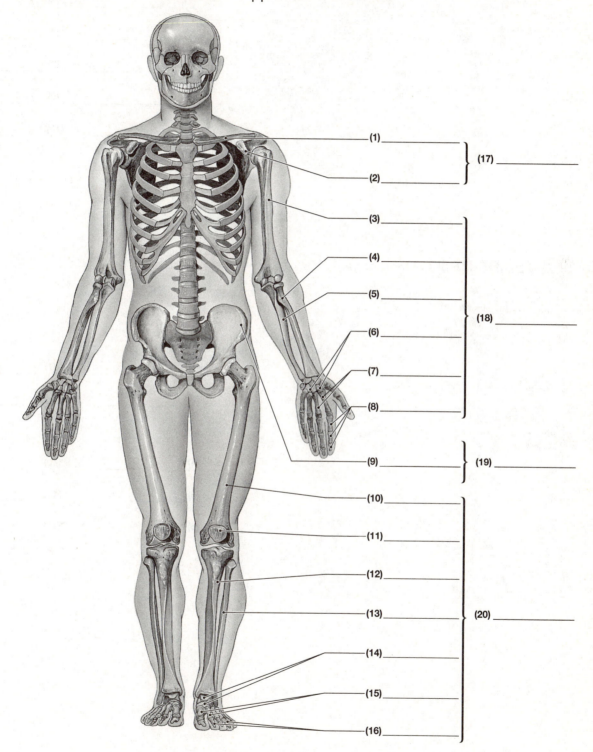

(1) _____

(2) _____

(17) _____

(3) _____

(4) _____

(5) _____

(6) _____

(7) _____

(8) _____

(18) _____

(9) _____

(19) _____

(10) _____

(11) _____

(12) _____

(13) _____

(20) _____

(14) _____

(15) _____

(16) _____

OUTCOME 8-1   **FIGURE 8-2**   The Scapula

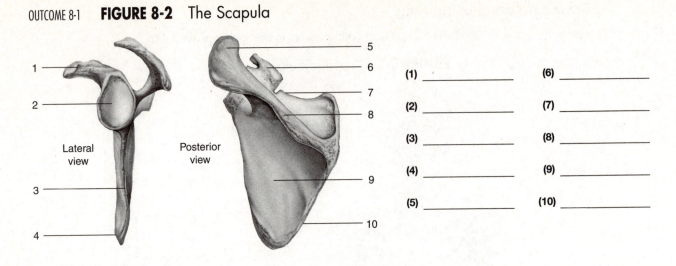

Lateral view

Posterior view

(1) _____   (6) _____

(2) _____   (7) _____

(3) _____   (8) _____

(4) _____   (9) _____

(5) _____   (10) _____

OUTCOME 8-2   **FIGURE 8-3**   The Humerus

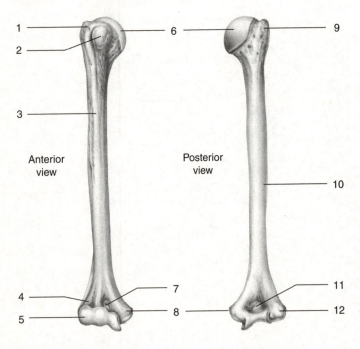

Anterior view

Posterior view

(1) _____   (7) _____

(2) _____   (8) _____

(3) _____   (9) _____

(4) _____   (10) _____

(5) _____   (11) _____

(6) _____   (12) _____

OUTCOME 8-2   **FIGURE 8-4**   The Radius and Ulna

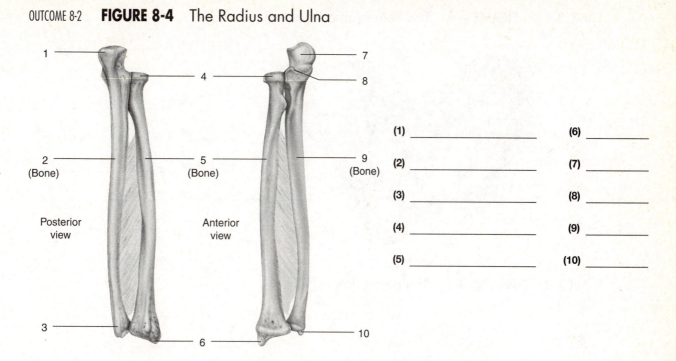

Posterior view

Anterior view

(1) _____   (6) _____

(2) _____   (7) _____

(3) _____   (8) _____

(4) _____   (9) _____

(5) _____   (10) _____

OUTCOME 8-2   **FIGURE 8-5**   Bones of the Wrist and Hand

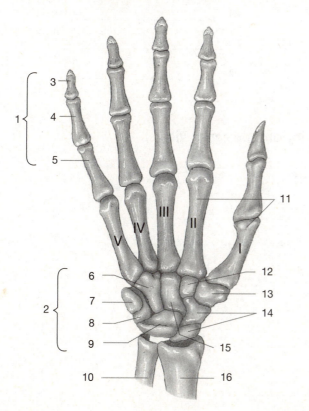

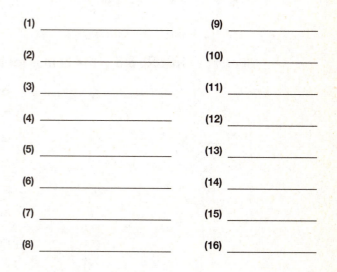

(1) _____   (9) _____

(2) _____   (10) _____

(3) _____   (11) _____

(4) _____   (12) _____

(5) _____   (13) _____

(6) _____   (14) _____

(7) _____   (15) _____

(8) _____   (16) _____

OUTCOME 8-3 **FIGURE 8-6** The Pelvis (anterior view)

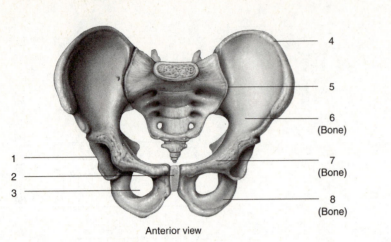
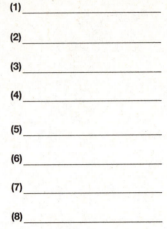

Anterior view

(1) _____

(2) _____

(3) _____

(4) _____

(5) _____

(6) _____

(7) _____

(8) _____

OUTCOME 8-3 **FIGURE 8-7** The Pelvis (lateral view)

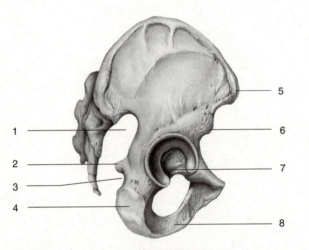

Lateral view

(1) _____

(2) _____

(3) _____

(4) _____

(5) _____

(6) _____

(7) _____

(8) _____

OUTCOME 8-4 **FIGURE 8-8** The Femur (anterior and posterior views)

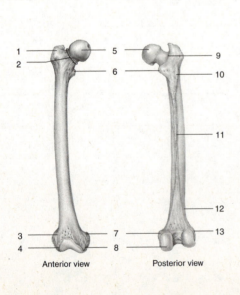
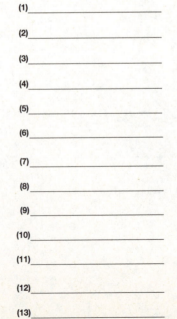

Anterior view   Posterior view

(1) _____

(2) _____

(3) _____

(4) _____

(5) _____

(6) _____

(7) _____

(8) _____

(9) _____

(10) _____

(11) _____

(12) _____

(13) _____

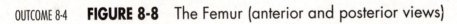

OUTCOME 8-4   **FIGURE 8-9**   The Tibia and Fibula

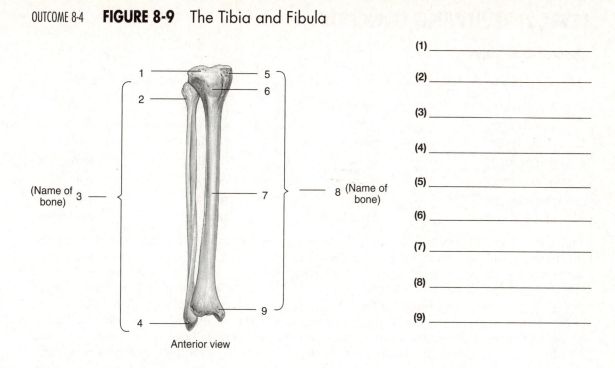

Anterior view

(1) _____

(2) _____

(3) _____

(4) _____

(5) _____

(6) _____

(7) _____

(8) _____

(9) _____

OUTCOME 8-4   **FIGURE 8-10**   Bones of the Ankle and Foot

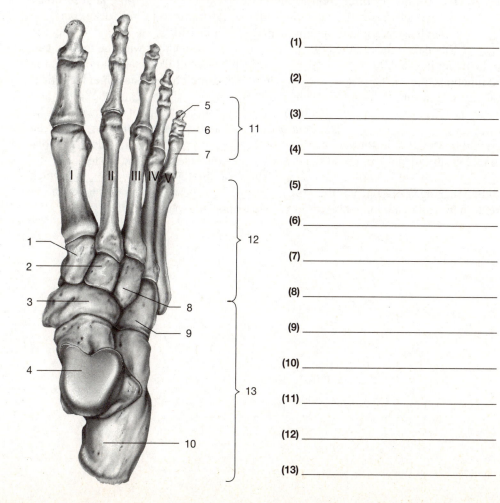

(1) _____

(2) _____

(3) _____

(4) _____

(5) _____

(6) _____

(7) _____

(8) _____

(9) _____

(10) _____

(11) _____

(12) _____

(13) _____

LEVEL 2: REVIEWING CONCEPTS

Chapter Overview

Using the terms below, fill in the blanks to complete the chapter overview of the appendicular skeleton.

acetabulum	humerus	scapulae	pectoral
glenohumeral	ischium	girdles	ilium
metacarpals	carpals	patella	knee
metatarsals	limbs	tarsals	toes
pubic symphysis	ulna	glenoid	tibia
clavicles	femur	fibula	pubis
ball and socket	coxal bones	radius	phalanges

Appended to the 80 bones of the axial skeleton are the remaining 126 bones that make up the appendicular skeleton, which includes the bones of the (1) _____ and the supporting elements, the (2) _____. Two S-shaped (3) _____ and two broad, flat (4) _____ make up the (5) _____ girdle. The shoulder joint known as the (6) _____ joint consists of the (7) _____ cavity, which articulates with the ball of the (8) _____, the single prominent bone of the upper arm. At the elbow joint, this single long bone of the upper arm is joined with the two parallel bones of the lower arm, the (9) _____, which lies medial to the (10) _____. The distal ends of the lower arm bones are articulated with the wrist bones, the (11) _____, which join with the bones of the hands, the (12) _____, which are connected to the bones of the fingers, the (13) _____. The pelvic girdle consists of the paired hip bones; the (14) _____, which is formed by the fusion of three bones; an (15) _____, the largest bone; the (16) _____, which forms the posterior, inferior portion of the hip socket; and the (17) _____, which is interconnected by a pad of fibrocartilage at a joint called the (18) _____. The pelvic socket, the (19) _____, articulates with the longest and heaviest bone in the upper leg, the (20) _____, which has a ball-like head that fits into the pelvic socket and functions like a (21) _____ joint. The large bone of the upper leg articulates at the distal end with a long, lateral thin bone, the (22) _____, and the larger and thicker medial bone, the (23) _____, both of which make up the bones of the lower leg. The kneecap, called the (24) _____, is a bone that glides across the articular surface of the femur embedded within a tendon in the region of the (25) _____ joint, which functions like the elbow joint. Distally, the long bones of the lower leg are connected with the ankle bones, the (26) _____, which are joined to the bones of the foot, the (27) _____, which are connected to the phalanges, or bones of the (28) _____. The girdles and limbs of the appendicular skeleton let you manipulate objects and move from place to place.

Concept Map I

Using the following terms, fill in the circled, numbered, blank spaces to correctly complete the concept map. Use each term only once.

fibula	ischium	femur
humerus	pectoral girdle	metacarpals
phalanges	radius	tarsals

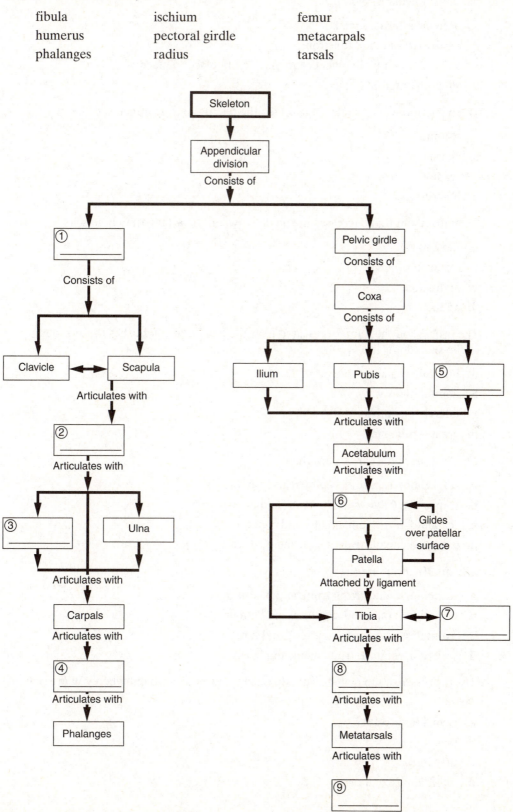

Multiple Choice

Place the letter corresponding to the best answer in the space provided.

_____ 1. The appendicular skeleton

    a. protects internal organs.

    b. participates in vital functions.

    c. lets you manipulate objects and move from place to place.

    d. supports internal organs.

_____ 2. The bone of the axial skeleton to which the pectoral girdle attaches is the

    a. sternum.

    b. clavicle.

    c. scapula.

    d. humerus.

_____ 3. The movement that decreases the angle between the articulating bones is

    a. flexion.

    b. extension.

    c. abduction.

    d. adduction.

_____ 4. The process on the humerus located near the head that establishes the contour of the shoulder is the

    a. intertubercular groove.

    b. deltoid tuberosity.

    c. lateral epicondyle.

    d. greater tubercle.

_____ 5. The four *proximal* carpals are the

    a. trapezium, trapezoid, capitate, and hamate.

    b. trapezium, triquetrum, capitate, and lunate.

    c. scaphoid, trapezium, lunate, and capitate.

    d. scaphoid, lunate, triquetrum, and pisiform.

_____ 6. The *distal* carpals are the

    a. scaphoid, lunate, triquetrum, and pisiform.

    b. trapezium, trapezoid, capitate, and hamate.

    c. trapezium, scaphoid, trapezoid, and lunate.

    d. scaphoid, capitate, triquetrum, and hamate.

_____ 7. The processes of the pelvic girdle that bear your body's weight when you are seated are the

    a. ischial tuberosities.

    b. iliac spines.

    c. iliac fossa.

    d. pubic and ischial rami.

_____ 8. The only ankle bone that articulates with the tibia and the fibula is the

    a. calcaneus.

    b. talus.

    c. navicular.

    d. cuboid.

_____ 9. The primary responsibility for stabilizing, positioning, and bracing the pectoral girdle is that of the

    a. joint shape.

    b. tendons.

    c. muscles.

    d. ligaments.

_____ 10. From the following selections, which pair of terms is correctly associated?

    a. Pubic symphysis; pectoral girdle

    b. Patella; acetabulum

    c. Radius; phalanges

    d. Femur; linea aspera

_____ 11. The femur in older individuals, particularly small women, can be distinguished by

    a. being stronger than other bones of the pelvis.

    b. having a greater number of trabeculae than those of younger individuals.

    c. a greater tendency to bend with the stress of weight bearing.

    d. being the most likely site of a breach when the person suffers a broken hip.

_____ 12. Of the two bones of the forearm, the one that articulates with the carpal bones of the wrist is the

    a. ulna.

    b. humerus.

    c. radius.

    d. tibia.

_____ 13. Processes that develop proximal to an articulation and provide additional surface area for muscle attachment are

    a. tuberosities.

    b. fossae.

    c. epicondyles.

    d. spinae.

_____ 14. The tarsal bone that accepts weight and distributes it to the heel or the toes is the

    a. navicular.

    b. talus.

    c. cuneiform.

    d. calcaneous.

Completion

Using the terms below, complete the following statements. Use each term only once.

calcaneus	hallux	pubic symphysis	thumb
clavicle	fibula	ilium	metacarpals
femur	perineum	pelvis	coxal bone
ulna	scapula	acetabulum	

1. The bone that cannot resist strong forces and provides the only fixed support for the pectoral girdle is the _____.
2. At its proximal end, the round head of the humerus articulates with the _____.
3. The bone that forms the medial support of the forearm is the _____.
4. The bones that form the palm of the hand are the _____.
5. The ilium, ischium, and a pubis fuse to form each _____.
6. Anteriorly, the medial surfaces of the hip bones are interconnected by a pad of fibrous cartilage at a joint called the _____.
7. At the proximal end of the femur, the head articulates with the curved surface of the _____.
8. The largest coxal bone is the _____.
9. The coxae, the sacrum, and the coccyx form a composite structure called the _____.
10. The longest and heaviest bone in the body is the _____.
11. The bone in the lower leg that is completely excluded from the knee joint is the _____.
12. The large heel bone that receives weight transmitted to the ground by the inferior surface of the talus is the _____.
13. If a person has dislocated his pollex, he has injured his _____.
14. The first toe is anatomically referred to as the _____.
15. The region bounded by the inferior edges of the pelvis is called the _____.

Short Essay

Briefly answer the following questions in the spaces provided below.

1. What are the primary functions of the appendicular skeleton?

2. What are the components of the appendicular skeleton?

3. What bones comprise the pectoral girdle? The pelvic girdle?

4. What are the structural and functional similarities and differences between the ulna and the radius?

5. List all the bones of the upper limbs.

6. What are the structural and functional similarities and differences between the tibia and the fibula?

7. Why is it necessary for the bones of the pelvic girdle to be more massive than those of the pectoral girdle?

8. What skeletal adaptations in the female pelvis make childbearing possible?

9. List all the bones of the lower limbs.

LEVEL 3: CRITICAL THINKING AND CLINICAL APPLICATIONS

Using principles and concepts learned in Chapter 8, answer the following questions. Write your answers on a separate sheet of paper.

1. What effect does running on hard or slanted surfaces, such as an inter-tidal area of a beach or the shoulder of a road, have on the knees?

2. Matthew is the midfielder on his high school soccer team. He is injured in a game and is taken to the hospital emergency room, where he is diagnosed with a Pott's fracture. What causes this type of fracture?

3. In a fitness program that includes street running, why is it essential to have proper support for the bones of the foot?

4. Janet is experiencing pain, weakness, and loss of wrist mobility. After a professional evaluation and diagnosis, her condition is identified as carpal tunnel syndrome. What causes this condition?

5. How does a comprehensive study of a human skeleton reveal important information about an individual?

Articulations

OVERVIEW

Ask the average individual what things are necessary to maintain life and usually the response will be oxygen, food, and water. Few people relate the importance of movement as one of the factors necessary to maintain life, yet the human body doesn't survive very long without the ability to produce body movements. One of the functions of the skeletal system is to permit body movements; however, it is not the rigid bones that allow movement, but the articulations, or joints, between the bones.

The elegant movements of the ballet dancer or gymnast, as well as the everyday activities of walking, talking, eating, and writing, require the coordinated activity of all the joints, some of which are freely movable and flexible while others remain rigid to stabilize the body and serve to maintain balance.

Chapter 9 classifies the joints of the body according to structure and function. Classification by structure is based on the type of connective tissue between the articulating bones, while the functional classification depends on the degree of movement permitted within the joint.

The exercises in this chapter are designed to encourage you to organize and conceptualize the principles of kinetics and biomechanics in the science of arthrology—the study of joints. Demonstrating the various movements permitted at each of the movable joints will facilitate and help you to visualize the adaptive advantages as well as the limitations of each type of movement.

LEVEL 1: REVIEWING FACTS AND TERMS

Review of Learning Outcomes

After completing this chapter, you should be able to do the following:

OUTCOME 9-1   Contrast the major categories of joints, and explain the relationship between structure and function for each category.

OUTCOME 9-2   Describe the basic structure of a synovial joint, and describe common synovial joint accessory structures and their functions.

OUTCOME 9-3   Describe how the anatomical and functional properties of synovial joints permit dynamic movements of the skeleton.

OUTCOME 9-4   Describe the articulations between the vertebrae of the vertebral column.

OUTCOME 9-5   Describe the structure and function of the shoulder joint and the elbow joint.

OUTCOME 9-6   Describe the structure and function of the hip joint and the knee joint.

OUTCOME 9-7   Describe the effects of aging on articulations, and discuss the most common age-related clinical problems for articulations.

OUTCOME 9-8   Explain the functional relationships between the skeletal system and other body systems.

Multiple Choice

Place the letter corresponding to the best answer in the space provided.

OUTCOME 9-1   _____   1. Joints, or articulations, are classified on the basis of their degree of movement. From the following selections, choose the one that identifies, in correct order, the following joints on the basis of: no movement; slightly movable; freely movable.

a. Amphiarthrosis, diarthrosis, synarthrosis

b. Diarthrosis, synarthrosis, amphiarthrosis

c. Amphiarthrosis, synarthrosis, diarthrosis

d. Synarthrosis, amphiarthrosis, diarthrosis

OUTCOME 9-1   _____   2. The amphiarthrotic articulation that limits movements between the two pubic bones is the

a. pubic symphysis.

b. obturator foramen.

c. greater sciatic notch.

d. pubic tubercle.

OUTCOME 9-1   _____   3. The type of synarthrosis that binds each tooth to the surrounding bony socket is a

a. synchondrosis.

b. syndesmosis.

c. gomphosis.

d. symphysis.

OUTCOME 9-1   _____   4. Which joint is correctly matched with its type of joint?

a. Symphysis pubis—fibrous

b. Knee—synovial

c. Sagittal suture—cartilaginous

d. Intervertebral disc—synovial

OUTCOME 9-2   _____   5. The function(s) of synovial fluid that fills the joint cavity is (are)

a. it nourishes the chondrocytes.

b. it provides lubrication.

c. it acts as a shock absorber.

d. a, b, and c are correct.

OUTCOME 9-2   _____   6. The primary function(s) of menisci in synovial joints is (are)

a. to subdivide a synovial cavity.

b. to channel the flow of synovial fluid.

c. to allow for variations in the shapes of the articular surfaces.

d. a, b, and c are correct.

OUTCOME 9-3 _____ 7. Flexion is defined as movement that
  a. increases the angle between articulating elements.
  b. decreases the angle between articulating elements.
  c. moves a limb from the midline of the body.
  d. moves a limb toward the midline of the body.

OUTCOME 9-3 _____ 8. Abduction and adduction always refer to movements of the
  a. axial skeleton.
  b. vertebral column.
  c. appendicular skeleton.
  d. a, b, and c are correct.

OUTCOME 9-3 _____ 9. Movements that occur at the shoulder and the hip represent the actions that occur at a _____ joint.
  a. gliding
  b. ball-and-socket
  c. pivot
  d. hinge

OUTCOME 9-3 _____ 10. The reason that the elbow and knee are called hinge joints is
  a. the articulator surfaces are able to slide across one another.
  b. all combinations of movement are possible.
  c. they permit angular movement in a single plane.
  d. sliding and rotation are prevented, and angular motion is restricted to two directions.

OUTCOME 9-3 _____ 11. The type of joint that connects the fingers and toes with the metacarpals and metatarsals is a(n)
  a. condylar joint.
  b. biaxial joint.
  c. synovial joint.
  d. a, b, and c are correct.

OUTCOME 9-3 _____ 12. Examples of angular motion include
  a. flexion and extension.
  b. abduction and adduction.
  c. circumduction.
  d. a, b, and c are correct.

OUTCOME 9-4 _____ 13. The part(s) of the vertebral column that do not contain intervertebral discs is (are) the
  a. sacrum.
  b. coccyx.
  c. first and second cervical vertebrae.
  d. a, b, and c are correct.

OUTCOME 9-4 _____ 14. Movements of the vertebral column are limited to
  a. flexion and extension.
  b. lateral flexion.
  c. rotation.
  d. a, b, and c are correct.

OUTCOME 9-5 _____ 15. The joint that permits the greatest range of motion of any joint in the body is the
    a. hip joint.
    b. shoulder joint.
    c. elbow joint.
    d. knee joint.

OUTCOME 9-5 _____ 16. The elbow joint is quite stable because the
    a. bony surfaces of the humerus and ulna interlock.
    b. articular capsule is very thick.
    c. capsule is reinforced by stout ligaments.
    d. a, b, and c are correct.

OUTCOME 9-5, 9-6 _____ 17. Examples of monaxial joints, which permit angular movements in a single plane, are the
    a. shoulder and hip joints.
    b. intercarpal and intertarsal joints.
    c. elbow and knee joints.
    d. intervertebral joints.

OUTCOME 9-6 _____ 18. The knee joint functions as a
    a. hinge joint.
    b. ball-and-socket joint.
    c. saddle joint.
    d. gliding joint.

OUTCOME 9-6 _____ 19. The reason the points of contact in the knee joint are constantly changing is
    a. there is no single unified capsule or a common synovial cavity.
    b. the menisci conform to the shape of the surface of the femur.
    c. the rounded femoral condyles roll across the top of the tibia.
    d. a, b, and c are correct.

OUTCOME 9-7 _____ 20. The risk of fractures increases with age because
    a. of the cessation of continuous passive motion.
    b. bone mass increases and causes stress on bones.
    c. bone mass decreases and bones become weaker.
    d. of the excessive degree of calcium salt deposition.

OUTCOME 9-7 _____ 21. Arthritis encompasses all of the rheumatic diseases that affect
    a. mistaken immune responses.
    b. synovial joints.
    c. collagen formation.
    d. a, b, and c are correct.

OUTCOME 9-8 _____ 22. The digestive and urinary systems are intimately associated with the skeletal system because they
    a. provide areas of attachment for muscles.
    b. are extensively connected to the skeletal system.
    c. provide calcium and phosphate minerals needed for bone growth.
    d. a, b, and c are correct.

Completion

Using the terms below, complete the following statements. Use each term only once.

synovial	nervous	suture	synostosis
bursae	osteoarthritis	flexion	knee
glenohumeral	elbow	continuous passive motion	intrinsic ligaments
supination	anulus fibrosus	hip	gliding

OUTCOME 9-1　　1. A synarthrotic joint found only between the bones of the skull is a(n) _____.

OUTCOME 9-1　　2. A totally rigid immovable joint resulting from fusion of bones is a(n) _____.

OUTCOME 9-1　　3. Joints that permit a wide range of motion are called diarthrotic, or _____ joints.

OUTCOME 9-2　　4. Localized thickenings of joint capsule are called _____.

OUTCOME 9-2　　5. Small, synovial-filled pockets that form where a tendon or ligament rubs against other tissues are called _____.

OUTCOME 9-3　　6. A movement that reduces the angle between the articulating elements is _____.

OUTCOME 9-3　　7. Movement in the wrist and hand in which the palm is turned forward is _____.

OUTCOME 9-4　　8. The joints between the superior and inferior articulations of adjacent vertebrae are _____.

OUTCOME 9-4　　9. The tough outer layer of fibrocartilage on intervertebral discs is the _____.

OUTCOME 9-5　　10. The joint that permits the greatest range of motion of any joint in the body is the _____ joint.

OUTCOME 9-5　　11. The radial collateral, annular, and ulnar collateral ligaments provide stability for the _____ joint.

OUTCOME 9-6　　12. The extremely stable joint that is almost completely enclosed in a bony socket is the _____ joint.

OUTCOME 9-6　　13. The joint that contains three separate articulations with an incomplete capsule is the _____.

OUTCOME 9-7　　14. The condition resulting from cumulative wear and tear at joint surfaces or from genetic factors affecting collagen formation is _____.

OUTCOME 9-7　　15. The process that appears to encourage the repair of an injured joint by improving circulation of synovial fluid is called _____.

OUTCOME 9-8　　16. The system that regulates bone position by controlling muscle contractions is the _____ system.

Matching

Match the terms in column B with the terms in column A. Use letters for answers in the spaces provided.

Part I

		Column A	**Column B**
OUTCOME 9-1	_____	1. synarthrosis	A. ball and socket
OUTCOME 9-1	_____	2. amphiarthrosis	B. forearm bone movements
OUTCOME 9-2	_____	3. articular discs	C. intervertebral articulation
OUTCOME 9-3	_____	4. acetabulum; head of femur	D. glenohumeral joint
OUTCOME 9-3	_____	5. pronation-supination	E. hinge joint
OUTCOME 9-4	_____	6. gliding joint	F. slightly movable joint
OUTCOME 9-5	_____	7. elbow	G. menisci
OUTCOME 9-5	_____	8. shoulder	H. immovable joint

Part II

		Column A	**Column B**
OUTCOME 9-5	_____	9. partial dislocation	I. knee joint posterior
OUTCOME 9-5	_____	10. shoulder joint	J. articular cartilage damage
OUTCOME 9-6	_____	11. hinge joint	K. varies with age
OUTCOME 9-6	_____	12. acetabulum	L. triaxial movement
OUTCOME 9-6	_____	13. reinforce knee joint	M. nursemaid's elbow
OUTCOME 9-6	_____	14. popliteal ligaments	N. deep fossa
OUTCOME 9-7	_____	15. gout	O. cruciate ligaments
OUTCOME 9-7	_____	16. arthritis	P. knee
OUTCOME 9-8	_____	17. bone recycling	Q. uric acid

Drawing/Illustration Labeling

Identify each numbered structure. Place your answers in the spaces provided.

OUTCOME 9-2 **FIGURE 9-1** Synovial Joint—Sagittal Section

(1)_____   (5)_____

(2)_____   (6)_____

(3)_____   (7)_____

(4)_____   (8)_____

OUTCOME 9-5 **FIGURE 9-2** The Shoulder Joint—Anterior View—Frontal Section

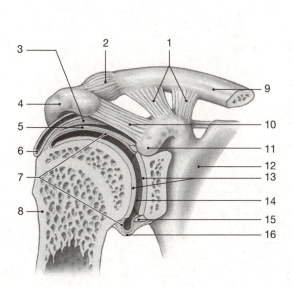

(1)_____

(2)_____

(3)_____

(4)_____

(5)_____

(6)_____

(7)_____

(8)_____

(9)_____

(10)_____

(11)_____

(12)_____

(13)_____

(14)_____

(15)_____

(16)_____

Identify each skeletal movement. Place your answers in the spaces provided.

OUTCOME 9-3 **FIGURE 9-3** Movements of the Skeleton

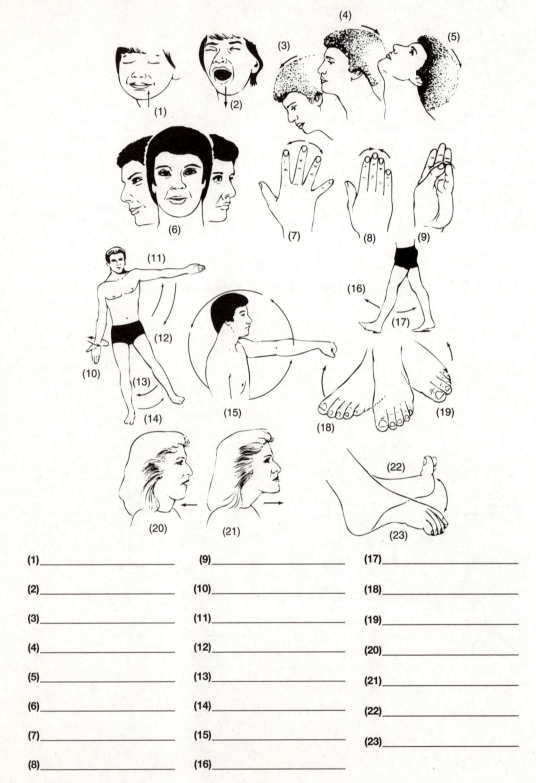

(1)_____	(9)_____	(17)_____
(2)_____	(10)_____	(18)_____
(3)_____	(11)_____	(19)_____
(4)_____	(12)_____	(20)_____
(5)_____	(13)_____	(21)_____
(6)_____	(14)_____	(22)_____
(7)_____	(15)_____	(23)_____
(8)_____	(16)_____	

Instructions

Match the joint movements pictured at the top of the page with the appropriate type of joint in Figure 9-4. Use letters for answers in spaces labeled *movement*.

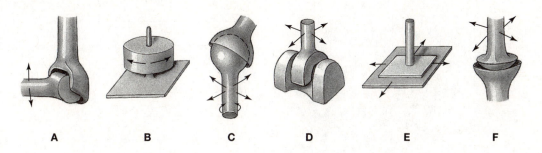

A   B   C   D   E   F

Identify the *types* of joints in Figure 9-4. Place the answers in the spaces labeled *type*.

OUTCOME 9-2, 9-3   **FIGURE 9-4**   Types of Synovial Joints

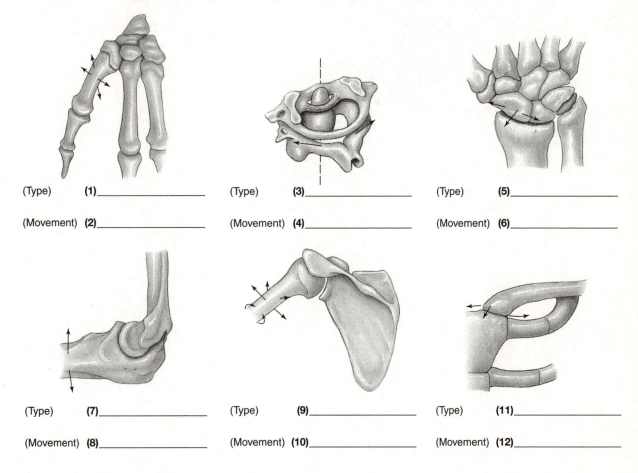

(Type)   **(1)**_____

(Movement) **(2)**_____

(Type)   **(3)**_____

(Movement) **(4)**_____

(Type)   **(5)**_____

(Movement) **(6)**_____

(Type)   **(7)**_____

(Movement) **(8)**_____

(Type)   **(9)**_____

(Movement) **(10)**_____

(Type)   **(11)**_____

(Movement) **(12)**_____

OUTCOME 9-6   **FIGURE 9-5**   Anterior View of the Knee Joint

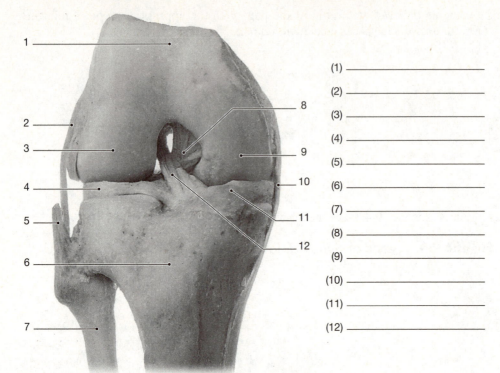

(1) _____

(2) _____

(3) _____

(4) _____

(5) _____

(6) _____

(7) _____

(8) _____

(9) _____

(10) _____

(11) _____

(12) _____

LEVEL 2: REVIEWING CONCEPTS

Chapter Overview

Using the terms below, fill in the blanks to complete the chapter overview of the ways bones interact at joints, or articulations.

capsule	fibula	tibial collateral	gomphoses
articulations	arthritis	menisci	fibular collateral
lubrication	symphysis	diarthrosis	popliteal
femoral	amphiarthrosis	lateral	syndesmosis
synostosis	seven	synovial	patellar
sutures	synchondroses	synarthrosis	cruciate

Because the bones of the skeleton are relatively inflexible, movements can only occur at (1) _____, or joints where two bones interconnect. Based on a functional scheme, the types of joints include: (a) an immovable joint, which is a(n) (2) _____; (b) a slightly movable joint, a(n) (3) _____; and (c) a freely movable joint, a(n) (4) _____. The four major types of immovable joints are: those found between the bones of the skull called (5) _____; those that bind the teeth to bony sockets in the maxillary and mandible, the (6) _____; those that form a rigid, cartilaginous bridge between two articulating bones, a(n) (7) _____; and a totally rigid, immovable joint, a(n) (8) _____. The two types of slightly movable joints are a(n) (9) _____, where bones are connected by a ligament, and a(n) (10) _____, where the articulating bones are separated by a wedge or pad of fibrous cartilage. The freely movable joints contain (11) _____ fluid and permit a wider range of motion than do other types of joints. The fluid has three primary functions, (a) nutrient distribution, (b) shock absorption, and (c) (12) _____. The knee joint is a complex synovial joint with three separate articulations, two between the femur and the tibia, and one between the patella and the patellar surface of the femur. There is no single unified (13) _____, and no common synovial cavity. A pair of fibrocartilage pads, the medial and lateral (14) _____, lie between the femoral and tibial surfaces. The pads cushion the area and provide some (15) _____ stability to the joint. There are (16) _____ major ligaments that stabilize the knee. The (17) _____ ligament provides support to the anterior surface of the knee joint, while two (18) _____ ligaments on the posterior surface extend between the femur and the heads of the tibia and the (19) _____, providing reinforcement to the back of the knee joint. Inside the joint capsule, the crossed anterior and posterior (20) _____ ligaments limit movement of the femur and maintain alignment of the (21) _____ and tibial condyles. Due to the tightness of the (22) _____ ligament reinforcing the medial surface of the knee joint, and the (23) _____ ligament reinforcing the lateral surface, the leg may be fully extended and stabilized, allowing the region to be free from being crushed or damaged due to knee flexion. Joints are subject to heavy wear and tear throughout the lifetime, and problems with joints are relatively common, one of which is (24) _____, which encompasses all the diseases that affect synovial joints.

Concept Map I

Using the following terms, fill in the circled, numbered, blank spaces to correctly complete the concept map. Use each term only once.

amphiarthrosis   sutures   wrist   monaxial
symphysis   cartilaginous   synovial   fibrous
no movement

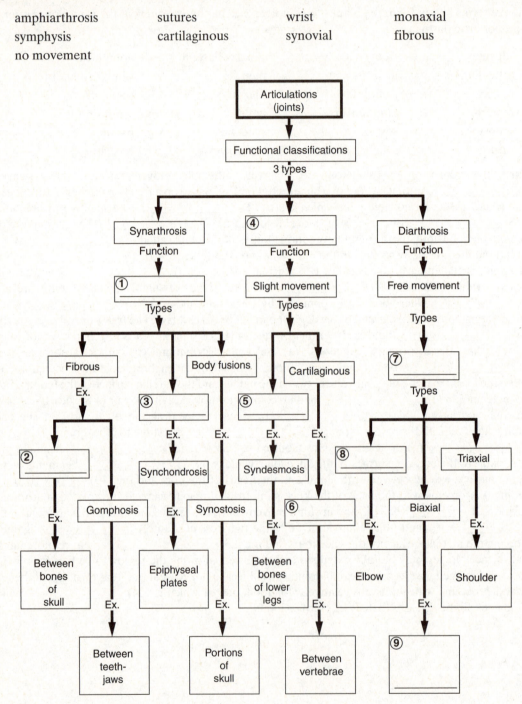

Multiple Choice

Place the letter corresponding to the best answer in the space provided.

_____   1. Localized masses of adipose tissue that protect articular cartilages and act as packing material for the joint are

    a. bursae.

    b. menisci.

    c. fat pads.

    d. articular capsules.

_____   2. A movement away from the longitudinal axis of the body in the frontal plane is

    a. adduction.

    b. hyperextension.

    c. abduction.

    d. circumduction.

_____   3. The opposing movement of supination is

    a. rotation.

    b. pronation.

    c. protraction.

    d. opposition.

_____   4. The special movement of the thumb that enables it to grasp and hold an object is

    a. supination.

    b. inversion.

    c. circumduction.

    d. opposition.

_____   5. Twiddling your thumbs during a lecture demonstrates the action that occurs at a

    a. hinge joint.

    b. ball-and-socket joint.

    c. saddle joint.

    d. gliding joint.

_____   6. A characteristic decrease in height with advanced age may result from

    a. decreased water content of the nucleus pulposus in an intervertebral disc.

    b. a decrease in the anulus fibrosus of an intervertebral disc.

    c. a decrease in the collagen fibers in the bodies of the vertebrae.

    d. a, b, and c are correct.

_____   7. When the nucleus pulposus breaks through the anulus and enters the vertebral canal, the result is a(n)

    a. slipped disc.

    b. anulus fibrosus.

    c. herniated disc.

    d. ligamentous nuchae.

_____ 8. Contraction of the biceps brachii muscle produces

    a. pronation of the forearm and extension of the elbow.

    b. supination of the forearm and extension of the elbow.

    c. supination of the forearm and flexion of the elbow.

    d. pronation of the forearm and flexion of the elbow.

_____ 9. The unique compromise of the articulations in the appendicular skeleton is

    a. the stronger the joint, the less restricted the range of motion.

    b. the weaker the joint, the more restricted the range of motion.

    c. the stronger the joint, the more restricted the range of motion.

    d. the strength of the joint and range of motion are unrelated.

_____ 10. Even though the specific cause may vary, arthritis always involves damage to the

    a. articular cartilages.

    b. bursae.

    c. accessory ligaments.

    d. epiphyseal discs.

_____ 11. In a shoulder separation, the

    a. clavicle and scapula separate.

    b. head of the femur separates from the clavicle.

    c. muscles attached to the clavicle and scapula are torn.

    d. acromioclavicular joint undergoes partial or complete separation.

_____ 12. If you run your fingers along the superior surface of the shoulder joint, you will feel a process called the

    a. coracoid.

    b. acromion.

    c. coronoid.

    d. styloid.

_____ 13. Structures found in synovial joints that reduce friction where large muscles and tendons pass across the joint capsule are referred to as

    a. capsular ligaments.

    b. menisci.

    c. bursae.

    d. articular condyles.

_____ 14. A complete dislocation of the knee is extremely unlikely because of the

    a. pair of cartilaginous pads that surround the knee.

    b. presence of the medial and lateral muscle.

    c. presence of bursae, which serve to reduce friction.

    d. seven major ligaments that stabilize the knee joint.

_____ 15. The mechanism that allows standing for prolonged periods without continually contracting the extensor muscles is

    a. the popliteal ligaments extend between the femur and the heads of the tibia and fibula.

    b. the limitations of the ligaments and the movement of the femur.

    c. the patellar ligaments providing support for the front of the knee, helping to maintain posture.

    d. a slight lateral rotation of the tibia tightens the anterior cruciate ligament and jams the meniscus between the tibia and femur.

_____ 16. Continuous passive motion (CPM) of any injured joint appears to encourage the repair process by

    a. improving the circulation of synovial fluid.

    b. taking drugs that reduce inflammation.

    c. engaging in a vigorous exercise program.

    d. joint immobilization.

_____ 17. A meniscus, or articular disc within a synovial joint, may

    a. subdivide a synovial cavity.

    b. channel the flow of synovial fluid.

    c. allow for variations in the shapes of the articular surfaces.

    d. a, b, and c are correct.

_____ 18. Rotation of the forearm that makes the palm face posteriorly is

    a. projection.

    b. pronation.

    c. supination.

    d. angulation.

Completion

Using the terms below, complete the following statements. Use each term only once.

gomphosis	synchondrosis	luxation	articular cartilage
hyperextension	fat pads	intervertebral discs	syndesmosis
menisci	bunion	arthritis	rheumatism
tendons			

1. The most common pressure-related bursitis, involving a tender nodule around bursae over the base of the great toe, is a(n) _____.

2. Rheumatic diseases that affect synovial joints result in the development of _____.

3. The synarthrosis that binds each tooth to the surrounding bony socket is a(n) _____.

4. A rigid cartilaginous connection such as an epiphyseal plate is called a(n) _____.

5. The amphiarthrotic distal articulation between the tibia and fibula is a(n) _____.

6. The joint accessory structures that may subdivide a synovial cavity are _____.

7. The accessory structures that provide protection for the articular cartilages are the _____.

8. Arthritis always involves damage to the _____.

9. A movement that allows you to gaze at the ceiling is _____.

10. A general term that indicates pain and stiffness affecting the skeletal and/or muscular systems is _____.

11. The structures that pass across or around a joint that may limit the range of motion and provide mechanical support are _____.

12. When articulating surfaces are forced out of position, the displacement is called a(n) _____.

13. The anulus fibrosus and nucleus pulposus are structures composing the _____.

Short Essay

1. What three major categories of joints are identified based on the range of motion permitted?

2. What is the difference between a *synostosis* and a *symphysis*?

3. What is the functional difference between a ligament and a tendon?

4. What are the structural and functional differences between: (a) a bursa and (b) a meniscus?

5. What is the functional role of synovial fluid in a diarthrotic joint?

6. Give examples of the types of joints and their location in the body that exhibit the following types of movements: (a) monaxial, (b) biaxial, (c) triaxial.

7. What six types of synovial joints are found in the human body? Give an example of each type.

8. Functionally, what is the similarity between the elbow and the knee joint?

9. Functionally, what is the similarity between the shoulder joint and the hip joint?

10. How are the articulations of the carpals of the wrist similar to those of the tarsals of the ankle?

11. Based on an individual in anatomical position, what types of movements can be used to illustrate angular motion?

12. Why is the glenohumeral joint the most frequently dislocated joint in the body?

13. What regions of the vertebral column do not contain intervertebral discs? Why are they unnecessary in these regions?

14. Identify the unusual types of movements that apply to the following examples:
    a. twisting motion of the foot that turns the sole inward
    b. grasping and holding an object with the thumb
    c. standing on tiptoes
    d. crossing the arms
    e. opening the mouth
    f. shrugging the shoulders

15. How does the stability of the elbow joint compare with that of the hip joint? Why?

16. What structural damage would most likely occur if a "locked knee" was struck from the side while in a standing position?

LEVEL 3: CRITICAL THINKING AND CLINICAL APPLICATIONS

Using principles and concepts learned in Chapter 9, answer the following questions. Write your answers on a separate sheet of paper.

1. As a teacher of anatomy, what structural characteristics would you identify for the students to substantiate that the hip joint is stronger and more stable than the shoulder joint?

2. While playing football, Steve is involved in a "pile on" and receives a momentary shock to his right knee. Even though he experienced a brief period of pain, he re-entered the game after a short period of time. How would you explain his quick recovery?

3. Stairmasters (stair-climbing machines) are popular in health clubs and gyms for a cardiovascular workout. How might this activity lead to chronic knee problems and/or the development of a "trick knee"?

4. Greg is a pitcher on the high school baseball team. He spends many hours practicing to improve his pitching skills. Recently, he has been complaining about persistent pain beneath his right shoulder blade (scapula). What do you think is causing the pain?

5. When excessive forces are applied, why is it easier to dislocate a shoulder than it is to dislocate a hip?

6. Why are hip fractures more common in individuals over age 60 than in those who are younger?

7. Joan works for a large corporation. Her normal attire for work includes high-heeled shoes, sometimes with pointed toes, resulting in the formation of bunions on her big toes. What are bunions and what causes them to form?

CHAPTER 10

Muscle Tissue

OVERVIEW

It would be impossible for human life to exist without muscle tissue. This is because many vital processes and all of the dynamic interactions internally and externally with our environment depend on movement. In the human body, muscle tissue is necessary for movement to take place. The body consists of various types of muscle tissue, including cardiac, which is found in the heart; smooth, which forms a substantial part of the walls of hollow organs; and skeletal, which is attached to the skeleton. Smooth and cardiac muscles are involuntary and are responsible for transport of materials within the body. Skeletal muscle is voluntary and allows us to maneuver and manipulate in the environment; it also supports soft tissues, guards entrances and exits, and serves to maintain body temperature.

The activities in this chapter review the basic structural and functional characteristics of skeletal muscle tissue. Emphasis is placed on muscle cell structure, muscle contraction and muscle mechanics, the energetics of muscle activity, muscle performance, and integration with other systems.

LEVEL 1: REVIEWING FACTS AND TERMS

Review of Learning Outcomes

After completing this chapter, you should be able to do the following:

OUTCOME 10-1   Specify the functions of skeletal muscle tissue.

OUTCOME 10-2   Describe the organization of muscle at the tissue level.

OUTCOME 10-3   Explain the characteristics of skeletal muscle fibers, and identify the structural components of a sarcomere.

OUTCOME 10-4   Identify the components of the neuromuscular junction, and summarize the events involved in the neural control of skeletal muscle contraction and relaxation.

OUTCOME 10-5   Describe the mechanism responsible for tension production in a muscle fiber, and compare the different types of muscle contraction.

OUTCOME 10-6   Describe the mechanisms by which muscle fibers obtain the energy to power contractions.

OUTCOME 10-7   Relate the types of muscle fibers to muscle performance, and distinguish between aerobic and anaerobic endurance.

OUTCOME 10-8   Identify the structural and functional differences between skeletal muscle fibers and cardiac muscle cells.

OUTCOME 10-9   Identify the structural and functional differences between skeletal muscle fibers and smooth muscle cells, and discuss the roles of smooth muscle tissue in systems throughout the body.

Multiple Choice

Place the letter corresponding to the best answer in the space provided.

OUTCOME 10-1 _____ 1. Muscle tissue consists of cells that are highly specialized for the function of

    a. excitability.

    b. contraction.

    c. extensibility.

    d. a, b, and c are correct.

OUTCOME 10-1 _____ 2. The primary function(s) performed by skeletal muscles is (are) to

    a. produce skeletal movement.

    b. guard entrances and exits.

    c. maintain body temperature.

    d. a, b, and c are correct.

OUTCOME 10-1 _____ 3. Skeletal muscles move the body by

    a. using the energy of ATP to form ADP.

    b. activation of the excitation–coupling reaction.

    c. means of neural stimulation.

    d. pulling on the bones of the skeleton.

OUTCOME 10-2 _____ 4. Skeletal muscles are often called voluntary muscles because

    a. ATP activates skeletal muscles for contraction.

    b. the skeletal muscles contain myoneural junctions.

    c. they contract only when stimulated by motor neurons of the central nervous system.

    d. connective tissue harnesses generated forces voluntarily.

OUTCOME 10-2 _____ 5. The three layers of connective tissue supporting each muscle are the

    a. skeletal, cardiac, and smooth.

    b. epimysium, perimysium, and endomysium.

    c. voluntary, involuntary, and resting.

    d. myosatellite cells, tendons, and aponeurosis.

OUTCOME 10-2 _____ 6. Nerves and blood vessels that supply the muscle fibers are contained within the connective tissues of the

    a. epimysium and endomysium.

    b. endomysium only.

    c. endomysium and perimysium.

    d. perimysium only.

OUTCOME 10-3 _____ 7. The smallest functional unit of the muscle fiber is

    a. thick filaments.

    b. thin filaments.

    c. the Z line.

    d. the sarcomere.

OUTCOME 10-3 _____ 8. The *thin* filaments consist of a

    a. pair of protein strands wound together to form chains of actin molecules.

    b. helical array of actin molecules.

    c. pair of protein strands wound together to form chains of myosin molecules.

    d. helical array of myosin molecules.

OUTCOME 10-3 _____ 9. The *thick* filaments consist of a

    a. pair of protein strands wound together to form chains of myosin molecules.

    b. helical array of myosin molecules.

    c. pair of protein strands wound together to form chains of actin molecules.

    d. helical array of actin molecules.

OUTCOME 10-3 _____ 10. The *sliding filament theory* explains that the *physical* change that takes place during contraction is the

    a. thick filaments are sliding toward the center of the sarcomere alongside the thin filaments.

    b. thick and thin filaments are sliding toward the center of the sarcomere together.

    c. Z lines are sliding toward the H zone.

    d. thin filaments are sliding toward the center of the sarcomere alongside the thick filaments.

OUTCOME 10-3 _____ 11. Troponin and tropomyosin are two proteins that can prevent the contractile process by

    a. combining with calcium to prevent active site binding.

    b. causing the release of calcium from the sacs of the sarcoplasmic reticulum.

    c. covering the active site and blocking the actin–myosin interaction.

    d. inactivating the myosin to prevent cross-bridging.

OUTCOME 10-4 _____ 12. The first step in excitation–contraction coupling is the

    a. stimulation of the sarcolemma.

    b. neuronal activation in the CNS.

    c. release of calcium ions from the cisternae of the sarcoplasmic reticulum.

    d. production of tension at the neuromuscular junction.

OUTCOME 10-4 _____ 13. Muscle relaxation begins with

    a. release of calcium ions by the sarcoplasmic reticulum.

    b. binding of ACh at neuromuscular junctions.

    c. breakdown of ACh by AChE.

    d. excitation–contraction coupling.

OUTCOME 10-5 _____ 14. All of the muscle fibers controlled by a single motor neuron constitute a
    a. motor unit.
    b. sarcomere.
    c. myoneural junction.
    d. cross-bridge.

OUTCOME 10-5 _____ 15. The tension in a muscle fiber will vary depending on the
    a. structure of individual sarcomeres.
    b. initial length of muscle fibers.
    c. number of cross-bridge interactions within a muscle fiber.
    d. a, b, and c are correct.

OUTCOME 10-5 _____ 16. There is *less* precise control over leg muscles compared to the muscles of the eye because
    a. single muscle fibers are controlled by many motor neurons.
    b. many muscle fibers are controlled by many motor neurons.
    c. a single muscle fiber is controlled by a single motor neuron.
    d. many muscle fibers are controlled by a single motor neuron.

OUTCOME 10-5 _____ 17. Skeletal muscle fibers contract most forcefully when stimulated over a
    a. narrow range of resting lengths.
    b. wide range of resting lengths.
    c. decrease in the resting sarcomere length.
    d. resting sarcomere that is as short as it can be.

OUTCOME 10-5 _____ 18. The amount of tension produced by a skeletal muscle is controlled by
    a. a push applied to an object.
    b. the amount of troponin released.
    c. the total number of muscle fibers stimulated.
    d. the number of calcium ions released.

OUTCOME 10-5 _____ 19. Peak tension production occurs when all motor units in the muscle contract in a state of
    a. treppe.
    b. twitch.
    c. wave summation.
    d. complete tetanus.

OUTCOME 10-5 _____ 20. In a concentric *isotonic* contraction,
    a. the tension in the muscle varies as it shortens.
    b. the muscle length doesn't change due to the resistance.
    c. the cross-bridges must produce enough tension to overcome the resistance.
    d. tension in the muscle decreases as the resistance increases.

OUTCOME 10-5 _____ 21. In an *isometric* contraction,
    a. tension rises but the length of the muscle remains constant.
    b. tension rises and the muscle shortens.
    c. the tension produced by the muscle is greater than the resistance.
    d. the tension of the muscle increases as the resistance decreases.

OUTCOME 10-6   _____ 22. A high blood concentration of the enzyme creatine kinase (CK) usually indicates
    a. the release of stored energy.
    b. serious muscle damage.
    c. that an excess of energy is being produced.
    d. that the mitochondria are malfunctioning.

OUTCOME 10-6   _____ 23. Mitochondrial activities are relatively efficient, but their rate of ATP generation is limited by the
    a. presence of enzymes.
    b. availability of carbon dioxide and water.
    c. energy demands of other organelles.
    d. availability of oxygen.

OUTCOME 10-6   _____ 24. Of the following selections, the one that has been correlated with muscle fatigue is
    a. an increase in metabolic reserves within the muscle fibers.
    b. a decline in pH within the muscle, altering enzyme activities.
    c. an increase in pH within the muscle fibers, affecting storage of glycogen.
    d. increased muscle performance due to an increased pain threshold.

OUTCOME 10-6   _____ 25. During the recovery period, the body's oxygen demand is
    a. elevated above normal resting levels.
    b. decreased below normal resting levels.
    c. unchanged.
    d. an irrelevant factor during recovery.

OUTCOME 10-7   _____ 26. The three major types of skeletal muscle fibers in the human body are
    a. slow, fast resistant, and fast fatigue.
    b. slow, intermediate, and fast.
    c. red muscle, slow twitch, and slow oxidative fibers.
    d. Type I, Type II, and Type III fibers.

OUTCOME 10-7   _____ 27. Extensive blood vessels, mitochondria, and myoglobin are found in the greatest concentration in
    a. fast fibers.
    b. slow fibers.
    c. intermediate fibers.
    d. Type II fibers.

OUTCOME 10-7   _____ 28. The length of time a muscle can continue to contract while supported by mitochondrial activities is referred to as
    a. anaerobic endurance.
    b. aerobic endurance.
    c. hypertrophy.
    d. recruitment.

OUTCOME 10-7 _____ 29. Altering the characteristics of muscle fibers and improving the performance of the cardiovascular system results in improving
   a. thermoregulatory adjustment.
   b. hypertrophy.
   c. anaerobic endurance.
   d. aerobic endurance.

OUTCOME 10-8 _____ 30. The property of cardiac muscle that allows it to contract without neural stimulation is
   a. intercalation.
   b. automaticity.
   c. plasticity.
   d. pacesetting.

OUTCOME 10-9 _____ 31. Structurally, smooth muscle cells differ from skeletal muscle cells because smooth muscle cells
   a. contain many nuclei.
   b. contain a network of T tubules.
   c. lack myofibrils and sarcomeres.
   d. possess striations.

OUTCOME 10-9 _____ 32. Smooth muscle tissue differs from other muscle tissue in
   a. excitation–contraction coupling.
   b. length–tension relationships.
   c. control of contraction.
   d. a, b, and c are correct.

OUTCOME 10-9 _____ 33. Neural, hormonal, or chemical factors can stimulate smooth muscle contraction, producing
   a. a decrease in muscle tone.
   b. an increase in muscle tone.
   c. no effect on muscle tone.
   d. variable effects on muscle tone.

Completion

Using the terms below, complete the following statements. Use each term only once.

anaerobic endurance	sarcomeres	treppe
twitch	smooth muscle	oxygen debt
tendon	T tubules	plasticity
contraction	pacemaker	lactic acid
sphincter	incomplete tetanus	action potential
troponin	tension	sarcolemma
motor unit	skeletal muscle	red muscles
Z lines	glycolysis	recruitment
epimysium	ATP	cross-bridges
white muscles	fascicles	complete tetanus

OUTCOME 10-1   1. Muscle tissues are highly specialized for producing _____.

OUTCOME 10-2   2. The dense layer of collagen fibers surrounding a muscle is called the _____.

OUTCOME 10-2   3. Bundles of muscle fibers are called _____.

OUTCOME 10-2   4. The dense regular connective tissue that attaches skeletal muscle to bones is known as a(n) _____.

OUTCOME 10-3   5. The cell membrane that surrounds the cytoplasm of a muscle fiber is called the _____.

OUTCOME 10-3   6. Structures that distribute the command to contract within a muscle fiber are called _____.

OUTCOME 10-3   7. Muscle cells contain contractible units called _____.

OUTCOME 10-3   8. Because they connect thick and thin filaments, the myosin heads are also known as _____.

OUTCOME 10-3   9. The boundary between adjacent sarcomeres is marked by the _____.

OUTCOME 10-4   10. The conducted charge in the transmembrane potential is called a(n) _____.

OUTCOME 10-5   11. The increase in muscular tension produced by increasing the number of active motor units is called _____.

OUTCOME 10-4   12. Active site exposure during the contraction process occurs when calcium binds to _____.

OUTCOME 10-5   13. The interactions between the thick and the thin filaments produce _____.

OUTCOME 10-5   14. When the calcium ion concentration in the cytoplasm prolongs the contraction state, making it continuous, the contraction is called _____.

OUTCOME 10-5   15. A muscle producing almost peak tension during rapid cycles of contraction and relaxation is said to be in _____.

OUTCOME 10-5   16. An indication of how fine the control of movement can be is determined by the size of the _____.

OUTCOME 10-5   17. By controlling the number of activated muscle fibers, you can control the amount of tension produced by the _____.

OUTCOME 10-5   18. The "staircase" phenomenon during which the peak muscle tension rises in stages is called _____.

OUTCOME 10-5   19. A single stimulus–contraction–relaxation sequence in a muscle fiber is a(n) _____.

OUTCOME 10-6   20. When muscles are actively contracting, the process requires large amounts of energy in the form of _____.

OUTCOME 10-6   21. At peak activity levels, most of the ATP is provided by glycolysis, leading to the production of _____.

OUTCOME 10-6   22. A skeletal muscle continues to contract even when mitochondrial activity is limited by the availability of oxygen due to the process of _____.

OUTCOME 10-7   23. Muscles dominated by fast fibers are sometimes referred to as _____.

OUTCOME 10-7   24. Muscles dominated by slow fibers are sometimes referred to as _____.

OUTCOME 10-7   25. The amount of oxygen used in the recovery period to restore normal pre-exertion conditions is referred to as _____.

OUTCOME 10-7   26. The amount of time muscular contraction can continue to be supported by glycolysis and existing reserves of ATP and CP is _____.

OUTCOME 10-8 27. The timing of contractions in cardiac muscle tissues is determined by specialized muscle fibers called _____ cells.

OUTCOME 10-9 28. The ability of smooth muscle to function over a wide range of lengths is called _____.

OUTCOME 10-9 29. Spindle-shaped cells with a single, centrally located nucleus are characteristic of _____.

OUTCOME 10-9 30. In the digestive and urinary systems, the rings of smooth muscle that regulate the movement of materials along internal passageways are called _____.

Matching

Match the terms in column B with the terms in column A. Use letters for answers in the spaces provided.

		Column A	Column B
OUTCOME 10-1	_____	1. skeletal muscle	A. thick filaments
OUTCOME 10-2	_____	2. fascicles	B. resting tension
OUTCOME 10-2	_____	3. aponeurosis	C. synaptic cleft
OUTCOME 10-2	_____	4. tendons	D. cardiac muscle fibers
OUTCOME 10-3	_____	5. myoblasts	E. muscle bundles
OUTCOME 10-3	_____	6. myosin	F. lactic acid
OUTCOME 10-4	_____	7. neuromuscular junction	G. red muscles
OUTCOME 10-4	_____	8. cross-bridging	H. peak tension
OUTCOME 10-5	_____	9. contraction phase	I. smooth muscle cell
OUTCOME 10-5	_____	10. myogram	J. produce body movements
OUTCOME 10-5	_____	11. muscle tone	K. actin–myosin interaction
OUTCOME 10-6	_____	12. energy reserve	L. embryonic cells
OUTCOME 10-6	_____	13. lactic acid	M. creatine phosphate
OUTCOME 10-6	_____	14. anaerobic glycolysis	N. lowers intracellular pH
OUTCOME 10-7	_____	15. slow fibers	O. measures external tension
OUTCOME 10-8	_____	16. intercalated discs	P. timing of contractions
OUTCOME 10-9	_____	17. no striations	Q. bundle of collagen fibers
OUTCOME 10-9	_____	18. pacemaker cells	R. broad sheet

Drawing/Illustration Labeling

Identify each numbered structure. Place your answers in the spaces provided.

OUTCOME 10-2   **FIGURE 10-1**   Organization of Skeletal Muscles

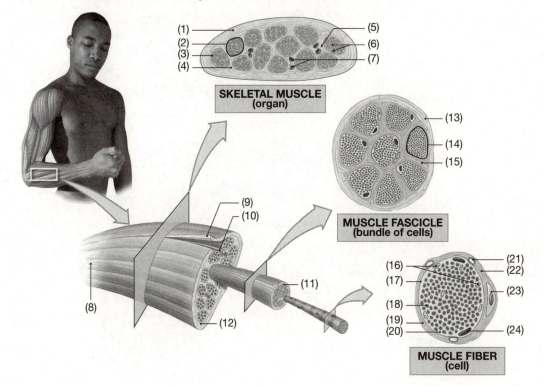

(1) _____	(13) _____
(2) _____	(14) _____
(3) _____	(15) _____
(4) _____	(16) _____
(5) _____	(17) _____
(6) _____	(18) _____
(7) _____	(19) _____
(8) _____	(20) _____
(9) _____	(21) _____
(10) _____	(22) _____
(11) _____	(23) _____
(12) _____	(24) _____

OUTCOME 10-3   **FIGURE 10-2**   The Structure of a Skeletal Muscle Fiber

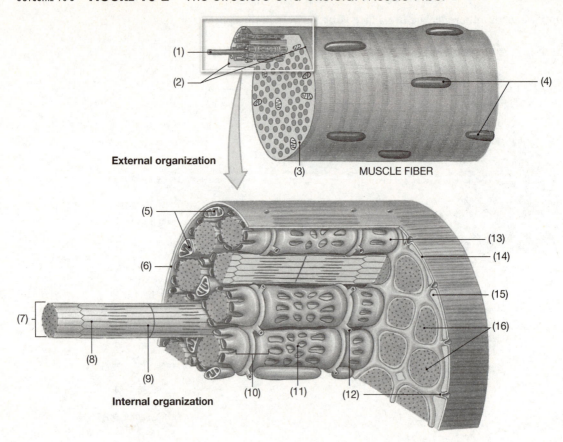

External organization

MUSCLE FIBER

Internal organization

(1) _____	(9) _____
(2) _____	(10) _____
(3) _____	(11) _____
(4) _____	(12) _____
(5) _____	(13) _____
(6) _____	(14) _____
(7) _____	(15) _____
(8) _____	(16) _____

OUTCOME 10-3, 10-8, 10-9   **FIGURE 10-3** Types of Muscle Tissue

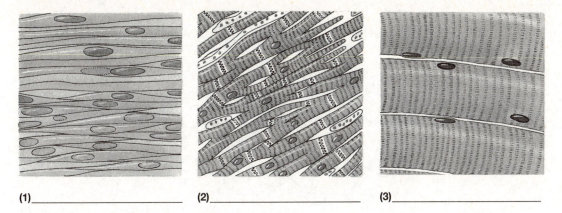

(1)_____   (2)_____   (3)_____

OUTCOME 10-3   **FIGURE 10-4** Structure of a Sarcomere

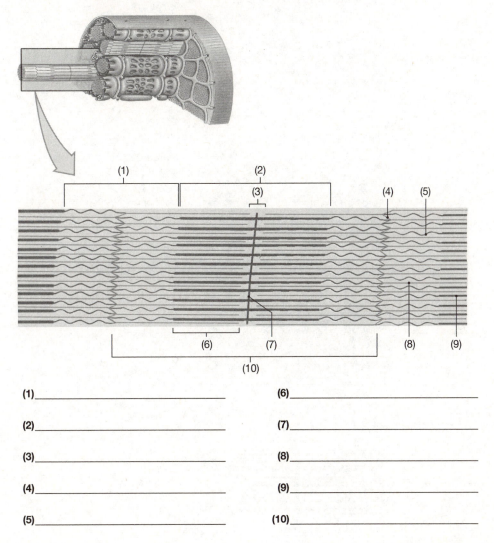

(1)_____   (6)_____

(2)_____   (7)_____

(3)_____   (8)_____

(4)_____   (9)_____

(5)_____   (10)_____

Identify each level of functional organization in a skeletal muscle. Place your answers in the spaces provided.

OUTCOME 10-2, 10-3 **FIGURE 10-5** Levels of Functional Organization in a Skeletal Muscle

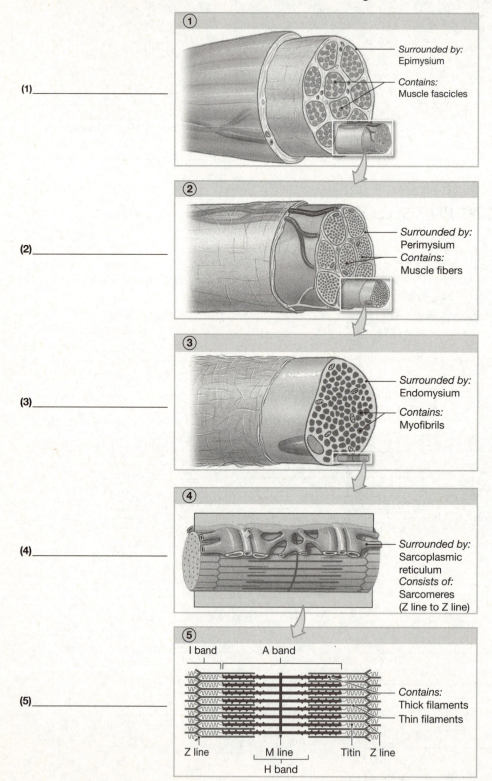

(1)_____

Surrounded by:
Epimysium

Contains:
Muscle fascicles

(2)_____

Surrounded by:
Perimysium
Contains:
Muscle fibers

(3)_____

Surrounded by:
Endomysium

Contains:
Myofibrils

(4)_____

Surrounded by:
Sarcoplasmic
reticulum
Consists of:
Sarcomeres
(Z line to Z line)

(5)_____

I band   A band

Contains:
Thick filaments
Thin filaments

Z line   M line   Titin Z line
H band

LEVEL 2: REVIEWING CONCEPTS

Chapter Overview

Using the terms below, fill in the blanks to correctly complete the chapter overview of muscle tissue. Use each term only once.

muscle fibers	myofibrils	nuclei
endomysium	myosin	skeletal
myofilaments	T tubules	fascicle
sarcolemma	perimysium	sarcomeres
sliding filament	Z line	A band
contraction	actin	muscle contraction
epimysium	myosatellite	thick

Muscle tissue, one of the four primary types of tissue, consists chiefly of muscle cells that are highly specialized for (1) _____. The three types of muscle tissue are (2) _____, cardiac, and smooth muscle. Three layers of connective tissue are part of each muscle. The entire muscle is surrounded by the (3) _____, a dense layer of collagen fibers. The connective tissue fibers of the (4) _____ divide the skeletal muscle into a series of compartments, which possess collagen and elastic fibers, blood vessels and nerves, and a bundle of muscle fibers called a(n) (5) _____. Within this bundle of muscle fibers, the delicate connective tissue of the (6) _____ surrounds the individual skeletal muscle cells, or (7) _____, and loosely interconnects adjacent muscle fibers. Between the surrounding layer and the muscle cell membrane, (8) _____ cells, the "doctors" that function in the repair of damaged muscle tissue, are sparsely scattered. The muscle cell membrane, called the (9) _____, is facilitated by a series of openings to "pipe-like" structures that are regularly arranged and that project into the muscle fibers. These "pipes" are the (10) _____ that serve as the transport system between the outside of the cell and the cell interior. Just inside the membrane, there are many (11) _____, a characteristic common to skeletal muscle fibers. Inside the muscle fiber, branches of the transverse tubules encircle cylindrical structures called (12) _____, which extend from one end of the muscle cell to the other. The cylindrical structures consist of bundles of protein filaments called (13) _____. The smaller filaments are arranged into repeating functional units called (14) _____, which are arranged into thin filaments called (15) _____, and thick filaments composed primarily of (16) _____. Each functional unit extends from one (17) _____ to another, forming a filamentous network of disc-like protein structures for the attachment of actin molecules. At the center of each of these functional units is a dark area called the (18) _____. In the center of this area, only (19) _____ filaments are present. The interactions between the filaments cause the sliding of the thin filaments of the unit, a process referred to as the (20) _____ theory, ultimately producing a function unique to muscle cells, the process of (21) _____.

Concept Map I

Using the following terms, fill in the circled, numbered, blank spaces to correctly complete the concept map. Use each term only once.

smooth   involuntary   striated
multinucleate   bones   nonstriated
heart

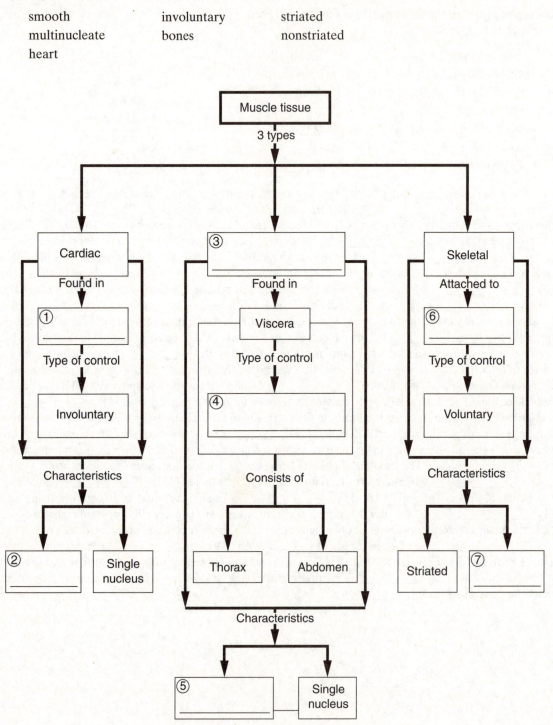

Concept Map II

Using the following terms, fill in the circled, numbered, blank spaces to correctly complete the concept map. Use each term only once.

Z lines   muscle bundles (fascicles)   myofibrils
actin   thick filaments   sarcomeres
H band

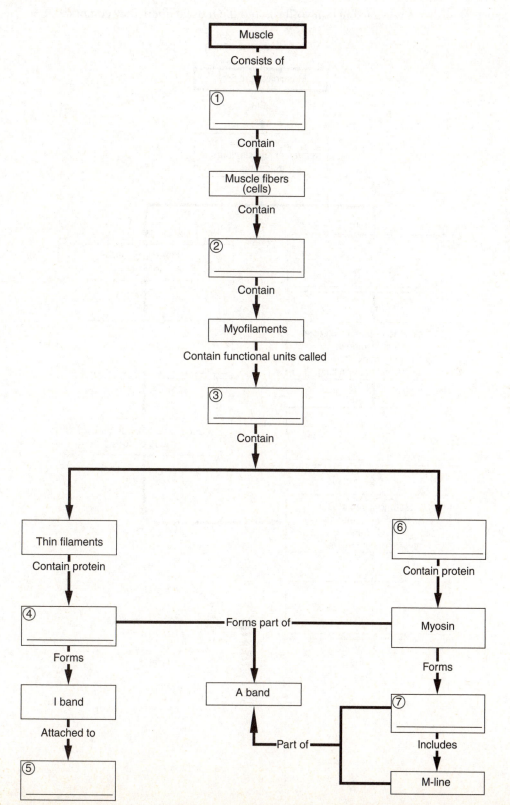

Concept Map III

Using the following phrases, fill in the circled, numbered, blank spaces to correctly complete the concept map. Use each phrase only once.

Cross-bridging (heads of myosin attach to turned-on thin filaments)

Energy + ADP + phosphate

Release of Ca^{2+} from sacs of sarcoplasmic reticulum

Shortening, that is, contraction of myofibrils and the muscle fibers they comprise

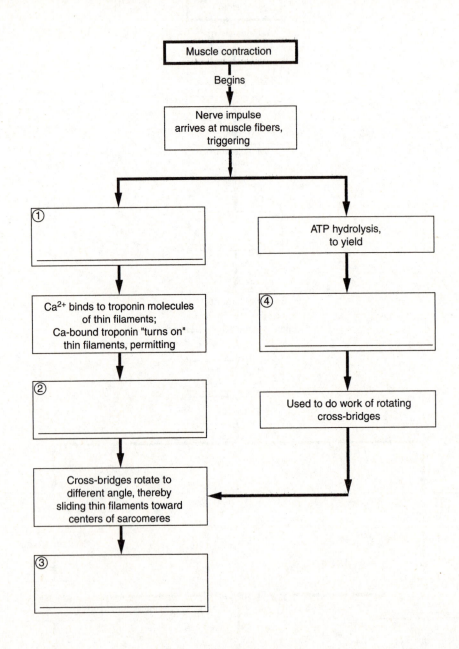

Multiple Choice

Place the letter corresponding to the best answer in the space provided.

_____ 1. Muscle contraction occurs as a result of

    a. interactions between the thick and thin filaments of the sarcomere.

    b. the interconnecting filaments that make up the Z lines.

    c. shortening of the A band, which contains thick and thin filaments.

    d. shortening of the I band, which contains thin filaments only.

_____ 2. The area of the A band in the sarcomere consists of

    a. Z line, H band, and M line.

    b. M line, H band, and zone of overlap.

    c. thin filaments only.

    d. overlapping thick and thin filaments.

_____ 3. The process of cross-bridging, which occurs at an active site, involves a series of sequential-cyclic reactions that include

    a. attach, return, pivot, and detach.

    b. attach, pivot, detach, and return.

    c. attach, detach, pivot, and return.

    d. attach, return, detach, and pivot.

_____ 4. Excitation–contraction coupling forms the link between

    a. the release of Ca^{2+} to bind with the troponin molecule.

    b. depolarization and repolarization.

    c. electrical activity in the sarcolemma and the initiation of a contraction.

    d. the neuromuscular junction and the sarcoplasmic reticulum.

_____ 5. When Ca^{2+} binds to troponin, it produces a change by

    a. initiating activity at the neuromuscular junction.

    b. causing the actin–myosin interaction to occur.

    c. decreasing the calcium concentration at the sarcomere.

    d. exposing the active site on the thin filaments.

_____ 6. The phases of a single twitch in sequential order are

    a. contraction phase, latent phase, relaxation phase.

    b. latent period, relaxation phase, contraction phase.

    c. latent period, contraction phase, relaxation phase.

    d. relaxation phase, latent phase, contraction phase.

_____ 7. After contraction, a muscle fiber returns to its original length through

    a. the active mechanism for fiber elongation.

    b. elastic forces and the movement of opposing muscles.

    c. the tension produced by the initial length of the muscle fiber.

    d. involvement of all the sarcomeres along the myofibrils.

_____ 8. A muscle producing almost peak tension during rapid cycles of contraction and relaxation is said to be in

    a. complete tetanus.

    b. incomplete tetanus.

    c. treppe.

    d. recruitment.

_____ 9. The process of reaching *complete tetanus* is obtained by

   a. applying a second stimulus before the relaxation phase has ended.
   b. decreasing the concentration of calcium ions in the cytoplasm.
   c. activation of additional motor units.
   d. increasing the rate of stimulation until the relaxation phase is completely eliminated.

_____ 10. The total force exerted by a muscle as a whole depends on

   a. the rate of stimulation.
   b. how many motor units are activated.
   c. the number of calcium ions released.
   d. a, b, and c are correct.

_____ 11. The principal energy reserves found in skeletal muscle cells are

   a. carbohydrates, fats, and proteins.
   b. DNA, RNA, and ATP.
   c. ATP, creatine phosphate, and glycogen.
   d. ATP, ADP, and AMP.

_____ 12. The two mechanisms used to generate ATP from glucose are

   a. aerobic respiration and anaerobic glycolysis.
   b. ADP and creatine phosphate.
   c. cytoplasm and mitochondria.
   d. a, b, and c are correct.

_____ 13. In *anaerobic glycolysis*, glucose is broken down to pyruvic acid, which is converted to

   a. glycogen.
   b. lactic acid.
   c. acetyl-CoA.
   d. citric acid.

_____ 14. The maintenance of normal body temperature is dependent upon

   a. the temperature of the environment.
   b. the pH of the blood.
   c. the production of heat by muscles.
   d. the amount of energy produced by anaerobic glycolysis.

_____ 15. Growth hormone from the pituitary gland and the male sex hormone, testosterone, stimulate

   a. the rate of energy consumption by resting and active skeletal muscles.
   b. muscle metabolism and increased force of contraction.
   c. synthesis of contractile proteins and the enlargement of skeletal muscles.
   d. the amount of tension produced by a muscle group.

_____ 16. The hormone responsible for stimulating muscle metabolism and increasing the force of contraction during a sudden crisis is

   a. epinephrine.
   b. thyroid hormone.
   c. growth hormone.
   d. testosterone.

_____ 17. The type of skeletal muscle fibers that have low fatigue resistance are

    a. fast fibers.

    b. slow fibers.

    c. intermediate fibers.

    d. Type I fibers.

_____ 18. An example of an activity that requires *anaerobic endurance* is

    a. a 50-yard dash.

    b. a 3-mile run.

    c. a 10-mile bicycle ride.

    d. running a marathon.

_____ 19. Athletes training to develop anaerobic endurance perform

    a. few, long, relaxing workouts.

    b. a combination of weight training and marathon running.

    c. frequent, brief, intensive workouts.

    d. stretching, flexibility, and relaxation exercises.

_____ 20. The major support that the muscular system gets from the cardiovascular system is

    a. a direct response by controlling the heart rate and the respiratory rate.

    b. constriction of blood vessels and decrease in heart rate for thermoregulatory control.

    c. nutrient and oxygen delivery and carbon dioxide removal.

    d. decreased volume of blood and rate of flow for maximal muscle contraction.

Completion

Using the terms below, complete the following statements. Use each term only once.

myosatellite	myoblasts	fatigue
concentric isotonic	complete tetanus	motor unit
plasticity	muscle tone	recovery period
A bands		

1. Specialized cells that function in the repair of damaged muscle tissue are called _____ cells.

2. During development, groups of embryonic cells that fuse together to create individual muscle fibers are called _____.

3. Resting tension in a skeletal muscle is called _____.

4. The ability of a stretched smooth muscle to function over a wide range of lengths is called _____.

5. The time during which conditions in muscle fibers are returned to normal, pre-exertion levels is the _____.

6. The condition that results when an active skeletal muscle can no longer perform at the required level of activity is referred to as _____.

7. In a sarcomere, the dark bands (anisotropic bands) are referred to as _____.

8. A single cranial or spinal motor neuron and the muscle fibers it innervates comprise a _____.

9. At sufficiently high stimulation frequencies, the overlapping twitches result in one strong, steady contraction referred to as _____.

10. When the muscle shortens but its tension remains the same, the contraction is a(n) _____ contraction.

Short Essay

Briefly answer the following questions in the spaces provided below.

1. What are the five functions performed by skeletal muscles?

2. What are the three layers of connective tissue that are part of each muscle?

3. Draw an illustration of a sarcomere and label the parts according to the unit organization.

4. Cite the five sequential steps involved in the contraction process.

5. Describe the major events in sequence that occur at the neuromuscular junction to initiate the contractive process.

6. Identify the types of muscle contractions illustrated at points A, B, and C on the following diagram.

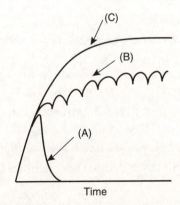

Time

7. What are the differences between an isometric and an isotonic contraction?

8. What is the relationship among fatigue, anaerobic glycolysis, and oxygen debt?

9. Why do fast fibers fatigue more rapidly than slow fibers?

LEVEL 3: CRITICAL THINKING AND CLINICAL APPLICATIONS

Using principles and concepts learned in Chapter 10, answer the following questions. Write your answers on a separate sheet of paper.

1. Why do athletes perform better if they warm up before a competitive event?
2. Suppose you are assigned the responsibility of developing training programs tailored to increase both aerobic and anaerobic endurance. What types of activities would be necessary to support these training programs?
3. Even though muscle tone does not produce active movements, what important purpose does it serve as a desirable condition of muscle tissue?
4. The Law of Use and Disuse states:

   "When skeletal muscles are forcefully exercised they tend to enlarge; conversely, when a muscle is not exercised, it undergoes atrophy . . . . that is, it decreases in size and strength."

   Explain.
5. A mortician is called to a scene where a 32-year-old male has overdosed on drugs, causing his death. He has been dead for about five hours, and is found sitting upright in a chair with an empty vial in his hand. The mortician has difficulty removing the vial from his hand and will need to wait for a few days to properly prepare the body for visitation and burial. Why?

CHAPTER 11

The Muscular System

OVERVIEW

The study of the muscular system includes the skeletal muscles, which make up about 40 percent of the body mass and can be controlled voluntarily. It is virtually impossible to master all the facts involved with the approximately 700 muscles that have been identified in the human body. Thus, representative muscles from all parts of the body are selected and surveyed. The survey includes the gross anatomy of muscles, anatomical arrangements, muscle attachments, and muscular performance relative to basic mechanical laws.

The activities for Chapter 11 focus on organizing the muscles into limited numbers of anatomical and functional groups, the general appearance of the muscles, and the factors that interact to determine the effects of muscle contraction. This chapter requires a great deal of memorization; therefore, arranging the information in an organized way and learning to recognize clues will make the task easier to manage and result in facilitating the memorization process.

LEVEL 1: REVIEWING FACTS AND TERMS

Review of Learning Outcomes

After completing this chapter, you should be able to do the following:

OUTCOME 11-1   Describe the arrangement of fascicles in the various types of muscles, and explain the resulting functional differences.

OUTCOME 11-2   Describe the classes of levers, and explain how they make muscles more efficient.

OUTCOME 11-3   Predict the actions of a muscle on the basis of the relative positions of its origin and insertion, and explain how muscles interact to produce or oppose movements.

OUTCOME 11-4   Explain how the name of a muscle can help identify its location, appearance, or function.

OUTCOME 11-5   Identify the principal axial muscles of the body, plus their origins, insertions, actions, and innervation.

OUTCOME 11-6   Identify the principal appendicular muscles of the body, plus their origins, insertions, actions, and innervation, and compare the major functional differences between the upper and lower limbs.

OUTCOME 11-7   Identify age-related changes of the muscular system.

OUTCOME 11-8   Explain the functional relationship between the muscular system and other body systems, and explain the role of exercise in producing various responses in other body systems.

Multiple Choice

Place the letter corresponding to the best answer in the space provided.

OUTCOME 11-1 _____ 1. The four types of muscles identified by different patterns of organization are

    a. skeletal, smooth, cardiac, and visceral.

    b. movers, synergists, antagonists, and agonists.

    c. parallel, convergent, pennate, and circular.

    d. flexors, extensors, adductors, and abductors.

OUTCOME 11-1 _____ 2. In a convergent muscle the muscle fibers are

    a. parallel to the long axis of the muscle.

    b. based over a broad area, but all the fibers come together at a common attachment site.

    c. arranged to form a common angle with the tendon.

    d. arranged concentrically around an opening or recess.

OUTCOME 11-2 _____ 3. A first-class lever is one in which

    a. the resistance is located between the applied force and the fulcrum.

    b. a force is applied between the resistance and the fulcrum.

    c. speed and distance traveled are increased at the expense of the force.

    d. the fulcrum lies between the applied force and the resistance.

OUTCOME 11-2 _____ 4. The effect of an arrangement where a force is applied between the resistance and the fulcrum illustrates the principles operating

    a. first-class levers.

    b. second-class levers.

    c. third-class levers.

    d. fourth-class levers.

OUTCOME 11-3 _____ 5. The site where the movable end of a muscle attaches to bone or other connective tissue is referred to as the

    a. origin.

    b. insertion.

    c. rotator.

    d. joint.

OUTCOME 11-3 _____ 6. A muscle whose contraction is chiefly responsible for producing a particular movement is called a(n)

    a. synergist.

    b. antagonist.

    c. originator.

    d. prime mover.

OUTCOME 11-3 _____ 7. Muscles are classified functionally as synergists when

    a. muscles perform opposite tasks and are located on opposite sides of the limb.

    b. their contraction is coordinated in a way that assists another muscle in affecting a particular movement.

    c. a muscle is responsible for a particular movement.

    d. the movement involves flexion and extension.

OUTCOME 11-4 _____ 8. Extrinsic muscles are those that

    a. operate within an organ.

    (b) position or stabilize an organ.

    c. are prominent and can be easily seen.

    d. are visible at the body surface.

OUTCOME 11-4 _____ 9. The reason we use the word *bicep* to describe a particular muscle is

    a. there are two areas in the body where biceps are found.

    b. there are two muscles in the body with the same characteristics.

    (c.) there are two tendons of origin.

    d. the man who named it was an Italian by the name of Biceppe Longo.

OUTCOME 11-5 _____ 10. Which of the following selections includes only muscles of *facial expression*?

    a. Lateral rectus, medial rectus, hypoglossus, stylohyoideus

    b. Splenius, masseter, scalenes, platysma

    c. Procerus, capitis, cervicis, zygomaticus

    (d.) Buccinator, orbicularis oris, risorius, occipitofrontalis

OUTCOME 11-5 _____ 11. The muscles that bring the teeth together during mastication are the

    (a.) temporalis, pterygoid, and masseter.

    b. procerus, capitis, and zygomaticus.

    c. mandibular, maxillary, and zygomaticus.

    d. glossus, platysma, and risorius.

OUTCOME 11-5 _____ 12. The names of the muscles of the tongue are readily identified because their descriptive names end in

    a. *genio.*

    b. *pollicus.*

    (c.) *glossus.*

    d. *hallucis.*

OUTCOME 11-5 _____ 13. The superficial muscles of the spine are identified by *subdivisions* that include the

    a. cervicis, thoracis, and lumborum.

    (b.) iliocostalis, longissimus, and spinalis.

    c. longissimus, transversus, and longus.

    d. capitis, splenius, and spinalis.

OUTCOME 11-5 _____ 14. The muscles in the abdominal region that compress the abdomen, depress the ribs, and flex or bend the spine are the

    a. rectus abdominis.

    b. internal intercostals.

    (c.) external obliques.

    d. transverse abdominis.

OUTCOME 11-5   _____   15. The muscular floor of the pelvic cavity is formed by muscles that make up the

        a. urogenital triangle and anal triangle.

        b. sacrum and the coccyx.

        c. ischium and the pubis.

        d. ilium and the ischium.

OUTCOME 11-6   _____   16. Which of the following selections includes only muscles that move the *shoulder girdle*?

        a. Teres major, deltoid, pectoralis major, triceps

        b. Procerus, capitis, pterygoid, brachialis

        c. Trapezius, serratus anterior, pectoralis minor, subclavius

        d. Internal oblique, thoracis, deltoid, pectoralis minor

OUTCOME 11-6   _____   17. Which of the following selections includes only muscles that move the *upper arm*?

        a. Deltoid, teres major, latissimus dorsi, pectoralis major

        b. Trapezius, pectoralis minor, subclavius, triceps

        c. Rhomboideus, serratus anterior, subclavius, trapezius

        d. Brachialis, brachioradialis, pronator, supinator

OUTCOME 11-6   _____   18. The muscles that arise on the humerus and the forearm and rotate the radius without producing either flexion or extension of the elbow are the

        a. pronator teres and supinator.

        b. brachialis and brachioradialis.

        c. triceps and biceps brachii.

        d. carpi ulnaris and radialis.

OUTCOME 11-6   _____   19. The pectoralis major produces _____ at the shoulder joint, and the latissimus dorsi muscle produces _____.

        a. abduction; adduction

        b. extension; flexion

        c. medial rotation; lateral rotation

        d. flexion; extension

OUTCOME 11-6   _____   20. The muscle *groups* that are responsible for movement of the thigh include the

        a. abductor, flexor, and extensor.

        b. adductor, gluteal, and lateral rotator.

        c. depressor, levator, and rotator.

        d. procerus, capitis, and pterygoid.

OUTCOME 11-6   _____   21. The *flexors* that move the lower leg, commonly known as the hamstrings, include the

        a. rectus femoris, vastus intermedius, vastus lateralis, and vastus medialis.

        b. sartorius, rectus femoris, gracilis, and vastus medialis.

        c. piriformis, lateral rotators, obturator, and sartorius.

        d. biceps femoris, semimembranosus, and semitendinosus.

OUTCOME 11-6   _____ 22. The *extensors* that move the lower leg, commonly known as the *quadriceps*, include the

    a. piriformis, lateral rotator, obturator, and sartorius.

    b. semimembranosus, semitendinosus, gracilis, and sartorius.

    c. popliteus, gracilis, rectus femoris, and biceps femoris.

    d. rectus femoris, vastus intermedius, vastus lateralis, and vastus medialis.

OUTCOME 11-6   _____ 23. Major muscles that produce plantar flexion involved with movement of the lower leg are the

    a. tibialis anterior, calcaneal, and popliteus.

    b. flexor hallucis, obturator, and gracilis.

    c. gastrocnemius, soleus, and tibialis posterior.

    d. sartorius, soleus, and flexor hallucis.

OUTCOME 11-6   _____ 24. The actions that the arm muscles produce that are not evident in the action of the leg muscles are

    a. abduction and adduction.

    b. flexion and extension.

    c. pronation and supination.

    d. rotation and adduction.

OUTCOME 11-6   _____ 25. Common functional actions of the muscles of the forearm and the upper leg involve

    a. flexion and extension.

    b. adduction and abduction.

    c. rotation and supination.

    d. pronation and supination.

OUTCOME 11-7   _____ 26. One of the effects of aging on the muscular system is

    a. skeletal muscle fibers become larger in diameter.

    b. tolerance for exercise decreases.

    c. skeletal muscles become more elastic.

    d. the ability to recover from muscular injury increases.

OUTCOME 11-7   _____ 27. The number of satellite cells steadily decreases with age and the

    a. number of myofibrils increases.

    b. tolerance for exercise increases.

    c. muscle fibers contain larger ATP, CP, and glycogen reserves.

    d. amount of fibrous tissue increases.

OUTCOME 11-8   _____ 28. For all the body systems, the muscular system

    a. generates heat that helps maintain normal body temperature.

    b. removes waste products of protein metabolism.

    c. provides oxygen and eliminates carbon dioxide.

    d. removes excess body heat.

OUTCOME 11-8 _____ 29. While the integumentary system helps the skin surface remove excess heat generated by muscle activity, _____ mechanisms are also coordinating responses to maintain homeostasis.

    a. respiratory

    b. nervous

    c. cardiovascular

    (d.) all of the above

Completion

Using the terms below, complete the following statements. Use each term only once.

cervicis	origin	popliteus	diaphragm
third-class	deltoid	biceps brachii	second-class
rectus femoris	sphincters	synergist	rotator cuff
nervous	fibrosis		

OUTCOME 11-1   1. An example of a parallel muscle with a central body or belly is the _____.

OUTCOME 11-1   2. Circular muscles that guard entrances and exits of internal passageways are called _____.

OUTCOME 11-2   3. The most common levers in the body are classified as _____ levers.

OUTCOME 11-2   4. The type of lever in which a small force can move a larger weight is classified as a(n) _____ lever.

OUTCOME 11-3   5. The stationary, immovable, or less movable attachment of a muscle is the _____.

OUTCOME 11-3   6. A muscle that assists the prime mover in performing a particular action is a(n) _____.

OUTCOME 11-4   7. The term that identifies the region of the body behind the knee is _____.

OUTCOME 11-4   8. The term that identifies the neck region of the body is _____.

OUTCOME 11-5   9. The muscle that separates the thoracic and abdominopelvic cavities is the _____.

OUTCOME 11-6   10. The major abductor of the arm is the _____.

OUTCOME 11-6   11. The large quadricep muscle that extends the leg and flexes the thigh is the _____.

OUTCOME 11-6   12. The supraspinatus, infraspinatus, subscapularis, and teres minor muscles and their associated tendons form the _____.

OUTCOME 11-7   13. Aging skeletal muscles develop increasing amounts of fibrous connective tissue, a process called _____.

OUTCOME 11-8   14. The system that controls skeletal muscle contractions is the _____ system.

Matching

Match the terms in column B with the terms in column A. Use letters for answers in the spaces provided.

Part I

		Column A	Column B
OUTCOME 11-1	_____	1. slender band of collagen fibers	A. wheelbarrow
OUTCOME 11-1	_____	2. tendon branches within muscles	B. agonist
OUTCOME 11-2	_____	3. first-class lever	C. antagonist
OUTCOME 11-2	_____	4. second-class lever	D. raphe
OUTCOME 11-3	_____	5. stationary muscle attachment	E. insertion
OUTCOME 11-3	_____	6. movable muscle attachment	F. see-saw
OUTCOME 11-3	_____	7. prime mover	G. multipennate
OUTCOME 11-3	_____	8. oppose action of prime mover	H. origin

Part II

		Column A	Column B
OUTCOME 11-4	_____	9. long and round muscles	I. Hyoids
OUTCOME 11-4	_____	10. tailor's muscle	J. adducts the arm
OUTCOME 11-5	_____	11. oculomotor muscles	K. teres
OUTCOME 11-5	_____	12. extrinsic laryngeal muscles	L. rotator cuff
OUTCOME 11-5	_____	13. "six pack"	M. gastrocnemius
OUTCOME 11-6	_____	14. pectoralis major	N. eye position
OUTCOME 11-6	_____	15. support shoulder joint	O. sartorius
OUTCOME 11-6	_____	16. "calf" muscle	P. refers to knee
OUTCOME 11-6	_____	17. popliteus	Q. rectus abdominis
OUTCOME 11-7	_____	18. cardiovascular performance	R. hormones coordinate homeostatic response
OUTCOME 11-8	_____	19. endocrine system	S. decreases with age

Drawing/Illustration Labeling

I. Match the muscle types (a–f) with the muscle type locations on the human body. Use letters for answers in the spaces provided.

II. Name the muscle.

OUTCOME 11-1 **FIGURE 11-1** Muscle Types Based on Patterns of Fascicle Organization

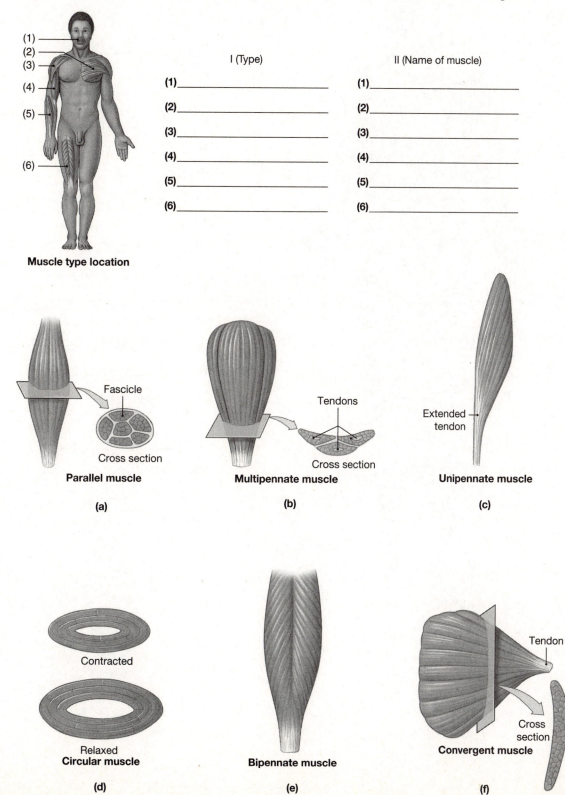

I (Type)

(1)_____

(2)_____

(3)_____

(4)_____

(5)_____

(6)_____

II (Name of muscle)

(1)_____

(2)_____

(3)_____

(4)_____

(5)_____

(6)_____

Muscle type location

Fascicle

Cross section

Parallel muscle

(a)

Tendons

Cross section

Multipennate muscle

(b)

Extended tendon

Unipennate muscle

(c)

Contracted

Relaxed
Circular muscle

(d)

Bipennate muscle

(e)

Tendon

Cross section

Convergent muscle

(f)

Identify each numbered structure in the following figures. Place your answers in the spaces provided.

OUTCOME 11-5, 11-6 **FIGURE 11-2** Major Superficial Skeletal Muscles (anterior view)

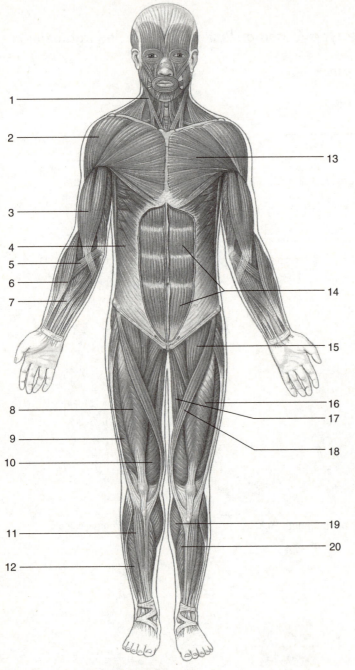

Anterior view

(1)_____

(2)_____

(3)_____

(4)_____

(5)_____

(6)_____

(7)_____

(8)_____

(9)_____

(10)_____

(11)_____

(12)_____

(13)_____

(14)_____

(15)_____

(16)_____

(17)_____

(18)_____

(19)_____

(20)_____

OUTCOME 11-5, 11-6 **FIGURE 11-3** Major Superficial Skeletal Muscles (posterior view)

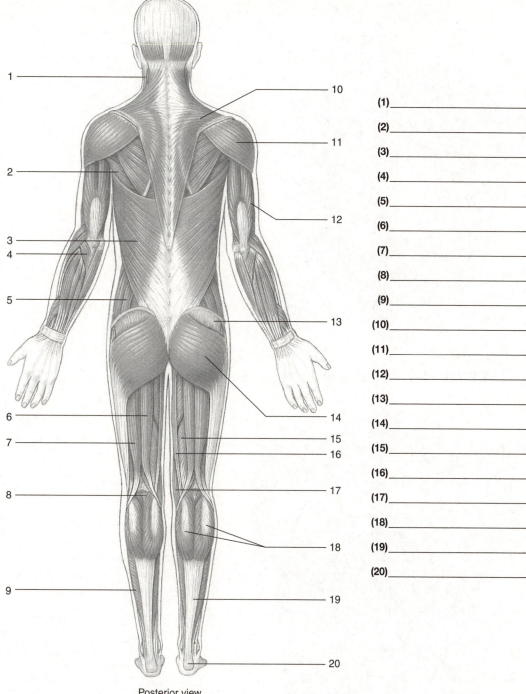

Posterior view

(1)_____

(2)_____

(3)_____

(4)_____

(5)_____

(6)_____

(7)_____

(8)_____

(9)_____

(10)_____

(11)_____

(12)_____

(13)_____

(14)_____

(15)_____

(16)_____

(17)_____

(18)_____

(19)_____

(20)_____

OUTCOME 11-5   **FIGURE 11-4**   Superficial View of Facial Muscles (lateral view)

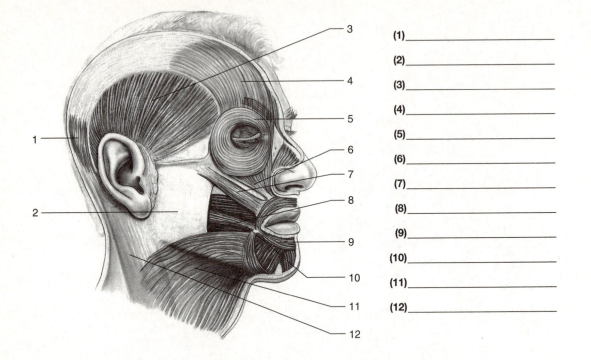

(1)_____

(2)_____

(3)_____

(4)_____

(5)_____

(6)_____

(7)_____

(8)_____

(9)_____

(10)_____

(11)_____

(12)_____

OUTCOME 11-5   **FIGURE 11-5**   Muscles of the Neck (anterior view)

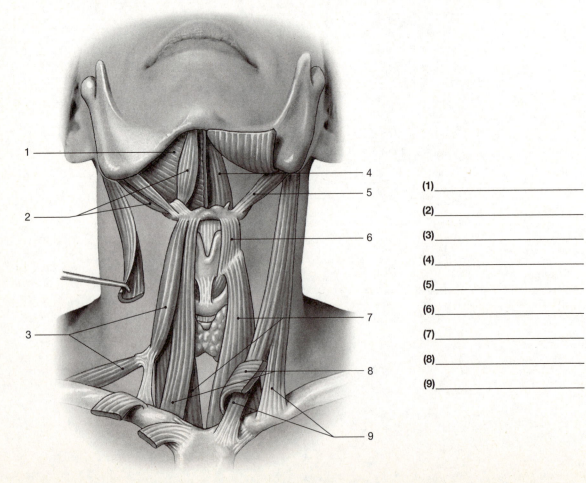

(1)_____

(2)_____

(3)_____

(4)_____

(5)_____

(6)_____

(7)_____

(8)_____

(9)_____

OUTCOME 11-6 **FIGURE 11-6** Superficial Muscles of the Forearm (anterior view)

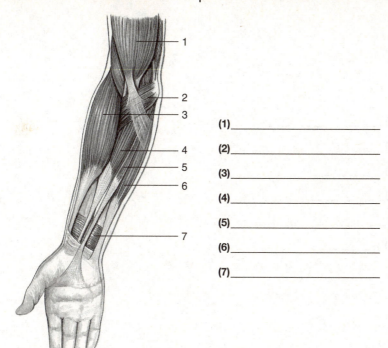

(1)_____

(2)_____

(3)_____

(4)_____

(5)_____

(6)_____

(7)_____

OUTCOME 11-6 **FIGURE 11-7** Muscles of the Hand (palmar group)

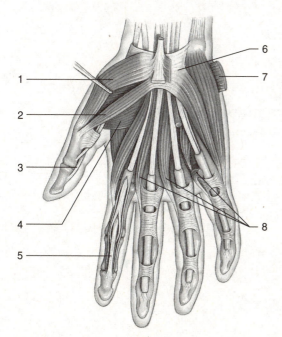
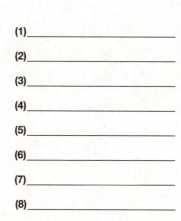

(1)_____

(2)_____

(3)_____

(4)_____

(5)_____

(6)_____

(7)_____

(8)_____

OUTCOME 11-6 **FIGURE 11-8** Superficial Muscles of the Thigh (anterior view)

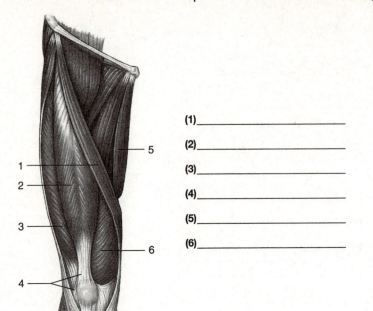

(1)_____

(2)_____

(3)_____

(4)_____

(5)_____

(6)_____

OUTCOME 11-6 **FIGURE 11-9** Superficial Muscles of the Thigh (posterior view)

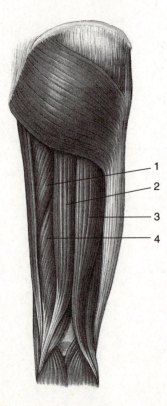

(1)_____

(2)_____

(3)_____

(4)_____

OUTCOME 11-6   **FIGURE 11-10**   Superficial Muscles of the Lower Leg (lateral view)

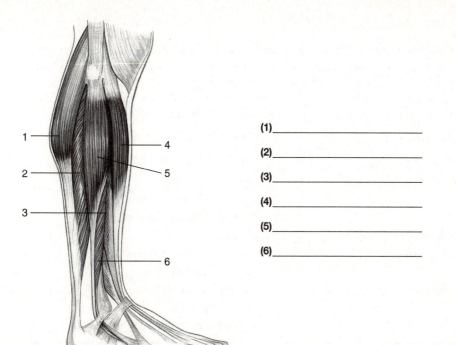

(1)_____

(2)_____

(3)_____

(4)_____

(5)_____

(6)_____

OUTCOME 11-6   **FIGURE 11-11**   Superficial Muscles of the Foot (plantar view)

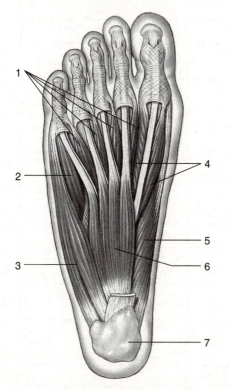

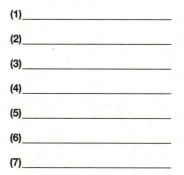

(1)_____

(2)_____

(3)_____

(4)_____

(5)_____

(6)_____

(7)_____

LEVEL 2: REVIEWING CONCEPTS

Chapter Overview

For the study of origins and insertions, the use of a muscular model that can be dissected along with a skeleton to locate and identify the muscles and the bone processes where they originate and insert is recommended. To complete the chapter overview, charts and diagrams will facilitate observation and identification. From the selections below, complete the table for each superficial muscle listed. The origins are randomly listed in Column 1 and the insertions in Column 2.

COLUMN 1—Origins

ischium, femur, ilium
medial margin of orbit
scapula
ribs
ribs (8–12), thoracic-lumbar vertebrae
humerus-lateral epicondyle
femoral condyles
manubrium and clavicle
lower eight ribs

COLUMN 2—Insertions

tibia
lips
deltoid tuberosity of humerus
humerus-greater tubercle
symphysis pubis
ulna-olecranon process
first metatarsal
iliotibial tract and femur
clavicle, scapula
skin of eyebrow, bridge of nose

ORIGINS–INSERTIONS TABLE		
Muscle—Location	**Origin**	**Insertion**
Frontal Belly of Occipitofrontalis (Forehead)	Epicranial Aponeurosis	(1)
Orbicularis Oculi (Eye)	(2)	Skin Around Eyelids
Orbicularis Oris (Mouth)	Maxilla and Mandible	(3)
Sternocleidomastoid (Neck)	(4)	Mastoid—Skull
Trapezius (Back—Neck)	Occipital Bone Thoracic Vertebrae	(5)
Latissimus Dorsi (Back)	(6)	Lesser Tubercle of Humerus
Deltoid (Shoulder)	Clavicle, Scapula	(7)
Biceps Brachii (Upper Arm)	(8)	Tuberosity of Radius
Triceps (Upper Arm—Posterior)	Humerus	(9)
Brachioradialis (Lower Arm)	(10)	Radius—Styloid Process
Pectoralis Major (Chest)	Ribs 2–6, Sternum, Clavicle	(11)
Intercostals (Ribs)	(12)	Ribs
Rectus Abdominis— (Abdomen—Anterior)	(13)	Xiphoid Process—Costal Cartilages (5–7)
External Obliques Abdomen—(Lateral—Anterior)	(14)	Linea Alba, Iliac Crest
Gluteus Maximus—(Posterior)	Ilium, Sacrum, Coccyx	(15)
Hamstrings (Thigh—Posterior)	(16)	Fibula, Tibia
Quadriceps—(Thigh—Anterior)	Ilium, Femur (Linea Aspera)	(17)
Gastrocnemius— (Posterior—Lower Leg)	(18)	Calcaneous via Achilles Tendon
Tibialis Anterior (Anterior—Lower Leg)	Tibia	(19)

Concept Map I

Using the following terms, fill in the circled, numbered, blank spaces to correctly complete the concept map. Use each term only once.

bulbospongiosus   capitis   urethral sphincter

splenius   masseter   external obliques

buccinator   thoracis   oculomotor muscles

rectus abdominis   diaphragm   superior constrictor

scalenes   external anal sphincter external-internal intercostals

hypoglossus   stylohyoid

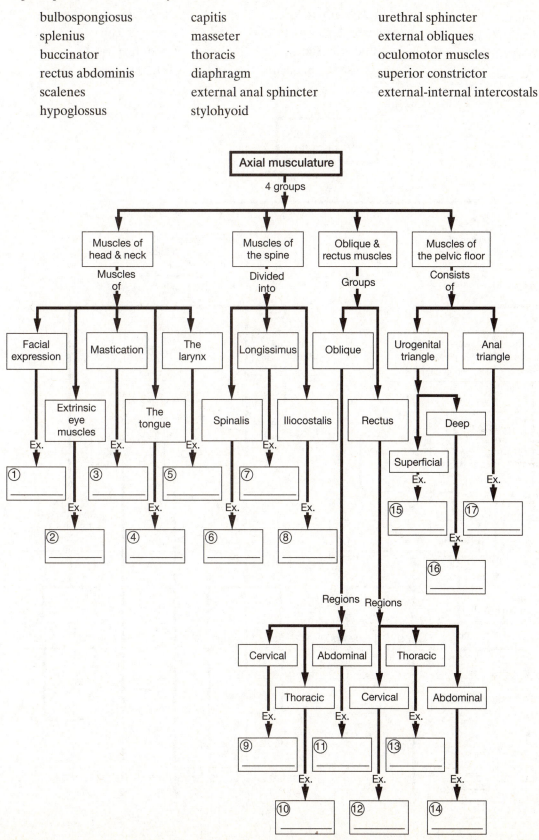

Concept Map II

Using the following terms, fill in the circled, numbered, blank spaces to correctly complete the concept map. Use each term only once.

gracilis
triceps brachii
obturators
deltoid
gastrocnemius

flexor carpi radialis
flexor digitorum longus
gluteus maximus
extensor carpi ulnaris
extensor digitorium

trapezius
biceps brachii
biceps femoris
rectus femoris

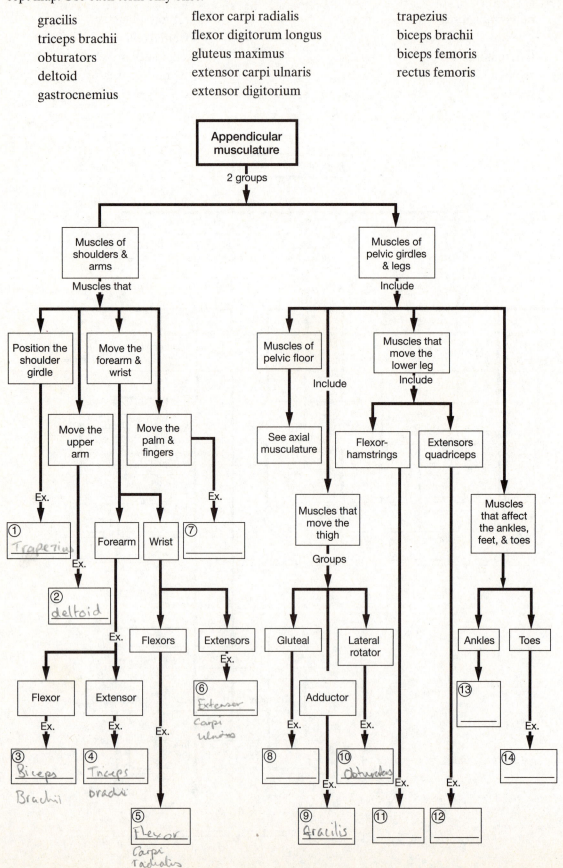

Multiple Choice

Place the letter corresponding to the best answer in the space provided.

_____ 1. The two factors that interact to determine the effects of *individual* skeletal muscle contraction are the

    a. degree to which the muscle is stretched and the amount of tension produced.

    b. anatomical arrangement of the muscle fibers and the way the muscle attaches to the skeletal system.

    c. strength of the stimulus and the speed at which the stimulus is applied.

    d. length of the fibers and the metabolic condition of the muscle.

_____ 2. In compartment syndrome, accumulated blood and fluid trapped within a musculoskeletal compartment results in

    a. ischemia.

    b. fibrosis.

    c. a hernia.

    d. carpel tunnel syndrome.

_____ 3. When a muscle contracts and its fibers shorten, the

    a. origin moves toward the insertion.

    b. origin and the insertion move in opposite directions.

    c. origin and insertion move in the same direction.

    d. insertion moves toward the origin.

_____ 4. The *rectus* muscles, which lie between the vertebral spines and the ventral midline, are important

    a. sphincters of the rectum.

    b. rotators of the spinal column.

    c. flexors of the spinal column.

    d. extensors of the spine.

_____ 5. Attaching a muscle to a lever can change the

    a. direction of an applied force.

    b. distance and speed of movement applied by force.

    c. strength of a force.

    d. a, b, and c are correct.

_____ 6. A collagenous sheet connecting two muscles is a(n)

    a. aponeurosis.

    b. ligament.

    c. tendon.

    d. fascicle.

_____ 7. Which of the following is the origin and insertion of the *genioglossus* muscle?

    a. Inserts at chin and originates in tongue

    b. Inserts and originates at chin

    c. Originates at chin and inserts at tongue

    d. Inserts and originates at tongue

_____ 8. The muscle(s) that is (are) synergistic with the diaphragm during inspiration is (are) the

    a. internal intercostals.

    b. rectus abdominis.

    c. pectoralis minor.

    d. external intercostals.

_____ 9. When playing an instrument such as a trumpet, the muscle used to purse the lips and blow forcefully is the

    a. masseter.

    b. buccinator.

    c. pterygoid.

    d. stylohyoid.

_____ 10. The most powerful and important muscle used when chewing food is the

    a. buccinator.

    b. pterygoid.

    c. masseter.

    d. zygomaticus.

_____ 11. If you are engaging in an activity in which the action involves the use of the _levator scapulae_, you are

    a. shrugging your shoulders.

    b. raising your hand.

    c. breathing deeply.

    d. looking up toward the sky.

_____ 12. The muscular elements that provide substantial support for the loosely built shoulder joint are collectively referred to as the

    a. levator scapulae.

    b. coracobrachialis.

    c. pronator quadratus.

    d. rotator cuff.

_____ 13. The biceps muscle makes a prominent bulge when

    a. extending the forearm pronated.

    b. flexing the forearm supinated.

    c. extending the forearm supinated.

    d. flexing the forearm pronated.

_____ 14. The carpal tunnels that are associated with the wrist bones are formed by the presence of

    a. muscles that extend from the lower arm.

    b. ligaments that attach the arm bones to the wrist.

    c. tendon sheaths crossing the surface of the wrist.

    d. muscular attachments to the phalanges.

_____ 15. If an individual complains because of shin splints, the affected muscles are located over the

    a. anterior surface of the leg.

    b. posterior surface of the leg.

    c. anterior and posterior surfaces of the leg.

    d. anterior surface of the thigh.

Completion

Using the terms below, complete the following statements. Use each term only once.

quadriceps   antagonist   sartorius
cervicis   sphincters   hernia
perineum   hamstrings   iliopsoas
fixator   risorius   masseter

1. Circular muscles in which the fascicles are concentrically arranged around an opening or a recess are called _____.
2. The term that identifies the neck region of the body is _____.
3. A muscle whose action opposes that of another particular muscle is a(n) _____.
4. The muscle that is active when crossing the legs is the _____.
5. The facial muscle that is active when laughing is the _____.
6. The muscles of the pelvic floor that extend between the sacrum and pelvic girdle form the muscular _____.
7. The flexors of the knees are commonly referred to as the _____.
8. The extensors of the knees are commonly referred to as the _____.
9. A _____ develops when a visceral organ or part of an organ protrudes abnormally through an opening in a surrounding muscular wall.
10. When a synergist assists an agonist by preventing movement at another joint and thereby stabilizes the origin of the agonist, it is called a(n) _____.
11. The strongest muscle of mastication is the _____ muscle.
12. The powerful flexors of the hip are the _____ muscles.

Short Essay

Briefly answer the following questions in the spaces provided below.

1. List the four types of skeletal muscles based on the pattern of fascicle organization. Give an example of each type.

2. What is the primary functional difference between an origin and an insertion?

3. List the three functionalities used to describe muscles.

4. What are the four groups of muscles that comprise the axial musculature?

5. What two major groups of muscles comprise the appendicular musculature?

6. List the three classes of levers, give a practical example of each, and identify a place where the action occurs in the body.

7. Name the three muscles that are included in the hamstrings.

8. Name the four muscles that are included in the quadriceps.

9. Why can a pennate muscle generate more tension than a parallel muscle of the same size?

LEVEL 3: CRITICAL THINKING AND CLINICAL APPLICATIONS

Using principles and concepts learned in Chapter 11, answer the following questions. Write your answers on a separate sheet of paper.

1. What effects does a lever have on modifying the contraction of a muscle?

2. Baseball players, especially pitchers, are susceptible to injuries that affect the *rotator cuff* muscles. In what area of the body are these muscles located, and what specific muscles may be affected?

3. One of the most common leg injuries in football involves pulling or tearing the "hamstrings." In what area of the body does this occur, and what muscles may be damaged?

4. How does damage to a musculoskeletal compartment produce the condition of ischemia or "blood starvation"?

5. Your responsibility as a nurse includes giving intramuscular (IM) injections. (a) Why is an IM preferred over an injection directly into circulation? (b) What muscles are best suited as sites for IM injections?

CHAPTER 12

Neural Tissue

OVERVIEW

The nervous system is the principal control center and communication network of the body, and its overall function is the maintenance of homeostasis. The nervous system and the endocrine system acting in a complementary way regulate and coordinate the activities of the body's organ systems. The nervous system generally affects short-term control, whereas endocrine regulation is slower and the general effect is long-term control.

In this chapter the introductory material begins with an overview of the nervous system and the cellular organization in neural tissue. The emphasis for the remainder of the chapter concerns the structure and function of neurons, information processing, and the functional patterns of neural organization.

The integration and interrelationship of the nervous system with the other organ systems is an integral part of understanding many of the body's activities, which must be controlled and adjusted to meet changing internal and external environmental conditions.

LEVEL 1: REVIEWING FACTS AND TERMS

Review of Learning Outcomes

After completing this chapter, you should be able to do the following:

OUTCOME 12-1   Describe the anatomical and functional divisions of the nervous system.

OUTCOME 12-2   Sketch and label the structure of a typical neuron, describe the functions of each component, and classify neurons on the basis of their structure and function.

OUTCOME 12-3   Describe the locations and functions of the various types of neuroglia.

OUTCOME 12-4   Explain how the resting potential is created and maintained.

OUTCOME 12-5   Describe the events involved in the generation and propagation of an action potential.

OUTCOME 12-6   Discuss the factors that affect the speed with which action potentials are propagated.

OUTCOME 12-7   Describe the structure of a synapse, and explain the mechanism involved in synaptic activity.

OUTCOME 12-8   Describe the major types of neurotransmitters and neuromodulators, and discuss their effects on postsynaptic membranes.

OUTCOME 12-9   Discuss the interactions that enable information processing to occur in neural tissue.

Multiple Choice

Place the letter corresponding to the best answer in the space provided.

OUTCOME 12-1   _____   1. The two major anatomical subdivisions of the nervous system are the
       a. central nervous system (CNS) and peripheral nervous system (PNS).
       b. somatic nervous system and autonomic nervous system.
       c. neurons and neuroglia.
       d. afferent division and efferent division.

OUTCOME 12-1   _____   2. The central nervous system (CNS) consists of the
       a. afferent and efferent division.
       b. somatic and visceral division.
       c. brain and spinal cord.
       d. autonomic and somatic division.

OUTCOME 12-1   _____   3. The primary function(s) of the nervous system include
       a. providing sensation of the internal and external environments.
       b. integrating sensory information.
       c. regulating and controlling peripheral structures and systems.
       d. a, b, and c are correct.

OUTCOME 12-2   _____   4. Neurons are responsible for
       a. creating a three-dimensional framework for the CNS.
       b. performing repairs in damaged neural tissue.
       c. information transfer and processing in the nervous system.
       d. controlling the interstitial environment.

OUTCOME 12-2   _____   5. The region of a neuron with voltage-gated sodium channels is the
       a. soma.
       b. dendrite.
       c. axon.
       d. perikaryon.

OUTCOME 12-2   _____   6. Neurons are classified on the basis of their *structure* as
       a. astrocytes, oligodendrocytes, microglia, and ependymal.
       b. efferent, afferent, and interneurons.
       c. motor, sensory, and interneurons.
       d. anaxonic, unipolar, bipolar, and multipolar.

OUTCOME 12-2   _____   7. Neurons are classified on the basis of their *function* as
       a. unipolar, bipolar, and multipolar.
       b. motor, sensory, and interneurons.
       c. somatic, visceral, and autonomic.
       d. central, peripheral, and somatic.

OUTCOME 12-2   _____   8. Sensory neurons are responsible for carrying impulses
       a. to the CNS.
       b. away from the CNS.
       c. to the PNS.
       d. from the CNS to the PNS.

OUTCOME 12-2   _____   9. Efferent pathways consist of axons that carry impulses
   - a. toward the CNS.
   - b. from the PNS to the CNS.
   - c. away from the CNS.
   - d. to the spinal cord and into the brain.

OUTCOME 12-3   _____   10. The two major cell types in neural tissue are
   - a. astrocytes and oligodendrocytes.
   - b. microglia and ependymal cells.
   - c. satellite cells and Schwann cells.
   - d. neurons and neuroglia.

OUTCOME 12-3   _____   11. The types of glial cells in the central nervous system are
   - a. astrocytes, oligodendrocytes, microglia, and ependymal cells.
   - b. unipolar, bipolar, and multipolar cells.
   - c. efferent, afferent, and interneuron cells.
   - d. motor, sensory, and interneuron cells.

OUTCOME 12-3   _____   12. The neuroglia that play a role in structural organization by tying clusters of axons together are the
   - a. astrocytes.
   - b. oligodendrocytes.
   - c. microglia.
   - d. ependymal cells.

OUTCOME 12-4   _____   13. Depolarization of the membrane will shift the membrane potential toward
   - a. -90 mV.
   - b. -85 mV.
   - c. -70 mV.
   - d. 0 mV.

OUTCOME 12-4   _____   14. The resting membrane potential (RMP) of a typical neuron is
   - a. -85 mV.
   - b. -60 mV.
   - c. -70 mV.
   - d. 0 mV.

OUTCOME 12-5   _____   15. If resting membrane potential is -70 mV and the threshold is -60 mV, a membrane potential of -62 mV will
   - a. produce an action potential.
   - b. depolarize the membrane to 0 mV.
   - c. repolarize the membrane to -80 mV.
   - d. not produce an action potential.

OUTCOME 12-5   _____   16. An action potential is triggered when
   - a. local current reaches a voltage that opens voltage-gated Na^+ channels in the axon hillock.
   - b. Na^+ leaks out of ion channels in the axon.
   - c. the transmembrane potential becomes hyperpolarized.
   - d. K^+ leaks out of the cell body.

OUTCOME 12-5 _____ 17. If the resting membrane potential is −70 mV, a hyperpolarized membrane is
   a. 0 mV.
   b. +30 mV.
   c. −80 mV.
   d. −65 mV.

OUTCOME 12-6 _____ 18. A node along the axon represents an area where there is
   a. a layer of fat.
   b. interwoven layers of myelin and protein.
   c. a gap in the cell membrane.
   d. an absence of myelin.

OUTCOME 12-6 _____ 19. The larger the diameter of the axon, the
   a. slower an action potential is conducted.
   b. greater the number of action potentials.
   c. faster an action potential will be conducted.
   d. less effect it will have on action potential propagation.

OUTCOME 12-6 _____ 20. The two most important factors that determine the rate of action potential propagation are the
   a. number of neurons and the length of their axons.
   b. strength of the stimulus and the rate at which the stimulus is applied.
   c. presence or absence of a myelin sheath and the diameter of the axon.
   d. a, b, and c are correct.

OUTCOME 12-6 _____ 21. The reason(s) that active neurons need ATP is to support the
   a. synthesis, release, and recycling of neurotransmitter molecules.
   b. recovery from action potentials.
   c. movement of materials to and from the soma via axoplasmic flow.
   d. a, b, and c are correct.

OUTCOME 12-7 _____ 22. At an electrical synapse, the presynaptic and postsynaptic membranes are locked together at
   a. gap junctions.
   b. synaptic vesicles.
   c. myelinated axons.
   d. neuromuscular junctions.

OUTCOME 12-7 _____ 23. Chemical synapses differ from electric synapses, because chemical synapses
   a. involve direct physical contact between cells.
   b. involve a neurotransmitter.
   c. contain integral proteins called connexons.
   d. propagate action potentials quickly and efficiently.

OUTCOME 12-7 _____ 24. The effect of a neurotransmitter on the postsynaptic membrane depends on the
    a. nature of the neurotransmitter.
    b. properties of the receptor.
    c. number of synaptic vesicles.
    d. width of the synaptic cleft.

OUTCOME 12-7 _____ 25. Exocytosis and the release of acetylcholine into the synaptic cleft is triggered by
    a. calcium ions leaving the cytoplasm.
    b. calcium ions flooding into the axoplasm.
    c. reabsorption of calcium into the endoplasmic reticulum.
    d. active transport of calcium into synaptic vesicles.

OUTCOME 12-7 _____ 26. The normal stimulus for neurotransmitter release is the depolarization of the synaptic terminal by the
    a. release of calcium ions.
    b. binding of ACh receptor sites.
    c. opening of voltage-regulated calcium channels.
    d. arrival of an action potential.

OUTCOME 12-8 _____ 27. Compounds that alter presynaptic or postsynaptic function, thereby affecting information processing, are
    a. neurotransmitters.
    b. neuromodulators.
    c. EPSPs.
    d. IPSPs.

OUTCOME 12-8 _____ 28. Compounds that have an indirect effect on membrane potential work through intermediaries known as
    a. second messengers.
    b. neurotransmitters.
    c. first messengers.
    d. neuromodulators.

OUTCOME 12-9 _____ 29. An excitatory postsynaptic potential (EPSP) is
    a. an action potential complying with the all-or-none principle.
    b. a result of a stimulus strong enough to produce threshold.
    c. the same as a nerve impulse along an axon.
    d. a depolarization produced by the arrival of a neuro-transmitter.

OUTCOME 12-9 _____ 30. An inhibitory postsynaptic potential (IPSP) is a
    a. depolarization produced by the effect of a neurotransmitter.
    b. graded hyperpolarization of the postsynaptic membrane.
    c. repolarization produced by the addition of multiple stimuli.
    d. reflection of the activation of an opposing transmembrane potential.

OUTCOME 12-9 _____ 31. Graded potentials that develop in the postsynaptic membrane in response to a neurotransmitter are

    a. presynaptic facilitators.

    b. presynaptic inhibitors.

    c. presynaptic potentials.

    d. postsynaptic potentials.

OUTCOME 12-9 _____ 32. The addition of stimuli occurring in rapid succession is

    a. temporal summation.

    b. spatial summation.

    c. facilitation.

    d. the absolute refractory period.

Completion

Using the terms below, complete the following statements. Use each term only once.

cholinergic	spatial summation	facilitated	threshold
proprioceptors	temporal summation	electrical	adrenergic
afferent	autonomic nervous system	microglia	IPSP
collaterals	neuromodulators	saltatory	skeletal muscle fiber
electrochemical gradient			

OUTCOME 12-1   1. The visceral motor system that provides automatic, involuntary regulation of smooth and cardiac muscle and glandular secretions is the _____.

OUTCOME 12-2   2. The "branches" that enable a single neuron to communicate with several other cells are called _____.

OUTCOME 12-2   3. Sensory information is brought to the CNS by means of the _____ fibers.

OUTCOME 12-2   4. Sensory neurons that monitor the position of skeletal muscles and joints are called _____.

OUTCOME 12-3   5. In times of infection or injury, the type of neuroglia that will increase in numbers is _____.

OUTCOME 12-4   6. For a particular ion, the sum of all the chemical and electrical forces active across the cell membrane is known as the _____ for that ion.

OUTCOME 12-5   7. An action potential occurs only if the membrane is depolarized to the level known as _____.

OUTCOME 12-6   8. The process that conducts impulses along a myelineated axon at a high rate of speed is called _____ propagation.

OUTCOME 12-7   9. The type of synapse where direct physical contact between the cells occurs is a(n) _____.

OUTCOME 12-7   10. The neuromuscular junction is a synapse where the postsynaptic cell is a(n) _____.

OUTCOME 12-7   11. Chemical synapses that release the neurotransmitter acetylcholine are known as _____ synapses.

OUTCOME 12-8   12. Chemical synapses that release the neurotransmitter norepinephrine are known as _____ synapses.

OUTCOME 12-8   13. Compounds that influence the postsynaptic cells' response to a neurotransmitter are called _____.

OUTCOME 12-9   14. Addition of stimuli occurring in rapid succession at a single synapse is called _____.

OUTCOME 12-9   15. Addition of stimuli arriving at different locations of the nerve cell membrane is called _____.

OUTCOME 12-9   16. A neuron whose transmembrane potential shifts closer to the threshold is said to be _____.

OUTCOME 12-9   17. A graded hyperpolarization of the postsynaptic membrane is referred to as a(n) _____.

Matching

Match the terms in column B with the terms in column A. Use letters for answers in the spaces provided.

Part I

		Column A	Column B
OUTCOME 12-1	_____	1. somatic nervous system	A. astrocytes
OUTCOME 12-1	_____	2. autonomic nervous system	B. −70 mV
OUTCOME 12-2	_____	3. axons	C. interoceptors
OUTCOME 12-2	_____	4. visceral sensory neurons	D. transmit action potentials
OUTCOME 12-2	_____	5. somatic sensory neurons	E. +30 mV
OUTCOME 12-3	_____	6. neuroglia	F. involuntary control
OUTCOME 12-3	_____	7. maintain blood–brain barriers	G. supporting brain cells
OUTCOME 12-4	_____	8. sodium channel inactivation	H. voluntary control
OUTCOME 12-4	_____	9. resting membrane	I. exteroceptors potential (neuron)

Part II

		Column A	Column B
OUTCOME 12-5	_____	10. potassium ion movement	J. gap junctions
OUTCOME 12-6	_____	11. unmyelinated axons	K. cAMP
OUTCOME 12-6	_____	12. nodes of Ranvier	L. continuous propagation
OUTCOME 12-7	_____	13. electrical synapses	M. serotonin
OUTCOME 12-8	_____	14. norepinephrine	N. depolarization
OUTCOME 12-8	_____	15. GABA	O. saltatory propagation
OUTCOME 12-8	_____	16. CNS neurotransmitter	P. simultaneous multiple synapses
OUTCOME 12-8	_____	17. second messenger	Q. repolarization
OUTCOME 12-9	_____	18. EPSP	R. inhibitory effect
OUTCOME 12-9	_____	19. spatial summation	S. adrenergic synapse

Drawing/Illustration Labeling

Identify each numbered structure in the following figures. Place your answers in the spaces provided.

OUTCOME 12-2 **FIGURE 12-1** Anatomy of a Multipolar Neuron

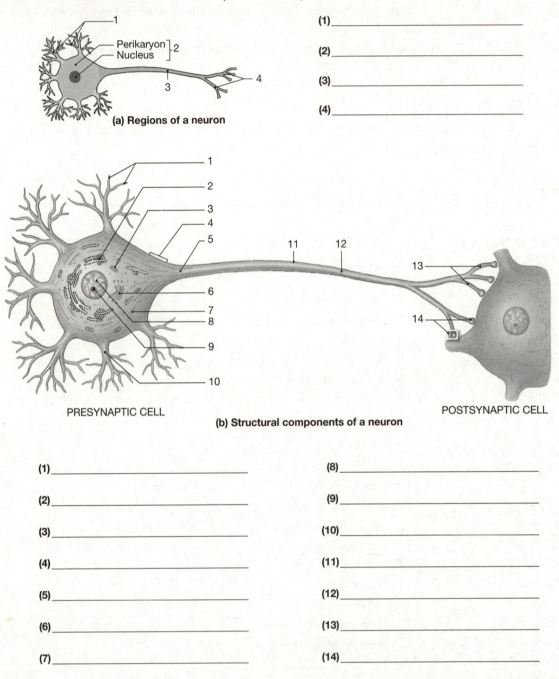

(a) Regions of a neuron

(1)_____

(2)_____

(3)_____

(4)_____

PRESYNAPTIC CELL

POSTSYNAPTIC CELL

(b) Structural components of a neuron

(1)_____   (8)_____

(2)_____   (9)_____

(3)_____   (10)_____

(4)_____   (11)_____

(5)_____   (12)_____

(6)_____   (13)_____

(7)_____   (14)_____

Identify the types of neurons. Place your answers in the spaces provided.

OUTCOME 12-2 **FIGURE 12-2** Neuron Classification (Based on Structure)

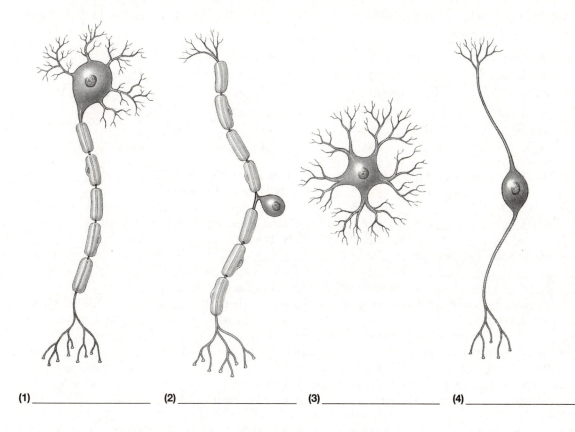

(1) _____ (2) _____ (3) _____ (4) _____

OUTCOME 12-2, 12-7 **FIGURE 12-3** The Structure of a Typical Synapse

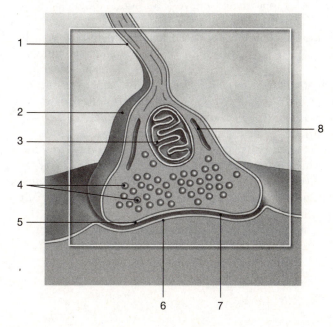

(1) _____

(2) _____

(3) _____

(4) _____

(5) _____

(6) _____

(7) _____

(8) _____

LEVEL 2: REVIEWING CONCEPTS

Chapter Overview

Using the terms below, fill in the blanks to correctly complete the chapter overview of neural tissue. Use each term only once.

Schwann	synaptic terminal	satellite	neuromuscular
dendrites	axoplasmic transport	neurons	anterograde
kinesin	oligodendrocytes	presynaptic	astrocytes
microglia	neurotransmitters	retrograde	dynein
synapse	neuroglandular	ependymal	neuroglia
axon	action potential	postsynaptic	cell body

The basic functional units of the nervous system are individual cells called (1) _____. Each functional unit has a large (2) _____ that is connected to a single, elongated cytoplasmic process, a(n) (3) _____, which is capable of propagating an electrical impulse known as a(n) (4) _____. A variable number of slender, highly branched processes known as (5) _____ extend out from the cell body. The specialized site where the neuron communicates with another cell is the (6) _____, which involves two cells: (a) the (7) _____ cell, which includes the synaptic terminal and sends a message, and (b) the (8) _____ cell, which receives the message. The communication between cells at a synapse most commonly involves the release of chemicals called (9) _____ by the synaptic terminal. This release is triggered by electrical events, such as the arrival of an action potential. A synapse between a neuron and a muscle cell is called a(n) (10) _____ junction, and if a neuron controls or regulates the activity of a secretory cell it is called a(n) (11) _____ junction. Where the postsynaptic cell is another neuron, the (12) _____ is simple, round, and contains mitochondria, portions of the endoplasmic reticulum, and thousands of vesicles filled with neurotransmitter molecules. The movement of materials between the cell body and synaptic terminals is called (13) _____. Known as the "slow stream," it occurs in both directions. The flow of materials from the cell body to the synaptic terminal is (14) _____ flow, pulled along by a "molecular motor" called (15) _____. At the same time, other substances are being transported toward the cell body in (16) _____ flow, carried by (17) _____. Other kinds of cells in the nervous system, which provide a supportive framework for neural tissue and protect neurons, are the (18) _____. There are four types of these cells in the central nervous system: (a) (19) _____ cells, which line the central canal of the spinal cord and the ventricles in several regions of the brain; (b) (20) _____, which are the largest and most numerous of these cell types in the CNS; (c) (21) _____, which play a role in structural organization by tying clusters of axons together, and improve the functional performance of neurons by wrapping axons within a myelin sheath; and (d) (22) _____, the smallest of these cell types in the CNS, which are capable of migrating through neural tissue, where in effect they act as a wandering police force and janitorial surface by engulfing cellular debris, waste products, and pathogens. There are two cell types of neuroglia in the peripheral nervous system: (a) (23) _____ cells or amphicytes, which surround neuron cell bodies in ganglia and regulate the environment around the neurons; and (b) (24)_____ cells, or neurilemmocytes, which form a sheath around peripheral axons, and shield most axons in the PNS from contact with interstitial fluids. The neural tissue, with supporting blood vessels and connective tissues, forms the organs of the nervous system, the receptors in complex sense organs, and the nerves that link the nervous system with other systems.

Concept Map I

Using the following terms, fill in the circled, numbered, blank spaces to correctly complete the concept map. Use each term only once.

surround peripheral ganglia   transmit nerve impulses   astrocytes

central nervous system   Schwann cells   microglia

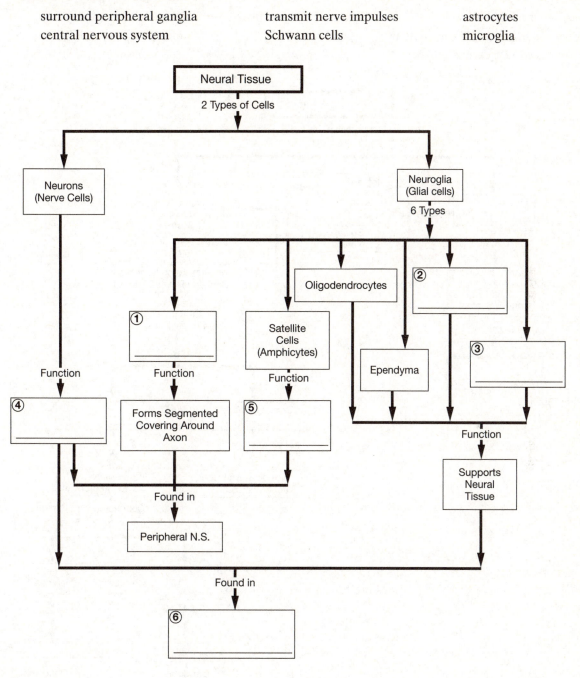

Concept Map II — Levels of Organization of the Nervous System

Using the following terms, fill in the circled, numbered, blank spaces to correctly complete the concept map. Use each term only once.

cranial nerves   parasympathetic division   voluntary nervous system

autonomic nervous system   spinal cord   motor (efferent) division

central nervous system

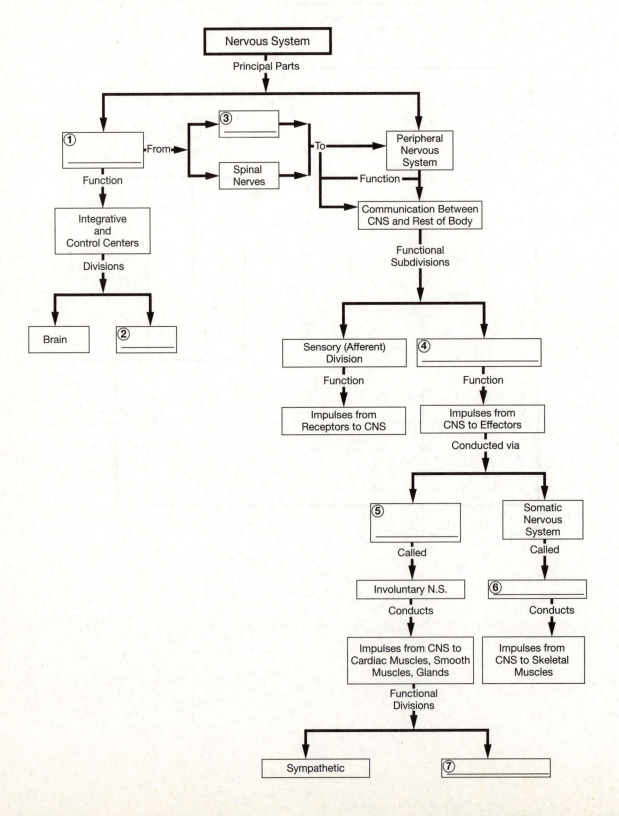

Multiple Choice

Place the letter corresponding to the best answer in the space provided.

_____ 1. The anatomical division of the nervous system responsible for integrating, processing, and coordinating sensory information is the

    a. PNS.

    b. ANS.

    c. CNS.

    d. SNS.

_____ 2. Interneurons are responsible for

    a. carrying instructions from the CNS to peripheral effectors.

    b. delivery of information to the CNS.

    c. collecting information from the external or internal environment.

    d. analysis of sensory inputs and coordination of motor outputs.

_____ 3. A long cytoplasmic process capable of propagating an action potential is the

    a. axon.

    b. dendrite.

    c. synaptic vesicle.

    d. a, b, and c are correct.

_____ 4. The type of cells that surround the nerve cell bodies in peripheral ganglia are

    a. Schwann cells.

    b. satellite cells.

    c. microglia.

    d. oligodendrocytes.

_____ 5. Schwann cells are glial cells responsible for

    a. producing a myelin layer around peripheral axons.

    b. secretion of cerebrospinal fluid.

    c. phagocytic activities in the neural tissue of the PNS.

    d. surrounding nerve cell bodies in peripheral ganglia.

_____ 6. When a barrier prevents the movement of opposite charges toward one another, a(n)

    a. action potential occurs.

    b. current is produced.

    c. potential difference may exist.

    d. generation potential is produced.

_____ 7. The membranous wrapping of electrical insulation, called myelin, around an axon is responsible for

    a. decreasing the speed at which action potentials travel along an axon.

    b. providing janitorial services for the axon by engulfing cellular debris.

    c. regulating the environment around the neuron.

    d. increasing the speed at which an action potential travels along an axon.

_____ 8. The simplest form of information processing in the nervous system is

    a. the integration of stimuli through the CNS.

    b. peripheral nervous system stimulation and integration.

    c. neurotransmission of a stimulus across a synaptic cleft.

    d. the integration of stimuli at the level of the individual cell.

_____ 9. During the relative refractory period, a larger-than-normal depolarizing stimulus can

    a. initiate a second action potential.

    b. cause the membrane to hyperpolarize.

    c. inhibit the production of an action potential.

    d. cause a membrane to reject a response to further stimulation.

_____ 10. Saltatory propagation conducts impulses along an axon

    a. two to three times more slowly than continuous propagation.

    b. five to seven times faster than continuous propagation.

    c. at a rate determined by the strength of the stimulus.

    d. at a velocity determined by the rate at which the stimulus is applied.

_____ 11. In type C fibers, action potentials are conducted at speeds of approximately

    a. 2 mph.

    b. 40 mph.

    c. 150 mph.

    d. 500 mph.

_____ 12. The larger the diameter of the axon, the

    a. slower the rate of transmission.

    b. greater the resistance.

    c. faster the rate of transmission.

    d. size of the axon doesn't affect rate of transmission or resistance.

_____ 13. *Facilitation* in the neuron's transmembrane potential refers to

    a. a shift closer to threshold.

    b. repolarization produced by the addition of multiple stimuli.

    c. transient hyperpolarization of a postsynaptic membrane.

    d. a, b, and c are correct.

_____ 14. Sensory neurons that provide information about the external environment through the sense of sight, smell, hearing, and touch are called

    a. proprioceptors.

    b. exteroceptors.

    c. enviroceptors.

    d. interoceptors.

_____ 15. The main functional difference between the autonomic nervous system and the somatic nervous system is that the activities of the ANS are

    a. primarily voluntary controlled.

    b. primarily involuntary or under "automatic" control.

    c. involved with affecting skeletal muscle activity.

    d. involved with carrying impulses to the CNS.

_____ 16. EPSPs and IPSPs reflect the activation of different types of chemically gated channels, producing

    a. the same effects on the transmembrane potential.

    b. summation resulting in facilitation.

    c. opposing effects on the transmembrane potential.

    d. temporal summation at a single synapse.

_____ 17. If one EPSP depolarizes the initial segment from a resting potential of -70 mV to -65 mV, and threshold is at -60 mV, a(n)

    a. spatial summation will occur.

    b. IPSP will occur.

    c. action potential will not be generated.

    d. action potential will be generated.

_____ 18. An example of presynaptic facilitation is

    a. calcium channels remaining open for a longer period, due to the influence of axoaxonic synapse activity, thus increasing the amount of neurotransmitter released.

    b. reduction of the amount of neurotransmitter released due to the closing of calcium channels in the synaptic terminal.

    c. a larger-than-usual depolarizing stimulus necessary to bring the membrane potential to threshold.

    d. shift in the transmembrane potential toward threshold, which makes the cell more sensitive to further stimulation.

Completion

Using the terms below, complete the following statements. Use each term only once.

interneurons	neurotransmitter	preganglionic fibers	current
voltage-gated	tracts	postganglionic fibers	nuclei
perikaryon	voltage	hyperpolarization	temporal summation
microglia	ganglia	nerve impulse	

1. The cytoplasm that surrounds a neuron's nucleus is referred to as the _____.

2. Nerve cell bodies in the PNS are clustered together in masses called _____.

3. Movement of charges, such as ions, is referred to as _____.

4. The potential difference that exists across a membrane or other barrier is expressed as a(n) _____.

5. Ion channels that open or close in response to changes in transmembrane potential are called _____ channels.

6. The loss of positive ions, which causes a shift in the resting potential to -80 mV or more, is referred to as _____.

7. An action potential traveling along an axon is called a(n) _____.

8. The small phagocytic cells that occur in increased numbers in infected and damaged areas of the CNS are called _____.

9. Axons extending from the CNS to a ganglion are called _____.

10. Axons connecting the ganglionic cells with peripheral effectors are known as _____.

11. Neurons that may be situated between sensory and motor neurons are called _____ neurons.

12. If a synapse involves direct physical contact between cells, it is termed electrical; if the synapse is termed chemical, it involves a(n) _____.

13. The addition of stimuli that arrive at a single synapse in rapid succession is called _____.

14. Collections of nerve cell bodies in the CNS are termed _____.

15. The axonal bundles that make up the white matter of the CNS are called _____.

Short Essay

Briefly answer the following questions in the spaces provided below.

1. What are the two anatomical divisions of the nervous system?

2. What are the major components of the central nervous system and the peripheral nervous system?

3. a. What four types of neuroglia (glial cells) are found in the central nervous system?

   b. What two types of neuroglia are found in the peripheral nervous system?

4. Functionally, what is the major difference between neurons and neuroglia?

5. Using a generalized model, list and describe the four steps that describe an action potential.

6. What is the primary difference between continuous propagation of an action potential and saltatory propagation of an action potential?

7. What are the primary differences with action potential in neural tissue compared to those in muscle tissue?

8. What is the difference between an EPSP and an IPSP, and how does each type affect the generation of an action potential?

9. How are neurons classified functionally, and how does each group function?

10. Compare the actions that occur at an electrical synapse to those that occur at a chemical synapse.

LEVEL 3: CRITICAL THINKING AND CLINICAL APPLICATIONS

Using principles and concepts learned in Chapter 12, answer the following questions. Write your answers on a separate sheet of paper.

1. How does the nicotine in cigarettes initiate the process of facilitation, and ultimately trigger action potentials, and/or lead to addiction?

2. Guillain-Barré syndrome is a degeneration of the myelin sheath that ultimately may result in paralysis. What is the relationship between degeneration of the myelin sheath and muscular paralysis?

3. Even though microglia are found in the CNS, why might these specialized cells be considered a part of the body's immune system?

4. Substantiate the following statement concerning neurotransmitter function: "The effect on the postsynaptic membrane depends on the properties of the receptor, not on the nature of the neurotransmitter."

5. What physiological mechanisms operate to reach threshold when a single stimulus is not strong enough to initiate an action potential?

The Spinal Cord, Spinal Nerves, and Spinal Reflexes

OVERVIEW

The central nervous system (CNS) consists of the brain and the spinal cord. Both of these parts of the CNS are covered with meninges, both are bathed in cerebrospinal fluid, and both areas contain gray and white matter. Even though there are structural and functional similarities, the brain and the spinal cord show significant independent structural and functional differences.

Chapter 13 focuses on the structure and function of the spinal cord and the spinal nerves. Emphasis is placed on the importance of the spinal cord as a communication link between the brain and the peripheral nervous system (PNS), and as an integrating center that can be independently involved with somatic reflex activity.

The questions and exercises for this chapter will help you to identify the integrating sites and the conduction pathways that make up the reflex mechanisms necessary to maintain homeostasis throughout the body.

LEVEL 1: REVIEWING FACTS AND TERMS

Review of Learning Outcomes

After completing this chapter, you should be able to do the following:

OUTCOME 13-1   Describe the basic structural and organizational characteristics of the nervous system.

OUTCOME 13-2   Discuss the structure and functions of the spinal cord, and describe the three meningeal layers that surround the central nervous system.

OUTCOME 13-3   Explain the roles of white matter and gray matter in processing and relaying sensory information and motor commands.

OUTCOME 13-4   Describe the major components of a spinal nerve, and relate the distribution pattern of spinal nerves to the regions they innervate.

OUTCOME 13-5   Discuss the significance of neuronal pools, and describe the major patterns of interaction among neurons within and among these pools.

OUTCOME 13-6   Describe the steps in a neural reflex, and classify the types of reflexes.

OUTCOME 13-7   Distinguish among the types of motor responses produced by various reflexes, and explain how reflexes interact to produce complex behaviors.

OUTCOME 13-8   Explain how higher centers control and modify reflex responses.

Multiple Choice

Place the letter corresponding to the best answer in the space provided.

OUTCOME 13-1   _____   1. The central nervous system (CNS) consists of the
- a. neuron cell bodies located in ganglia.
- b. axons bundled together in nerves.
- c. spinal and cranial nerves.
- d. brain and spinal cord.

OUTCOME 13-1   _____   2. In the peripheral nervous system (PNS),
- a. neuron cell bodies are located in ganglia.
- b. spinal nerves connect to the spinal cord.
- c. cranial nerves connect to the brain.
- d. a, b, and c are correct.

OUTCOME 13-2   _____   3. The axons of motor neurons that extend into the periphery to control somatic and visceral effectors are contained in a pair of
- a. ventral roots.
- b. dorsal roots.
- c. motor nuclei.
- d. sensory nuclei.

OUTCOME 13-2   _____   4. The regions of the spinal cord that are based on the regions of the vertebral column are
- a. cervical, thoracic, lumbar, and sacral.
- b. pia mater, dura mater, and arachnoid mater.
- c. axillary, radial, median, and ulnar.
- d. cranial, visceral, autonomic, and spinal.

OUTCOME 13-2   _____   5. The cervical enlargement of the spinal cord supplies nerves to the
- a. shoulder girdle and arms.
- b. pelvis and legs.
- c. thorax and abdomen.
- d. back and lumbar region.

OUTCOME 13-2   _____   6. If cerebrospinal fluid was withdrawn during a spinal tap, a needle would be inserted into the
- a. pia mater.
- b. subdural space.
- c. subarachnoid space.
- d. epidural space.

OUTCOME 13-2   _____   7. The meninx that is firmly bound to neural tissue and deep to the other meninges is the
- a. pia mater.
- b. arachnoid membrane.
- c. dura mater.
- d. epidural space.

OUTCOME 13-3   8. The white matter of the spinal cord contains

    a. cell bodies of neurons and glial cells.

    b. somatic and visceral sensory nuclei.

    c. large numbers of myelinated and unmyelinated axons.

    d. sensory and motor nuclei.

OUTCOME 13-3   9. The area of the spinal cord that surrounds the central canal and is dominated by the cell bodies of neurons and glial cells is the

    a. white matter.

    b. gray matter.

    c. ascending tracts.

    d. descending tracts.

OUTCOME 13-3   10. The posterior gray horns of the spinal cord contain

    a. somatic and visceral sensory nuclei.

    b. somatic and visceral motor nuclei.

    c. ascending and descending tracts.

    d. anterior and posterior columns.

OUTCOME 13-4   11. The delicate connective tissue fibers that surround individual axons of spinal nerves comprise a layer called the

    a. perineurium.

    b. epineurium.

    c. endoneurium.

    d. commissures.

OUTCOME 13-4   12. The branches of the cervical plexus innervate the muscles of the

    a. shoulder girdle and arm.

    b. pelvic girdle and leg.

    c. neck and extend into the thoracic cavity to control the diaphragm.

    d. back and lumbar region.

OUTCOME 13-4   13. The brachial plexus innervates the

    a. neck and shoulder girdle.

    b. shoulder girdle and arm.

    c. neck and arm.

    d. thorax and arm.

OUTCOME 13-5   14. Divergence is the "neural circuit" that permits

    a. several neurons to synapse on the postsynaptic neuron.

    b. the broad distribution of a specific input.

    c. the processing of the same information at one time.

    d. utilizing positive feedback to stimulate presynaptic neurons.

OUTCOME 13-5   15. When sensory information is relayed from one part of the brain to another, the pattern is called

    a. parallel processing.

    b. reverberation.

    c. convergence.

    d. serial processing.

OUTCOME 13-6 _____ 16. The final step involved in a neural reflex is
    a. information processing.
    b. the activation of a motor neuron.
    c. a response by an effector.
    d. the activation of a sensory neuron.

OUTCOME 13-6 _____ 17. The goals of information processing during a neural reflex are the selection of
    a. appropriate sensor selections and specific motor responses.
    b. an appropriate motor response and the activation of specific motor neurons.
    c. appropriate sensory selections and activation of specific sensory neurons.
    d. appropriate motor selections and specific sensory responses.

OUTCOME 13-6, 13-7 _____ 18. Reflex arcs in which sensory stimulus and the motor response occur on the same side of the body are
    a. monosynaptic.
    b. contralateral.
    c. ipsilateral.
    d. paraesthetic.

OUTCOME 13-6 _____ 19. The basic motor patterns of innate reflexes are
    a. learned responses.
    b. genetically programmed.
    c. a result of environmental conditioning.
    d. usually unpredictable and extremely complex.

OUTCOME 13-6 _____ 20. A professional skier making a rapid, automatic adjustment in body position while racing is an example of a(n) _____ reflex.
    a. innate
    b. cranial
    c. patellar
    d. acquired

OUTCOME 13-7 _____ 21. When a sensory neuron synapses directly on a motor neuron, which itself serves as the processing center, the reflex is called a(n) _____ reflex.
    a. polysynaptic
    b. innate
    c. monosynaptic
    d. acquired

OUTCOME 13-7 _____ 22. The sensory receptors in the stretch reflex are the
    a. muscle spindles.
    b. Golgi tendon organs.
    c. Pacinian corpuscles.
    d. Meissner's corpuscles.

OUTCOME 13-7 _____ 23. When one set of motor neurons is stimulated, those controlling antagonistic muscles are inhibited. This statement illustrates the principle of

    a. contralateral reflex.

    b. reciprocal inhibition.

    c. crossed extensor reflex.

    d. ipsilateral reflex.

OUTCOME 13-7 _____ 24. All polysynaptic reflexes share the same basic characteristic(s), which include

    a. they involve pools of interneurons.

    b. they are intersegmental in distribution.

    c. they involve reciprocal inhibition.

    d. a, b, and c are correct.

OUTCOME 13-8 _____ 25. As descending inhibitory synapses develop,

    a. the Babinski response disappears.

    b. there is a positive Babinski response.

    c. there will be a noticeable fanning of the toes in adults.

    d. there will be a decrease in the facilitation of spinal reflexes.

OUTCOME 13-8 _____ 26. The highest level of motor control involves a series of interactions that occur

    a. as monosynaptic reflexes that are rapid but stereotyped.

    b. in centers in the brain that can modulate or build upon reflexive motor patterns.

    c. in the descending pathways that provide facilitation and inhibition.

    d. in the gamma efferents to the intrafusal fibers.

Completion

Using the terms below, complete the following statements. Use each term only once.

conus medullaris	acquired reflexes	dermatome	somatic reflexes
receptor	ganglia	innate reflexes	filum terminale
flexor reflex	polysynaptic reflex	dorsal ramus	crossed extensor
Babinski sign	nuclei	reinforcement	reflex
columns	cranial reflexes	epineurium	convergence

OUTCOME 13-1   1. In the peripheral nervous system (PNS), neuron cell bodies are located in _____.

OUTCOME 13-2   2. The terminal portion of the spinal cord is called the _____.

OUTCOME 13-2   3. The supportive fibrous strand of the pia mater is the _____.

OUTCOME 13-3   4. The cell bodies of neurons in the gray matter of the spinal cord form groups called _____.

OUTCOME 13-3   5. The white matter of the spinal cord is divided into regions called _____.

OUTCOME 13-4   6. The outermost layer of a spinal nerve is called the _____.

OUTCOME 13-4   7. The branch of each spinal nerve that provides sensory and motor innervation to the skin and muscles of the back is the _____.

OUTCOME 13-4   8. The specific region of the body surface monitored by a single pair of spinal nerves is known as a(n) _____.

OUTCOME 13-5   9. When two neuronal pools synapse on the same motor neurons, the process is called _____.

OUTCOME 13-6   10. _____ provide a mechanism for involuntary control of the muscular system.

OUTCOME 13-6   11. A specialized cell that monitors conditions in the body or the external environment is called a(n) _____.

OUTCOME 13-6   12. Connections that form between neurons during development produce _____.

OUTCOME 13-6   13. Complex, learned motor patterns are called _____.

OUTCOME 13-6   14. Reflexes processed in the brain are called _____.

OUTCOME 13-7   15. A reflex arc that has at least one interneuron between the sensory neuron and the motor neuron is a(n) _____.

OUTCOME 13-7   16. A motor response that occurs on the side opposite the stimulus is referred to as a(n) _____.

OUTCOME 13-7   17. The withdrawal reflex affecting the muscles of a limb is a(n) _____.

OUTCOME 13-8   18. Elevated facilitation leading to an enhancement of spinal reflexes is called _____.

OUTCOME 13-8   19. Stroking an infant's foot on the side of the sole produces a fanning of the toes known as the _____.

Matching

Match the terms in column B with the terms in column A. Use letters for answers in the spaces provided.

Part I

		Column A	Column B
OUTCOME 13-1	_____	1. nerve	A. specialized membranes
OUTCOME 13-2	_____	2. dorsal roots	B. outermost layer of a peripheral nerve
OUTCOME 13-2	_____	3. ventral roots	
OUTCOME 13-2	_____	4. spinal meninges	C. spinal nerves C_1–C_5
OUTCOME 13-3	_____	5. white matter	D. spinal nerves C_5–T_1
OUTCOME 13-4	_____	6. epineurium	E. ascending, descending tracts
OUTCOME 13-4	_____	7. cervical plexus	F. contains axons of motor neurons
OUTCOME 13-4	_____	8. brachial plexus	G. sensory information to spinal cord
			H. bundles of axons

Part II

		Column A	Column B
OUTCOME 13-5	_____	9. reverberation	I. monosynaptic reflex
OUTCOME 13-6	_____	10. neural "wiring"	J. controls activities of muscular system
OUTCOME 13-6	_____	11. peripheral effector	
OUTCOME 13-6	_____		K. muscle or gland cells
OUTCOME 13-7	_____	12. somatic reflexes	L. reflex arc
OUTCOME 13-7	_____	13. stretch reflex	M. monitors muscle tension
OUTCOME 13-7	_____	14. tendon reflex	N. provides facilitation, inhibition
OUTCOME 13-7	_____	15. crossed extensor reflex	O. contralateral response
OUTCOME 13-8	_____	16. same-side reflex	P. ipsilateral response
		17. descending pathways	Q. positive feedback

Drawing/Illustration Labeling

Identify each numbered structure in the following figures. Place your answers in the spaces provided.

OUTCOME 13-2 **FIGURE 13-1**   Anterior (sectional) View of the Adult Spinal Cord

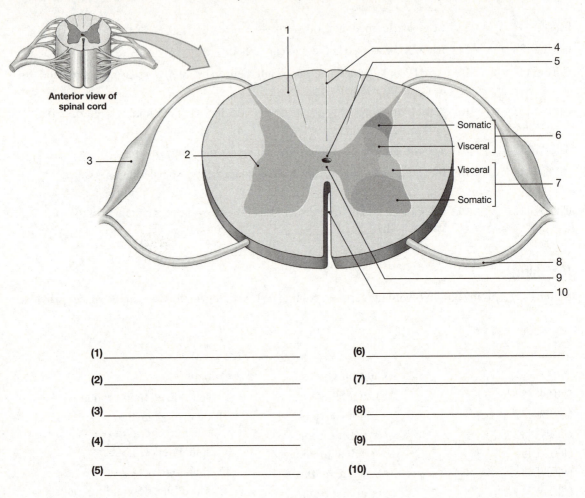

Anterior view of
spinal cord

Somatic
Visceral
Visceral
Somatic

(1)_____   (6)_____

(2)_____   (7)_____

(3)_____   (8)_____

(4)_____   (9)_____

(5)_____   (10)_____

OUTCOME 13-6   **FIGURE 13-2**   Organization of the Reflex Arc

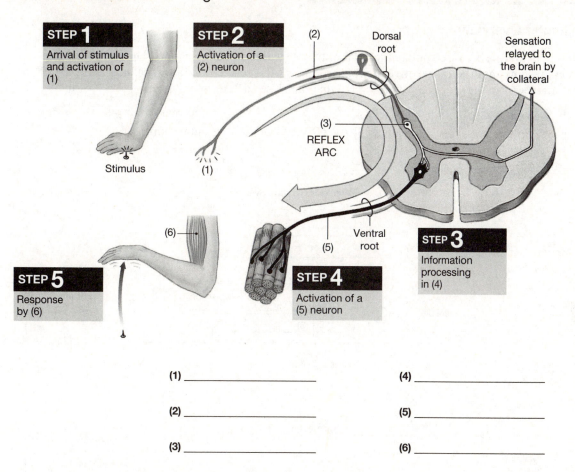

STEP **1**
Arrival of stimulus and activation of (1)

Stimulus

STEP **2**
Activation of a (2) neuron

(1)

(2)

Dorsal root

Sensation relayed to the brain by collateral

(3)

REFLEX ARC

(6)

STEP **5**
Response by (6)

(5)

Ventral root

STEP **4**
Activation of a (5) neuron

STEP **3**
Information processing in (4)

(1) _____   (4) _____

(2) _____   (5) _____

(3) _____   (6) _____

OUTCOME 13-5   **FIGURE 13-3**   The Organization of Neuronal Pools

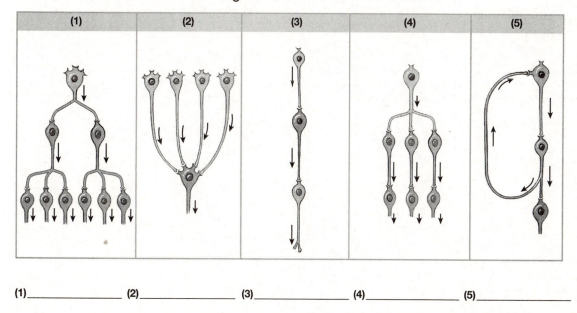

(1)   (2)   (3)   (4)   (5)

(1)_____ (2)_____ (3)_____ (4)_____ (5)_____

LEVEL 2: REVIEWING CONCEPTS

Chapter Overview

Using the terms below, complete the chapter overview through a typical body neural processing system. Assume that the consecutive numbers represent the correct neural processing pathways. Place your answers in the spaces provided on the diagram. Use each term only once.

effectors   afferent   motor
receptors   postganglionic efferent fiber   interneurons
smooth muscles   sensory   visceral motor neurons

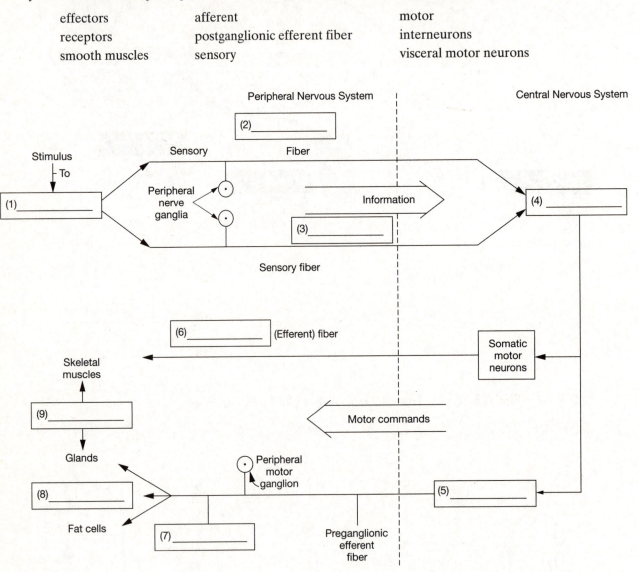

Concept Map I

Using the following terms, fill in the circled, numbered, blank spaces to correctly complete the concept map. Use each term only once.

viscera	glial cells	sensory information
motor commands	posterior gray horns	anterior white columns
gray matter	skeletal muscles	somatic motor neurons
brain	ascending tracts	

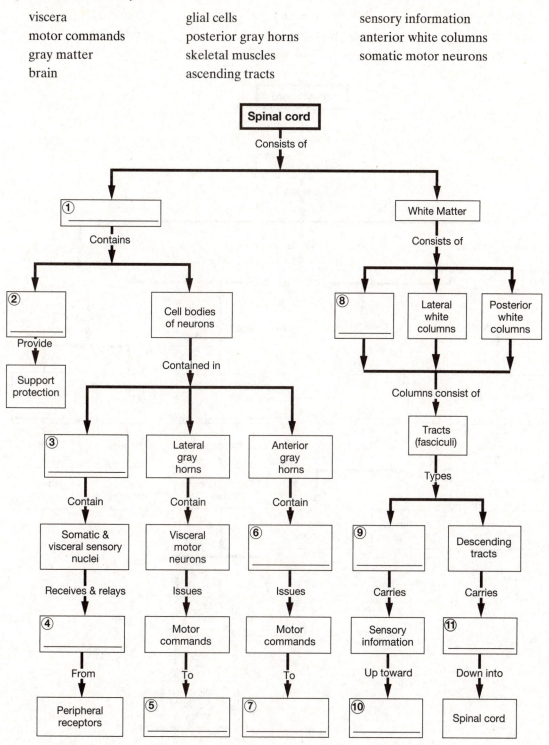

Concept Map II

Using the following terms, fill in the circled, numbered, blank spaces to correctly complete the concept map. Use each term only once.

stability, support   diffusion medium   pia mater

shock absorber   subarachnoid space   dura mater

blood vessels   simple squamous epithelium

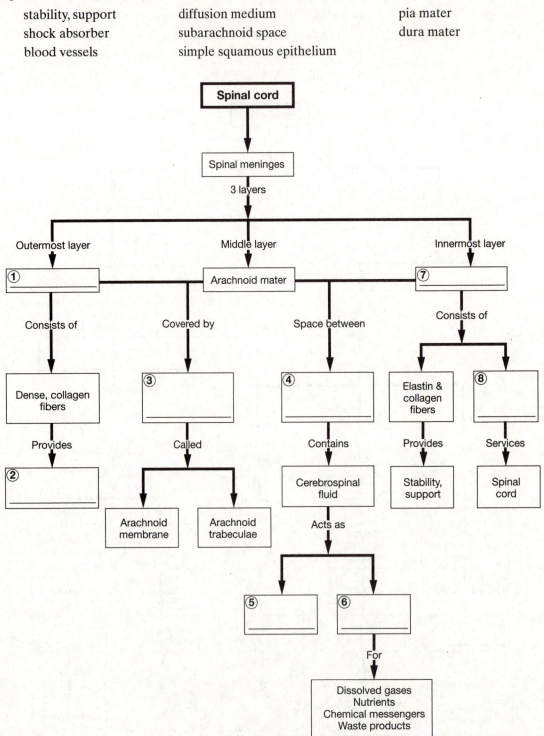

Multiple Choice

Place the letter corresponding to the best answer in the space provided.

_____ 1. The designation C_3 relative to the spinal cord refers to the third

    a. cranial nerve.

    b. coccygeal ligament.

    c. cervical segment.

    d. cerebrospinal layer.

_____ 2. The specialized membranes that provide physical stability and shock absorption are the

    a. conus medullaris.

    b. spinal meninges.

    c. filum terminale.

    d. denticulate ligaments.

_____ 3. Spinal nerves are classified as mixed nerves because they contain

    a. both sensory and motor fibers.

    b. both dorsal and ventral roots.

    c. white matter and gray matter.

    d. ascending and descending pathways.

_____ 4. The three meningeal layers of the spinal cord are the

    a. spinal meninges, cranial meninges, and conus medullaris.

    b. cauda equina, conus medullaris, and filum terminale.

    c. white matter, gray matter, and central canal.

    d. dura mater, pia mater, and arachnoid mater.

_____ 5. The *cauda equina* is a complex of the spinal cord that includes the

    a. enlargements of the cervical and thoracic regions of the cord.

    b. posterior and anterior median sulci.

    c. coccygeal and denticulate ligaments.

    d. filum terminale and the ventral and dorsal roots caudal to the conus medullaris.

_____ 6. The epidural space is an area that contains

    a. blood vessels and sensory and motor fibers.

    b. loose connective tissue, blood vessels, and adipose tissue.

    c. a delicate network of collagen and elastic fibers.

    d. the spinal fluid.

_____ 7. In the spinal cord, the cerebrospinal fluid is found within the

    a. central canal and subarachnoid space.

    b. subarachnoid and epidural spaces.

    c. central canal and epidural space.

    d. subdural and epidural spaces.

_____ 8. The meninx in contact with the brain and the spinal cord is the
   a. dura mater.
   b. arachnoid mater.
   c. pia mater.
   d. subdural space.

_____ 9. The division between the left and right sides of the spinal cord is the
   a. anterior median fissure and the posterior median sulcus.
   b. anterior gray horns and the posterior white columns.
   c. white matter and the gray matter.
   d. anterior and posterior white columns.

_____ 10. The axons in the white matter of the spinal cord that carry sensory information up toward the brain are organized into
   a. descending tracts.
   b. anterior white columns.
   c. ascending tracts.
   d. posterior white columns.

_____ 11. In reference to the vertebral column, C_2 refers to the cervical nerve that
   a. is between vertebrae C_2 and C_3.
   b. precedes vertebra C_2.
   c. follows vertebra C_2.
   d. precedes vertebra C_3.

_____ 12. The white ramus is the branch of a spinal nerve that consists of
   a. myelinated postganglionic axons.
   b. unmyelinated preganglionic axons.
   c. unmyelinated postganglionic axons.
   d. myelinated preganglionic axons.

_____ 13. Unmyelinated fibers that innervate glands and smooth muscles in the body wall or limbs form the
   a. autonomic ganglion.
   b. gray ramus.
   c. white ramus.
   d. ramus communicans.

_____ 14. The lumbar and sacral plexuses that supply the pelvic girdle and the leg include spinal nerves
   a. T_{12}–L_4.
   b. L_4–S_4.
   c. T_1–T_{12}.
   d. T_{12}–S_4.

_____ 15. The subarachnoid space contains
   a. cerebrospinal fluid.
   b. lymph.
   c. blood vessels and connective tissues.
   d. denticulate ligaments.

_____ 16. Reflexive removal of a hand from a hot stove and blinking when the eyelashes are touched are examples of _____ reflexes.

    a. acquired

    b. cranial

    c. innate

    d. synaptic

_____ 17. The most complicated responses are produced by polysynaptic responses because

    a. the interneurons can control several different muscle groups.

    b. a delay between stimulus and response is minimized.

    c. the postsynaptic motor neuron serves as the processing center.

    d. there is an absence of interneurons, which minimizes delay.

_____ 18. The duration of synaptic delay is

    a. determined by the length of the interneuron.

    b. proportional to the amount of neurotransmitter substance released.

    c. determined by the number of sensory neurons.

    d. proportional to the number of synapses involved.

_____ 19. The reflexive contraction of the quadriceps muscle in response to a distortion of stretch receptors is the _____ reflex.

    a. monosynaptic

    b. polysynaptic

    c. stretch

    d. patellar

_____ 20. Examples of somatic reflexes include

    a. pupillary, ciliospinal, salivary, and micturition reflexes.

    b. contralateral, ipsilateral, pupillary, and defecation reflexes.

    c. stretch, crossed extensor, corneal, and gag reflexes.

    d. micturition, Hering-Breuer, defecation, and carotid sinus reflexes.

_____ 21. In this type of neuronal processing, many responses to a single stimulus can occur simultaneously.

    a. Parallel

    b. Convergence

    c. Reverberation

    d. Serial

_____ 22. In an adult, CNS injury is suspected if there is a(n)

    a. plantar reflex.

    b. increase in facilitating synapses.

    c. positive Babinski reflex.

    d. negative Babinski reflex.

Completion

Using the terms below, complete the following statements. Use each term only once.

brachial plexus	pia mater	cauda equina	reflexes
dura mater	sensory nuclei	gray commissures	descending tracts
motor nuclei	dorsal ramus	spinal meninges	dermatome
gray ramus	gamma efferents	nerve plexus	perineurium

1. The connective tissue partition that separates adjacent bundles of nerve fibers in a peripheral nerve is referred to as the _____.

2. Axons convey motor commands down into the spinal cord by way of _____.

3. The outermost covering of the spinal cord is the _____.

4. The specialized protective membrane that surrounds the spinal cord is the _____.

5. The blood vessels servicing the spinal cord run along the surface of the _____.

6. The nerves that collectively radiate from the conus medullaris are known as the _____.

7. The cell bodies of neurons that receive and relay information from peripheral receptors are called _____.

8. The cell bodies of neurons that issue commands to peripheral effectors are called _____.

9. The specific bilateral region of the skin surface monitored by a single pair of spinal nerves is known as a(n) _____.

10. The areas of the spinal cord where axons cross from one side of the cord to the other are the _____.

11. Unmyelinated, postganglionic fibers that innervate glands and smooth muscles in the body wall or limbs form the _____.

12. Spinal nerve sensory and motor innervation to the skin and muscles of the back is provided by the _____.

13. A complex, interwoven network of nerves is called a _____.

14. The radial, median, and ulnar nerves originate in the _____.

15. Programmed, automatic, involuntary motor responses and motor patterns are called _____.

16. Contraction of the myofibrils within intrafusal muscle occurs when impulses arrive over _____.

Short Essay

Briefly answer the following questions in the spaces provided below.

1. What three membranes make up the spinal meninges?

2. Why are spinal nerves classified as *mixed* nerves?

3. What structural components make the white matter different from the gray matter in the spinal cord?

4. What is a dermatome, and why are dermatomes clinically important?

5. List the five steps involved in a neural reflex.

6. What four criteria are used to classify reflexes?

7. Why can polysynaptic reflexes produce more complicated responses than monosynaptic reflexes?

8. What is reciprocal inhibition?

9. What is the difference between a contralateral reflex arc and an ipsilateral reflex arc?

10. What are the clinical implications of a positive Babinski reflex in an adult?

LEVEL 3: CRITICAL THINKING AND CLINICAL APPLICATIONS

Using principles and concepts learned in Chapter 13, answer the following questions. Write your answers on a separate sheet of paper.

1. What structural features make it possible to do a spinal tap without damaging the spinal cord?

2. What is spinal meningitis, what causes the condition, and how does it ultimately affect the victim?

3. Why is lumbar epidural anesthesia often preferred over caudal anesthesia to control pain during labor and delivery?

4. Why are reflexes such as the patellar reflex an important part of a routine physical examination?

5. How might damage to the following areas of the vertebral column affect the body?

    a. damage to the 4th or 5th vertebrae

    b. damage to C_3 to C_5

    c. damage to the thoracic vertebrae

    d. damage to the lumbar vertebrae

14

The Brain and Cranial Nerves

OVERVIEW

In Chapter 13 we ventured into the central nervous system (CNS) via the spinal cord. Chapter 14 considers the other division of the CNS, the brain. This large, complex organ is located in the cranial cavity completely surrounded by cerebrospinal fluid (CSF), the cranial meninges, and the bony structures of the cranium.

The development of the brain is one of the primary factors that have been used to classify us as "human." Our range of thinking and reasoning skills is a key identifying characteristic of our species.

The brain consists of billions of neurons organized into hundreds of neuronal pools with extensive interconnections that provide great versatility and variability. The same basic principles of neural activity and information processing that occur in the spinal cord also apply to the brain; however, because of the increased amount of neural tissue, there is an expanded versatility and complexity in response to neural stimulation and neural function.

Included in Chapter 14 are exercises and test questions that focus on the major regions and structures of the brain and their relationships with the cranial nerves. The concept maps of the brain will be particularly helpful in making the brain come "alive" structurally.

LEVEL 1: REVIEWING FACTS AND TERMS

Review of Learning Outcomes

After completing this chapter, you should be able to do the following:

OUTCOME 14-1   Name the major brain regions, vesicles, and ventricles, and describe the locations and functions of each.

OUTCOME 14-2   Explain how the brain is protected and supported, and discuss the formation, circulation, and function of cerebrospinal fluid.

OUTCOME 14-3   Describe the anatomical differences between the medulla oblongata and the spinal cord, and identify the medulla oblongata's main components and functions.

OUTCOME 14-4   List the main components of the pons, and specify the functions of each.

OUTCOME 14-5   List the main components of the cerebellum, and specify the functions of each.

OUTCOME 14-6   List the main components of the midbrain, and specify the functions of each.

OUTCOME 14-7   List the main components of the diencephalon, and specify the functions of each.

OUTCOME 14-8   Identify the main components of the limbic system, and specify the locations and functions of each.

OUTCOME 14-9   Identify the major anatomical subdivisions and functions of the cerebrum, and discuss the origin and significance of the major types of brain waves seen in an electroencephalogram.

OUTCOME 14-10   Describe representative examples of cranial reflexes that produce somatic responses or visceral responses to specific stimuli.

Multiple Choice

Place the letter corresponding to the best answer in the space provided.

OUTCOME 14-1   _____   1. The major region of the brain responsible for conscious thought processes, sensations, intellectual functions, memory, and complex motor patterns is the

    a. cerebellum.

    b. medulla.

    c. pons.

    d. cerebrum.

OUTCOME 14-1   _____   2. The region of the brain that adjusts voluntary and involuntary motor activities on the basis of sensory information and stored memories of previous movements is the

    a. cerebrum.

    b. cerebellum.

    c. medulla.

    d. diencephalon.

OUTCOME 14-1   _____   3. The brain stem consists of the

    a. midbrain, pons, medulla oblongata.

    b. cerebrum, cerebellum, medulla, and pons.

    c. thalamus, hypothalamus, cerebellum, and medulla.

    d. diencephalon, spinal cord, cerebellum, and medulla.

OUTCOME 14-1   _____   4. The expansion of the neurocoel enlarges to create three primary brain vesicles called the

    a. cerebrum, cerebellum, and medulla oblongata.

    b. diencephalon, mesencephalon, and pons.

    c. telencephalon, metencephalon, and myelencephalon.

    d. prosencephalon, mesencephalon, and rhombencephalon.

OUTCOME 14-1   _____   5. The cerebrum of the adult brain ultimately forms from the

    a. telencephalon.

    b. mesencephalon.

    c. diencephalon.

    d. myelencephalon.

OUTCOME 14-1   _____   6. The lateral ventricles of the cerebral hemisphere communicate with the third ventricle of the diencephalon through the

    a. septum pellucidum.

    b. cerebral aqueduct.

    c. mesencephalic aqueduct.

    d. interventricular foramen.

OUTCOME 14-1   _____   7. The slender canal that connects the third ventricle with the fourth ventricle is the

   a. cerebral aqueduct.

   b. foramen of Monro.

   c. septum pellucidum.

   d. diencephalic chamber.

OUTCOME 14-2   _____   8. The delicate tissues of the brain are protected from mechanical forces by the

   a. bones of the cranium.

   b. cranial meninges.

   c. cerebrospinal fluid.

   d. a, b, and c are correct.

OUTCOME 14-2   _____   9. The cranial meninges offer protection to the brain by

   a. protecting the brain from extremes of temperature.

   b. affording mechanical protection to the brain tissue.

   c. providing a barrier against invading pathogenic organisms.

   d. preventing contact with surrounding bones.

OUTCOME 14-2   _____ 10. The neural tissue of the brain is biochemically isolated from the general circulation by the

   a. cerebrospinal fluid.

   b. blood–brain barrier.

   c. choroid plexus.

   d. subarachnoid space.

OUTCOME 14-2   _____ 11. The most important function(s) of the cerebrospinal fluid is (are)

   a. to cushion delicate neural structures.

   b. to support the brain.

   c. the transport of nutrients, chemical messengers, and waste products.

   d. all of the above.

OUTCOME 14-2   _____ 12. A combination of specialized ependymal cells and permeable capillaries involved in the production of cerebrospinal fluid forms a network called the

   a. falx cerebri.

   b. choroid plexus.

   c. tentorium cerebelli.

   d. arachnoid granulations.

OUTCOME 14-2   _____ 13. Excess cerebrospinal fluid is returned to venous circulation by

   a. diffusion across the arachnoid villi.

   b. active transport across the choroid plexus.

   c. diffusion through the lateral and medial apertures.

   d. passage through the subarachnoid space.

OUTCOME 14-3   _____ 14. The inferior portion of the medulla oblongata resembles the spinal cord, with the presence of a

   a. large exterior prominence.

   b. small central canal.

   c. large clump of "olives."

   d. a, b, and c are correct.

OUTCOME 14-3 _____ 15. The similarity of the medulla oblongata to the spinal cord disappears with the formation of the

    a. third ventricle.

    b. pons.

    c. fourth ventricle.

    d. small central canal.

OUTCOME 14-3 _____ 16. The cardiovascular centers and the respiratory rhythmicity centers are located in the

    a. spinal cord.

    b. medulla oblongata.

    c. pons.

    d. cerebellum.

OUTCOME 14-3 _____ 17. The medulla oblongata contains sensory and motor nuclei associated with cranial nerves

    a. I, II, III, IV, and V.

    b. II, IV, VI, VIII, and X.

    c. VIII, IX, X, XI, and XII.

    d. III, IV, IX, X, and XII.

OUTCOME 14-3 _____ 18. In the medulla oblongata, the relay stations along sensory or motor pathways include the

    a. vasomotor and rhythmicity centers.

    b. arbor vitae and cerebellar peduncles.

    c. cerebral peduncles and red nuclei.

    d. nucleus gracilis and nucleus cuneatus.

OUTCOME 14-4 _____ 19. The pons links the cerebellum with the

    a. diencephalon.

    b. cerebrum.

    c. spinal cord.

    d. a, b, and c are correct.

OUTCOME 14-4 _____ 20. The centers in the pons that modify the activity of the rhythmicity centers in the medulla oblongata are the

    a. inferior and superior peduncles.

    b. apneustic and pneumotaxic centers.

    c. nucleus gracilis and nucleus cuneatus.

    d. cardiac and vasomotor centers.

OUTCOME 14-5 _____ 21. Huge, highly branched Purkinje cells are found in the

    a. cerebral cortex.

    b. primary fissures.

    c. flocculonodular lobe.

    d. cerebellar cortex.

OUTCOME 14-5   _____ 22. Coordination and refinement of learned movement patterns at the subconscious level are performed by the
    a. cerebellum.
    b. hypothalamus.
    c. pons.
    d. association fibers.

OUTCOME 14-6   _____ 23. The corpora quadrigemina of the midbrain are responsible for processing
    a. sensations of taste and smell.
    b. complex coordinated movements.
    c. visual and auditory sensations.
    d. balance and equilibrium.

OUTCOME 14-6   _____ 24. The nerve fiber bundles on the ventrolateral surfaces of the midbrain are the
    a. cerebral peduncles.
    b. substantia nigra.
    c. red nuclei.
    d. superior and inferior colliculi.

OUTCOME 14-7   _____ 25. The epithalamus, thalamus, and hypothalamus are anatomical structures of the
    a. cerebellum.
    b. diencephalon.
    c. midbrain.
    d. metencephalon.

OUTCOME 14-7   _____ 26. Relay and processing centers for sensory information are found in the
    a. hypothalamus.
    b. epithalamus.
    c. pineal gland.
    d. thalamus.

OUTCOME 14-7   _____ 27. The hypothalamus contains centers involved with
    a. voluntary somatic motor responses.
    b. somatic and visceral motor control.
    c. emotions, autonomic function, and hormone production.
    d. maintenance of consciousness.

OUTCOME 14-8   _____ 28. The sea-horse-like structure in the limbic system responsible for storage and retrieval of new long-term memories is the
    a. corpus callosum.
    b. amygdaloid body.
    c. hippocampus.
    d. cingulate gyrus.

OUTCOME 14-8   _____ 29. As a functional group, the limbic system behaves as a
    a. mini-neural system.
    b. motivational system.
    c. subendocrine system.
    d. complex task-performing system.

OUTCOME 14-9 _____ 30. Even though almost completely separated, the two cerebral hemispheres are connected by a thick band of white matter called the

a. corpus callosum.

b. corpora quadrigemina.

c. infundibulum.

d. intermediate mass.

OUTCOME 14-9 _____ 31. The cerebral cortex is linked to the diencephalon, brain stem, cerebellum, and spinal cord by _____ fibers.

a. commissural

b. projection

c. association

d. arcuate

OUTCOME 14-9 _____ 32. The lobes of the cerebral cortex involved in processing sensory data and motor activities are the

a. commissural, projection, association, and arcuate.

b. sulci, gyri, fissures, and capsules.

c. frontal, parietal, occipital, and temporal.

d. prefrontal, interpretive, integrative, and association.

OUTCOME 14-9 _____ 33. The masses of gray matter that lie within each cerebral hemisphere deep to the floor of the lateral ventricle are the

a. commissural fibers.

b. pyramidal cells.

c. gyri and sulci.

d. basal nuclei.

OUTCOME 14-9 _____ 34. The general interpretive area that receives information from all of the sensory association areas is called

a. Brodmann's area.

b. Wernicke's area.

c. Broca's area.

d. the limbic system.

OUTCOME 14-9 _____ 35. Pyramidal cells are cortical neurons that

a. direct voluntary movements.

b. link the cerebral cortex to the brain stem.

c. permit communication between the cerebral hemispheres.

d. monitor and interpret sensory information.

OUTCOME 14-9 _____ 36. Higher-frequency beta waves are typical of an individual who is

a. resting with his or her eyes closed.

b. in a deep sleep.

c. brain damaged.

d. under stress or in a state of psychological tension.

OUTCOME 14-9 _____ 37. Very-large-amplitude, low-frequency waves normally seen during deep sleep are _____ waves.
 a. delta
 b. theta
 c. alpha
 d. beta

OUTCOME 14-9 _____ 38. The brains of healthy, awake adults who are resting with their eyes closed produce _____ waves.
 a. alpha
 b. beta
 c. delta
 d. theta

OUTCOME 14-10 _____ 39. A loud noise produces a tympanic reflex that results in
 a. unstabilized eye movements.
 b. unusual head movements.
 c. reduced movement of auditory ossicles.
 d. increased olfactory activity.

OUTCOME 14-10 _____ 40. Constriction of the contralateral pupil due to light striking photoreceptor cells results from the _____ reflex.
 a. direct light
 b. consensual light
 c. vestibulo-ocular
 d. corneal

OUTCOME 14-10 _____ 41. A loud noise produces an auditory reflex that results in
 a. constriction of the ipsilateral pupil.
 b. constriction of the contralateral pupil.
 c. increased olfactory activity.
 d. eye and/or head movements.

Completion

Using the terms below, complete the following statements. Use each term only once.

ependymal	third ventricle	tentorium cerebelli	corneal reflex
cerebral aqueduct	red nuclei	respiratory centers	pons
ataxia	cranial nerves	choroid plexus	lentiform nucleus
cerebrum	falx cerebri	white matter	direct light reflex
spinal cord	brain stem	diencephalon	basal nuclei
cerebrospinal fluid	thalamus	seizures	cerebellum
hypothalamus	myelencephalon	theta waves	vestibulo-ocular
amygdaloid body			

OUTCOME 14-1   1. Conscious thoughts, sensations, intellect, memory, and complex movements all originate in the _____.

OUTCOME 14-1   2. The region of the brain that would allow you to perform the same movements over and over is the _____.

OUTCOME 14-1   3. The midbrain, pons, and medulla oblongata are components of the _____.

OUTCOME 14-1   4. During embryological development, the ventral portion of the metencephalon will become the _____.

OUTCOME 14-1   5. The medulla oblongata forms from the _____.

OUTCOME 14-1   6. The ventricles in the cerebral hemispheres are filled with _____.

OUTCOME 14-1   7. The slender canal in the midbrain that connects the third ventricle with the fourth ventricle is the _____.

OUTCOME 14-2   8. The fold of dura mater projecting between the cerebral hemispheres in the midsagittal plane that provides stabilization and support for the brain is the _____.

OUTCOME 14-2   9. The dural sheet that separates and protects the cerebellar hemispheres from those of the cerebrum is the _____.

OUTCOME 14-2   10. The site of cerebrospinal fluid (CSF) production is the _____.

OUTCOME 14-2   11. The specialized cells that remove waste products from the CSF and adjust its composition over time are the _____ cells.

OUTCOME 14-3   12. The ascending and descending tracts of the medulla oblongata link the brain with the _____.

OUTCOME 14-3   13. The medulla oblongata contains sensory and motor nuclei associated with five of the _____.

OUTCOME 14-3   14. The nucleus gracilis and the nucleus cuneatus pass somatic sensory information to the _____.

OUTCOME 14-4   15. The reticular formation within the pons contains two _____.

OUTCOME 14-5   16. The arbor vitae, or "tree of life," in the cerebellum is a branching array of _____.

OUTCOME 14-5   17. A disturbance of balance caused by a damaged cerebellum due to trauma or stroke results in a condition called _____.

OUTCOME 14-6   18. Involuntary control of background muscle tone and limb position is the function of components in the midbrain called _____.

OUTCOME 14-7   19. The thalamus and hypothalamus are the main components of the _____.

OUTCOME 14-7   20. The left thalamus is separated from the right thalamus by the _____.

OUTCOME 14-8   21. The cerebral contribution to the limbic system is the _____.

OUTCOME 14-8   22. The diencephalic component of the limbic system concerned with emotions, thirst, and hunger is the _____.

OUTCOME 14-9   23. Cerebral processing of sensory information and issuing of motor commands at the subconscious level are directed by the _____.

OUTCOME 14-9   24. A medial globus pallidus and a lateral putamen comprise the _____.

OUTCOME 14-9   25. In an EEG, the waves that may appear transiently during sleep in normal adults but are most often observed in children and in intensely frustrated adults are _____.

OUTCOME 14-9   26. Marked changes in the pattern of electrical activity monitored in an electroencephalogram are an indication of _____.

OUTCOME 14-10   27. Constriction of the ipsilateral pupil results from the _____.

OUTCOME 14-10   28. Blinking of the eyelids results from the _____.

OUTCOME 14-10   29. Rotation of the head initiates the _____ reflexes.

Matching

Match the terms in column B with the terms in column A. Use letters for answers in the spaces provided.

Part I

		Column A	Column B
OUTCOME 14-1	_____	1. cerebrum	A. contain cerebrospinal fluid
OUTCOME 14-1	_____	2. rhombencephalon	B. CSF absorbed into venous circulation
OUTCOME 14-1	_____	3. ventricles	C. receives visceral sensory information
OUTCOME 14-2	_____	4. falx cerebri	D. medulla oblongata
OUTCOME 14-2	_____	5. arachnoid granulations	E. conscious thoughts
OUTCOME 14-3	_____	6. "olives"	F. apneustic and pneumotaxic centers
OUTCOME 14-3	_____	7. solitary nucleus	G. hindbrain
OUTCOME 14-4	_____	8. pons	H. fold of dura mater

Part II

		Column A	Column B
OUTCOME 14-5	_____	9. arbor vitae	I. secretes melatonin
OUTCOME 14-6	_____	10. cerebral peduncles	J. speech center
OUTCOME 14-7	_____	11. pineal gland	K. links cerebral hemispheres
OUTCOME 14-8	_____	12. limbic system	L. electrical activity of the brain
OUTCOME 14-9	_____	13. corpus callosum	M. cerebellum
OUTCOME 14-9	_____	14. Broca's area	N. autonomic nervous system
OUTCOME 14-9	_____	15. EEG	O. processing of memories
OUTCOME 14-10	_____	16. reflex motor output	P. carry voluntary motor commands

Drawing/Illustration Labeling

Identify each numbered structure. Place your answers in the spaces provided.

OUTCOME 14-1, 14-9 **FIGURE 14-1**   Lateral View of the Human Brain

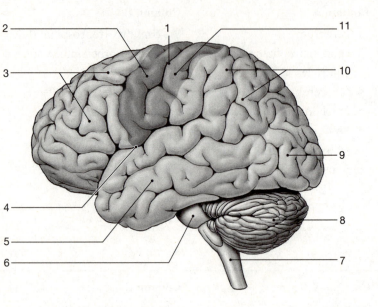

(1) _____

(2) _____

(3) _____

(4) _____

(5) _____

(6) _____

(7) _____

(8) _____

(9) _____

(10) _____

(11) _____

Identify the major brain regions and landmarks labeled 1–8.

OUTCOME 14-1   **FIGURE 14-2**   An Introduction to Brain Structures and Functions

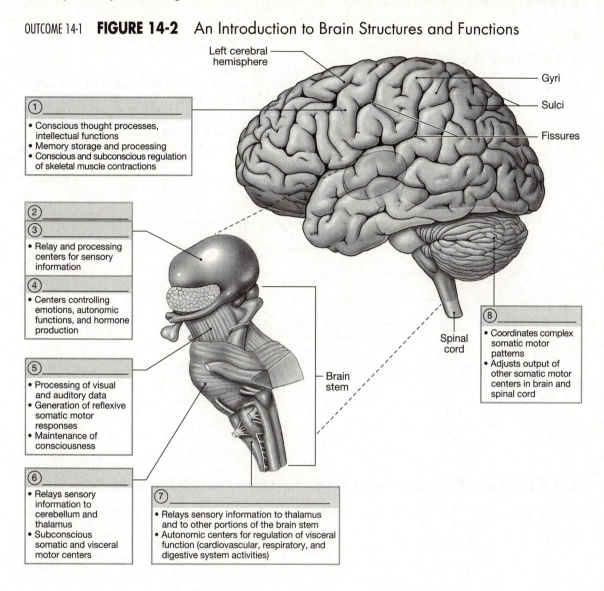

Left cerebral hemisphere

Gyri

Sulci

Fissures

1 _____
- Conscious thought processes, intellectual functions
- Memory storage and processing
- Conscious and subconscious regulation of skeletal muscle contractions

2 _____
3 _____
- Relay and processing centers for sensory information

4 _____
- Centers controlling emotions, autonomic functions, and hormone production

5 _____
- Processing of visual and auditory data
- Generation of reflexive somatic motor responses
- Maintenance of consciousness

6 _____
- Relays sensory information to cerebellum and thalamus
- Subconscious somatic and visceral motor centers

7 _____
- Relays sensory information to thalamus and to other portions of the brain stem
- Autonomic centers for regulation of visceral function (cardiovascular, respiratory, and digestive system activities)

Brain stem

Spinal cord

8 _____
- Coordinates complex somatic motor patterns
- Adjusts output of other somatic motor centers in brain and spinal cord

LEVEL 2: REVIEWING CONCEPTS

Chapter Overview

PART I: Review of Cranial Nerves

On the following chart, fill in the blanks to complete the names of the cranial nerves and their functions. If the function is motor, use *M*; if sensory, use *S*; if both motor and sensory, use *B*.

Number	Name	Function
I	(1)	S
II	Optic	(7)
III	(2)	M
IV	Trochlear	(8)
V	(3)	B
VI	Abducens	(9)
VII	(4)	B
VIII	Vestibulocochlear	(10)
IX	(5)	B
X	Vagus	(11)
XI	(6)	M
XII	Hypoglossal	(12)

PART II: Origins of Cranial Nerves—Inferior View of the Human Brain—Partial

Identify the numbered structures in the following figure and write your answers in the spaces provided.

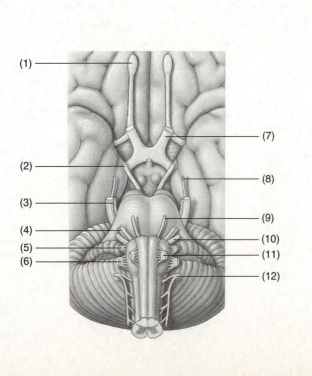

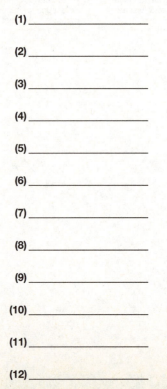

(1) _____

(2) _____

(3) _____

(4) _____

(5) _____

(6) _____

(7) _____

(8) _____

(9) _____

(10) _____

(11) _____

(12) _____

Concept Map I

Using the following terms, fill in the circled, numbered, blank spaces to correctly complete the concept map. Use each term only once.

2 cerebellar hemispheres   pons   hypothalamus

corpora quadrigemina   diencephalon   medulla oblongata

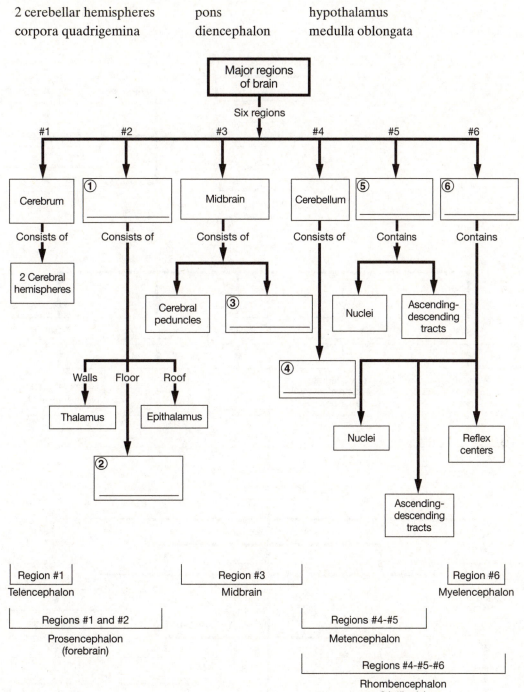

Concept Map II

Using the following terms, fill in the circled, numbered, blank spaces to correctly complete the concept map. Use each term only once.

occipital   cerebral cortex   temporal   commissural
basal nuclei   projections   somatic sensory cortex

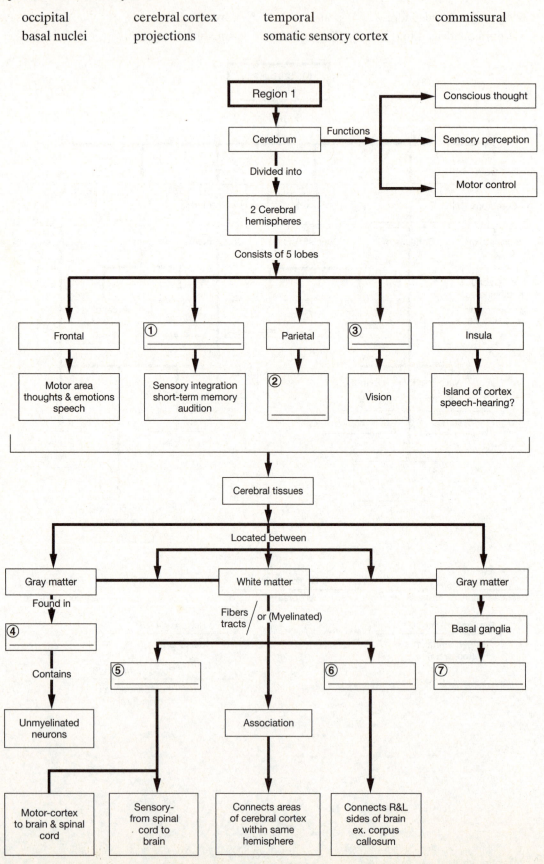

Concept Map III

Using the following terms, fill in the circled, numbered, blank spaces to correctly complete the concept map. Use each term only once.

autonomic centers   ventral   cerebrospinal fluid   preoptic area
thalamus   geniculates   anterior   pineal gland

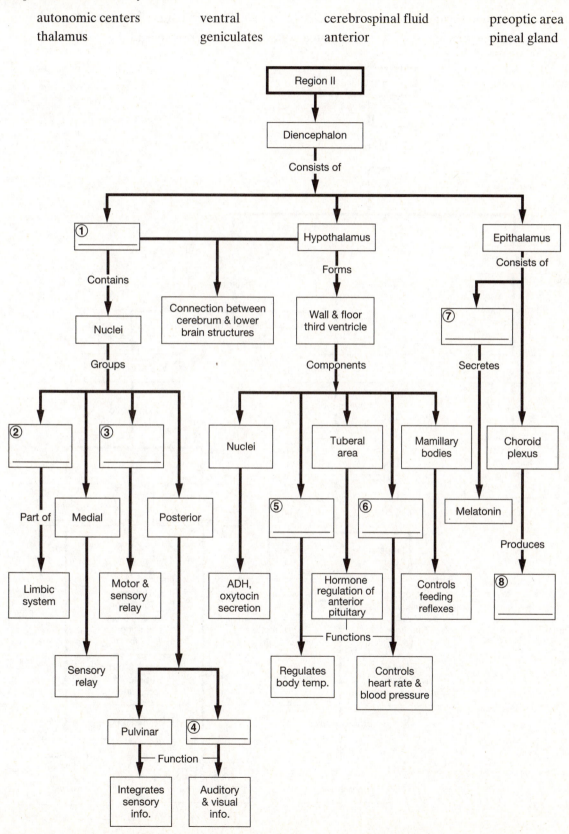

Concept Map IV

Using the following terms, fill in the circled, numbered, blank spaces to correctly complete the concept map. Use each term only once.

corpora quadrigemina   red nucleus   gray matter
cerebral peduncles   substantia nigra   inferior colliculi

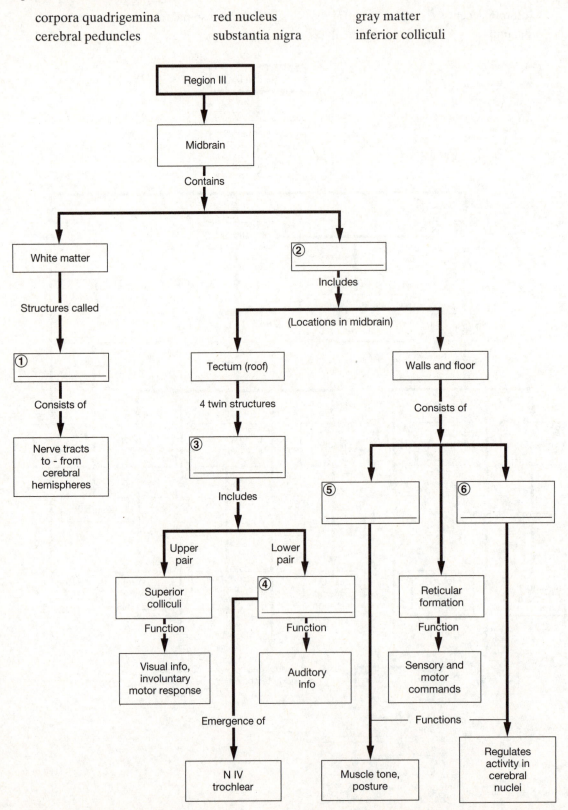

Concept Map V

Using the following terms, fill in the circled, numbered, blank spaces to correctly complete the concept map. Use each term only once.

Purkinje cells   cerebellar nuclei   gray matter
vermis   arbor vitae

```
                        ┌──────────────────┐
                        │    Region IV     │
                        └──────────────────┘
                                  │
                                  ▼
                        ┌──────────────────┐   Membrane     ┌──────────────┐
                        │    Cerebellum    │───covering────▶│  Tentorium   │
                        └──────────────────┘                │  cerebelli   │
                                  │                          └──────────────┘
                             Consists of
                                  │
                                  ▼
                ┌──────────────────────┐   Separated    ┌──────────────┐
                │ 2 Lateral cerebellar │────by────────▶ │ ①            │
                │     hemispheres      │                │ _____ │
                └──────────────────────┘                └──────────────┘
                                  │
                             Consists of
                                  │
            ┌─────────────────────┴─────────────────────┐
            ▼                                            ▼
    ┌──────────────┐                           ┌──────────────┐
    │ White matter │                           │ ③           │
    └──────────────┘                           │ _____   │
            │                                   └──────────────┘
        Contains                                     Contains
            │                                            │
            ▼                              ┌─────────────┴─────────────┐
    ┌──────────────┐                       ▼                           ▼
    │ ②           │              ┌──────────────┐           ┌──────────────┐
    │ _____   │              │  Cerebellar  │           │ ⑤           │
    └──────────────┘              │    cortex    │           │ _____   │
            │                     └──────────────┘           └──────────────┘
        Connects                         │                           │
            │                        Contains               Embedded within
            ▼                            │                           │
    ┌──────────────┐            ┌──────────────┐           ┌──────────────┐
    │  Cerebellar  │            │ ④           │           │    Arbor     │
    │    cortex    │            │              │           │    vitae     │
    │  and nuclei  │            │ _____   │           └──────────────┘
    └──────────────┘            └──────────────┘                   │
            │                            │                         │
          With                           └───────────┬─────────────┘
            │                                    Functions
            ▼                                         │
    ┌──────────────┐                                  ▼
    │  Cerebellar  │                      ┌────────────────────────┐
    │  peduncles   │                      │ Involuntary coordination│
    └──────────────┘                      │  and control of ongoing │
                                          │  movements of body      │
                                          │         parts           │
                                          └────────────────────────┘
```

Concept Map VI

Using the following terms, fill in the circled, numbered, blank spaces to correctly complete the concept map. Use each term only once.

inferior   transverse fibers   superior
apneustic   white matter   respiratory centers

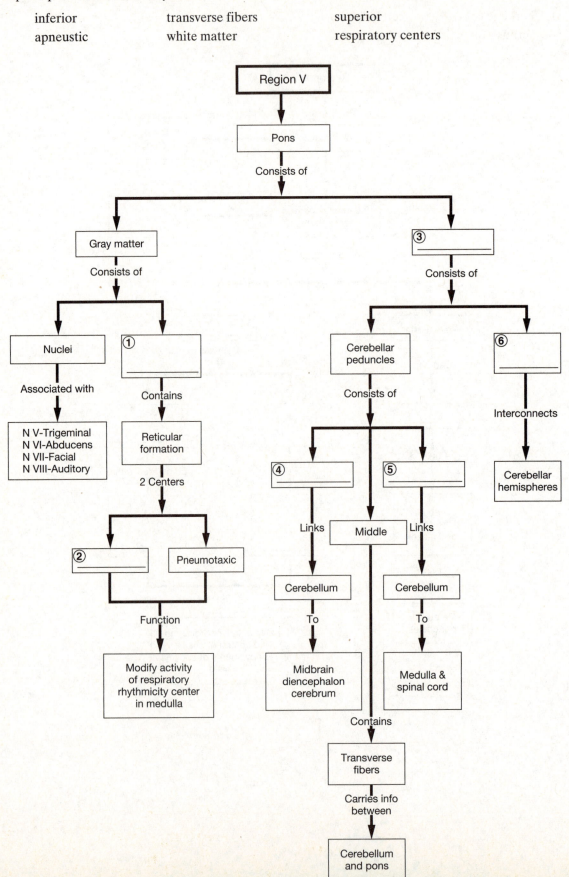

Concept Map VII

Using the following terms, fill in the circled, numbered, blank spaces to correctly complete the concept map. Use each term only once.

cardiac   cuneatus   reflex centers
brain and spinal cord   olivary   white matter
distribution of blood flow   respiratory rhythmicity centers

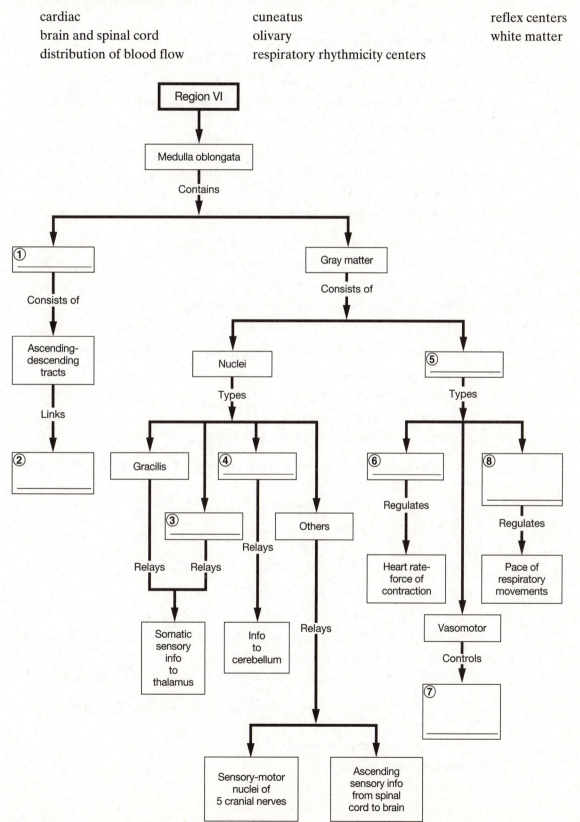

Multiple Choice

Place the letter corresponding to the best answer in the space provided.

_____ 1. The versatility of the brain to respond to stimuli is greater than that of the spinal cord because of the

    a. location of the brain in the cranial vault.

    b. size of the brain and the number of myelinated neurons.

    c. fast-paced processing centers located in the brain.

    d. number of neurons and complex interconnections between the neurons.

_____ 2. The midbrain processes

    a. visual and auditory information and generates involuntary motor responses.

    b. emotions, sensations, memory, and visceral motor control.

    c. voluntary motor activities and sensory information.

    d. sensory information from the cerebellum to the cerebrum.

_____ 3. An individual with a damaged *visual association area*

    a. is incapable of receiving somatic sensory information.

    b. cannot read because he or she is blind.

    c. is red-green color blind and experiences glaucoma.

    d. can see letters clearly, but cannot recognize or interpret them.

_____ 4. The three major groups of axons that comprise the central white matter are

    a. motor, sensory, and interneurons.

    b. unipolar, bipolar, and multipolar.

    c. reticular, caudate, and arcuate fibers.

    d. association, commissural, and projection fibers.

_____ 5. The masses of gray matter that lie between the bulging surface of the insula and the lateral wall of the diencephalon are the

    a. amygdaloid, lentiform, and caudate.

    b. putamen and globus pallidus.

    c. formix, gyrus, and corpus striatum.

    d. cingulate, reticular, and hippocampus.

_____ 6. The functions of the limbic system involve

    a. unconscious processing of visual and auditory information.

    b. adjusting muscle tone and body position to prepare for a voluntary movement.

    c. emotional states and related behavioral drives.

    d. controlling reflex movements and processing of conscious thought.

_____ 7. The mamillary bodies in the floor of the hypothalamus contain motor nuclei that control the reflex movements involved with

    a. chewing, licking, and swallowing.

    b. touch, pressure, and pain.

    c. eye movements and auditory responses.

    d. taste and temperature responses.

_____   8. Hypothalamic or thalamic stimulation that depresses reticular formation activity in the brain stem results in

  a. heightened alertness and a generalized excitement.

  b. emotions of fear, rage, and pain.

  c. sexual arousal and pleasure.

  d. generalized lethargy or actual sleep.

_____   9. The pineal gland, an endocrine structure that secretes the hormone melatonin, is found in the

  a. hypothalamus.

  b. thalamus.

  c. epithalamus.

  d. brain stem.

_____   10. The pulvinar nuclei of the thalamus integrate

  a. visual information from the eyes and auditory signals from the ears.

  b. sensory information for projection to the association areas of the cerebral cortex.

  c. sensory data concerning touch, pressure, and pain.

  d. sensory information for relay to the frontal lobes.

_____   11. The part(s) of the diencephalon responsible for coordination of activities of the central nervous system and the endocrine system is (are) the

  a. thalamus.

  b. hypothalamus.

  c. epithalamus.

  d. a, b, and c are correct.

_____   12. The hypothalamus produces and secretes the hormones

  a. ADH and oxytocin.

  b. adrenalin and growth hormone.

  c. estrogen and testosterone.

  d. prolactin and FSH.

_____   13. The preoptic area of the hypothalamus controls the

  a. visual responses of the eye.

  b. dilation and constriction of the iris in the eye.

  c. emotions of rage and aggression.

  d. physiological responses to changes in body temperature.

_____   14. The nerve fiber bundles on the ventrolateral surfaces of the midbrain are the

  a. corpora quadrigemina.

  b. substantia nigra.

  c. cerebral peduncles.

  d. inferior and superior colliculi.

_____   15. The effortless serve of a tennis player is a result of establishing

  a. cerebral motor responses.

  b. cerebellar motor patterns.

  c. sensory patterns within the medulla.

  d. motor patterns in the pons.

_____ 16. Huge, highly branched Purkinje cells are found in the

    a. cerebellar cortex.

    b. pons.

    c. medulla oblongata.

    d. cerebral cortex.

_____ 17. The centers in the pons that modify the activity of the respiratory rhythmicity center in the medulla oblongata are the

    a. vasomotor and cardiac centers.

    b. ascending and descending tracts.

    c. apneustic and pneumotaxic centers.

    d. nucleus gracilis and nucleus cuneatus.

_____ 18. The cardiovascular centers and the respiratory rhythmicity center are found in the

    a. cerebrum.

    b. cerebellum.

    c. pons.

    d. medulla oblongata.

_____ 19. The large veins found between the inner and outer layers of the dura mater are called the

    a. dural venules.

    b. dural sinuses.

    c. dural falx.

    d. venous sinuses.

_____ 20. Neural tissue in the CNS is isolated from the general circulation by the

    a. blood–brain barrier.

    b. falx cerebri.

    c. corpus striatum.

    d. anterior commissures.

_____ 21. The CSF reaches the subarachnoid space via

    a. the subdural sinuses.

    b. the choroid plexus.

    c. three holes in the fourth ventricle.

    d. the superior sagittal sinus.

_____ 22. The blood–brain barrier remains intact throughout the CNS, except in

    a. the pons and the medulla.

    b. the thalamus and the midbrain.

    c. portions of the cerebrum and the cerebellum.

    d. portions of the hypothalamus and the choroid plexus.

_____ 23. The ventricles in the brain form hollow chambers that serve as passageways for the circulation of

    a. blood.

    b. cerebrospinal fluid.

    c. interstitial fluid.

    d. lymph.

_____ 24. The central white matter of the cerebrum is found

    a. in the superficial layer of the neural cortex.

    b. beneath the neural cortex and around the basal nuclei.

    c. in the deep basal nuclei and the neural cortex.

    d. in the cerebral cortex and in the basal nuclei.

_____ 25. The series of elevated ridges that increase the surface area of the cerebral hemispheres and the number of neurons in the cortical area are called

    a. sulci.

    b. fissures.

    c. gyri.

    d. a, b, and c are correct.

_____ 26. The cranial nerves responsible for all aspects of eye function include

    a. I, II, and VII.

    b. V, VII, and VIII.

    c. II, III, IV, V, and VI.

    d. IX, XI, and XII.

_____ 27. If a person has damage to the trochlear nerve of the right eye, the person would be unable to

    a. roll the eye upward and to the right (laterally).

    b. look upward or to the right or left.

    c. look downward.

    d. blink the eye.

_____ 28. If a person has Bell's palsy, a cranial nerve disorder that results from an inflammation of a facial nerve, a typical sign and symptom is

    a. a loss of balance and equilibrium.

    b. the inability to blink the eyes.

    c. loss of taste sensations from the anterior two-thirds of the tongue.

    d. the inability to control the amount of food that can be swallowed.

_____ 29. A lesion in the hypothalamus may cause a person to have a difficult time

    a. remembering past events.

    b. regulating the rate of respiratory movements.

    c. controlling the expression of emotions.

    d. processing visual and auditory information.

Completion

Using the terms below, complete the following statements. Use each term only once.

fissures	hypothalamus	corpus striatum	aphasia
drives	hippocampus	fornix	dyslexia
thalamus	pituitary gland	spinal cord	global aphasia
commissural	cerebral aqueduct	third ventricle	dopamine
sulci	arcuate	Wernicke	arousal

1. The walls of the diencephalon are formed by the _____.

2. Integration of the nervous and endocrine systems results from the link between the hypothalamus and the _____.

3. The shortest association fibers in the central white matter are called _____ fibers.

4. The fibers that permit communication between the two cerebral hemispheres are called _____ fibers.

5. The caudate and lentiform nuclei are collectively referred to as the _____.

6. The part of the limbic system that appears to be important in learning and storage of long-term memory is the _____.

7. The tract of white matter that connects the hippocampus with the hypothalamus is the _____.

8. Unfocused "impressions" originating in the hypothalamus are called _____.

9. Because the cerebrum has a pair of lateral ventricles, the diencephalic chamber is called the _____.

10. Instead of a ventricle, the midbrain has a slender canal known as the _____.

11. In the caudal half of the medulla, the fourth ventricle narrows and becomes continuous with the central canal of the _____.

12. The shallow depressions that separate the cortical surface of the cerebral hemispheres are called _____.

13. Deep grooves separating the cortical surface of the cerebral hemispheres are called _____.

14. The interpretive area in the left cerebral hemisphere that plays an essential role in your personality by integrating sensory information and coordinating access to complex visual and auditory memories is called the _____ area.

15. The component of the brain that integrates with the endocrine system is the _____.

16. Extensive damage to the general interpretive area or to the associated sensory tracts results in _____.

17. A disorder affecting the ability to speak or read is _____.

18. A disorder affecting the comprehension and use of words is referred to as _____.

19. One of the functions of the reticular formation is _____.

20. Parkinson's disease is associated with inadequate production of the neurotransmitter _____.

Short Essay

Briefly answer the following questions in the spaces provided below.

1. Why is the brain more versatile than the spinal cord when responding to stimuli?

2. List the six major regions in the adult brain.

3. Where is the limbic system, and what are its functions?

4. What are the hypothalamic contributions to the endocrine system?

5. What are the two primary functions of the cerebellum?

6. What are the four primary structural components of the pons?

7. What is the primary function of the blood–brain barrier?

8. What are the three major vital functions of the cerebrospinal fluid?

9. What cranial nerves are classified as "special sensory"?

10. Why is examination of the cranial reflexes clinically important?

LEVEL 3: CRITICAL THINKING AND CLINICAL APPLICATIONS

Using principles and concepts learned in Chapter 14, answer the following questions. Write your answers on a separate sheet of paper.

1. How does a blockage of cerebrospinal fluid (CSF) in an arachnoid granulation cause irreversible brain damage?
2. During the action in a football game, one of the players was struck with a blow to the head, knocking him unconscious. The first medics to arrive at his side immediately subjected him to smelling salts. Why?
3. After taking a walk you return home and immediately feel the urge to drink water because you are thirsty. What part of the brain is involved in the urge to drink because of thirst?
4. Steve has been stopped by a highway patrolman on suspicion of driving while intoxicated. It appears that he cannot stand or sit without assistance. What part of the brain is affected, and what results from using a drug like alcohol?
5. Suppose you are participating in a contest where you are blindfolded and numerous objects are placed in your hands to identify. The person in the contest who identifies the most objects while blindfolded receives a prize. What neural activities and areas in the CNS are necessary in order for you to name the objects correctly and win the prize?

CHAPTER 15

Neural Integration I: Sensory Pathways and the Somatic Nervous System

OVERVIEW

You probably have heard the statement that "No man is an island; no man can stand alone," implying that humans are societal creatures, depending on and interacting with other people in the world in which they live. So it is with the parts of the human nervous system.

The many diverse and complex physical and mental activities require that all component parts of the nervous system function in an integrated and coordinated way if homeostasis is to be maintained within the human body.

Chapters 13 and 14 considered the structure and functions of the spinal cord and the brain. Chapter 15 focuses on how communication processes take place via pathways, nerve tracts, and nuclei that relay sensory information to the CNS and the output of somatic motor commands. The following exercises are designed to test your understanding of patterns and principles of the brain's sensory pathways and the somatic nervous system.

LEVEL 1: REVIEWING FACTS AND TERMS

Review of Learning Outcomes

After completing this chapter, you should be able to do the following:

OUTCOME 15-1   Specify the components of the afferent and efferent divisions of the nervous system, and explain what is meant by the somatic nervous system.

OUTCOME 15-2   Explain why receptors respond to specific stimuli, and how the organization of a receptor affects its sensitivity.

OUTCOME 15-3   Identify the receptors for the general senses, and describe how they function.

OUTCOME 15-4   Identify the major sensory pathways, and explain how it is possible to distinguish among sensations that originate in different areas of the body.

OUTCOME 15-5   Describe the components, processes, and functions of the somatic motor pathways, and the levels of information processing involved in motor control.

Multiple Choice

Place the letter corresponding to the best answer in the space provided.

OUTCOME 15-1 _____ 1. The primary sensory cortex of the cerebral hemispheres or areas of the cerebellar hemispheres receive _____ information.

    a. visceral sensory

    b. visceral motor

    c. somatic sensory

    d. somatic motor

OUTCOME 15-1 _____ 2. Visceral sensory information is distributed primarily to reflex centers in the

    a. cerebral cortex.

    b. brain stem and diencephalon.

    c. cerebellar hemispheres.

    d. skeletal muscles.

OUTCOME 15-1 _____ 3. The somatic nervous system (SNS) consists of

    a. motor neurons and pathways that control skeletal muscles.

    b. ascending, descending, and transverse tracts.

    c. sensory nuclei of the cranial nerves.

    d. mamillary bodies that process sensory information.

OUTCOME 15-1 _____ 4. The somatic motor portion of the efferent division of the nervous system is responsible for control of

    a. visceral sensory information.

    b. somatic sensory information.

    c. higher-order function.

    d. skeletal muscles.

OUTCOME 15-2 _____ 5. Regardless of the nature of the stimulus, sensory information must be sent to the CNS in the form of

    a. perceptions.

    b. action potentials.

    c. sensations.

    d. receptor specificity.

OUTCOME 15-2 _____ 6. Free nerve endings can be stimulated by many different stimuli because they exhibit

    a. few accessory structures.

    b. little information about the location of a stimulus.

    c. little receptor specificity.

    d. little ability to initiate transduction.

OUTCOME 15-2 _____ 7. The receptors that provide information about the intensity and rate of change of a stimulus are called

    a. generator potentials.

    b. receptor potentials.

    c. tonic receptors.

    d. phasic receptors.

OUTCOME 15-2 _____ 8. Each receptor has a characteristic sensitivity.
  a. True
  b. False

OUTCOME 15-3 _____ 9. Nociceptors, common in the skin, joint capsules, and around the walls of blood vessels, are sensory receptors for
  a. temperature.
  b. pain.
  c. chemical concentration.
  d. physical distortion.

OUTCOME 15-3 _____ 10. The general sensory receptors that respond to physical distortion are the
  a. nociceptors.
  b. thermoreceptors.
  c. mechanoreceptors.
  d. chemoreceptors.

OUTCOME 15-3 _____ 11. Thermoreceptors are phasic receptors because they are
  a. slow when adapting to a stable temperature.
  b. inactive when the temperature is changing.
  c. unaffected by temperature sensations.
  d. very active when the temperature is changing.

OUTCOME 15-3 _____ 12. Tactile receptors provide sensations of
  a. joint and muscle movement.
  b. touch, pressure, and vibration.
  c. conscious perception of pain.
  d. pressure changes in the walls of blood vessels.

OUTCOME 15-3 _____ 13. Pressure changes in the wall of a blood vessel are monitored by
  a. Ruffini corpuscles.
  b. Pacinian corpuscles.
  c. baroreceptors.
  d. Merkel's discs.

OUTCOME 15-3 _____ 14. The pH, carbon dioxide, and oxygen concentrations of arterial blood are monitored by
  a. proprioceptors.
  b. mechanoreceptors.
  c. chemoreceptors.
  d. nociceptors.

OUTCOME 15-4 _____ 15. The three major somatic sensory pathways are the
  a. posterior column, spinothalamic, and spinocerebellar.
  b. first-, second-, and third-order pathways.
  c. nuclear, cerebellar, and thalamic.
  d. anterior, posterior, and lateral spinothalamic.

OUTCOME 15-4 _____ 16. The axons of the posterior column ascend within the

    a. fasciculus gracilis and fasciculus cuneatus.

    b. posterior and lateral spinothalamic tracts.

    c. posterior and interior spinocerebellar tracts.

    d. a, b, and c are correct.

OUTCOME 15-4 _____ 17. The spinothalamic pathway consists of

    a. anterior and posterior tracts.

    b. gracilis and cuneatus nuclei.

    c. proprioceptors and interneurons.

    d. lateral and anterior tracts.

OUTCOME 15-5 _____ 18. Proprioceptive data from peripheral structures, visual information from the eyes, and equilibrium-related sensations are processed and integrated in the

    a. cerebral cortex.

    b. cerebellum.

    c. red nucleus.

    d. basal nuclei.

OUTCOME 15-5 _____ 19. The integrative activities performed by neurons in the cerebellar cortex and basal nuclei are essential to the

    a. involuntary regulation of posture and muscle tone.

    b. precise control of voluntary and involuntary movements.

    c. involuntary regulation of autonomic functions.

    d. voluntary control of smooth and cardiac muscle.

OUTCOME 15-5 _____ 20. The somatic nervous system issues somatic motor commands that direct the

    a. activities of the autonomic nervous system.

    b. contractions of smooth and cardiac muscles.

    c. activities of glands and fat cells.

    d. control of skeletal muscles.

OUTCOME 15-5 _____ 21. The upper motor neuron of a somatic motor pathway has a cell body that lies in a(n)

    a. nucleus of the brain stem.

    b. nucleus of the spinal cord.

    c. area outside the CNS.

    d. CNS processing center.

OUTCOME 15-5 _____ 22. The three integrated pathways controlling conscious and subconscious motor commands in skeletal muscle are the

    a. vestibulospinal, tectospinal, and reticulospinal.

    b. corticospinal, medial, and lateral.

    c. corticobulbar, tectospinal, and rubrospinal.

    d. cerebellum, corticospinal, and reticulospinal.

OUTCOME 15-5 _____ 23. The motor tracts in the spinal cord controlling subconscious regulation of balance and muscle tone are the _____ tracts.

     a. vestibulospinal

     b. tectospinal

     c. reticulospinal

     d. corticobulbar

OUTCOME 15-5 _____ 24. The background patterns of movement involved in voluntary motor activities are controlled by

     a. superior and inferior colliculi.

     b. vestibular nuclei.

     c. basal nuclei.

     d. the pyramidal system.

OUTCOME 15-5 _____ 25. The center of somatic motor control that plans and initiates voluntary motor activity is the

     a. cerebellum.

     b. hypothalamus.

     c. basal nuclei.

     d. cerebral cortex.

OUTCOME 15-5 _____ 26. The center of somatic motor control that makes the movement efficient, smooth, and precisely controlled is the

     a. cerebral cortex.

     b. cerebellum.

     c. medulla oblongata.

     d. thalamus.

OUTCOME 15-5 _____ 27. The center of somatic motor control that controls basic respiratory reflexes is the

     a. cerebellum.

     b. medulla oblongata.

     c. cerebrum.

     d. midbrain.

Completion

Using the terms below, complete the following statements. Use each term only once.

cerebral hemispheres	generator potential	proprioceptors	diencephalon
transduction	adaptation	solitary nucleus	lateral
hypothalamus	first order	Merkel's discs	sensory homunculus
corticospinal	substance P	cerebellum	Purkinje cells
thalamus	corticobulbar	spinothalamic	lamellated corpuscles
interoceptors	spinal cord	muscle spindles	cerebral peduncles
sensory receptors			

OUTCOME 15-1   1. Specialized cells that monitor specific conditions in the body or the external environment are called _____.

OUTCOME 15-1   2. Visceral sensory information is distributed primarily to reflex centers in the brain stem and _____.

OUTCOME 15-2   3. The translation of a stimulus into an action potential that can be conducted to the CNS is called _____.

OUTCOME 15-2   4. A receptor potential large enough to produce an action potential is called a(n) _____.

OUTCOME 15-2   5. A reduction in sensitivity in the presence of a constant stimulus is _____.

OUTCOME 15-3   6. Receptors that monitor visceral organs and internal conditions are classified as _____.

OUTCOME 15-3   7. In addition to the use of the amino acid glutamate, afferent sensory neurons relaying pain sensations utilize the neuropeptide neurotransmitter _____.

OUTCOME 15-3   8. Receptors that provide awareness about skeletal muscles and joints are the _____.

OUTCOME 15-3   9. Fine touch and pressure receptors called tactile discs are also called _____.

OUTCOME 15-3   10. Fast-adapting receptors sensitive to deep pressure are called _____.

OUTCOME 15-3   11. Proprioceptors that monitor skeletal muscle length and trigger stretch reflexes are the _____.

OUTCOME 15-4   12. A sensory neuron that delivers the sensation to the CNS is referred to as a(n) _____ neuron.

OUTCOME 15-4   13. If a sensation is to reach our awareness, a second-order neuron must synapse with a third-order neuron in the _____.

OUTCOME 15-4   14. Conscious sensations of poorly localized "crude" touch, pressure, pain, and temperature are provided by the _____ pathway.

OUTCOME 15-4   15. The information carried by the spinocerebellar pathway ultimately arrives at the _____ of the cerebellar cortex.

OUTCOME 15-4   16. The medulla oblongata contains a major processing and sorting center for visceral sensory information called the _____.

OUTCOME 15-4   17. A functional map of the primary sensory cortex is called a _____.

OUTCOME 15-5   18. Conscious control over skeletal muscles that move the eye, jaw, face, and neck is provided by the _____ tracts.

OUTCOME 15-5   19. The "pyramids" are a pair of thick bands of axons comprising the _____ tract.

OUTCOME 15-5   20. The three pairs of descending tracts of the corticospinal pathway emerge on either side of the midbrain as the _____.

OUTCOME 15-5   21. The components that help control the distal limb muscles that perform more precise movements are those of the _____ pathway.

OUTCOME 15-5   22. Motor patterns that are related to eating, drinking, and sexual activity are controlled by the _____.

OUTCOME 15-5   23. Simple cranial and spinal reflexes are controlled by the brain stem and _____.

OUTCOME 15-5   24. Most complex and variable motor activities are directed by the primary motor cortex of the _____.

OUTCOME 15-5   25. Coordination and feedback control over muscle contractions is directed by the basal nuclei and _____.

Matching

Match the terms in column B with the terms in column A. Use letters for answers in the spaces provided.

Part I

		Column A	Column B
OUTCOME 15-1	_____	1. afferent division	A. crossing of an axon
OUTCOME 15-1	_____	2. efferent division	B. link from receptor to cortical neuron
OUTCOME 15-2	_____	3. sensation	C. crude touch and pressure
OUTCOME 15-2	_____	4. labeled line	D. physical distortion
OUTCOME 15-2	_____	5. phasic receptors	E. map of primary sensory cortex
OUTCOME 15-3	_____	6. nociceptors	F. motor pathways
OUTCOME 15-3	_____	7. mechanoreceptors	G. arriving information
OUTCOME 15-4	_____	8. decussation	H. fast adapting
OUTCOME 15-4	_____	9. sensory homunculus	I. sensory pathways
OUTCOME 15-4	_____	10. anterior spinothalamic tract	J. pain receptors

Part II

		Column A	Column B
OUTCOME 15-4	_____	11. Purkinje cells	K. gross movements of trunk and proximal limb muscles
OUTCOME 15-4	_____	12. solitary nucleus	L. stimulating neurotransmitter in basal nuclei
OUTCOME 15-5	_____	13. pyramidal system	M. visual and auditory sensations
OUTCOME 15-5	_____	14. medial pathway	N. coordinates complex motor patterns
OUTCOME 15-5	_____	15. vestibular nuclei	O. corticospinal pathways
OUTCOME 15-5	_____	16. tectospinal tracts	P. not fully functional at birth
OUTCOME 15-5	_____	17. acectylcholine (ACh)	Q. cerebellar cortex
OUTCOME 15-5	_____	18. gamma aminobutyric acid (GABA)	R. maintain posture and balance
OUTCOME 15-5	_____	19. cerebellum	S. inhibitory neurotransmitter in basal nuclei
OUTCOME 15-5	_____	20. cerebral cortex	T. medulla oblongata

Drawing/Illustration Labeling

Identify each numbered structure by labeling the following figure. Select your answers from the following choices:

posterior spinocerebellar	fasciculus gracilis	vestibulospinal
tectospinal	reticulospinal	rubrospinal
lateral corticospinal	anterior corticospinal	lateral spinothalamic
anterior spinothalamic	anterior spinocerebellar	fasciculus cuneatus

OUTCOME 15-4, 15-5   **FIGURE 15-1**   Ascending and Descending Tracts of the Spinal Cord

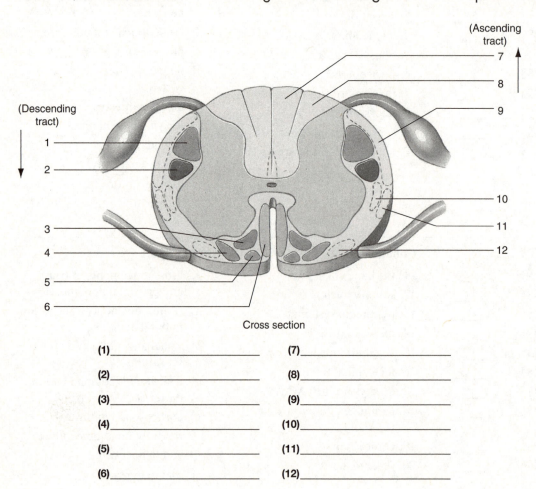

Cross section

(1)_____	(7)_____
(2)_____	(8)_____
(3)_____	(9)_____
(4)_____	(10)_____
(5)_____	(11)_____
(6)_____	(12)_____

From the labels on the drawing of the skin, identify the tactile receptors by placing the answers in spaces 1–6.

OUTCOME 15-3 **FIGURE 15-2** Tactile Receptors in the Skin

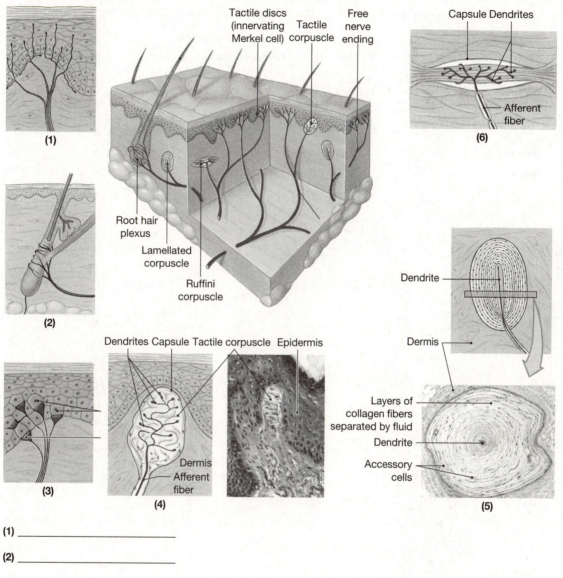

(1) _____

(2) _____

(3) _____

(4) _____

(5) _____

(6) _____

LEVEL 2: REVIEWING CONCEPTS

Chapter Overview

Using the terms below, identify the numbered areas to correctly complete the chapter overview of one of the principal descending motor pathways: the corticospinal pathway. Use each term only once.

decussation of pyramids   to skeletal muscles   midbrain

left cerebral hemisphere   medulla oblongata   spinal cord

motor nuclei of cranial nerves

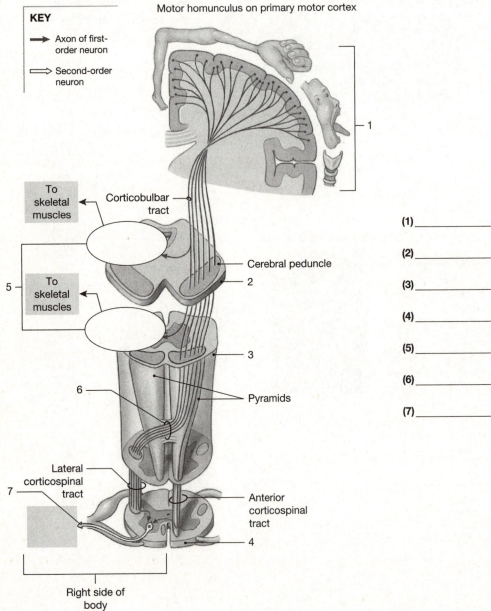

KEY

→ Axon of first-order neuron

⇒ Second-order neuron

Motor homunculus on primary motor cortex

1

To skeletal muscles

Corticobulbar tract

Cerebral peduncle

2

To skeletal muscles

5

3

6

Pyramids

Lateral corticospinal tract

7

Anterior corticospinal tract

4

Right side of body

(1)_____

(2)_____

(3)_____

(4)_____

(5)_____

(6)_____

(7)_____

Concept Map I

Using the following terms, fill in the circled, numbered, blank spaces to correctly complete the concept map. Use each term only once.

spinothalamic
fasciculus cuneatus
posterior column
medulla (nucleus gracilis)

posterior tract
thalamus (ventral nuclei)
proprioceptors
anterior tract

interneurons
cerebellum (cortex)

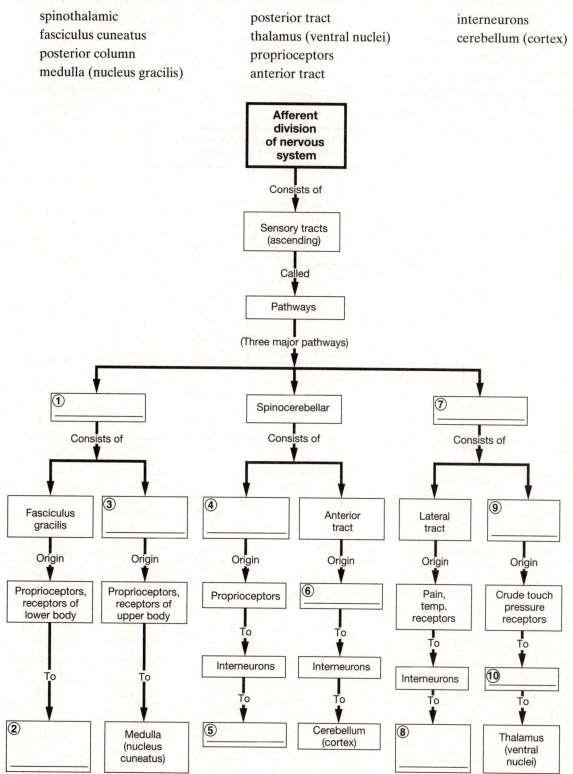

Concept Map II

Using the following terms, fill in the circled, numbered, blank spaces to correctly complete the concept map. Use each term only once.

vestibular nuclei

anterior corticospinal tracts

pyramids

brain stem

control of muscle tone and pre-
    cise distal limb movements

reticulospinal tracts

control of skeletal
    muscles

cranial nerves

lateral pathway

basal nuclei

tectospinal tracts

uncrossed

control of muscle tone
    and balance

regulation of reflex activity

regulation of eye, head, neck,
    and upper limb position

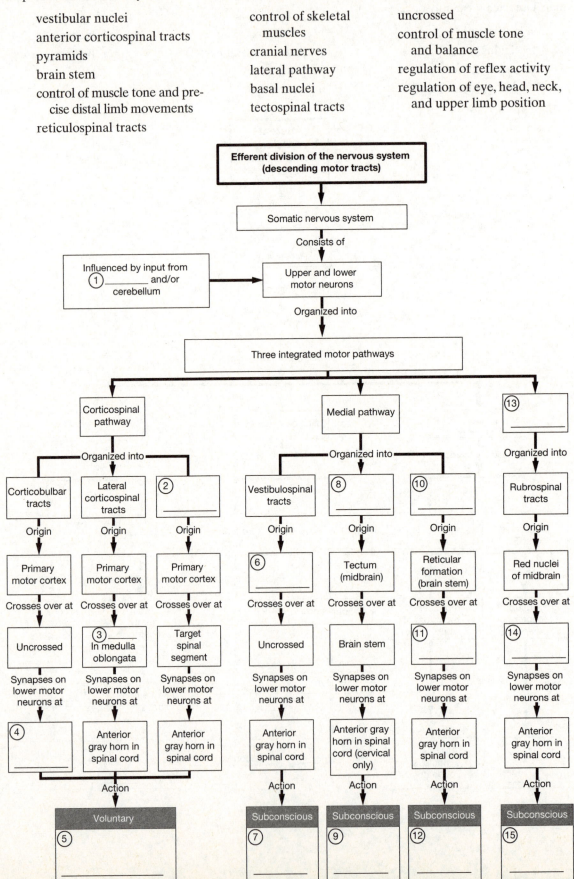

Multiple Choice

Place the letter corresponding to the best answer in the space provided.

_____ 1. If a tract name begins with *spino-*, it must start in the

    a. brain and end in the PNS, bearing motor commands.

    b. spinal cord and end in the brain, bearing motor commands.

    c. spinal cord and end in the PNS, carrying sensory information.

    d. spinal cord and end in the brain, carrying sensory information.

_____ 2. If the name of a tract ends in *-spinal*, its axons must start in the

    a. higher centers and end in the spinal cord, bearing motor commands.

    b. higher centers and end in the spinal cord, bearing sensory information.

    c. spinal cord and end in the brain, carrying sensory information.

    d. spinal cord and end in the PNS, carrying motor commands.

_____ 3. The posterior column pathway receives sensations associated with

    a. pain and temperature.

    b. crude touch and pressure.

    c. highly localized fine touch, pressure, vibration, and position.

    d. a, b, and c are correct.

_____ 4. In the thalamus, data arriving over the posterior column pathway are integrated, sorted, and projected to the

    a. cerebellum.

    b. spinal cord.

    c. primary sensory cortex.

    d. peripheral nervous system (PNS).

_____ 5. If a sensation arrives at the wrong part of the sensory cortex, we will

    a. experience pain in the posterior column pathway.

    b. reach an improper conclusion about the source of the stimulus.

    c. be incapable of experiencing pain or pressure.

    d. lose all capability of receiving and sending information.

_____ 6. The presence of abnormal sensations such as the pins-and-needles sensation when an arm or leg "falls asleep" as a result of pressure on a peripheral nerve is called

    a. paresthesia.

    b. hypesthesia.

    c. two-point discrimination.

    d. decussation.

_____ 7. The spinothalamic pathway relays impulses associated with

    a. "fine" touch, pressure, vibration, and proprioception.

    b. "crude" sensations of touch, pressure, pain, and temperature.

    c. the position of muscles, tendons, and joints.

    d. proprioceptive information and vibrations.

_____ 8. The spinocerebellar pathway includes

    a. lateral and anterior tracts.

    b. gracilis and cuneatus nuclei.

    c. lateral and posterior tracts.

    d. anterior and posterior tracts.

_____ 9. The spinocerebellar pathway carries information concerning the

    a. pressure on each side of the body.

    b. sensations that cause referred and phantom limb pain.

    c. position of muscles, tendons, and joints to the cerebellum.

    d. sensations of touch, pain, and temperature.

_____ 10. Somatic motor pathways always involve

    a. an upper and lower motor neuron.

    b. a ganglionic and preganglionic neuron.

    c. sensory and motor fibers.

    d. anterior and lateral nuclei.

_____ 11. Voluntary and involuntary somatic motor commands issued by the brain reach peripheral targets by traveling over the

    a. sensory and motor fibers.

    b. corticospinal, medial, and lateral pathways.

    c. ganglionic and preganglionic fibers.

    d. spinothalamic tracts.

_____ 12. The primary goal of the vestibular nuclei is the

    a. monitoring of muscle tone throughout the body.

    b. triggering of visual and auditory stimuli.

    c. maintenance of posture and balance.

    d. control of involuntary eye movements.

_____ 13. The reticulospinal tract is involved with regulation of

    a. involuntary reflex activity and autonomic functions.

    b. involuntary regulation of balance and posture.

    c. balance and muscle tone.

    d. voluntary motor control of skeletal muscles.

_____ 14. An individual whose primary motor cortex has been destroyed retains the ability to walk and maintain balance but the movements

    a. are restricted and result in partial paralysis.

    b. are under involuntary control and are poorly executed.

    c. are characteristic involuntary motor commands.

    d. lack precision and are awkward and poorly controlled.

_____ 15. Basal nuclei create background patterns of movement by

    a. modulation of upper motor neuron activity.

    b. indirectly changing the sensitivity of pyramidal cells in the primary motor cortex.

    c. altering the excitatory or inhibitory output of the reticulospinal tracts.

    d. a, b, and c are correct.

_____ 16. When someone touches a hot stove, the rapid, automatic preprogrammed response that preserves homeostasis is provided by

    a. the cerebral cortex.

    b. the primary sensory cortex.

    c. a spinal reflex.

    d. the cerebellum.

_____ 17. At the highest level of processing, the complex, variable, and voluntary motor patterns are dictated by the

    a. cerebral cortex.

    b. frontal lobe.

    c. cerebellum.

    d. diencephalon.

_____ 18. Changes in pressure or movement stimulate

    a. pain receptors.

    b. mechanoreceptors.

    c. photoreceptors.

    d. chemoreceptors.

_____ 19. The CNS interprets the type of stimulus involved in a sensory transmission on the basis of the

    a. labeled line over which it traveled.

    b. type of receptor activated.

    c. speed at which the impulse travels.

    d. context within which the impulse is generated.

_____ 20. Sensory receptors for all of the following adapt to repeated stimulation by sending fewer and fewer impulses, except those for

    a. heat.

    b. light.

    c. pain.

    d. touch.

_____ 21. Receptors for the somatic senses are located in the

    a. skin.

    b. muscles.

    c. joints.

    d. all of these.

_____ 22. Tactile corpuscles and lamellated corpuscles are sensitive to

    a. touch and pressure.

    b. pain.

    c. heat.

    d. light.

Completion

Using the terms below, complete the following statements. Use each term only once.

cerebellum	thalamus	basal nuclei	homunculus
deep pressure	anencephaly	descending	nociceptors
GABA	posterior	Meissner's	primary sensory cortex
interoceptors	medial lemniscus	ascending	

1. The vestibulospinal tracts in the spinal cord consist of _____ fibers.

2. The spinothalamic tracts in the spinal cord consist of _____ fibers.

3. Basal nuclei inhibit motor neurons by releasing _____.

4. A rare condition in which the brain fails to develop at levels above the midbrain or lower diencephalon is called _____.

5. Motor patterns associated with walking and body positioning are controlled by _____.

6. Tactile corpuscles are also sometimes called _____ corpuscles.

7. Lamellated corpuscles are responsible for sensations of _____.

8. Pain receptors are more appropriately called _____.

9. The tract that relays information from the nuclei gracilis and cuneatus to the thalamus is the _____.

10. Receptors responsible for maintaining the homeostasis of the blood belong to the class called _____.

11. The general organization of the spinal cord is such that motor tracts are anterior and sensory tracts are _____.

12. The destination of the impulses that travel along the spinocerebellar pathways is the _____.

13. The destination of the impulses that travel along the spinothalamic pathways is the _____.

14. A sensory map created by electricity stimulating the cortical surface is called a sensory _____.

15. The ability to localize a specific stimulus depends on the organized distribution of sensory information to the _____.

Short Essay

Briefly answer the following questions in the spaces provided below.

1. What are the three major somatic sensory pathways in the body?

2. Name the three pairs of descending tracts that make up the corticospinal pathway.

3. List the descending tracts comprising the medial and lateral pathways of the somatic nervous system.

4. What are the two primary functions of the cerebellum?

5. Distinguish between tonic and phasic receptors. Which type is likely to be a fast-adapting receptor? Which type is likely to be a slow-adapting receptor?

6. For sensory information coming from peripheral receptors, how does the CNS determine (a) stimulus type, (b) stimulus location, and (c) stimulus strength, duration, and/or variation?

7. List the general senses and the special senses.

LEVEL 3: CRITICAL THINKING AND CLINICAL APPLICATIONS

Using principles and concepts learned in Chapter 15, answer the following questions. Write your answers on a separate sheet of paper.

1. A friend of yours complains of pain radiating down through the left arm. After having an EKG, it was discovered that there were signs of a mild heart attack. Why would the pain occur in the arm instead of the cardiac area?

2. While on a hunting trip, Sam was accidentally shot in his lower right leg just below the knee. The damage was severe enough to necessitate amputation of his leg below the knee. After removal of the lower limb and foot, he says he can feel pain in his toes even though his foot is gone. What causes these sensations and feelings?

3. Basal nuclei do not exert direct control over CNS motor neurons; however, they do adjust the motor commands issued in other processing centers. What two major pathways do they use to accomplish this?

4. An individual whose primary motor cortex has been destroyed retains the ability to walk, maintain balance, and perform other voluntary and involuntary movements. Even though the movements lack precision and may be awkward and poorly controlled, why is the ability to walk and maintain balance possible?

5. The proud parents of a newborn take the infant home and experience all the joy of watching the baby suckle, stretch, yawn, cry, kick, track movements with the eyes, and stick its fingers in its mouth. Suddenly, after a few weeks, the baby dies for no *apparent* reasons. What do you suspect and how do you explain the suspicious cause of death?

CHAPTER 16

Neural Integration II: The Autonomic Nervous System and Higher-Order Functions

OVERVIEW

Can you imagine what life would be like if you were responsible for coordinating the involuntary activities that occur in the cardiovascular, respiratory, digestive, excretory, and reproductive systems? Your entire life would be consumed with monitoring and giving instructions to maintain homeostasis in the body. The autonomic nervous system (ANS) does this involuntary work with meticulous efficiency.

The ANS consists of efferent motor fibers that innervate smooth and cardiac muscle as well as glands. The controlling centers of the ANS are found in the brain, whereas the nerve fibers of the ANS, which are subdivided into the sympathetic and parasympathetic fibers, belong to the peripheral nervous system. The sympathetic nervous system originates in the thoracic and lumbar regions of the spinal cord, whereas the parasympathetic nervous system originates in the brain and sacral region of the spinal cord. Both divisions terminate on the smooth muscle, cardiac muscle, or glandular tissue of numerous organs.

The primary focus of Chapter 16 is on the structural and functional properties of the sympathetic and parasympathetic divisions of the ANS. Integration and control of autonomic functions are included to show how the ANS coordinates involuntary activities throughout the body and how interactions between the nervous system and other systems make homeostatic adjustments necessary for survival.

LEVEL 1: REVIEWING FACTS AND TERMS

Review of Learning Outcomes

After completing this chapter, you should be able to do the following:

OUTCOME 16-1   Compare the organization of the autonomic nervous system with that of the somatic nervous system.

OUTCOME 16-2   Describe the structures and functions of the sympathetic division of the autonomic nervous system.

OUTCOME 16-3   Describe the mechanisms of sympathetic neurotransmitter release and their effects on target organs and tissues.

OUTCOME 16-4   Describe the structures and functions of the parasympathetic division of the autonomic nervous system.

OUTCOME 16-5   Describe the mechanisms of parasympathetic neurotransmitter release and their effects on target organs and tissues.

OUTCOME 16-6   Discuss the functional significance of dual innervation and autonomic tone.

OUTCOME 16-7   Describe the hierarchy of interacting levels of control in the autonomic nervous system, including the significance of visceral reflexes.

OUTCOME 16-8   Explain how memories are created, stored, and recalled, and distinguish among the levels of consciousness and unconsciousness.

OUTCOME 16-9   Describe some of the ways in which the interactions of neurotransmitters influence brain function.

OUTCOME 16-10   Summarize the effects of aging on the nervous system and give examples of interactions between the nervous system and each of the other organ systems.

Multiple Choice

Place the letter corresponding to the best answer in the space provided.

OUTCOME 16-1   _____   1. A distinct similarity between the organization of the somatic and autonomic nervous systems is that both are
    a. afferent divisions that carry sensory commands.
    b. efferent divisions that carry sensory commands.
    c. efferent divisions that carry motor commands.
    d. afferent divisions that carry motor commands.

OUTCOME 16-1   _____   2. The lower motor neurons of the somatic nervous system (SNS) exert direct control over skeletal muscles. By contrast, in the ANS there is
    a. no synapse between the efferent neuron and peripheral effector.
    b. no CNS integration of peripheral response.
    c. a greater degree of voluntary control.
    d. a synapse on visceral motor neurons, which lie between the CNS and the peripheral effector.

OUTCOME 16-1   _____   3. The system that coordinates cardiovascular, respiratory, digestive, urinary, and reproductive functions is the
    a. somatic nervous system (SNS).
    b. autonomic nervous system (ANS).
    c. central nervous system (CNS).
    d. enteric nervous system (ENS).

OUTCOME 16-2   _____   4. The division of the nervous system that "kicks in" during periods of exertion, stress, or emergency is the
    a. enteric division of the CNS.
    b. sympathetic division of the ANS.
    c. somatic motor division of the PNS.
    d. parasympathetic division.

OUTCOME 16-2 _____ 5. Preganglionic fibers from the thoracic and lumbar segments form part of the

    a. sympathetic division of the ANS.

    b. parasympathetic division of the ANS.

    c. processing centers of the CNS.

    d. cell bodies in the CNS.

OUTCOME 16-2 _____ 6. The nerve bundle that carries preganglionic fibers to a nearby sympathetic chain ganglion is the

    a. collateral ganglion.

    b. autonomic nerve.

    c. gray ramus.

    d. white ramus.

OUTCOME 16-2 _____ 7. Sympathetic postganglionic fibers entering the thoracic cavity can create

    a. an accelerated heart rate.

    b. increased force of cardiac contractions.

    c. dilated respiratory passageways.

    d. a, b, and c are correct.

OUTCOME 16-3 _____ 8. The effect of secretions from the adrenal medulla is that they

    a. do not last as long as those produced by direct sympathetic stimulation.

    b. are limited to peripheral tissues and CNS activity.

    c. serve to dilate blood vessels and elevate blood pressure.

    d. last longer than those produced by direct sympathetic stimulation.

OUTCOME 16-3 _____ 9. At their synapses with ganglionic neurons, all preganglionic neurons in the sympathetic division release

    a. epinephrine.

    b. norepinephrine.

    c. acetylcholine.

    d. dopamine.

OUTCOME 16-3 _____ 10. At neuroeffector junctions, typical sympathetic postganglionic fibers release

    a. epinephrine.

    b. norepinephrine.

    c. acetylcholine.

    d. dopamine.

OUTCOME 16-3 _____ 11. The sympathetic division can change tissue and organ activities by

    a. releasing norepinephrine at peripheral synapses.

    b. distribution of norepinephrine by the bloodstream.

    c. distribution of epinephrine by the bloodstream.

    d. a, b, and c are correct.

OUTCOME 16-3 _____ 12. Cholinergic postganglionic sympathetic fibers that innervate the sweat glands of the skin and the blood vessels of the skeletal muscles are stimulated during exercise to

a. constrict the blood vessels and inhibit sweat gland secretion.

b. keep the body cool and provide oxygen and nutrients to active skeletal muscles.

c. increase the smooth muscle activity in the digestive tract for better digestion.

d. decrease the body temperature and decrease the pH in the blood.

OUTCOME 16-4 _____ 13. In general, the parasympathetic division of the ANS predominates

a. during periods of exertion or stress.

b. during musculoskeletal activity.

c. during heightened levels of somatic activity.

d. under resting condition.

OUTCOME 16-4 _____ 14. Preganglionic neurons in the parasympathetic division of the ANS originate in the

a. peripheral ganglia adjacent to the target organ.

b. thoracolumbar area of the spinal cord.

c. walls of the target organ.

d. brain stem and sacral segments of the spinal cord.

OUTCOME 16-5 _____ 15. At synapses and neuroeffector junctions, all preganglionic and postganglionic fibers in the parasympathetic division release

a. epinephrine.

b. norepinephrine.

c. acetylcholine.

d. a, b, and c are correct.

OUTCOME 16-5 _____ 16. Whether a neurotransmitter produces stimulation or inhibition of activity depends on the

a. target organ that is affected.

b. arrangement of the postganglionic fibers.

c. rate at which the neurotransmitter is released to the receptor.

d. response of the membrane receptor to the presence of the neurotransmitter.

OUTCOME 16-5 _____ 17. The two types of parasympathetic receptors that occur on the postsynaptic membranes are

a. adrenergic and cholinergic.

b. nicotinic and muscarinic.

c. alpha one and beta one.

d. alpha two and beta two.

OUTCOME 16-5 _____ 18. Compared to sympathetic stimulation, the effects of parasympathetic stimulation are usually

a. brief and restricted to specific sites.

b. more specific and localized.

c. specific for controlling a variety of visceral effectors.

d. controlled by alpha and beta receptors.

OUTCOME 16-5 _____ 19. The major effect(s) produced by the parasympathetic division include

    a. secretion by digestive glands.

    b. increased smooth muscle activity along the digestive tract.

    c. constriction of the respiratory passageway.

    d. a, b, and c are correct.

OUTCOME 16-6 _____ 20. Where dual innervation exists, the two divisions of the ANS commonly have

    a. stimulatory effects.

    b. inhibitory effects.

    c. opposing effects.

    d. no effect.

OUTCOME 16-6 _____ 21. Autonomic tone is present in the heart because

    a. ACh (acetylcholine) released by the parasympathetic division decreases the heart rate, and norepinephrine (NE) released by the sympathetic division accelerates the heart rate.

    b. ACh released by the parasympathetic division accelerates the heart rate, and NE released by the sympathetic division decreases the heart rate.

    c. NE released by the parasympathetic division accelerates the heart rate, and ACh released by the sympathetic division decreases the heart rate.

    d. NE released by the sympathetic division and ACh released by the parasympathetic division accelerate the heart rate.

OUTCOME 16-6 _____ 22. The primary factor that determines an individual's autonomic tone is the

    a. autonomic motor neuron activity.

    b. resting levels of acetylcholinesterase.

    c. level of CNS involvement.

    d. increase in demand for ANS activity.

OUTCOME 16-7 _____ 23. The lowest level of integration in the ANS consists of

    a. centers in the medulla that control visceral functions.

    b. regulatory centers in the posterior and lateral hypothalamus.

    c. regulatory centers in the midbrain that control the viscera.

    d. visceral motor neurons that participate in cranial and spinal visceral reflexes.

OUTCOME 16-7 _____ 24. Coordination and regulation of sympathetic function occurs in centers in the medulla oblongata, which can be influenced directly at an additional level by activity of the

    a. cerebral cortex.

    b. spinal cord.

    c. hypothalamus.

    d. cerebellum.

OUTCOME 16-7 _____ 25. A visceral reflex arc consists of a receptor, a sensory neuron, a processing center, and

a. two or more interneurons.

b. two visceral motor neurons.

c. the medulla oblongata.

d. the hypothalamus.

OUTCOME 16-7 _____ 26. Visceral reflexes have the same basic components as somatic reflexes, but all visceral reflexes are

a. monosynaptic.

b. adrenergic.

c. cholinergic.

d. polysynaptic.

OUTCOME 16-7 _____ 27. Visceral reflexes provide

a. automatic motor responses.

b. automatic sensory stimulations.

c. voluntary sensory movements.

d. voluntary motor control.

OUTCOME 16-7 _____ 28. When a bright light shines in the eyes, a parasympathetic reflex causes

a. dilation of the pupils in both eyes.

b. the eyes to begin watering.

c. blindness.

d. constriction of the pupils in both eyes.

OUTCOME 16-8 _____ 29. The cellular mechanism(s) that seem to be involved in memory formation and storage include

a. increased neurotransmitter release.

b. facilitation of synapses.

c. formation of additional synaptic connections.

d. a, b, and c are correct.

OUTCOME 16-8 _____ 30. The two components of the limbic system essential to memory consolidation are the

a. fornix and mamillary body.

b. intermediate mass and pineal gland.

c. amygdaloid body and hippocampus.

d. cingulate gyrus and corpus callosum.

OUTCOME 16-8 _____ 31. A state of awareness of and attention to external events and stimuli implies

a. amnesia.

b. rapid eye movement (REM).

c. consciousness.

d. dual innervation.

OUTCOME 16-8   _____   32. An important brain stem component involved in the conscious state is the
    a. reticular activating system.
    b. pons.
    c. medulla oblongata.
    d. hypothalamus.

OUTCOME 16-8   _____   33. When the heart rate, blood pressure, respiratory rate, and energy utilization decline by up to 30 percent, it indicates
    a. rapid eye movement sleep (REM).
    b. slow wave or non-REM sleep.
    c. production of alpha waves in awake adults.
    d. oscillation between sleep stages.

OUTCOME 16-9   _____   34. Compounds that inhibit serotonin production or block its action cause
    a. hallucinations.
    b. exhilaration.
    c. schizophrenia.
    d. severe depression and anxiety.

OUTCOME 16-9   _____   35. Compounds that stimulate norepinephrine release in certain brain pathways can cause
    a. the motor problems of Parkinson's disease.
    b. hallucinations.
    c. exhilaration.
    d. severe depression and anxiety.

OUTCOME 16-10   _____   36. A common age-related anatomical change in the nervous system is
    a. increased brain size and weight.
    b. an increased number of neurons.
    c. increased blood flow to the brain.
    d. none of the above.

OUTCOME 16-10   _____   37. The most common and incapacitating form of senile dementia is
    a. retrograde amnesia.
    b. Alzheimer's disease.
    c. Parkinson's disease.
    d. Down's syndrome.

Completion

Using the terms below, complete the following statements. Use each term only once.

CNS	sympathetic activation	vagus nerve	norepinephrine
memory consolidation	rapid eye movement	schizophrenia	limbic
"fight or flight"	acetylcholine	muscarinic	collateral ganglia
cholinergic	hypothalamus	acetylcholinesterase	plaques
excitatory	opposing	anabolic	
epinephrine	receptor	visceral reflexes	
"rest and digest"	autonomic tone	involuntary	

OUTCOME 16-1   1. The lower motor neurons of the SNS exert voluntary control over skeletal muscles, while in the ANS the control of smooth and cardiac muscle is _____.

OUTCOME 16-2   2. Because the sympathetic division of the ANS stimulates tissue metabolism and increases alertness, it is called the _____ subdivision.

OUTCOME 16-2   3. The three unpaired sympathetic ganglia that lie anterior to the vertebral bodies are the _____.

OUTCOME 16-2   4. The response of the entire sympathetic division in a crisis causes an event called _____.

OUTCOME 16-3   5. At their synaptic terminals, preganglionic autonomic fibers release _____.

OUTCOME 16-3   6. The two neurotransmitters released into circulation by sympathetic stimulation of the adrenal medulla are norepinephrine and _____.

OUTCOME 16-3   7. Preganglionic fibers always have a(n) _____ effect on postganglionic neurons.

OUTCOME 16-3   8. The neurotransmitter that is released by sympathetic postganglionic fibers and that causes constriction of most peripheral blood vessels is _____.

OUTCOME 16-4   9. Because the parasympathetic division of the ANS conserves energy and promotes sedentary activity, it is known as the _____ subdivision.

OUTCOME 16-4   10. Approximately 75 percent of all parasympathetic outflow is provided by the _____.

OUTCOME 16-4   11. Due to its functions, the parasympathetic division is referred to as the _____ system.

OUTCOME 16-4   12. All parasympathetic neurons are _____.

OUTCOME 16-5   13. Most of the ACh released during parasympathetic stimulation is inactivated by _____.

OUTCOME 16-5   14. The effects of a neurotransmitter substance on a postsynaptic cell can vary widely due to the type of _____.

OUTCOME 16-5   15. The G protein-based cholinergic receptors in the parasympathetic division are the _____ receptors.

OUTCOME 16-6   16. In an organ with dual innervation, sympathetic and parasympathetic actions would likely be described as _____.

OUTCOME 16-6   17. The release of small amounts of acetylcholine and norepinephrine on a continual basis in an organ with dual innervation is referred to as _____.

OUTCOME 16-7   18. The system in the brain that would most likely exert an influence on autonomic control if an emotional condition is present is the _____ system.

OUTCOME 16-7   19. Centers and nuclei in the medulla oblongata are subject to regulation by the _____.

OUTCOME 16-7   20. The simplest functional units in the ANS are _____.

OUTCOME 16-8   21. The conversion from short-term to long-term memory is called _____.

OUTCOME 16-8   22. The degree of wakefulness at any moment is an indication of the level of ongoing _____ activity.

OUTCOME 16-8   23. Active dreaming occurs during _____ sleep.

OUTCOME 16-9   24. Excessive secretion of dopamine may be associated with _____.

OUTCOME 16-10   25. Extracellular accumulations of fibrillar proteins, surrounded by abnormal dendrites and axons, are called _____.

Matching

Match the terms in column B with the terms in column A. Use letters for answers in the spaces provided. Use each term only once.

Part I

		Column A	Column B
OUTCOME 16-1	_____	1. somatic nervous system	A. secretes norepinephrine and epinephrine
OUTCOME 16-1	_____	2. autonomic nervous system	
OUTCOME 16-2	_____	3. thoracolumbar axons and ganglia	B. alpha and beta
OUTCOME 16-2	_____	4. craniosacral axons and ganglia	C. vasodilation
OUTCOME 16-2	_____	5. adrenal medulla	D. splanchnic neurons
OUTCOME 16-2	_____	6. collateral ganglia	E. homeostatic adjustments
OUTCOME 16-3, 16-5	_____	7. cholinergic receptors	F. sympathetic division
OUTCOME 16-3	_____	8. adrenergic receptors	G. vasoconstriction
OUTCOME 16-3	_____	9. nitric oxide	H. conscious control
OUTCOME 16-3	_____	10. norepinephrine	I. parasympathetic activity
OUTCOME 16-4	_____	11. sexual arousal	J. nicotinic, muscarinic
OUTCOME 16-5	_____	12. parasympathetic neurotransmitter	K. parasympathetic division
OUTCOME 16-5	_____	13. tissue cholinesterase	L. acetylcholine
			M. inactivates ACh

Part II

		Column A	Column B
OUTCOME 16-5	_____	14. parasympathetic effect	N. localized, short-lived
OUTCOME 16-5	_____	15. parasympathetic division	O. skill memories
OUTCOME 16-6	_____	16. dual innervation	P. defecation, urination
OUTCOME 16-6	_____	17. resting level of spontaneous activity	Q. cardiac centers
OUTCOME 16-7	_____	18. medulla oblongata	R. reduced blood flow
OUTCOME 16-7	_____	19. visceral reflexes	S. autonomic tone
OUTCOME 16-7	_____	20. swallowing reflex	T. REM sleep
OUTCOME 16-8	_____	21. learned motor behaviors	U. inadequate dopamine
OUTCOME 16-8	_____	22. coma	V. state of unconsciousness
OUTCOME 16-8	_____	23. dreaming	W. parasympathetic reflex
OUTCOME 16-9	_____	24. Parkinson's disease	X. stimulates visceral activity
OUTCOME 16-10	_____	25. arteriosclerosis	Y. opposing effects

Drawing/Illustration Labeling

Using the terms below, fill in the blank, numbered spaces to correctly complete the organization of the sympathetic division of the ANS. Place your answers in the spaces provided below.

organs and systems throughout body   visceral effectors in abdominopelvic cavity

paired sympathetic chain ganglia   unpaired collateral ganglia

paired adrenal medullae   preganglionic neurons

OUTCOME 16-2   **FIGURE 16-1**   Organization of the Sympathetic Division of the ANS

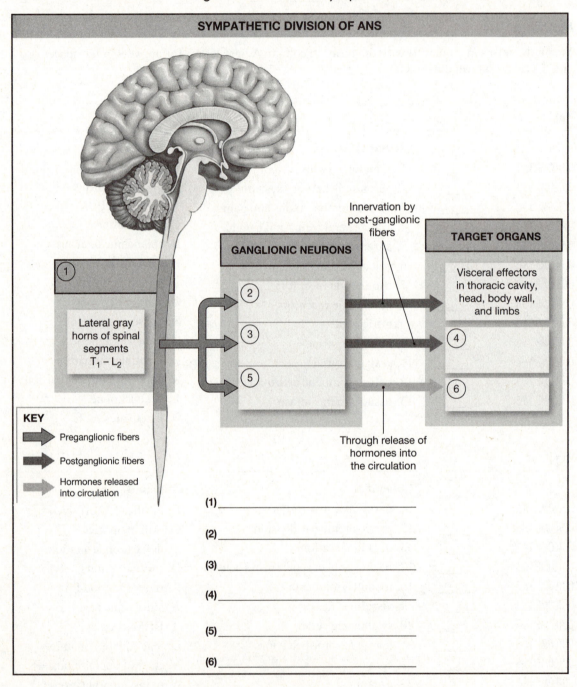

(1) _____

(2) _____

(3) _____

(4) _____

(5) _____

(6) _____

Using the terms below, fill in the blank, numbered spaces to correctly complete the organization of the parasympathetic division of the ANS. Place your answers in the spaces provided below.

parotid salivary gland

otic ganglion

ciliary ganglion

nasal, tear, and salivary glands

intramural ganglia

intrinsic eye muscles (pupil and lens shape)

nuclei in brain stem

N X

N VII

OUTCOME 16-4   **FIGURE 16-2**   Organization of the Parasympathetic Division of the ANS

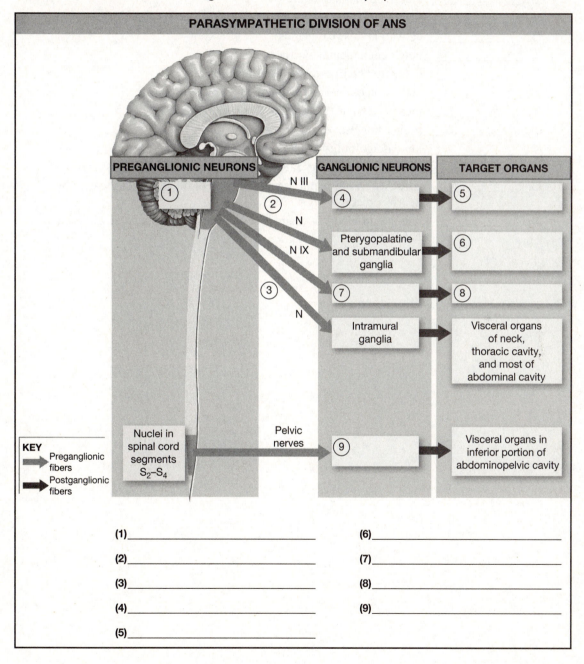

(1)_____   (6)_____

(2)_____   (7)_____

(3)_____   (8)_____

(4)_____   (9)_____

(5)_____

LEVEL 2: REVIEWING CONCEPTS

Chapter Overview

To complete the chapter overview, identify the major effects produced by the autonomic nervous system. Use the letter *S* for the sympathetic division and the letter *P* for the parasympathetic division. Place your answers in the spaces provided next to each effect.

1. _____ decreased metabolic rate
2. _____ increased salivary and digestive secretion
3. _____ increased metabolic rate
4. _____ stimulation of urination and defecation
5. _____ activation of sweat glands
6. _____ heightened mental alertness
7. _____ decreased heart rate and blood pressure
8. _____ activation of energy reserves
9. _____ increased heart rate and blood pressure
10. _____ reduced digestive and urinary functions
11. _____ increased motility of blood flow in the digestive tract
12. _____ increased respiratory rate and dilation of respiratory passageways
13. _____ constriction of the pupils and focusing of the lenses of the eyes on nearby objects
14. _____ changes in blood flow and glandular activity associated with sexual arousal

Concept Map I

Using the following terms, fill in the circled, numbered, blank spaces to correctly complete the concept map. Use each term only once.

adrenal medulla (paired)

ganglionic neurons

thoracolumbar

sympathetic chain of ganglia (paired)

spinal segments T_1–L_2

visceral effectors

general circulation

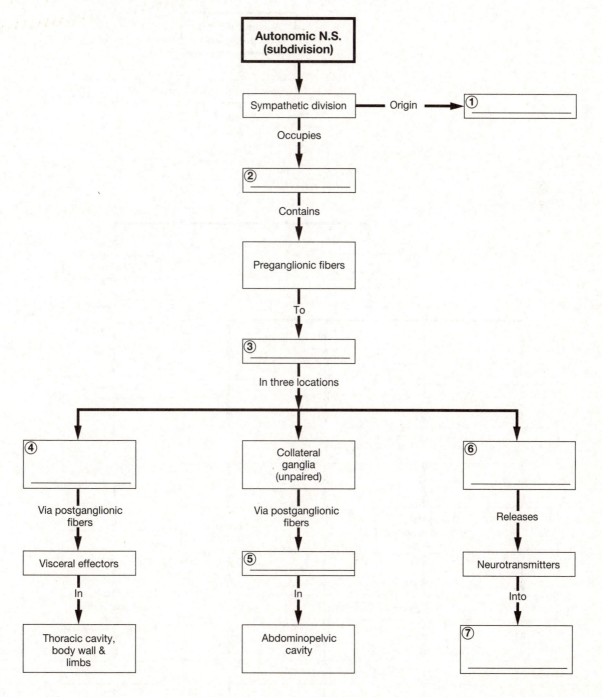

Concept Map II

Using the following terms, fill in the circled, numbered, blank spaces to correctly complete the concept map. Use each term only once.

lower abdominopelvic cavity	segments S_2–S_4	ciliary ganglion
nasal, tear, salivary glands	otic ganglia	N X
craniosacral	intramural ganglia	N VII
		brain stem

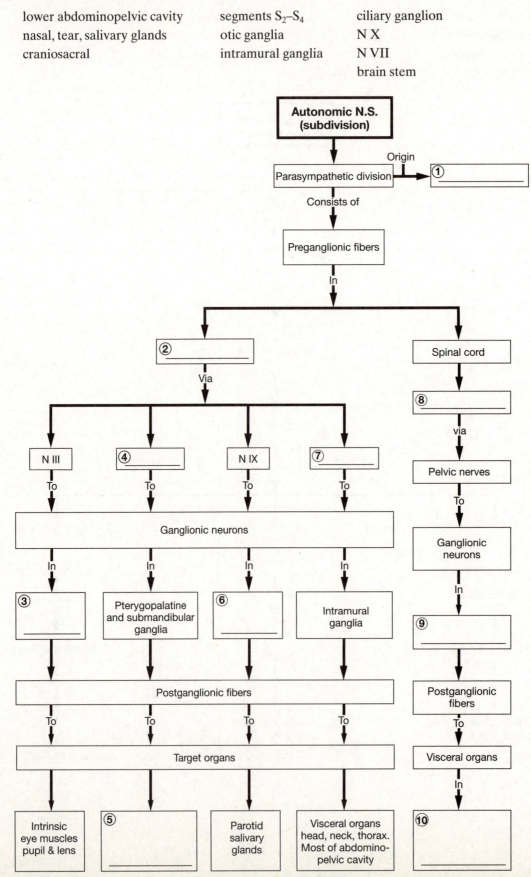

Concept Map III

Using the following terms, fill in the circled, numbered, blank spaces to correctly complete the concept map. Unless otherwise noted, use each term only once.

medulla oblongata   spinal cord   respiratory

parasympathetic (use twice)   cardiovascular   hypothalamus

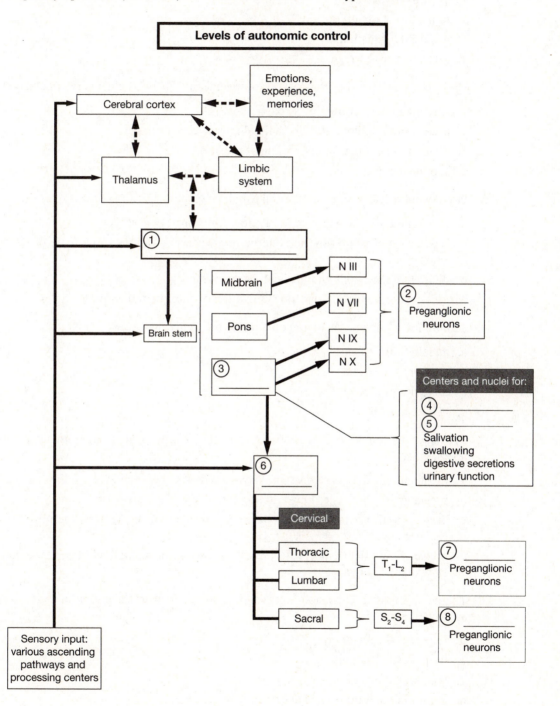

Multiple Choice

Place the letter corresponding to the best answer in the space provided.

_____ 1. The axon of a ganglionic neuron is called a postganglionic fiber because it carries impulses

   a. away from the ganglion.
   b. to the target organ.
   c. toward the ganglion.
   d. a, b, and c are correct.

_____ 2. The effect(s) produced by sympathetic postganglionic fibers in spinal nerves include(s)

   a. stimulation of secretion by sweat glands.
   b. acceleration of blood flow to skeletal muscles.
   c. dilation of the pupils and focusing of the eyes.
   d. a, b, and c are correct.

_____ 3. The summary effects of the collateral ganglia include

   a. accelerating the heart rate and dilation of respiratory passageways.
   b. redirection of blood flow and energy use by visceral organs and release of stored energy.
   c. dilation of the pupils and focusing of the eyes.
   d. acceleration of blood flow and stimulation of energy production.

_____ 4. Stimulation of alpha-1 receptors on a target cell generally causes

   a. an inhibitory effect.
   b. a refractory period.
   c. an excitatory effect.
   d. no effect at all.

_____ 5. The major structural difference between sympathetic preganglionic and postganglionic fibers is that

   a. preganglionic fibers are short and postganglionic fibers are long.
   b. preganglionic fibers are long and postganglionic fibers are short.
   c. preganglionic fibers are close to target organs and postganglionic fibers are close to the spinal cord.
   d. preganglionic fibers innervate target organs and postganglionic fibers originate from cranial nerves.

_____ 6. The parasympathetic division innervates areas serviced by the cranial nerves and innervates

   a. the general circulation.
   b. the collateral ganglia.
   c. the adrenal medulla.
   d. organs in the thoracic and abdominopelvic cavities.

_____ 7. Vasomotor reflexes are responsible for

   a. changes in the diameter of peripheral blood vessels to maintain a normal range of blood pressure.
   b. a sudden decline in blood pressure in the carotid artery.
   c. smooth muscle contractions in the digestive tract.
   d. increase in heart rate and force of contractions.

_____ 8. The functions of the parasympathetic division center on

    a. accelerating the heart rate and the force of contraction.

    b. dilation of the respiratory passageways.

    c. relaxation, food processing, and energy absorption.

    d. a, b, and c are correct.

_____ 9. During a crisis, the event necessary for the individual to cope with stressful and potentially dangerous situations is called

    a. splanchnic innervation.

    b. sympathetic activation.

    c. parasympathetic activation.

    d. the effector response.

_____ 10. The two classes of sympathetic receptors are

    a. alpha and beta receptors.

    b. alpha-1 and alpha-2 receptors.

    c. beta-1 and beta-2 receptors.

    d. ganglionic and preganglionic receptors.

_____ 11. Intramural ganglia are components of the parasympathetic division that are located

    a. in the nasal and salivary glands.

    b. in the intrinsic eye muscles.

    c. inside the tissues of visceral organs.

    d. in the adrenal medulla.

_____ 12. Parasympathetic preganglionic fibers of the vagus nerve entering the abdominopelvic cavity join the _____ plexus.

    a. cardiac

    b. pulmonary

    c. hypogastric

    d. celiac

_____ 13. Sensory nerves deliver information to the CNS along

    a. spinal nerves.

    b. cranial nerves.

    c. autonomic nerves that innervate visceral effectors.

    d. a, b, and c are correct.

_____ 14. Which of the following selections includes only sympathetic reflexes?

    a. Swallowing, coughing, sneezing, and vomiting reflexes

    b. Baroreceptor, vasomotor, pupillary dilation, and ejaculation reflexes

    c. Defecation, urination, digestion, and secretion reflexes

    d. Light, consensual light, sexual arousal, and baroreceptor reflexes

_____ 15. In general, the primary result(s) of treatment with beta-blockers is (are)

    a. reducing the load on the heart and lowering peripheral blood pressure.

    b. decreasing the heart rate.

    c. decreasing the force of contraction of the heart.

    d. a, b, and c are correct.

_____ 16. A change or changes that occur in the synaptic organization of the brain due to aging is (are)

    a. the number of dendritic branches decreases.

    b. synaptic connections are lost.

    c. the rate of neurotransmitter production declines.

    d. all of the above.

Completion

Using the terms below, complete the following statements. Use each term only once.

hypothalamus	collateral	white ramus
postganglionic	gray ramus	norepinephrine
acetylcholine	blocker	splanchnic
adrenal medulla		

1. The autonomic fibers that innervate peripheral organs are called _____ fibers.

2. Adrenergic postganglionic sympathetic terminals primarily release the neurotransmitter _____.

3. Modified sympathetic ganglia containing ganglionic neurons are located in the _____.

4. Prevertebral ganglia found anterior to the vertebral bodies are called _____ ganglia.

5. The nerve bundle that carries sympathetic postganglionic fibers is known as the _____.

6. The nerve bundle containing the myelinated preganglionic axons of sympathetic motor neurons en route to the sympathetic chain or collateral ganglion is the _____.

7. In the dorsal wall of the abdominal cavity, preganglionic fibers that innervate the collateral ganglia form the _____ nerves.

8. Sympathetic activation is controlled by sympathetic centers in the _____.

9. Muscarinic and nicotine receptors are stimulated by the neurotransmitter _____.

10. Drugs that reduce the effects of autonomic stimulation by keeping the neurotransmitter from affecting the postsynaptic membranes are known as _____.

Short Essay

Briefly answer the following questions in the spaces provided below.

1. Why is the sympathetic division of the ANS called the "fight or flight" system?

2. Why is the parasympathetic division of the ANS known as the "rest and digest" system?

3. Identify the neurotransmitters used by preganglionic and postganglionic neurons in the sympathetic and parasympathetic subdivisions of the ANS. Describe the three major patterns observed.

4. What are the three major components of the sympathetic division?

5. What two general patterns summarize the effects of the functions of the collateral ganglia?

6. What are the two major components of the parasympathetic division of the ANS?

7. In what four cranial nerves do parasympathetic preganglionic fibers travel when leaving the brain?

8. Why is the parasympathetic division sometimes referred to as the anabolic system?

9. List five types of adrenergic receptors found in the human body.

10. Name two cholinergic receptors found in the human body.

11. What effect does dual innervation have on autonomic control throughout the body?

12. Why is autonomic tone an important aspect of ANS function?

13. a. What parasympathetic reflexes would distention of the rectum and urinary bladder initiate?

    b. What sympathetic reflexes would low light levels and changes in blood pressure in major arteries initiate?

LEVEL 3: CRITICAL THINKING AND CLINICAL APPLICATIONS

Using principles and concepts learned in Chapter 16, answer the following questions. Write your answers on a separate sheet of paper.

1. You have probably heard stories about "superhuman" feats that have been performed during times of crisis. What autonomic mechanisms are involved in supporting a sudden, intensive physical activity that, under ordinary circumstances, would be impossible?

2. Amanda is leisurely swimming in the ocean about 30 yards from the beach when she sees the dorsal fin of a shark approaching. What physiological changes would occur due to the activation of the sympathetic nervous system?

3. Recent surveys show that 30 to 35 percent of the American adult population is involved in some type of exercise program. What contributions does sympathetic activation make to help the body adjust to the changes that occur during exercise and still maintain homeostasis?

4. Paul is diagnosed with high blood pressure. His physician prescribes a medication that blocks beta receptors. How might a beta-blocker help correct his high blood pressure?

CHAPTER 4

The Special Senses

OVERVIEW

Do we really see with our eyes, hear with our ears, smell with our nose, and taste with our tongue? Ask the layperson and the answer will probably be "yes." Ask the student of anatomy and physiology and the response should be "not really." Our awareness of the world within and around us is aroused by sensory receptors that react to stimuli within the body or to stimuli in the environment outside the body. The previous chapters on the nervous system described the mechanisms by which the body perceives stimuli as neural events and not necessarily as environmental realities.

Sensory receptors receive stimuli and neurons transmit action potentials to the CNS, where the sensations are processed, resulting in motor responses that serve to maintain homeostasis.

Chapter 17 considers the *special senses* of olfaction, gustation, vision, equilibrium, and hearing. The activities in this chapter will reinforce your understanding of the structure and function of specialized receptor cells that are structurally more complex than those that serve the general senses.

LEVEL 1: REVIEWING FACTS AND TERMS

Review of Learning Outcomes

After completing this chapter, you should be able to do the following:

OUTCOME 17-1   Describe the sensory organs of smell, trace the olfactory pathways to their destinations in the brain, and explain the physiological basis of olfactory discrimination.

OUTCOME 17-2   Describe the sensory organs of taste, trace the gustatory pathways to their destinations in the brain, and explain the physiological basis of gustatory discrimination.

OUTCOME 17-3   Identify the internal and accessory structures of the eye, and explain the functions of each.

OUTCOME 17-4   Explain color and depth perception, describe how light stimulates the production of nerve impulses, and trace the visual pathways to their destinations in the brain.

OUTCOME 17-5   Describe the structures of the external, middle, and internal ear, explain their roles in equilibrium and hearing, and trace the pathways for equilibrium and hearing to their destinations in the brain.

Multiple Choice

Place the letter corresponding to the best answer in the space provided.

OUTCOME 17-1 _____ 1. The two tissue layers that make up the olfactory organs are the

    a. supporting cells and basal cells.

    b. olfactory epithelium and olfactory receptors.

    c. olfactory epithelium and lamina propria.

    d. olfactory glands and lamina propria.

OUTCOME 17-1 _____ 2. The first step in olfactory reception occurs on the exposed

    a. olfactory cilia.

    b. columnar cells.

    c. basal cells.

    d. olfactory glands.

OUTCOME 17-1 _____ 3. The ultimate destination(s) for interpreting the sense of smell is (are) the

    a. olfactory cortex.

    b. hypothalamus.

    c. limbic system.

    d. a, b, and c are correct.

OUTCOME 17-1 _____ 4. The factor that ensures that you quickly lose awareness of a new smell but retain sensitivity to others is called

    a. olfactory discrimination.

    b. central adaptation.

    c. olfactory sensitivity.

    d. neuronal replacement.

OUTCOME 17-1 _____ 5. The CNS interprets smell on the basis of the particular pattern of

    a. cortical arrangement.

    b. neuronal replacement.

    c. receptor activity.

    d. sensory impressions.

OUTCOME 17-1 _____ 6. Although the human olfactory organs can discriminate among many smells, acuity varies widely depending on the

    a. number of receptors.

    b. nature of the odorant.

    c. type of receptor cells.

    d. olfactory sensitivity.

OUTCOME 17-2 _____ 7. The three different types of papillae on the human tongue are

    a. kinocilia, cupula, and maculae.

    b. sweet, sour, and bitter.

    c. cuniculate, pennate, and circumvallate.

    d. filiform, fungiform, and circumvallate.

OUTCOME 17-2 _____ 8. Taste buds are monitored by cranial nerves
a. VII, IX, and X.
b. IV, V, and VI.
c. I, II, and III.
d. VIII, XI, and XII.

OUTCOME 17-2 _____ 9. After synapsing in the thalamus, gustatory information is projected to the appropriate portion of the
a. medulla.
b. medial lemniscus.
c. gustatory cortex.
d. VII, IX, and X cranial nerves.

OUTCOME 17-2 _____ 10. Taste buds in all portions of the tongue provide all four primary taste sensations.
a. True
b. False

OUTCOME 17-2 _____ 11. Gustatory reception begins when dissolved chemicals contacting the taste hairs bind to receptor proteins of the
a. filiform papillae.
b. circumvallate papillae.
c. basal cells.
d. gustatory cells.

OUTCOME 17-2 _____ 12. Protein complexes that use second messengers to produce their gustatory effects are called
a. gustducins.
b. umami.
c. phenylthiocarbamide.
d. primary taste sensations.

OUTCOME 17-2 _____ 13. An example of one inherited taste sensitivity is for the compound
a. glutamate.
b. gustducin.
c. phenylthiourea.
d. peptidase.

OUTCOME 17-3 _____ 14. A lipid-rich product that helps to keep the eyelids from sticking together is produced by the
a. gland of Zeis.
b. tarsal glands.
c. lacrimal glands.
d. conjunctiva.

OUTCOME 17-3 _____ 15. The fibrous layer, the outermost layer covering the eye, consists of the
a. iris and choroid.
b. pupil and ciliary body.
c. sclera and cornea.
d. lacrimal sac and orbital fat.

OUTCOME 17-3 _____ 16. The vascular layer consists of three distinct structures, the
    a. sclera, cornea, and iris.
    b. choroid, pupil, and lacrimal sac.
    c. retina, cornea, and iris.
    d. iris, ciliary body, and choroid.

OUTCOME 17-3 _____ 17. The function of the vitreous body in the eye is to
    a. provide a fluid cushion for protection of the eye.
    b. serve as a route for nutrient and waste transport.
    c. stabilize the shape of the eye.
    d. serve as a medium for cleansing the inner eye.

OUTCOME 17-3 _____ 18. The primary function of the lens of the eye is to
    a. absorb light after it passes through the retina.
    b. biochemically interact with the photoreceptors of the retina.
    c. focus the visual image on retinal photoreceptors.
    d. integrate visual information for the retina.

OUTCOME 17-3 _____ 19. When looking directly at an object, its image falls upon the portion of the retina called the
    a. fovea centralis.
    b. choroid layer.
    c. sclera.
    d. focal point.

OUTCOME 17-3 _____ 20. The transparent proteins in the lens fibers, responsible for both the clarity and the focusing power of the lens, are
    a. amacrines.
    b. crystallins.
    c. opsins.
    d. rhodopsins.

OUTCOME 17-4 _____ 21. When photons of all wavelengths stimulate both rods and all three types of cones, the eye perceives
    a. "black" objects.
    b. all the colors of the visible light spectrum.
    c. either "red" or "blue" light.
    d. "white" light.

OUTCOME 17-4 _____ 22. Axons converge on the optic disc, penetrate the wall of the eye, and proceed toward the
    a. retina at the posterior part of the eye.
    b. diencephalon as the optic nerve (N II).
    c. retinal processing areas below the choroid coat.
    d. cerebral cortex area of the parietal lobes.

OUTCOME 17-4   _____   23. The perception of a visual image reflects the integration of information arriving at the
  a. lateral geniculate of the left side.
  b. visual cortex of the cerebrum.
  c. lateral geniculate of the right side.
  d. reflex centers in the brain stem.

OUTCOME 17-5   _____   24. The bony labyrinth of the ear is subdivided into the
  a. auditory meatus, auditory canal, and ceruminous glands.
  b. saccule, utricle, and vestibule.
  c. vestibule, semicircular canals, and cochlea.
  d. ampulla, crista, and cupula.

OUTCOME 17-5   _____   25. The external ear is separated from the middle ear by the
  a. pharyngotympanic tube.
  b. tympanic membrane.
  c. sacculus.
  d. utriculus.

OUTCOME 17-5   _____   26. The auditory ossicles of the middle ear include the
  a. sacculus, utriculus, and ampulla.
  b. vestibule, cochlea, and spiral organ.
  c. malleus, stapes, and incus.
  d. otoliths, maculae, and otoconia.

OUTCOME 17-5   _____   27. The structure in the cochlea of the internal ear that provides information to the CNS is the
  a. scala tympani.
  b. spiral organ.
  c. tectorial membrane.
  d. basilar membrane.

OUTCOME 17-5   _____   28. The receptors that provide the sensation of hearing are located in the
  a. vestibule.
  b. ampulla.
  c. tympanic membrane.
  d. cochlea.

OUTCOME 17-5   _____   29. The senses of equilibrium and hearing are provided by receptors in the
  a. external ear.
  b. middle ear.
  c. internal ear.
  d. a, b, and c are correct.

OUTCOME 17-5   _____   30. Ascending auditory sensations synapse in the thalamus and then are delivered by projection fibers to the auditory cortex of the
  a. parietal lobe.
  b. temporal lobe.
  c. occipital lobe.
  d. frontal lobe.

Completion

Using the terms below, complete the following statements. Use each term only once.

olfactory discrimination	sclera	occipital	olfactory
rods	taste buds	saccule	pupil
cerebral cortex	round window	cones	midbrain
amacrine	gustducins	astigmatism	accommodation
endolymph	photoreception	convergence	

OUTCOME 17-1   1. The information provided by receptors of the special senses is distributed to specific areas of the _____.

OUTCOME 17-1   2. The only type of sensory information that reaches the cerebral cortex without synapsing in the thalamus is _____ stimuli.

OUTCOME 17-1   3. The olfactory system's ability to make subtle distinctions among thousands of chemical stimuli is referred to as _____.

OUTCOME 17-2   4. Gustatory receptors are clustered in individual _____.

OUTCOME 17-2   5. Receptors responding to stimuli that produce sweet, bitter, and umami sensations are linked to G proteins called _____.

OUTCOME 17-3   6. Most of the external surface of the eye is covered by the _____.

OUTCOME 17-3   7. The opening surrounded by the iris is called the _____.

OUTCOME 17-3   8. The photoreceptors that enable us to see in dimly lit rooms, at twilight, or in pale moonlight are the _____.

OUTCOME 17-3   9. _____ cells, which occur where bipolar cells synapse with ganglion cells, can facilitate or inhibit communication between these two cell populations.

OUTCOME 17-3   10. The process that focuses images on the retina by changing the shape of the lens to keep the focal length constant is called _____.

OUTCOME 17-3   11. Distortion of the visual image, if light passing through the cornea and lens is not refracted properly, causes a condition called _____.

OUTCOME 17-4   12. The photoreceptors that account for the perception of color are the _____.

OUTCOME 17-4   13. Visual information is integrated in the cortical area of the _____ lobe.

OUTCOME 17-4   14. In darkness (resting state), photoreceptors release a constant level of neurotransmitter. In the process of _____, those photoreceptors become hyperpolarized in response to light stimuli and decrease their rate of neurotransmitter release.

OUTCOME 17-4   15. In the retina, a significant amount of _____ happens as information passes from the level of the photoreceptors to the level of the ganglion cells.

OUTCOME 17-5   16. The tubes of the membranous labyrinth in the internal ear contain a fluid called _____.

OUTCOME 17-5   17. The thin membranous partition that separates the perilymph of the scala tympani from the air spaces of the middle ear is the _____.

OUTCOME 17-5   18. The receptors in the internal ear that provide sensations of gravity and linear acceleration are located in the _____ and the utricle.

OUTCOME 17-5   19. The reflex processing center responsible for changing the position of the head in response to a sudden loud noise is located in the inferior colliculus of the _____.

Matching

Match the terms in column B with the terms in column A. Use letters for answers in the spaces provided. Use each term only once.

Part I

		Column A	Column B
OUTCOME 17-1	_____	1. olfactory glands	A. tarsal cyst
OUTCOME 17-1	_____	2. olfactory receptors	B. sense of smell
OUTCOME 17-1	_____	3. cerebrum	C. use G proteins
OUTCOME 17-1	_____	4. olfaction	D. PTC
OUTCOME 17-2	_____	5. odorant	E. olfactory bulbs
OUTCOME 17-2	_____	6. gustatory receptors	F. stem cells
OUTCOME 17-2	_____	7. basal cells	G. mucus secretion
OUTCOME 17-2	_____	8. circumvallate papillae	H. chemical stimulus
OUTCOME 17-3	_____	9. inherited taste sensitivity	I. modified neurons
OUTCOME 17-3	_____	10. chalazion	J. umami

Part II

		Column A	Column B
OUTCOME 17-3	_____	11. eyelids	K. inner layer
OUTCOME 17-3	_____	12. lacrimal glands	L. circadian rhythm
OUTCOME 17-3	_____	13. sharp vision	M. three-dimensional relationships
OUTCOME 17-3	_____	14. retina	N. internal ear
OUTCOME 17-3	_____	15. sclera	O. palpebral
OUTCOME 17-4	_____	16. rhodopsin	P. ear wax
OUTCOME 17-4	_____	17. depth perception	Q. equilibrium
OUTCOME 17-4	_____	18. suprachiasmatic nucleus	R. tears
OUTCOME 17-4	_____	19. P cells	S. visual pigment
OUTCOME 17-5	_____	20. ceruminous glands	T. fovea centralis
OUTCOME 17-5	_____	21. membranous labyrinth	U. white of the eye
OUTCOME 17-5	_____	22. utricle, saccule	V. monitor cones

Drawing/Illustration Labeling

Identify each numbered structure by labeling the following figures.

OUTCOME 17-3 **FIGURE 17-1** Sagittal Sectional Anatomy of the Left Eye

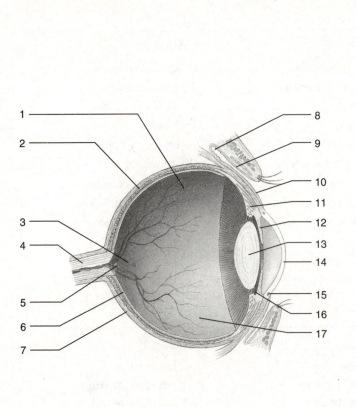

(1)_____

(2)_____

(3)_____

(4)_____

(5)_____

(6)_____

(7)_____

(8)_____

(9)_____

(10)_____

(11)_____

(12)_____

(13)_____

(14)_____

(15)_____

(16)_____

(17)_____

OUTCOME 17-5 **FIGURE 17-2** Anatomy of the Ear: External, Middle, and Internal Ear

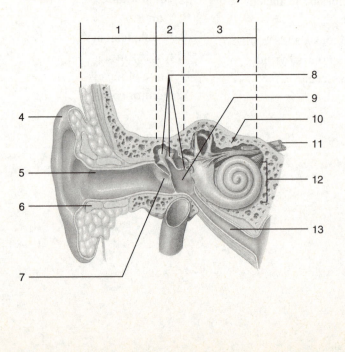

(1)_____

(2)_____

(3)_____

(4)_____

(5)_____

(6)_____

(7)_____

(8)_____

(9)_____

(10)_____

(11)_____

(12)_____

(13)_____

OUTCOME 17-5 **FIGURE 17-3** Anatomy of the Ear (Bony Labyrinth)

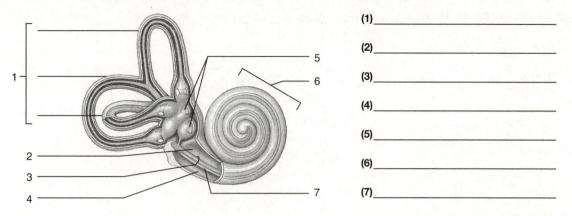

(1)_____

(2)_____

(3)_____

(4)_____

(5)_____

(6)_____

(7)_____

OUTCOME 17-5 **FIGURE 17-4** Anatomy of the Cochlea (Details Visible in Section)

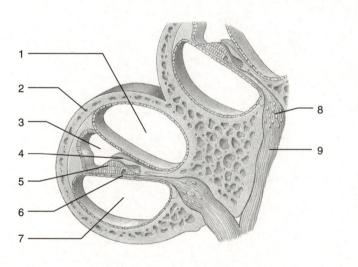

(1)_____

(2)_____

(3)_____

(4)_____

(5)_____

(6)_____

(7)_____

(8)_____

(9)_____

OUTCOME 17-5 **FIGURE 17-5** Three-Dimensional Structure of Tectorial Membrane and Hair Complex of the Spiral Organ

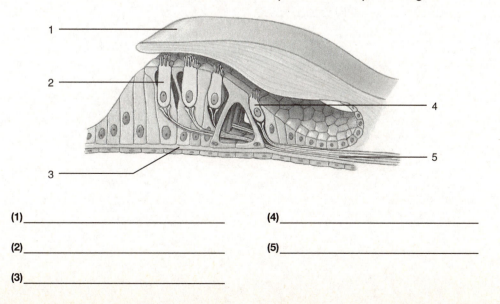

(1)_____   (4)_____

(2)_____   (5)_____

(3)_____

Identify the areas of taste on the tongue and the three types of lingual papillae.

OUTCOME 17-2 **FIGURE 17-6** Gustatory Receptors—Landmarks and Receptors on the Tongue

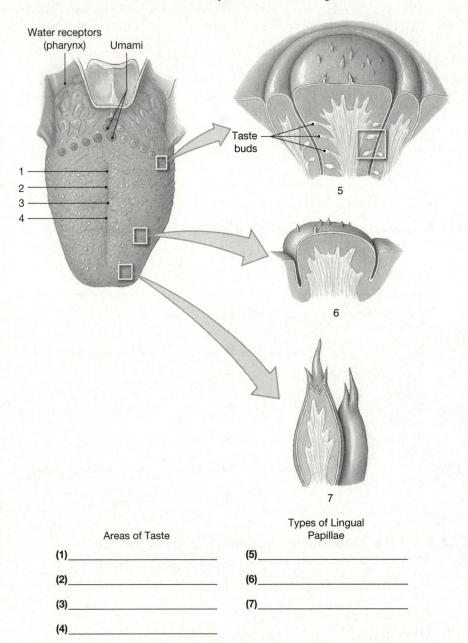

Areas of Taste

(1)_____

(2)_____

(3)_____

(4)_____

Types of Lingual Papillae

(5)_____

(6)_____

(7)_____

LEVEL 2: REVIEWING CONCEPTS

Chapter Overview

Using the terms provided, correctly complete the chapter overview for the special senses. Unless otherwise noted, each term should be used once.

internal ear	endolymph	floats
modified neurons	rotational planes	gravity
neurotransmitter	salty	green
olfactory cilia	semicircular ducts	spiral organ
open	umami	sterocilia
oval window	utricle and saccule	tympanic membrane
stimulates	vestibule	close
sweet	inhibits	cochlear duct
taste receptors	papillae	cones
tectorial	perilymph	rods
transducin	second messenger	bleaching
continually	accessory structures	sour
depolarize (use twice)	bipolar	water
distorts	bitter	olfactory epithelium

OLFACTION

Olfactory receptor cells are found in the (1) _____ in the nasal cavity. Odorant binding to receptor proteins on (2) _____ activates a G protein complex, which converts ATP to cAMP. cAMP, as a (3) _____, binds to and opens sodium channels to (4) _____ the receptor membrane. Because olfactory receptor cells are (5) _____, olfaction does not display a synaptic delay in the transmission of sensory information.

GUSTATION

(6) _____ clustered in taste buds are found in epithelial (7) _____ on the tongue. Also present are umami and water receptors. Each taste bud can provide four primary taste sensations. These include (8) _____ and (9) _____, in which a chemical binds directly to receptors and opens sodium channels to depolarize the membrane. The other taste sensations include (10) _____ and (11) _____, in which a chemical binds to G protein–mediated receptors, called gustducins, which activate a second messenger to stimulate the release of neurotransmitters. (12) _____ receptors use a similar G protein–based mechanism, but the distribution of these receptors is not known in full detail. (13) _____ receptors, found especially in the pharynx, have sensory output to the hypothalamus and play a role in fluid balance.

VISION

There are two types of photoreceptors in the retina of the eye: (14) _____ do not discriminate colors; (15) _____, of which there are three types—red, (16) _____, and blue—are sensitive to specific wavelengths to allow for color discrimination.

The basic mechanism of photoreception is the same for rods and cones. Differences in wavelength sensitivity depend on which type of opsin molecule is present.

In darkness, photoreceptors are tonically active, (17) _____ secreting neurotransmitter to bipolar cells. Gated sodium channels are kept (18) _____ by the presence of cGMP, and the transmembrane potential stays at about –40mV.

When photons strike the photoreceptor, the bound retinal changes configuration, which then activates the opsin, which activates (19) _____ (a G protein), which activates phosphodiesterase to break down the cGMP holding the sodium gate open. As cGMP levels decline, the gated sodium channels (20) _____ and the transmembrane potential hyperpolarizes toward –70mV. As a result, neurotransmitter release rate is altered, which in turn indicates to the (21) _____ cell that light has been detected.

In order to recover for the next photon, the entire opsin molecule must be broken down and reassembled to return retinal to its ready state (a process called [22] _____).

EQUILIBRIUM

The hair cells involved with our sense of equilibrium are found in the (23) _____. Hair cells located in the ampullae of the (24) _____ detect rotational movements of the head. Hair cells located in the maculae of the (25) _____ (in the [26] _____) detect position with respect to gravity and linear acceleration.

When an appropriate external force moves stereocilia or kinocilia of hair cells enough to distort their plasma membranes, the hair cells alter their rate of neurotransmitter release onto a sensory nerve, which then conveys information about that force to the CNS. Movement in one direction (27) _____ the hair cells; movement in the opposite direction (28) _____ the hair cells. The type of movement that will stimulate a hair cell is dependent on location of the hair cell and the nature of (29) _____ present.

In the semicircular ducts, which are organized into three (30) _____, hair cells are surrounded by a cupula that (31) _____ on the endolymph. When the head is rotated in the appropriate plane, the cupula moves and (32) _____ the kinocilia and stereocilia in the process.

In the utricle and saccule, located in the vestibule of the internal ear, hair cells are clustered in maculae that have surfaces covered with otolith. The otolith slides with (33) _____ when the head is tilted, and as it slides it moves the kinocilia and stereocilia.

HEARING

The hair cells involved with our sense of hearing are located in the (34) _____ of the cochlea, which is in the internal ear. Sound waves, which are pressure waves, are funneled through the external acoustic meatus to the (35) _____, where they are transduced into mechanical movements by the auditory ossicles of the middle ear. At the (36) _____ of the internal ear, these mechanical movements are then transduced again into pressure waves that are transmitted along the fluid-filled chambers of the cochlea.

The scala vestibuli and scala tympani form the (37) _____-filled chamber through which the transduced pressure waves travel. Nestled between the scalae is another fluid-filled chamber called the (38) _____ (filled with [39] _____).

The hair cells that respond to sound are located in a structure in the cochlear duct called the spiral organ, embedded in the basilar membrane that forms the boundary between the cochlear duct and the scala tympani. When a pressure wave moves through the scala tympani, it displaces the basilar membrane in proportion to the intensity and frequency of the external sound it represents. The basilar membrane has regional differences that make certain sections more responsive than others to certain sound frequencies.

Displacement of the basilar membrane vibrates the (40) _____ of the hair cells against the (41) _____ membrane, which causes an influx of ions. As a result, the hair cells (42) _____ and release (43) _____ onto the sensory nerves. Sensory nerves then relay this information to the CNS for processing.

Concept Map I

Using the following terms, fill in the circled, numbered, blank spaces to correctly complete the concept map. Use each term only once.

retina	audition	rods and cones
olfaction	smell	ears
taste buds	tongue	balance and hearing

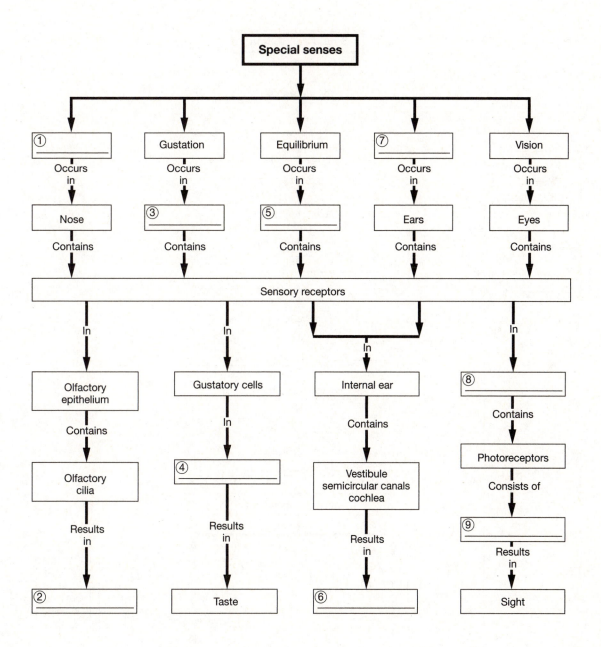

Concept Map il

Using the following terms, fill in the circled, numbered, blank spaces to correctly complete the concept map. Use each term only once.

round window

oval window

middle ear

vestibulocochlear nerve (VIII)

hair cells of organ of Corti

incus

endolymph in cochlear duct

external acoustic canal

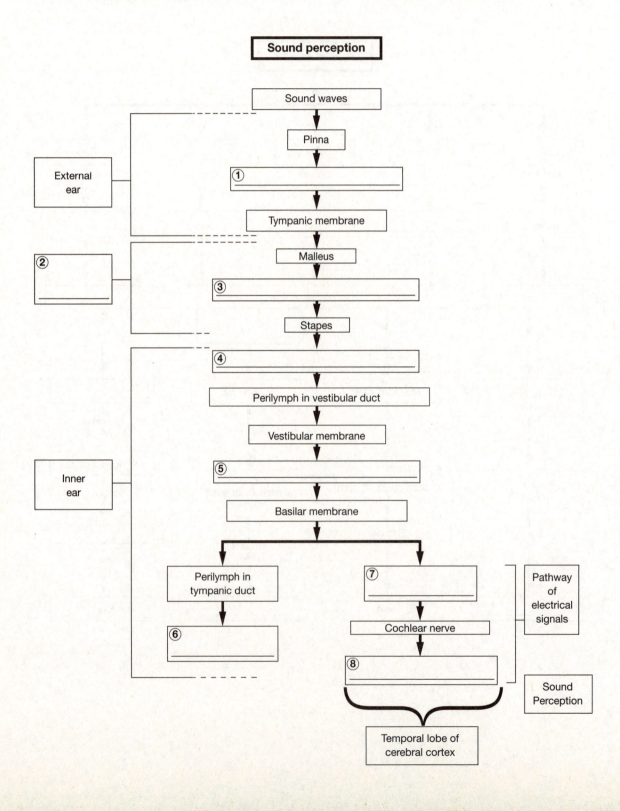

Multiple Choice

Place the letter corresponding to the best answer in the space provided.

_____ 1. Olfactory secretions that absorb water and form a thick, pigmented mucus are produced by

     a. basal cells.

     b. olfactory bulbs.

     c. olfactory glands.

     d. salivary glands.

_____ 2. During the olfaction process, the first synapse occurs at the

     a. olfactory bulbs of the cerebrum.

     b. initial receptor site.

     c. olfactory cortex of the cerebral hemisphere.

     d. olfactory epithelium.

_____ 3. One of the few examples of neuronal replacement occurs in the _____ population.

     a. gustatory receptor

     b. olfactory receptor

     c. optic receptor

     d. vestibuloreceptor

_____ 4. The tongue papillae that provide friction to move objects around in the mouth, but do not contain taste buds, are

     a. fungiform papillae.

     b. filiform papillae.

     c. circumvallate papillae.

     d. none of the above.

_____ 5. The ultimate higher-order olfactory destination is the

     a. thalamus.

     b. medulla oblongata.

     c. hypothalamus.

     d. olfactory cortex.

_____ 6. The mechanism that quickly reduces your sensitivity to a new taste is

     a. central adaptation.

     b. neurotransmitter release.

     c. receptor stimulation threshold.

     d. exposure to different papillae.

_____ 7. A taste receptor sensitive to dissolved chemicals but insensitive to pressure illustrates the concept of

     a. transduction.

     b. receptor specificity.

     c. receptor potential.

     d. phasic reception.

_____   8. During the focusing process, when light travels from the air into the relatively dense cornea,

    a. the sclera assumes an obvious color.

    b. the light path is bent.

    c. pupillary reflexes are triggered.

    d. reflexive adjustments occur in both pupils.

_____   9. Exposure to bright light produces a

    a. rapid reflexive increase in pupillary diameter.

    b. slow reflexive increase in pupillary diameter.

    c. very slow reflexive decrease in pupillary diameter.

    d. rapid reflexive decrease in pupillary diameter.

_____   10. The color of the eye is determined by

    a. light reflecting through the cornea onto the retina.

    b. the thickness of the iris and the number and distribution of pigment cells.

    c. the reflection of light from the aqueous humor.

    d. the number and distribution of pigment cells in the lens.

_____   11. In rating visual acuity, a person whose vision is rated 20/15 is better than normal since this person can

    a. read letters at 20 feet that are only discernible by the normal eye at 15 feet from the chart.

    b. read letters at 15 feet that are only discernible by the normal eye at 20 feet from the chart.

    c. read letters at 15 or 20 feet that normal individuals cannot read at all.

    d. read letters without the aid of a lens at 35 feet from the chart.

_____   12. After an intense exposure to light, a "ghost" image remains on the retina because a photoreceptor cannot respond to stimulation until

    a. the rate of neurotransmitter release declines at the receptor membrane.

    b. the receptor membrane channels close.

    c. opsin activates transducin.

    d. its rhodopsin molecules have been regenerated.

_____   13. The role that vitamin A plays in the eye is

    a. the visual pigment _retinal_ is synthesized from vitamin A.

    b. the protein part of rhodopsin is synthesized from vitamin A.

    c. it acts as a coenzyme for the activation of transducin.

    d. it activates opsin for rhodopsin-based photoreception.

_____   14. The reason everything appears to be shades of gray when we enter dimly lighted surroundings is

    a. rods and cones are stimulated.

    b. only cones are stimulated.

    c. only rods are stimulated.

    d. only the blue cones are stimulated.

_____ 15. When one or more classes of cones are nonfunctional, the result is

     a. a blind spot occurs.

     b. color blindness.

     c. image inversion.

     d. the appearance of "ghost" images.

_____ 16. The most detailed information about the visual image is provided by the

     a. cones.

     b. rods.

     c. optic disc.

     d. rods and cones.

_____ 17. The region of the retina called the "blind spot" is an area that structurally comprises the

     a. choroid coat.

     b. suprachiasmatic nucleus.

     c. optic disc.

     d. visual cortex.

_____ 18. The partial cross-over that occurs at the optic chiasm ensures that the visual cortex receives

     a. an image that is inverted.

     b. an image that is reversed before reaching the cortex.

     c. different images from the right and left eyes.

     d. a composite picture of the entire visual field.

_____ 19. The waxy material that slows the growth of microorganisms in the external acoustic canal and reduces the chances of infection is

     a. gustducin.

     b. phenylthiourea.

     c. cerumen.

     d. umami.

_____ 20. The auditory ossicles found in the middle ear resemble "tools" that would be found in a

     a. fitness center.

     b. horse stable.

     c. pig sty.

     d. mechanic's tool box.

_____ 21. Information about the direction and strength of mechanical stimuli is provided by the

     a. hair cells.

     b. endolymph.

     c. vestibular ganglia.

     d. auditory ossicles.

_____ 22. The receptors in the internal ear that provide sensations of gravity and linear acceleration are the

     a. ampulla and capula.

     b. otolith and statoconia.

     c. semicircular and cochlear ducts.

     d. saccule and utricle.

_____ 23. The intensity (volume) of a perceived sound is determined by

    a. which part of the cochlear duct is stimulated.

    b. how many of the hair cells in the cochlear duct are stimulated.

    c. pressure fluctuations in the endolymph of the vestibular complex.

    d. tectorial membrane vibrations.

_____ 24. Information about the region and intensity of auditory stimulation is relayed to the CNS over the cochlear branch of

    a. N IV.

    b. N VI.

    c. N VIII.

    d. N X.

_____ 25. The energy content of a sound determines its intensity, which is measured in

    a. millimeters.

    b. furlongs.

    c. hertz.

    d. decibels.

Completion

Using the terms below, complete the following statements. Use each term only once.

odorants	nystagmus	phasic
retina	olfactory stimulation	taste buds
receptive field	accommodation	afferent fiber
pupil	receptor potential	transduction
cataract	hyperopia	adaptation
hertz	sensory receptor	sensation
myopia	aqueous humor	ampulla

1. A specialized cell that monitors conditions in the body or the external environment is a(n) _____.

2. The information passed to the CNS is called a(n) _____.

3. The area monitored by a single receptor cell is its _____.

4. The process of translating a stimulus into an action potential is called _____.

5. The change in the transmembrane potential that accompanies receptor stimulation is called a(n) _____.

6. An action potential travels to the CNS over a(n) _____.

7. A reduction in sensitivity in the presence of a constant stimulus is _____.

8. Receptors that provide information about the intensity and rate of change of a stimulus are called _____ receptors.

9. Sensory receptors in the semicircular canals in the internal ear are located in the _____.

10. The condition in which short, jerky eye movements appear after damage to the brain stem or internal ear is referred to as _____.

11. The frequency of sound is measured in units called _____.

12. Chemicals that stimulate olfactory receptors are called _____.

13. The only type of sensory information that reaches the cerebral cortex without first synapsing in the thalamus is _____.

14. Taste receptors and specialized epithelial cells form sensory structures called _____.

15. When the muscles of the iris contract, they change the diameter of the central opening referred to as the _____.

16. The anterior cavity of the eye contains a fluid called _____.

17. The visual receptors and associated neurons in the eye are contained in the _____.

18. When the lens of the eye loses its transparency, this abnormality is known as a _____.

19. When the lens of the eye becomes rounder to focus the image of a nearby object on the retina, the mechanism is called _____.

20. If a person sees objects clearly at close range but not at a distance, the individual is said to be "nearsighted," a condition formally termed _____.

21. If a person sees distant objects more clearly than close objects, the individual is said to be "farsighted," a condition formally termed _____.

Short Essay

Briefly answer the following questions in the spaces provided below.

1. What sensations are included as *special senses*?

2. What three major steps are involved in the process of translating a stimulus into an action potential (transduction)?

3. Trace an olfactory sensation from the time it leaves the olfactory bulb until it reaches its final destinations in the higher centers of the brain.

4. What are the four *primary taste sensations*?

5. What sensations are provided by the *saccule* and *utricle* in the vestibule of the internal ear?

6. What are the four primary functions of the *fibrous layer*, which consists of the *sclera* and the *cornea*?

7. What are the four primary functions of the *vascular layer*, which consists of the iris, the ciliary body, and the choroid?

8. What are the primary functions of the eye's *inner layer*, which consists of an outer pigment layer and an inner retina that contains the visual receptors and associated neurons?

9. When referring to the eye, what is the purpose of *accommodation,* and how does it work?

10. Why are cataracts common in the elderly?

LEVEL 3: CRITICAL THINKING AND CLINICAL APPLICATIONS

Using principles and concepts learned in Chapter 17, answer the following questions. Write your answers on a separate sheet of paper.

1. The perfume industry expends considerable effort to develop scents that trigger sexual responses. How might this effort be anatomically related to the processes involved in olfaction?

2. Grandmother and Granddad are all "decked out" for a night of dining and dancing. Grandma has doused herself with her finest perfume and Grandpa has soaked his face with aftershave. The scents are overbearing to others; however, the old folks can barely smell the aromas. Why?

3. Suppose you were small enough to wander around in the middle ear. Describe the structures you would encounter as you moved from the external ear to the internal ear. Be sure to note any incident of signal transduction you might witness in this region.

4. We have all heard the statement, "Eat carrots because they are good for your eyes." What is the relationship between eating carrots and the "health" of the eye?

5. A bright flash of light temporarily blinds normal, well-nourished eyes. As a result, a "ghost" image remains on the retina. Why?

The Endocrine System

OVERVIEW

The nervous system and the endocrine system are the metabolic control systems in the body. Together they monitor and adjust physiological activities throughout the body to maintain homeostasis. The effects of nervous system regulation are usually rapid and short term, whereas the effects of endocrine regulation are ongoing and long term.

Endocrine cells are glandular secretory cells that release chemicals, called hormones, into the bloodstream for distribution to target tissues throughout the body. The influence of these chemical "messengers" results in facilitating processes that include growth and development, sexual maturation and reproduction, and the maintenance of homeostasis within other systems.

Chapter 18 consists of exercises that will test your knowledge of the endocrine organs and their functions. The integrative and complementary activities of the nervous and endocrine systems are presented to show the coordinated effort needed to properly regulate physiological activities in the body.

LEVEL 1: REVIEWING FACTS AND TERMS

Review of Learning Outcomes

After completing this chapter, you should be able to do the following:

OUTCOME 18-1   Explain the importance of intercellular communication, describe the mechanisms involved, and compare the modes of intercellular communication that occur in the endocrine and nervous systems.

OUTCOME 18-2   Compare the cellular components of the endocrine system with those of other systems, contrast the major structural classes of hormones, and explain the general mechanisms of hormonal action on target organs.

OUTCOME 18-3   Describe the location, hormones, and functions of the pituitary gland, and discuss the effects of abnormal pituitary hormone production.

OUTCOME 18-4   Describe the location, hormones, and functions of the thyroid gland, and discuss the effects of abnormal thyroid hormone production.

OUTCOME 18-5   Describe the location, hormone, and functions of the parathyroid glands, and discuss the effects of abnormal parathyroid hormone production.

OUTCOME 18-6   Describe the location, structure, hormones, and general functions of the adrenal glands, and discuss the effects of abnormal adrenal hormone production.

OUTCOME 18-7   Describe the location of the pineal gland, and discuss the functions of the hormone it produces.

OUTCOME 18-8   Describe the location, structure, hormones, and functions of the pancreas, and discuss the effects of abnormal pancreatic hormone production.

OUTCOME 18-9   Describe the functions of the hormones produced by the kidneys, heart, thymus, testes, ovaries, and adipose tissue.

OUTCOME 18-10   Explain how hormones interact to produce coordinated physiological responses and influence behavior, describe the role of hormones in the general adaptation syndrome, discuss how aging affects hormone production, and give examples of interactions between the endocrine system and other organ systems.

Multiple Choice

Place the letter corresponding to the best answer in the space provided.

OUTCOME 18-1   _____   1. Cellular activities are coordinated throughout the body by the endocrine system to
    a. preserve homeostasis.
    b. maintain cellular structure.
    c. maintain intracellular organization.
    d. all of the above.

OUTCOME 18-1   _____   2. The endocrine system regulates long-term processes such as growth and development by using
    a. DNA and RNA to control cellular metabolic activities.
    b. physical changes to control homeostatic regulation.
    c. chemical messengers to relay information and instructions between cells.
    d. nuclear control of cells to produce desired regulation.

OUTCOME 18-1   _____   3. Situations requiring split-second responses, or crisis management, are the job of the _____ system.
    a. nervous
    b. circulatory
    c. endocrine
    d. muscular

OUTCOME 18-1   _____   4. Gap junctions involved in direct communication between two cells of the same type serve to
    a. coordinate ciliary movement among epithelial cells.
    b. coordinate the contractions of cardiac muscle cells.
    c. facilitate the propagation of action potentials from one neuron to the next at electrical synapses.
    d. all of the above.

OUTCOME 18-1   _____   5. An example of a functional similarity between the nervous system and the endocrine system is
    a. both systems secrete hormones into the bloodstream.
    b. the cells of the endocrine and nervous systems are functionally the same.
    c. compounds used as hormones by the endocrine system may also function as neurotransmitters inside the CNS.
    d. both produce very specific responses to environmental stimuli.

OUTCOME 18-1   _____   6. The endocrine system expresses an organization that allows for
a. localized effects.
b. effects on distant targets.
c. simultaneous coordination of many widespread targets.
d. a, b, and c are correct.

OUTCOME 18-2   _____   7. The three different groups of hormones based on their chemical structure are
a. catecholamines, prehormones, and prohormones.
b. amino acid derivatives, peptide hormones, and lipid derivatives.
c. intercellular, intracellular, and G proteins.
d. peptides, dipeptides, and polypeptides.

OUTCOME 18-2   _____   8. The biogenic amino hormones include
a. E, NE, dopamine, melatonin, and thyroid hormones.
b. oxytocin, ADH, GH, and prolactin.
c. androgens, estrogens, corticosteroids, and calcitriol.
d. leukotrienes, prostaglandins, and thromboxanes.

OUTCOME 18-2   _____   9. Most peptide hormones are produced as
a. catecholamines.
b. eicosanoids.
c. prostacyclins.
d. prohormones.

OUTCOME 18-2   _____   10. The reason steroid hormones remain in circulation longer than the other groups of hormones is because they are
a. produced as prohormones.
b. bound to specific transport proteins in the plasma.
c. side chains attached to the basic ring structure.
d. attached to the hemoglobin molecule for cellular use.

OUTCOME 18-2   _____   11. Hormones alter the operations of target cells by changing the
a. cell membrane permeability properties.
b. types, quantities, or activities of important enzymes and structural proteins.
c. arrangement of the molecular complex of the cell membrane.
d. rate at which hormones affect the target organ cells.

OUTCOME 18-2   _____   12. Catecholamines and peptide hormones affect target organ cells by
a. binding to receptors in the cytoplasm.
b. binding to target receptors in the nucleus.
c. enzymatic reactions that occur in the ribosomes.
d. second messengers activated by G protein complexes when receptor binding occurs at the membrane surface.

OUTCOME 18-2   _____   13. Steroid hormones affect target organ cells by
a. target receptors in peripheral tissues.
b. releasing second messengers at cell membrane receptors.
c. binding to receptors in the cell membrane.
d. binding to target receptors in the nucleus.

OUTCOME 18-2   _____ 14. Endocrine cells responding directly to changes in the composition of the extracellular fluid are an example of a(n)

    a. neural reflex.

    b. reflex arc.

    c. endocrine reflex.

    d. hypothalamic control mechanism.

OUTCOME 18-2   _____ 15. The hypothalamus is a major coordinating and control center because it

    a. contains autonomic centers and acts as an endocrine organ.

    b. initiates endocrine and neural reflexes.

    c. stimulates appropriate responses by peripheral target cells.

    d. stimulates responses to restore homeostasis.

OUTCOME 18-3   _____ 16. The most notable effect of ADH produced in the neurohypophysis of the pituitary gland is to

    a. increase the amount of water lost at the kidneys.

    b. decrease the amount of water lost at the kidneys.

    c. stimulate the contraction of uterine muscles.

    d. increase or decrease calcium ion concentrations in body fluids.

OUTCOME 18-3   _____ 17. Stimulation of contractile cells in mammary tissue and uterine muscles in the female is initiated by secretion of

    a. oxytocin from the posterior pituitary.

    b. melatonin from the pineal gland.

    c. oxytocin from the adenohypophysis.

    d. melatonin from the neurohypophysis.

OUTCOME 18-3   _____ 18. The structure that locks the pituitary gland in position and isolates it from the cranial cavity is the

    a. infundibulum.

    b. sellar diaphragm.

    c. pars intermedia.

    d. pars distalis.

OUTCOME 18-3   _____ 19. The pituitary gland is connected to the hypothalamus by a slender, funnel-shaped structure called the

    a. hypophysis.

    b. sella turcica.

    c. adenohypophysis.

    d. infundibulum.

OUTCOME 18-3   _____ 20. The small, oval pituitary gland lies nestled within a depression of the sphenoid bone called the

    a. sella turcica.

    b. infundibulum.

    c. diaphragma sellae.

    d. pars tuberalis.

OUTCOME 18-3 _____ 21. The two classes of hypothalamic regulatory hormones are
   a. humoral and neural.
   b. releasing and inhibiting.
   c. cyclic-AMP and cyclic-GMP.
   d. G proteins and kinase C (PKC)

OUTCOME 18-3 _____ 22. The two *gonadotropic* hormones secreted by the *anterior pituitary* are
   a. TSH and ACTH.
   b. PRL and GH.
   c. LH and FSH.
   d. MSH and FSH.

OUTCOME 18-3 _____ 23. The hormonal effect of ACTH produced by the adenohypophysis is
   a. secretion of thyroid hormones by the thyroid glands.
   b. milk production in the female by the mammary glands.
   c. protein synthesis by body cells.
   d. secretion of glucocorticoids by the adrenal cortex.

OUTCOME 18-3 _____ 24. The gonadotropins FSH and LH exert their hormonal effects in the
   _____ of the male and female.
   a. reproductive system
   b. excretory system
   c. circulatory system
   d. integumentary system

OUTCOME 18-3 _____ 25. An abnormally low production of gonadotropin produces
   a condition called
   a. diabetogenesis.
   b. somatomedins.
   c. hypogonadism.
   d. follitropism.

OUTCOME 18-3 _____ 26. When the posterior lobe of the pituitary gland no longer releases
   adequate amounts of ADH, the result is development of
   a. hypergonadism.
   b. diabetes insipidus.
   c. decreased urine production.
   d. decreased blood glucose levels.

OUTCOME 18-3 _____ 27. If abnormal amounts of glucocorticoids are released by the adrenal
   cortex, it affects
   a. glucose metabolism.
   b. synthesis and release of somatomedins.
   c. production of gonadotropins.
   d. synthesis and release of sex hormones.

OUTCOME 18-4 _____ 28. The thyroid gland curves across the anterior surface of the trachea just
   inferior to the
   a. sternum.
   b. thyroid cartilage.
   c. scapula and clavicles.
   d. dorsal ascending aorta.

OUTCOME 18-4   _____ 29. Structurally, the thyroid gland is best described as

    a. a pea-sized gland divided into anterior and posterior lobes.

    b. pyramid-shaped glands embedded in the parathyroids.

    c. a two-lobed gland united by an isthmus.

    d. a slender, pear-shaped gland with a lumpy consistency.

OUTCOME 18-4   _____ 30. The hormones produced by the thyroid gland are

    a. thymosins, TSH, and PTH.

    b. calcitriol and TSH.

    c. thymosins, triiodothyronine, and calcitriol.

    d. thyroxine, triiodothyronine, and calcitonin.

OUTCOME 18-4   _____ 31. The overall effect of hormones secreted by the thyroid gland is

    a. a decrease in the rate of mitochondrial ATP production.

    b. maturation and functional competence of the immune system.

    c. an increase in cellular rates of metabolism and oxygen consumption.

    d. increased reabsorption of sodium ions and water from urine.

OUTCOME 18-4   _____ 32. Inadequate production of thyroid hormones in infants produces cretinism, a condition marked by

    a. inadequate skeletal and nervous system development.

    b. increased blood pressure and heart rate.

    c. irregular heart beat and flushed skin.

    d. shifts in mood and emotional states.

OUTCOME 18-4   _____ 33. Hypothyroidism, myxedema, and goiter may result from an inadequate amount of

    a. salt in the diet.

    b. dietary iodide.

    c. sodium and calcium.

    d. magnesium and zinc.

OUTCOME 18-5   _____ 34. The two pairs of parathyroid glands are embedded in the

    a. posterior surfaces of the thyroid glands.

    b. anterior surfaces of the thyroid glands.

    c. lateral surfaces of the thyroid cartilage.

    d. anterior surfaces of the thyroid cartilage.

OUTCOME 18-5   _____ 35. The net result of parathyroid hormone (PTH) secretion is that it

    a. inhibits the reabsorption of Ca^{2+} at the kidneys.

    b. increases the rate of calcium deposition in bone.

    c. increases Ca^{2+} concentration in body fluids.

    d. increases the urinary output and losses.

OUTCOME 18-6   _____ 36. The highly vascularized, pyramid-shaped adrenal gland is located on the

    a. superior surface of the thyroid gland.

    b. anterior surface of the trachea inferior to the thyroid cartilage.

    c. posterior surface of the hypothalamus in the sella turcica.

    d. superior border of each kidney.

OUTCOME 18-6 _____ 37. The adrenal cortex consists of a
    a. posterior, anterior, and lateral lobe.
    b. zona glomerulosa, zona fasciculata, and zona reticularis.
    c. pars tuberalis, pars distalis, and pars intermedia.
    d. C cell population, chief cells, and oxyphils.

OUTCOME 18-6 _____ 38. The hormones released by the adrenal medulla are
    a. epinephrine and norepinephrine.
    b. mineralocorticoids and glucocorticoids.
    c. cortisol and corticosterone.
    d. aldosterone and androgens.

OUTCOME 18-6 _____ 39. The hormonal effect(s) produced by adrenal epinephrine is (are)
    a. increased cardiac activity.
    b. increased blood glucose levels.
    c. increased glycogen breakdown.
    d. a, b, and c are correct.

OUTCOME 18-6 _____ 40. The mineralocorticoid aldosterone targets kidney cells and causes the
    a. loss of sodium ions and water, increasing body fluid losses.
    b. retention of sodium ions and water, reducing body fluid losses.
    c. production and release of angiotensin II by the kidneys.
    d. secretion of the enzyme renin by the kidneys.

OUTCOME 18-7 _____ 41. The pineal gland is located in the
    a. area between the kidneys and the diaphragm.
    b. neck region inferior to the thyroid gland.
    c. posterior portion of the roof of the third ventricle.
    d. sella turcica of the sphenoid bone.

OUTCOME 18-7 _____ 42. The pinealocytes of the pineal gland produce the hormone melatonin, which is believed to
    a. be an effective antioxidant.
    b. maintain basic circadian rhythms.
    c. inhibit reproductive function.
    d. a, b, and c are correct.

OUTCOME 18-8 _____ 43. The pancreas lies in the J-shaped loop between the stomach and the small intestine in the _____ cavity.
    a. cranial
    b. thoracic
    c. abdominopelvic
    d. pelvic

OUTCOME 18-8 _____ 44. The cells of the pancreatic islets of the endocrine pancreas include
    a. alpha, beta, delta, and F cells.
    b. lymphocytes, eosinophils, leukocytes, and neutrophils.
    c. epithelial, gland cells, goblet cells, and adipocytes.
    d. mast cells, melanocytes, mesenchymal cells, and macrophages.

OUTCOME 18-8 _____ 45. The hormones produced by the pancreatic islets that regulate blood glucose concentrations are
    a. cortisol and androgens.
    b. leptin and resistin.
    c. insulin and glucagon.
    d. calcitonin and calcitriol.

OUTCOME 18-8 _____ 46. The pancreatic hormone pancreatic polypeptide (PP)
    a. regulates production of some pancreatic enzymes.
    b. inhibits gallbladder contractions.
    c. may control the rate of nutrient absorption.
    d. a, b, and c are correct.

OUTCOME 18-8 _____ 47. Inadequate insulin production can lead to
    a. diabetes mellitus.
    b. hyperglycemia.
    c. glycosuria.
    d. a, b, and c are correct.

OUTCOME 18-8 _____ 48. A common diabetes-related medical problem is
    a. partial or complete blindness.
    b. low blood pressure.
    c. retarded growth.
    d. muscular weakness.

OUTCOME 18-9 _____ 49. The normal effect(s) of erythropoietin (EPO) produced by the kidney is (are) to
    a. stimulate calcium and phosphate absorption.
    b. stimulate red blood cell production.
    c. inhibit PTH secretion.
    d. a, b, and c are correct.

OUTCOME 18-9 _____ 50. One of the major effects of ANP produced by the heart is that it
    a. assists with the production of red blood cells.
    b. constricts peripheral blood vessels.
    c. decreases thirst.
    d. promotes water retention at the kidneys.

OUTCOME 18-9 _____ 51. The thymosins produced by the thymus affect the
    a. suppression of appetite.
    b. coordination of digestive activities.
    c. development and maturation of lymphocytes.
    d. suppression of insulin response.

OUTCOME 18-9 _____ 52. Testosterone produced in the interstitial cells of the testes is responsible for
    a. promoting the production of functional sperm.
    b. determining male secondary sexual characteristics.
    c. protein synthesis in skeletal muscles.
    d. a, b, and c are correct.

OUTCOME 18-9 _____ 53. The hormone leptin produced in adipose tissues targets the hypothalamus to affect the

    a. suppression of the insulin response.

    b. release of calcium salts from bone.

    c. suppression of appetite.

    d. suppression of ADH secretion.

OUTCOME 18-10 _____ 54. Of the following choices, the two hormones that have an antagonistic or opposing effect are

    a. mineralocorticoids and glucocorticoids.

    b. insulin and glucagon.

    c. GH and glucocorticoids.

    d. antidiuretic hormone and arginine vasopressin.

OUTCOME 18-10 _____ 55. The additive effects of growth hormone (GH) and glucocorticoids illustrate the _____ effect.

    a. antagonistic

    b. permissive

    c. integrative

    d. synergistic

OUTCOME 18-10 _____ 56. The differing but complementary effects of calcitriol and parathyroid hormone on tissues involved in calcium metabolism illustrate the _____ effect.

    a. antagonistic

    b. additive

    c. permissive

    d. integrative

OUTCOME 18-10 _____ 57. The hormones that are of primary importance to normal growth include

    a. TSH, ACTH, insulin, parathyroid hormone, and LH.

    b. GH, thyroid hormones, insulin, PTH, and reproductive hormones.

    c. ADH, AVP, E, NE, and PTH.

    d. mineralocorticoids, glucocorticoids, ACTH, LH, and insulin.

OUTCOME 18-10 _____ 58. Insulin is important to normal growth because it promotes

    a. adequate supplies of nutrients and energy for growing cells.

    b. absorption of calcium salts for deposition in bone.

    c. normal development of the nervous system.

    d. a, b, and c are correct.

OUTCOME 18-10 _____ 59. The body's response to stress is the general adaptation syndrome (GAS), which consists of the phase(s):

    a. alarm phase.

    b. resistance phase.

    c. exhaustion phase.

    d. a, b, and c are correct.

OUTCOME 18-10   _____ 60. When the body's temperature falls below normal, the homeostatic adjustment responses will include

    a. mobilization of glucose reserves.

    b. increases in heart rate and respiratory rates.

    c. shivering or changes in the pattern of blood flow.

    d. conservation of salts and body water.

OUTCOME 18-10   _____ 61. In children, when sex hormones are produced prematurely, the obvious behavioral changes will cause the child to be

    a. lethargic and docile.

    b. aggressive and assertive.

    c. regressive and introverted.

    d. overweight and detached.

OUTCOME 18-10   _____ 62. In adults, changes in the mixtures of hormones reaching the CNS can have significant effects on

    a. memory and learning.

    b. emotional states.

    c. intellectual capabilities.

    d. a, b, and c are correct.

Completion

Using the terms below, complete the following statements. Use each term only once.

TERMS FOR PART I

infundibulum	thyroid gland	isthmus
cortex	plasma membrane	Na^+ ions and water
calcitonin	bones	tropic hormones
hypothalamus	sella turcica	peritoneal cavity
diabetes insipidus	paracrine	gigantism
negative feedback	catecholamines	trachea
neurotransmitters	adrenal gland	Graves' disease
long-term	ADH; oxytocin	
parathyroid hormone (PTH)	aldosterone	

TERMS FOR PART II

androgens	antioxidant	general adaptation syndrome
GH	glucagons	synergism
exhaustion	epithalamus	thyroid hormone
glucocorticoids	leptin	integrative
islets of Langerhans	stress	abdominopelvic cavity
kidneys	thymosins	estrogens
diabetes mellitus	insulin	precocious puberty
Addison's disease	aldosteronism	PTH and calcitriol

PART I

OUTCOME 18-1   1. The type of regulation provided by the endocrine system is _____.

OUTCOME 18-1   2. The use of chemical messengers to transfer information from cell to cell within a single tissue is called _____ communication.

OUTCOME 18-1   3. Both the endocrine and nervous systems are regulated by _____ control mechanisms.

OUTCOME 18-1   4. Many compounds used as hormones by the endocrine system also function inside the nervous system as _____.

OUTCOME 18-2   5. An example of a gland in the body that functions as a neural tissue and an endocrine tissue is the _____.

OUTCOME 18-2   6. Epinephrine, norepinephrine, and dopamine are structurally similar and belong to a class of compounds sometimes called _____.

OUTCOME 18-2   7. The receptors for peptide hormones are most likely found in the _____ of target cells.

OUTCOME 18-2   8. The most complex endocrine responses are directed by the _____.

OUTCOME 18-3   9. The hypophysis lies nestled within a depression in the sphenoid bone called the _____.

OUTCOME 18-3   10. The pituitary gland is connected to the hypothalamus by a slender, funnel-shaped structure called the _____.

OUTCOME 18-3   11. The hypothalamic neurons of the posterior lobe of the pituitary gland manufacture _____ and _____.

OUTCOME 18-3   12. Because they "turn on" endocrine glands or support the functions of other organs, the hormones of the anterior lobe are called _____.

OUTCOME 18-3   13. Underproduction of ADH by the posterior lobe of the pituitary gland results in a condition called _____.

OUTCOME 18-3   14. Overproduction of GH by the anterior lobe of the pituitary gland results in a condition called _____.

OUTCOME 18-4   15. The two lobes of the thyroid gland are united by a slender connection, the _____.

OUTCOME 18-4   16. The thyroid gland curves across the anterior surface of the _____.

OUTCOME 18-4   17. The hormone produced by the C cells of the thyroid gland that causes a decrease in Ca^{2+} concentrations in body fluids is _____.

OUTCOME 18-4   18. Overproduction of T_3 and T_4 by the thyroid gland results in a condition called _____.

OUTCOME 18-5   19. The two pairs of parathyroid glands are embedded in the posterior surfaces of the _____.

OUTCOME 18-5   20. The hormone secreted by the chief cells of the parathyroid gland that causes an increase in Ca^{2+} concentrations in body fluids is _____.

OUTCOME 18-5   21. Hyperparathyroidism, an overproduction of PTH, may cause weak and brittle _____.

OUTCOME 18-6   22. The adrenal glands project into the _____.

OUTCOME 18-6   23. The superficial portion of the adrenal glands is the _____.

OUTCOME 18-6   24. The principal mineralocorticoid produced by the adrenal cortex is _____.

OUTCOME 18-6   25. The hormonal effect of aldosterone in the kidneys is to increase reabsorption of _____.

PART II

OUTCOME 18-6   26. The adrenal hormones that affect the release of amino acids from skeletal muscles and lipids from adipose tissues are the _____.

OUTCOME 18-6   27. An endocrine condition caused by an underproduction of glucocorticoids is _____.

OUTCOME 18-6   28. An overproduction of mineralocorticoids by the adrenal glands causes the condition called _____.

OUTCOME 18-7   29. The pineal gland is a part of the _____.

OUTCOME 18-7   30. Melatonin produced by the pineal gland may protect the CNS neurons from free radicals due to its identification as a very effective _____.

OUTCOME 18-8   31. The endocrine pancreas consists of cell clusters known as the _____.

OUTCOME 18-8   32. The pancreas lies within the _____.

OUTCOME 18-8   33. A peptide hormone released by beta cells when glucose levels exceed normal is _____.

OUTCOME 18-8   34. The hormone released by alpha cells that raises blood glucose levels is _____.

OUTCOME 18-8   35. An underproduction of insulin by the endocrine pancreas is one of the causes of a condition known as _____.

OUTCOME 18-9   36. The hormones released by the thymus that regulate the immune response are the _____.

OUTCOME 18-9   37. The hormones produced and released by the interstitial cells of the testes that produce male secondary sex characteristics are the _____.

OUTCOME 18-9   38. The hormones produced and released by the follicular cells of the ovaries that produce female secondary sex characteristics are the _____.

OUTCOME 18-9   39. The hormone produced in adipose tissue throughout the body that causes suppression of appetite is _____.

OUTCOME 18-10   40. When two hormones have an additive effect, the phenomenon is called _____.

OUTCOME 18-10   41. When hormones produce different but complementary results in specific tissues and organs, the effect is _____.

OUTCOME 18-10   42. The hormone that supports muscular and skeletal development in children is _____.

OUTCOME 18-10   43. Signs of cretinism such as mental retardation and an underdeveloped nervous system result from low levels of _____.

OUTCOME 18-10   44. The hormones that promote the absorption of calcium for well-developed, mineralized bones are _____.

OUTCOME 18-10   45. The stress responses caused by a wide variety of stress-causing factors are part of the _____.

OUTCOME 18-10   46. The phases of the GAS include alarm, resistance, and _____.

OUTCOME 18-10   47. Any physical or emotional condition that threatens homeostasis is a form of _____.

OUTCOME 18-10   48. When sex hormones are produced prematurely, causing development of adult secondary sex characteristics and significant behavioral change, it is referred to as _____.

OUTCOME 18-10   49. Aldosterone, ADH, ANP, and BNP adjust rates of fluid and electrolyte reabsorption in the _____.

Matching

Match the terms in column B with the terms in column A. Use letters for answers in the spaces provided. Use each term only once.

Part I

		Column A	Column B
OUTCOME 18-1	_____	1. gap junctions	A. cyclic-AMP
OUTCOME 18-1	_____	2. synaptic communication	B. tyrosine
OUTCOME 18-2	_____	3. endocrine cells	C. tropic hormones
OUTCOME 18-2	_____	4. eicosanoids	D. FSH
OUTCOME 18-2	_____	5. second messenger	E. glandular; secretory
OUTCOME 18-2	_____	6. hypothalamus	F. acromegaly
OUTCOME 18-3	_____	7. pituitary gland	G. calorigenic effect
OUTCOME 18-3	_____	8. gonadotropin	H. direct communication
OUTCOME 18-3	_____	9. excessive GH	I. parathyroid glands
OUTCOME 18-4	_____	10. thyroglobulin molecule	J. epinephrine
OUTCOME 18-4	_____	11. thyroid hormones	K. leukotrienes
OUTCOME 18-5	_____	12. chief cells	L. secretes releasing hormones
OUTCOME 18-6	_____	13. adrenal medulla	M. neural communication

Part II

		Column A	Column B
OUTCOME 18-6	_____	14. zona reticularis	N. acini (gland cells)
OUTCOME 18-6	_____	15. excessive glucocorticoids	O. integumentary system
OUTCOME 18-7	_____	16. pineal gland	P. sympathetic activation
OUTCOME 18-8	_____	17. exocrine pancreas	Q. secretes inhibin
OUTCOME 18-8	_____	18. pancreatic peptide	R. Cushing's disease
OUTCOME 18-8	_____	19. glucose in urine	S. protein synthesis
OUTCOME 18-9	_____	20. nurse cells	T. precocious puberty
OUTCOME 18-10	_____	21. antagonistic effect	U. androgens
OUTCOME 18-10	_____	22. growth hormone (GH)	V. PTH and calcitonin
OUTCOME 18-10	_____	23. alarm phase	W. F cells
OUTCOME 18-10	_____	24. behavioral changes	X. circadian rhythms
OUTCOME 18-10	_____	25. MSH stimulation	Y. glycosuria

Drawing/Illustration Labeling

Identify each numbered structure in the following figures. Place your answers in the spaces provided.

OUTCOME 18-2   **FIGURE 18-1**   The Endocrine System—Organs and Tissues

(1)_____

Production of ADH, oxytocin, and regulatory hormones

(2)_____

Adenohypophysis (anterior lobe):
    ACTH, TSH, GH, PRL, FSH, LH, and MSH
Neurohypophysis (posterior lobe):
    Release of oxytocin and ADH

(3)_____

Thyroxine (T_4)
Triiodothyronine (T_3)
Calcitonin (CT)

(4)_____

(Undergoes atrophy during adulthood)
Thymosins *(Chapter 22)*

(5)_____

Each suprarenal gland is subdivided into:

Suprarenal medulla:
    Epinephrine (E)
    Norepinephrine (NE)
Suprarenal cortex:
    Cortisol, corticosterone, aldosterone, androgens

(6)_____

Melatonin

(7)_____

(on posterior surface of thyroid gland)

Parathyroid hormone (PTH)

(8)_____

Natriuretic peptides:
Atrial natriuretic peptide (ANP)
Brain natriuretic peptide (BNP)
(Chapter 21)

(9)_____

Erythropoietin (EPO)
Calcitriol
(Chapters 19 and 26)

(10)_____

Leptin

(11)_____

Numerous hormones
(detailed in Chapter 25)

(12)_____

Insulin, glucagon

(13)_____

Testes (male):
    Androgens (especially testosterone), inhibin
Ovaries (female):
    Estrogens, progestins, inhibin

Testis

Ovary

OUTCOME 18-2 **FIGURE 18-2** Classification of Hormones by Chemical Structure

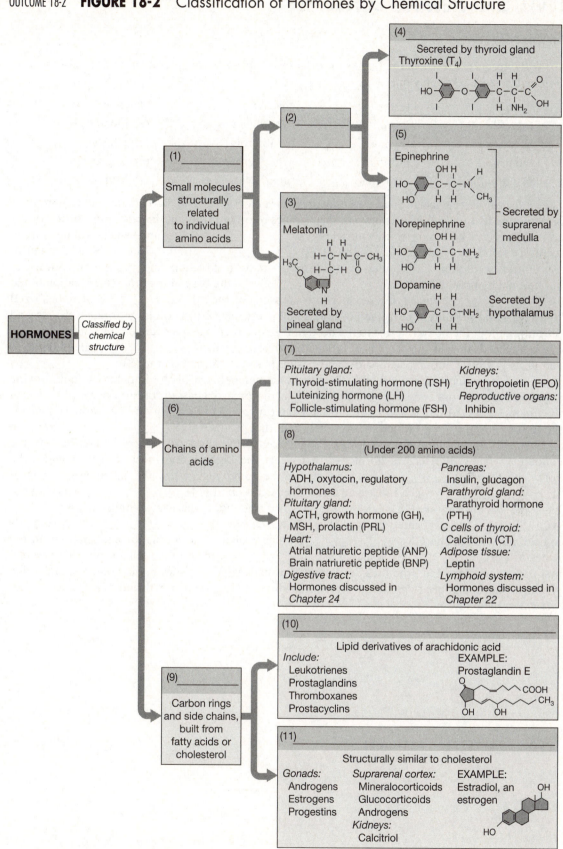

LEVEL 2: REVIEWING CONCEPTS

Chapter Overview

Using the terms below, fill in the blanks to correctly complete the chapter overview of the endocrine system. Use each term only once.

ovaries	testes	adrenal	pancreas
thyroid	insulin	hypothalamus	red blood cells
heart	thymus	parathyroid	testosterone
kidneys	pituitary	progesterone	pineal gland
melatonin	estrogen	sella turcica	erythropoietin

The endocrine system includes all the endocrine cells and tissues of the body that produce hormones or paracrine factors with effects beyond their tissues of origin. Lying in the posterior portion of the epithalamus attached to the third ventricle is the (1) _____ gland, which contains special secretory cells that synthesize the hormone (2) _____ from the molecules of the neurotransmitter serotonin. Continuing in an anterior–inferior direction and proceeding down the infundibulum, the stalk of the (3) _____ leads to the base of the brain, the location of the (4) _____, which is housed in a bony encasement, the (5) _____. This pea-sized "master" gland secretes nine tropic hormones that exert their effects on target organs, tissues, and cells throughout the body. Inferior to the head in the neck region and located on the anterior surface of the trachea is the (6) _____ gland, with four small (7) _____ glands embedded in its posterior surface. Just below and posterior to the sternum is the (8) _____, a gland that is primarily involved with the functional competence of the immune system. Moving slightly downward and away from the midline of the thorax, atrial natriuretic hormone is produced in the tissues of the (9) _____. Moving inferiorly and to the position behind the stomach is the (10) _____, which contains beta cells that produce (11) _____, a hormone necessary to lower blood glucose levels. Progressing to the superior borders of the kidneys, the endocrine organs that secrete hormones that affect the kidneys and most cells throughout the body are the (12) _____ glands. In the same region are the much larger pair of (13) _____, which release an endocrine secretion of (14) _____, which stimulates the production of (15) _____ by the bone marrow. Within the female, the (16) _____, located in the pelvic cavity, release (17) _____, which supports follicle maturation, and (18) _____, which prepares the uterus for implantation. In the male, the (19) _____ release (20) _____, the hormone that supports functional maturation of sperm and promotes development of male characteristics in the body. The endocrine system adjusts metabolic rates and substrate utilization and regulates growth and development for all systems in the body.

Concept Map I

Using the following terms, fill in the circled, numbered, blank spaces to correctly complete the concept map. Use each term only once.

male/female gonads   epinephrine   pineal

hormones   bloodstream   testosterone

heart   parathyroids   peptide hormones

pituitary

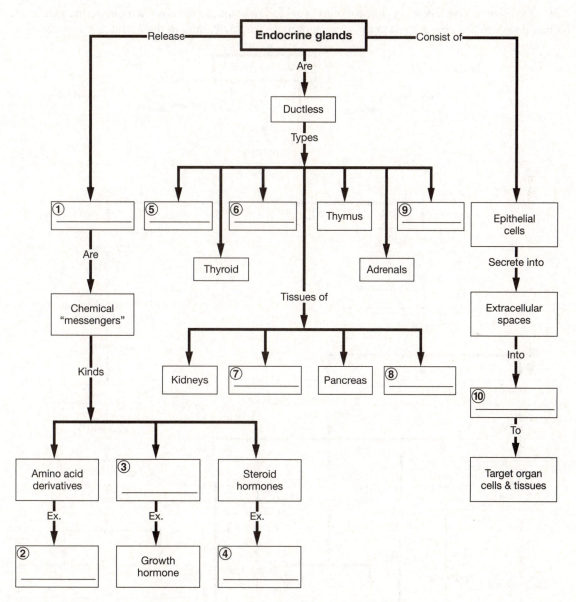

Concept Map II

Using the following terms, fill in the circled, numbered, blank spaces to correctly complete the concept map. Use each term only once.

pars nervosa	thyroid	adenohypophysis
mammary glands	melanocytes	posterior pituitary
sella turcica	ADH (vasopressin)	adrenal cortex
oxytocin	gonadotropic hormones	pars distalis

Note to student: The following concept map provides an example of how each endocrine gland can be mapped. You are encouraged to draw concept maps for each gland in the endocrine system.

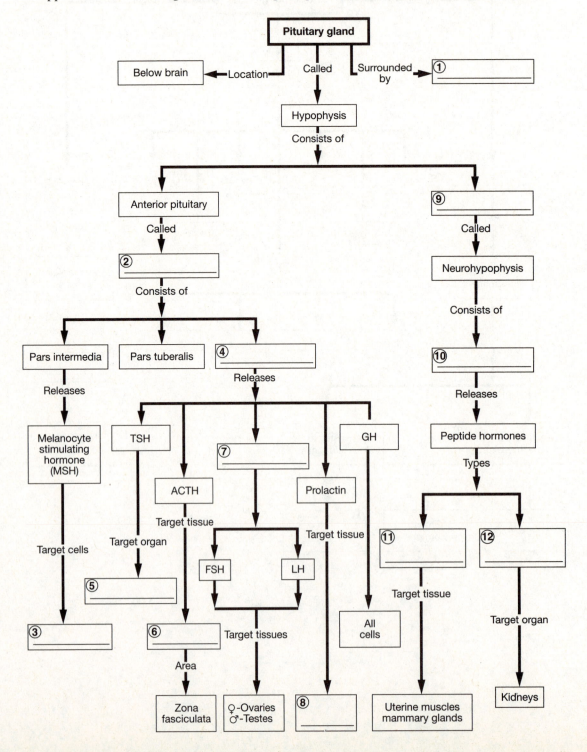

Concept Map III

Using the following terms, fill in the circled, numbered, blank spaces to correctly complete the concept map. Use each term only once.

homeostasis   hormones   target cells
contraction   ion channel opening   cellular communication

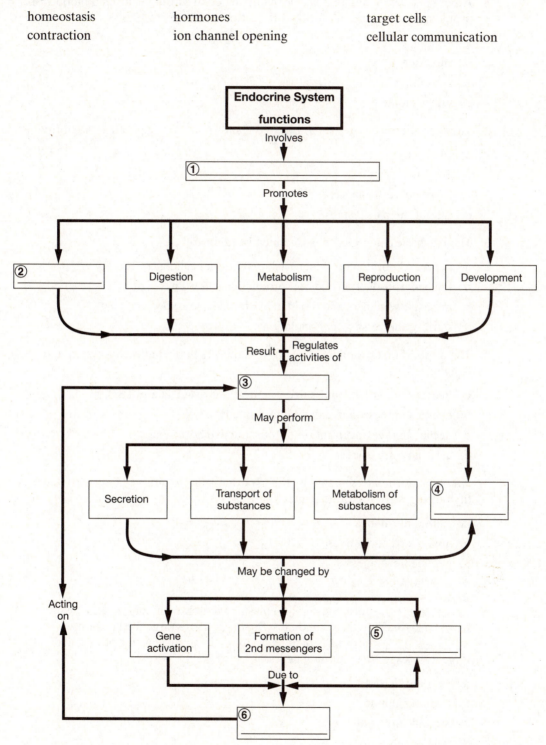

Multiple Choice

Place the letter corresponding to the best answer in the space provided.

_____ 1. An organic compound that has been implicated in causing heart and blood vessel problems, but is necessary for the production of corticosteroids, is

    a. sugar.

    b. cholesterol.

    c. unsaturated fat.

    d. saturated fat.

_____ 2. The specialized cells in the parathyroid glands that secrete parathyroid hormone are

    a. chief cells.

    b. C cells.

    c. follicular epithelial cells.

    d. cells of the pars distalis.

_____ 3. The two basic categories of endocrine disorders are

    a. abnormal hormone production and abnormal cellular sensitivity.

    b. the underproduction and overproduction syndromes.

    c. muscular weakness and the inability to tolerate stress.

    d. sterility and cessation of ovulation.

_____ 4. The most dramatic functional change that occurs in the endocrine system due to aging is a(n)

    a. decrease in blood and tissue concentrations of ADH and TSH.

    b. overall decrease in circulating hormone levels.

    c. decline in the concentration of reproductive hormones.

    d. a, b, and c are correct.

_____ 5. All of the hormones secreted by the hypothalamus, pituitary gland, heart, kidneys, thymus, digestive tract, and pancreas are

    a. steroid hormones.

    b. amino acid derivatives.

    c. peptide hormones.

    d. catecholamines.

_____ 6. _____ hormones bind to plasma membrane receptors attached to G proteins that activate second messengers to create target cell response; _____ hormones bind to cytoplasmic or nuclear receptors and create target cell response by turning genes on or off.

    a. Peptide; steroid

    b. Steroid; peptide

    c. Steroid; androgen

    d. Peptide; glycoprotein

_____ 7. The hypothalamus has a profound effect on endocrine functions through the secretion of

    a. neurotransmitters involved in reflex activity.

    b. anterior and posterior pituitary hormones.

    c. releasing and inhibiting hormones.

    d. a, b, and c are correct.

_____ 8. A pituitary endocrine cell that is stimulated by a releasing hormone is usually _____ by the peripheral hormone it controls.

    a. also stimulated

    b. not affected

    c. stimulated or inhibited

    d. inhibited

_____ 9. Endocrine reflexes, the functional counterparts of neural reflexes, can be triggered by _____ stimuli.

    a. humoral

    b. hormonal

    c. neural

    d. a, b, and c, are correct

_____ 10. The body system that adjusts metabolic rates and substrate utilization is the _____ system.

    a. nervous

    b. digestive

    c. lymphoid

    d. endocrine

_____ 11. Nurse cells in the testes support the

    a. production of male hormones known as androgens.

    b. development of CNS structures.

    c. differentiation and physical maturation of sperm.

    d. secretion of FSH at the adenohypophysis.

_____ 12. The protein hormone prolactin is involved with

    a. melanin synthesis.

    b. production of milk.

    c. production of testosterone.

    d. labor contractions and milk ejection.

_____ 13. The amino acid derivative hormone epinephrine is responsible for

    a. increased cardiac activity.

    b. glycogen breakdown.

    c. release of lipids by adipose tissues.

    d. a, b, and c are correct.

_____ 14. The zona fasciculata of the adrenal cortex produces steroid hormones that are known as glucocorticoids because of their

     a. anti-inflammatory activities.

     b. effects on glucose metabolism.

     c. effect on the electrolyte composition of body fluids.

     d. effect on glucagon production in the pancreas.

_____ 15. Corticosteroids are hormones that are produced and secreted by the

     a. medulla of the adrenal glands.

     b. cortex of the cerebral hemisphere.

     c. somatosensory cortex of the cerebrum.

     d. cortex of the adrenal glands.

_____ 16. The endocrine functions of the kidney and the heart include the production and secretion of the hormones

     a. renin and angiotensinogen.

     b. insulin and glucagon.

     c. erythropoietin and atrial natriuretic peptide.

     d. epinephrine and norepinephrine.

_____ 17. The testis of the male produces the endocrine hormones

     a. FSH and LH.

     b. inhibin and testosterone.

     c. LH and TSH.

     d. a, b, and c are correct.

_____ 18. The endocrine tissue of the ovaries in the female produce

     a. estrogens, inhibin, and progesterone.

     b. androgens and estrogens.

     c. FSH and LH.

     d. a, b, and c are correct.

Completion

Using the terms below, complete the following statements. Use each term only once.

reticularis	leptin	target cells
PTH	portal	cyclic-AMP
somatomedins	nucleus	fenestrated
hypothalamus	calmodulin	tyrosine
glucocorticoids	epinephrine	aldosterone
G protein	cytoplasm	

1. An important site of neural and endocrine integration is the _____.

2. The peripheral cells that are modified by the activities of a hormone are referred to as _____.

3. Some receptors for thyroid hormones are in the target cell's _____.

4. The hormone that promotes the absorption of calcium salts for bone deposition is _____.

5. Catecholamines and thyroid hormones are synthesized from molecules of the amino acid _____.

6. The feedback control of appetite, providing a sense of satiation and the suppression of appetite, results from the production and release of a peptide hormone by adipose tissue called _____.

7. Many hormones produce their effects by increasing intracellular concentrations of _____.

8. An intracellular protein that acts as an intermediary for oxytocin and several regulatory hormones secreted by the hypothalamus is _____.

9. Liver cells respond to the presence of growth hormone by releasing hormones called _____.

10. Capillaries that have a "swiss cheese" appearance and are found only where relatively large molecules enter or leave the circulatory system are described as _____.

11. Blood vessels that link two capillary networks are called _____ vessels.

12. The principal mineralocorticoid produced by the human adrenal cortex is _____.

13. The zona of the adrenal cortex involved in the secretion of androgens is the _____.

14. The dominant hormone of the alarm phase of the general adaptation syndrome is _____.

15. The dominant hormones of the resistance phase of the GAS are _____.

16. The link between the first messenger and the second messenger is a(n) _____.

17. Steroid hormones bind to receptors in the nucleus or the _____.

Short Essay

Briefly answer the following questions in the spaces provided below.

1. List the three groups of hormones on the basis of chemical structure.

2. What are the four possible events set in motion by activation of a G protein, each of which leads to the appearance of a second messenger in the cytoplasm?

3. What three mechanisms does the hypothalamus use to regulate the activities of the endocrine system?

4. What hypothalamic control mechanisms are used to effect endocrine activity in the anterior pituitary?

5. How does the calorigenic effect of thyroid hormones generate heat?

6. How do you explain the relationship between the pineal gland and circadian rhythms?

7. What are the possible results when a cell receives instructions from two different hormones at the same time?

LEVEL 3: CRITICAL THINKING AND CLINICAL APPLICATIONS

Using principles and concepts learned in Chapter 18, answer the following questions. Write your answers on a separate sheet of paper.

1. Androgen abuse, consisting of the use of anabolic steroids, is seen in a number of amateur and professional athletes. Many of these athletes contend that it gives them a "competitive edge," and that the use of the steroids is not only safe but good for you. What are the health risks associated with the use of these natural or synthetic androgens?

2. a. As a diagnostic clinician, what endocrine malfunctions and conditions would you associate with the following symptoms in patients A, B, and C?

   Patient A: ↑ urine output, ↑ blood sugar concentration, dehydration, ↑ thrist, ↑ weight loss, metabolic acidosis.

   Patient B: Subcutaneous swelling, hair loss, dry skin, low body temperature, muscle weakness, and slowed reflexes.

   Patient C: ↑ Blood glucose level, retention of sodium, tissue fluid increases, skin appears puffy and thin and may bruise easily, growth of facial hair and deepening of voice in females.

   b. What glands are associated with each disorder?

3. An anatomy and physiology peer of yours makes a claim stating that the hypothalamus is exclusively a part of the CNS. What argument(s) would you make to substantiate that the hypothalamus is an interface between the nervous system and the endocrine system?

4. Topical steroid creams are used effectively to control irritating allergic responses or superficial rashes. Why are they ineffective and dangerous to use in the treatment of open wounds?

5. The recommended daily allowance (RDA) for iodine in adult males and females is 150 μg/day. Even though only a small quantity is needed each day, what symptoms are apparent if the intake of this mineral is insufficient or there is a complete lack of it in the diet?

6. If you have participated in athletic competitions, you have probably heard the statement, "I could feel the adrenaline flowing." Why does this feeling occur, and how is it manifested in athletic performance?

Blood

OVERVIEW

Fact: Every cell of the body needs a continuous supply of water, nutrients, and oxygen.

The cardiovascular system of the adult male contains 5–6 liters (5.3–6.4 quarts) of whole blood; that of an adult female contains 4–5 liters (4.2–5.3 quarts), making up 7 percent of the body's total weight. The blood circulates constantly through the body, bringing to each cell the oxygen, nutrients, and chemical substances necessary for its proper functioning, and, at the same time, removing waste products. This life-sustaining fluid, along with the heart and a network of blood vessels, comprises the cardiovascular system, which, acting in concert with other systems, plays an important role in the maintenance of homeostasis. This specialized fluid connective tissue consists of two basic parts: the formed elements and cell fragments, and the fluid plasma in which they are carried.

Functions of the blood, including transportation, regulation, protection, and coagulation, serve to maintain the integrity of cells in the human body.

Chapter 19 provides an opportunity for you to identify and study the components of blood and review the functional roles the components play in order for cellular needs to be met throughout the body. Blood types, hemostasis, hematopoiesis, and blood abnormalities are also considered.

LEVEL 1: REVIEWING FACTS AND TERMS

Review of Learning Outcomes

After completing this chapter, you should be able to do the following:

OUTCOME 19-1   Describe the components and major functions of blood, identify blood collection sites, and list the physical characteristics of blood.

OUTCOME 19-2   Specify the composition and functions of plasma.

OUTCOME 19-3   List the characteristics and functions of red blood cells, describe the structure and functions of hemoglobin, describe how red blood cell components are recycled, and explain erythropoiesis.

OUTCOME 19-4   Explain the importance of blood typing, and the basis for ABO and Rh incompatibilities.

OUTCOME 19-5   Categorize white blood cell types based on their structures and functions, and discuss the factors that regulate the production of each type.

OUTCOME 19-6   Describe the structure, function, and production of platelets.

OUTCOME 19-7   Discuss the mechanisms that control blood loss after an injury, and describe the reaction sequences responsible for blood clotting.

Multiple Choice

Place the letter corresponding to the best answer in the space provided.

OUTCOME 19-1 _____ 1. Venipuncture is a common sampling technique because

    a. superficial veins are easy to locate.

    b. the walls of the veins are thinner than the walls of arteries.

    c. the blood pressure in veins is relatively low.

    d. a, b, and c are correct.

OUTCOME 19-1 _____ 2. In adults, the circulatory blood

    a. provides nutrients, oxygen, and chemical instructions for cells.

    b. is a mechanism for waste removal for cells.

    c. transports special cells to defend tissues from infection and disease.

    d. a, b, and c are correct.

OUTCOME 19-1 _____ 3. An arterial puncture generally drawn from the radial artery at the wrist or the brachial artery at the elbow would primarily be used for

    a. checking the clotting system.

    b. monitoring glucose and cholesterol levels.

    c. preparing a blood smear.

    d. checking the efficiency of gas exchange at the lungs.

OUTCOME 19-1 _____ 4. The blood transports specialized cells that defend peripheral tissues from

    a. infection and disease.

    b. overproduction of red and white blood cells.

    c. decreased levels of O_2 and CO_2.

    d. all of the above.

OUTCOME 19-1 _____ 5. The formed elements of the blood consist of

    a. antibodies, metalloproteins, and lipoproteins.

    b. red and white blood cells, and platelets.

    c. albumins, globulins, and fibrinogen.

    d. electrolytes, nutrients, and organic wastes.

OUTCOME 19-1 _____ 6. The primary function(s) of blood is (are)

    a. transportation of dissolved gases, nutrients, wastes, and hormones.

    b. regulation of pH and ion composition of interstitial fluids.

    c. defense against toxins and pathogens.

    d. all of the above.

OUTCOME 19-1 _____ 7. Blood transports dissolved gases, bringing oxygen from the lungs to the tissues and carrying carbon dioxide from

    a. the lungs to the tissues.

    b. one peripheral cell to another.

    c. the interstitial fluid to the cell.

    d. the tissues to the lungs.

OUTCOME 19-1 _____ 8. The "patrol agents" in the blood that defend the body against toxins and pathogens are

    a. hormones and enzymes.

    b. albumins and globulins.

    c. white blood cells and antibodies.

    d. red blood cells and platelets.

OUTCOME 19-1 _____ 9. The best laboratory procedure for obtaining blood to determine the hemoglobin level is

    a. arterial puncture.

    b. venipuncture.

    c. puncturing the tip of a finger.

    d. a, b, or c could be used.

OUTCOME 19-1 _____ 10. Blood temperature is roughly _____, and the blood pH averages _____.

    a. 0°C; 6.8

    b. 32°C; 7.0

    c. 38°C; 7.4

    d. 98°C; 7.8

OUTCOME 19-2 _____ 11. Which one of the following statements is most correct?

    a. Plasma contributes approximately 92 percent of the volume of whole blood, and H_2O accounts for 55 percent of the plasma volume.

    b. Plasma contributes approximately 55 percent of the volume of whole blood, and H_2O accounts for 92 percent of the plasma volume.

    c. H_2O accounts for 99 percent of the volume of the plasma, and plasma contributes approximately 45 percent of the volume of whole blood.

    d. H_2O accounts for 45 percent of the volume of the plasma, and plasma contributes approximately 99 percent of the volume of whole blood.

OUTCOME 19-2 _____ 12. The three primary classes of plasma proteins are

    a. antibodies, metalloproteins, and lipoproteins.

    b. serum, fibrin, and fibrinogen.

    c. albumins, globulins, and fibrinogen.

    d. heme, porphyrin, and globin.

OUTCOME 19-2 _____ 13. In addition to water and proteins, the plasma consists of

    a. erythrocytes, leukocytes, and platelets.

    b. electrolytes, nutrients, and organic wastes.

    c. albumins, globulins, and fibrinogen.

    d. a, b, and c are correct.

OUTCOME 19-2 _____ 14. Loose connective tissue and cartilage contain a network of insoluble fibers, whereas plasma, a fluid connective tissue, contains

    a. a network of collagen and elastic fibers.

    b. elastic fibers only.

    c. collagen fibers only.

    d. dissolved proteins.

OUTCOME 19-3   _____   15. Circulating mature RBCs lack

      a. mitochondria.

      b. ribosomes.

      c. nuclei.

      d. a, b, and c are correct.

OUTCOME 19-3   _____   16. The important effect(s) on RBCs due to their unusual shape is (are) that it

      a. enables RBCs to form stacks.

      b. gives each RBC a large surface-area-to-volume ratio.

      c. enables RBCs to bend and flex when entering small capillaries and branches.

      d. all of the above.

OUTCOME 19-3   _____   17. The *primary* function of a mature red blood cell is

      a. transport of respiratory gases.

      b. delivery of enzymes to target tissues.

      c. defense against toxins and pathogens.

      d. a, b, and c are correct.

OUTCOME 19-3   _____   18. The part of the hemoglobin molecule that *directly* interacts with oxygen is

      a. globin.

      b. the porphyrin.

      c. the iron ion.

      d. the sodium ion.

OUTCOME 19-3   _____   19. Iron is necessary in the diet because it is involved with

      a. the prevention of sickle cell anemia.

      b. hemoglobin production.

      c. prevention of hematuria.

      d. a, b, and c are correct.

OUTCOME 19-3   _____   20. The iron extracted from heme molecules during hemoglobin recycling is stored in the protein–iron complexes

      a. transferrin and porphyrin.

      b. biliverdin and bilirubin.

      c. urobilin and stercobilin.

      d. ferritin and hemosiderin.

OUTCOME 19-3   _____   21. During RBC recycling, each heme unit is stripped of its iron and converted to

      a. biliverdin.

      b. urobilin.

      c. transferrin.

      d. ferritin.

OUTCOME 19-3   _____   22. The primary site of erythropoiesis in the adult is the

      a. myeloid tissue.

      b. kidney.

      c. liver.

      d. heart.

OUTCOME 19-3   _____   23. Erythropoietin appears in the plasma when peripheral tissues, especially the kidneys, are exposed to

      a. extremes of temperature.

      b. high urine volumes.

      c. excessive amounts of radiation.

      d. low oxygen concentrations.

OUTCOME 19-4   _____   24. Agglutinogens are contained (on, in) the _____, while the agglutinins are found (on, in) the _____.

      a. plasma; cell membrane of RBC

      b. nucleus of the RBC; mitochondria

      c. cell membrane of RBC; plasma

      d. mitochondria; nucleus of the RBC

OUTCOME 19-4   _____   25. If you have type A blood, your plasma holds circulating _____ that will attack _____ erythrocytes.

      a. anti-B agglutinins; Type B

      b. anti-A agglutinins; Type A

      c. anti-A agglutinogens; Type A

      d. anti-A agglutinins; Type B

OUTCOME 19-4   _____   26. A person with type O blood contains

      a. anti-A and anti-B agglutinins.

      b. anti-O agglutinins.

      c. anti-A and anti-B agglutinogens.

      d. no agglutinins—type O blood lacks agglutinins altogether.

OUTCOME 19-5   _____   27. The two types of *agranular* leukocytes found in the blood are

      a. neutrophils and eosinophils.

      b. leukocytes and lymphocytes.

      c. monocytes and lymphocytes.

      d. neutrophils and monocytes.

OUTCOME 19-5   _____   28. Based on their staining characteristics, the types of *granular* leukocytes found in the blood are

      a. lymphocytes, monocytes, and erythrocytes.

      b. neutrophils, monocytes, and lymphocytes.

      c. eosinophils, basophils, and lymphocytes.

      d. neutrophils, eosinophils, and basophils.

OUTCOME 19-5 _____ 29. The number of eosinophils increases dramatically during

    a. an allergic reaction or a parasitic infection.

    b. an injury to a tissue or a bacterial infection.

    c. tissue degeneration or cellular deterioration.

    d. a, b, and c are correct.

OUTCOME 19-5 _____ 30. *Basophils* are specialized in that they

    a. contain microphages that engulf invading bacteria.

    b. contain histamine that exaggerates the inflammation response at the injury site.

    c. are enthusiastic phagocytes, often attempting to engulf items as large as or larger than themselves.

    d. produce and secrete antibodies that attack cells or proteins in distant portions of the body.

OUTCOME 19-6 _____ 31. *Megakaryocytes* are specialized cells of the bone marrow responsible for

    a. specific immune responses.

    b. engulfing invading bacteria.

    c. formation of platelets.

    d. production of scar tissue in an injured area.

OUTCOME 19-6 _____ 32. The rate of megakaryocyte activity and platelet formation is regulated by

    a. thrombopoietin.

    b. interleukin-6.

    c. multi-CSF.

    d. a, b, and c are correct.

OUTCOME 19-7 _____ 33. The process of hemostasis includes *five* phases. The correct order of the phases as they occur after injury is as follows:

    a. vascular, coagulation, platelet, clot retraction, fibrinolysis.

    b. coagulation, vascular, platelet, fibrinolysis, clot retraction.

    c. platelet, vascular, coagulation, clot retraction, fibrinolysis.

    d. vascular, platelet, coagulation, clot retraction, fibrinolysis.

OUTCOME 19-7 _____ 34. The *extrinsic* pathway involved in blood clotting involves the release of

    a. platelet factors and platelet thromboplastin.

    b. Ca^{2+} and clotting factors VIII, IX, X, XI, and XIII.

    c. tissue factor (TF).

    d. prothrombin and fibrinogen.

OUTCOME 19-7 _____ 35. Which of the relationships listed here correctly reflects events of the "common pathway"?

    a. Fibrinogen catalyzes the creation of fibrin, which catalyzes creation of thrombin.

    b. Prothrombinase catalyzes the creation of thrombin, which then catalyzes creation of fibrin.

    c. Factor X catalyzes the creation of prothrombin, which catalyzes creation of clotting factor VII.

    d. Fibrin catalyzes the creation of fibrinogen, which catalyzes creation of thrombin.

Completion

Using the terms below, complete the following statements. Use each term only once.

anticoagulants	venipuncture	vitamin B_{12}	rouleaux
hemoglobin	leukocytes	blood	formed elements
lymphocytes	diapedesis	vascular spasm	serum
plasma	globulins	arterial punctures	lymphopoiesis
erythroblasts	agglutinins	megakaryocytes	nonspecific
agglutinogens	plasmin	platelets	transferrin

OUTCOME 19-1   1. A specialized fluid connective tissue that contains cells suspended in a fluid matrix is the _____.

OUTCOME 19-1   2. In addition to RBCs and platelets, the blood carries WBCs, also called _____.

OUTCOME 19-1   3. The matrix of the blood is the _____.

OUTCOME 19-1   4. The blood cells and cell fragments suspended in the ground substance are referred to as _____.

OUTCOME 19-1   5. The most common clinical procedure for collecting blood for blood tests is the _____.

OUTCOME 19-1   6. To check the efficiency of gas exchange at the lungs, blood would be drawn from a(n) _____.

OUTCOME 19-2   7. Considering the most prevalent types of plasma proteins, a hormone-binding protein most likely would belong to the class called _____.

OUTCOME 19-2   8. When the clotting proteins are removed from the plasma, the remaining fluid is the _____.

OUTCOME 19-3   9. The flattened shape of RBCs enables them to form stacks called _____.

OUTCOME 19-3   10. The part of the red blood cell responsible for the cell's ability to transport oxygen and carbon dioxide is _____.

OUTCOME 19-3   11. During recycling, when iron is extracted from the heme molecules and released into the bloodstream, it binds to the plasma protein _____.

OUTCOME 19-3   12. For erythropoiesis to proceed normally, the myeloid tissues must receive adequate amounts of amino acids, iron, and _____.

OUTCOME 19-3   13. *Very* immature RBCs that actively synthesize hemoglobin are the _____.

OUTCOME 19-4   14. Antigens on the surfaces of RBCs whose presence and structure are genetically determined are called _____.

OUTCOME 19-4   15. Immunoglobulins in plasma that react with antigens on the surfaces of foreign red blood cells when donor and recipient differ in blood type are called _____.

OUTCOME 19-5   16. T cells and B cells are representative cell populations of WBCs identified as _____.

OUTCOME 19-5   17. Neutrophils, eosinophils, basophils, and monocytes are part of the body's _____ defense system.

OUTCOME 19-5   18. The process of WBCs migrating across the endothelial lining of a capillary, squeezing between adjacent endothelial cells, is called _____.

OUTCOME 19-5   19. The production of lymphocytes from stem cells is called _____.

OUTCOME 19-6   20. After a clot has formed, the clot shrinks due to the action of actin and myosin filaments contained in _____.

OUTCOME 19-6   21. Platelets originate from enormous cells called _____.

OUTCOME 19-7   22. Cutting the wall of a blood vessel triggers a local contraction, called a _____, in the smooth muscle fibers of the vessel wall.

OUTCOME 19-7   23. During fibrinolysis, the clot gradually dissolves through the action of _____.

OUTCOME 19-7   24. Normal plasma contains several _____, which inhibit clotting.

Matching

Match the terms in column B with the terms in column A. Use letters for answers in the spaces provided. Use each term only once.

Part I

		Column A	Column B
OUTCOME 19-1	_____	1. WBCs	A. matrix
OUTCOME 19-1	_____	2. plasma	B. heme
OUTCOME 19-1	_____	3. whole blood	C. buffer in blood
OUTCOME 19-1	_____	4. pH regulation	D. hormone
OUTCOME 19-1	_____	5. median cubital vein	E. enucleated
OUTCOME 19-2	_____	6. antibodies	F. jaundice
OUTCOME 19-2	_____	7. globulin	G. fight infections
OUTCOME 19-3	_____	8. mature red blood cells	H. immunoglobulin
OUTCOME 19-3	_____	9. pigment complex	I. transport protein
OUTCOME 19-3	_____	10. bilirubin	J. plasma and formed elements
OUTCOME 19-3	_____	11. ferritin	K. venipuncture
OUTCOME 19-3	_____	12. erythropoietin	L. protein–iron complex

Part II

		Column A	Column B
OUTCOME 19-3	_____	13. reticulocytes	M. specific immunity
OUTCOME 19-4	_____	14. agglutinogens	N. prevents clotting
OUTCOME 19-4	_____	15. cross-reaction	O. A, B, and Rh
OUTCOME 19-4	_____	16. agglutinins	P. neutrophils, eosinophils
OUTCOME 19-5	_____	17. microphages	Q. syneresis
OUTCOME 19-5	_____	18. neutrophils	R. polymorphonuclear leukocytes
OUTCOME 19-5	_____	19. lymphocytes	S. immature RBCs
OUTCOME 19-5	_____	20. monocyte	T. agglutination
OUTCOME 19-6	_____	21. platelets	U. macrophage
OUTCOME 19-6	_____	22. platelet production	V. antibodies
OUTCOME 19-7	_____	23. heparin	W. blood clotting
OUTCOME 19-7	_____	24. clot retraction	X. thrombocytopoiesis

Drawing/Illustration Labeling

Identify each numbered structure by labeling the following formed elements of the blood.

OUTCOME 19-1, 19-5 **FIGURE 19-1** Formed Elements of the Blood

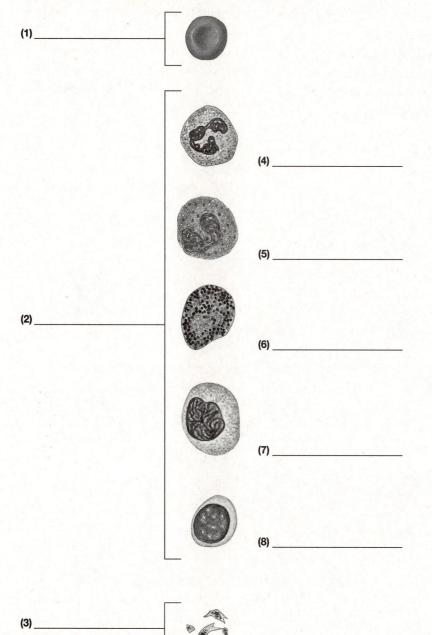

(1) _____

(2) _____

(3) _____

(4) _____

(5) _____

(6) _____

(7) _____

(8) _____

LEVEL 2: REVIEWING CONCEPTS

Chapter Overview

Using the terms below, correctly complete the chapter overview on the origins and differentiation of formed elements. Place your answers in the spaces below the flow diagram. Use each term only once.

eosinophil	lymphocyte	reticulocyte	platelets
blast cells	myeloid stem cells	hemocytoblasts	monocyte
erythrocyte	lymphoid stem cells	progenitor cells	neutrophil
basophil	formed elements of blood	proerythroblast	myelocytes
monoblast	megakaryocyte		

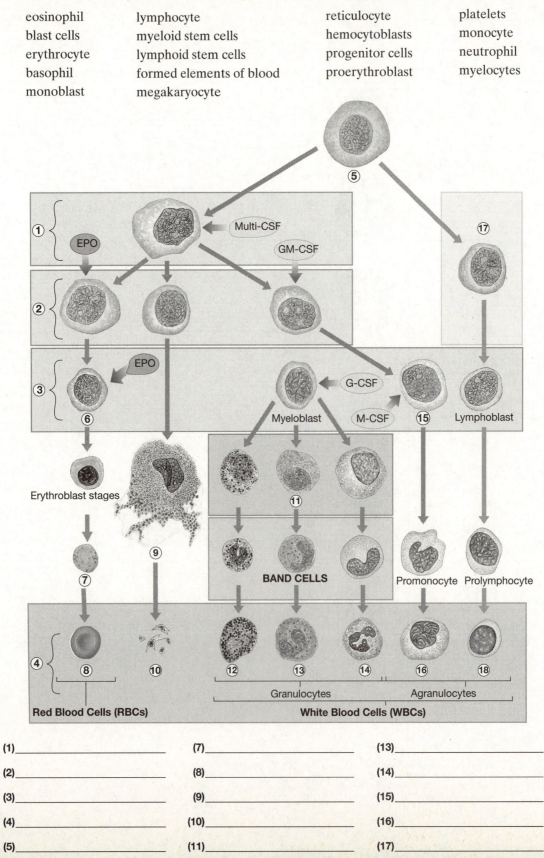

(1)_____   (7)_____   (13)_____

(2)_____   (8)_____   (14)_____

(3)_____   (9)_____   (15)_____

(4)_____   (10)_____   (16)_____

(5)_____   (11)_____   (17)_____

Concept Map I

Using the following terms, fill in the circled, numbered, blank spaces to correctly complete the concept map. Use each term only once.

neutrophils   monocytes   metalloproteins
albumins   solutes   plasma
leukocytes

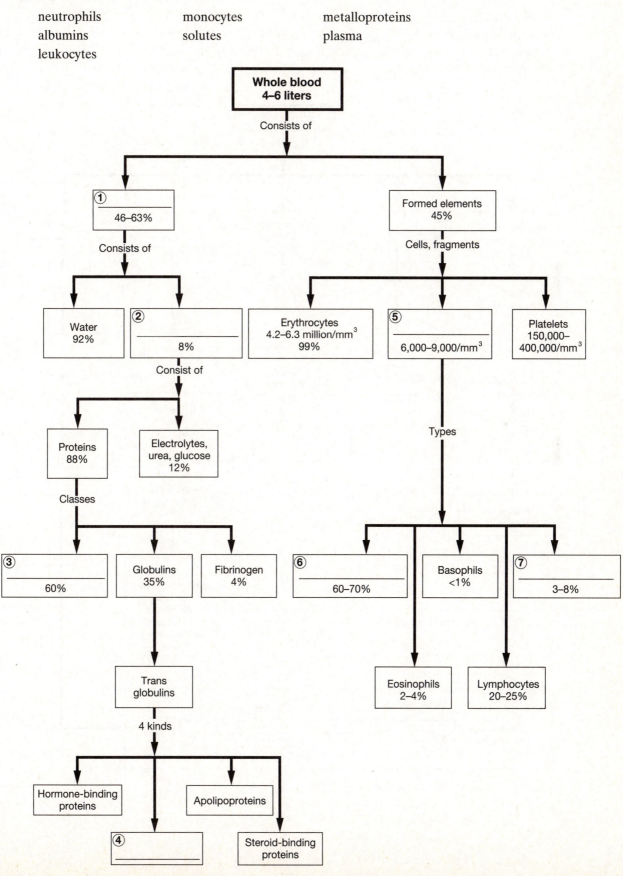

Concept Map II

Using the following terms, fill in the circled, numbered, blank spaces to correctly complete the concept map. Use each term only once.

clot dissolves   plasminogen   plug

platelet phase   plasmin   clot retraction

blood clot   vascular spasm

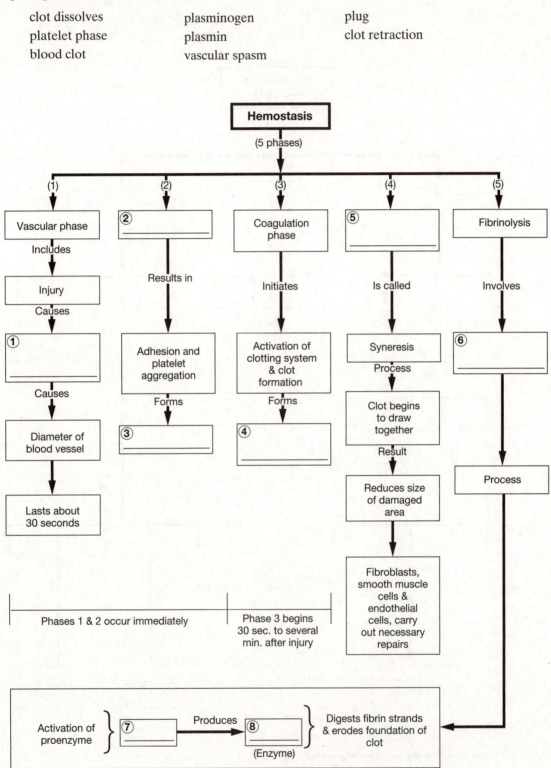

Concept Map III

Using the following terms, fill in the circled, numbered, blank spaces to correctly complete the concept map. Use each term only once.

factor VII & Ca^{2+} PF-3 (platelet factor)   fibrinogen

extrinsic pathway   intrinsic pathway   factor X

thrombin   fibrin

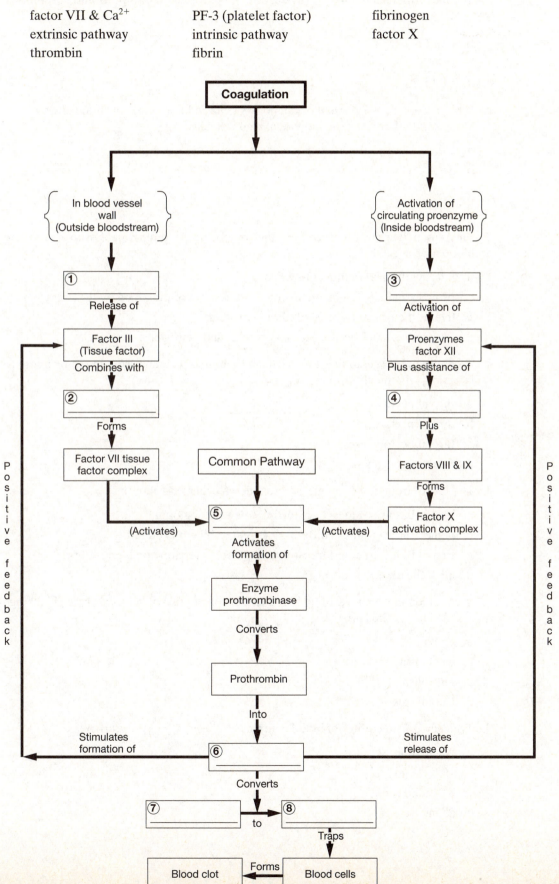

Multiple Choice

Place the letter corresponding to the best answer in the space provided.

_____ 1. The factor that stimulates the production of neurophils, eosinophils, and basophils is

    a. G-CSF.

    b. Gm-CSF.

    c. M-CSF.

    d. Multi-CSF.

_____ 2. If blood comprises 7 percent of the body weight in kilograms, how many liters of blood would there be in an individual who weighs 85 kg?

    a. 5.25

    b. 5.95

    c. 6.25

    d. 6.95

_____ 3. The reason liver disorders can alter the composition and functional properties of the blood is because the

    a. liver synthesizes immunoglobulins.

    b. proteins synthesized by the liver are filtered out of the blood by the kidneys.

    c. liver is the primary source of plasma proteins.

    d. liver serves as a filter for plasma proteins and pathogens.

_____ 4. A condition that generally results from accelerated platelet formation in response to infection, inflammation, or cancer is

    a. thrombocytosis.

    b. thrombocytopenia.

    c. thrombocytopoieses.

    d. megakaryocytosis.

_____ 5. The fate of iron after hemoglobin breakdown occurs as follows:

    a. iron → bloodstream → transferrin → new hemoglobin.

    b. heme → biliverdin → bilirubin → (liver) → bile → excreted.

    c. iron → bloodstream → transferrin → (stored) ferritin and hemosiderin.

    d. all of the above.

_____ 6. A condition in which a drifting blood clot becomes stuck in a blood vessel, blocking circulation, is called a(n)

    a. embolism.

    b. catabolism.

    c. anabolism.

    d. thrombosis.

_____ 7. Typically, one microliter of blood contains _____ erythrocytes.

    a. 260 million

    b. 10^{13}

    c. 5.2 million

    d. 25 million

_____ 8. In a mature RBC, energy is obtained exclusively by anaerobic respiration (i.e., the breakdown of glucose in the cytoplasm).

   a. True

   b. False

_____ 9. Protein synthesis in a circulating RBC occurs primarily in

   a. ribosomes.

   b. mitochondria.

   c. the nucleus.

   d. none of these.

_____ 10. If agglutinogen "B" meets with agglutinin "anti-A," the result would be

   a. a cross-reaction.

   b. no agglutination would occur.

   c. the patient would be comatose.

   d. the patient would die.

_____ 11. The precursor of all blood cells in the human body is the

   a. myeloblast.

   b. hemocytoblast.

   c. megakaryocyte.

   d. none of these.

_____ 12. *Reticulocytes* are nucleated immature cells that develop into mature

   a. lymphocytes.

   b. platelets.

   c. leukocytes.

   d. erythrocytes.

_____ 13. The blood cells that may originate in the thymus, spleen, and lymph nodes as well as in the bone marrow are the

   a. erythrocytes.

   b. lymphocytes.

   c. leukocytes.

   d. monocytes.

_____ 14. Rh-negative blood indicates the

   a. presence of the Rh agglutinogen.

   b. presence of agglutinogen A.

   c. absence of the Rh agglutinogen.

   d. absence of agglutinogen B.

_____ 15. In hemolytic disease of the newborn, the

   a. newborn's agglutinins cross the placental barrier and cause the newborn's RBCs to degenerate.

   b. mother's agglutinins cross the placental barrier and destroy fetal red blood cells.

   c. mother's agglutinogens destroy her own RBCs, causing deoxygenation of the newborn.

   d. all of the above occur.

_____ 16. *Circulating* leukocytes represent a small fraction of the total population, since most WBCs are found in peripheral tissues.

    a. True

    b. False

_____ 17. A typical microliter of blood contains _____ leukocytes.

    a. 1000–2000

    b. 3000–5000

    c. 6000–9000

    d. 10,000–12,000

_____ 18. Under "normal" conditions, neutrophils comprise _____ of the circulatory white blood cells.

    a. 10–20 percent

    b. 20–40 percent

    c. 50–70 percent

    d. 85–95 percent

_____ 19. Leukemia is a condition characterized by

    a. inadequate numbers of white blood cells in the bloodstream.

    b. sickle-shaped cells due to low oxygen levels.

    c. low tissue oxygen levels.

    d. extremely elevated levels of circulating white blood cells.

_____ 20. Platelets are unique formed elements of the blood because they

    a. initiate the immune response and destroy bacteria.

    b. are cytoplasmic fragments rather than individual cells.

    c. can reproduce because they contain a nucleus.

    d. none of the above apply to platelets.

_____ 21. Fibrinolysis involves a process that begins with activation of

    a. the proenzyme fibrinogen, which initiates the production of fibrin.

    b. prothrombin, which initiates the production of thrombin.

    c. Ca^{2+} to produce tissue plasmin.

    d. the proenzyme plasminogen, which leads to the production of plasmin.

_____ 22. A vitamin K deficiency in the body will eventually cause

    a. a breakdown of the common pathway, deactivating the clotting system.

    b. the body to become insensitive to situations that would necessitate the clotting mechanism.

    c. an overactive clotting system, which might necessitate thinning of the blood.

    d. all of the above are correct.

_____ 23. A vitamin B_{12} deficiency results in the type of anemia known as

    a. pernicious.

    b. aplastic.

    c. hemorrhagic.

    d. sickle cell.

_____ 24. An anemia resulting from the production of an abnormal form of hemoglobin results in a condition known as

    a. hematuria.

    b. leukemia.

    c. sickle cell anemia.

    d. leukopenia.

_____ 25. The activity of the common pathway stimulates both the intrinsic and extrinsic pathways by

    a. a negative feedback loop that accelerates the clotting process.

    b. removal of clotting factors from the blood.

    c. deactivation of stimulatory agents from the blood.

    d. a positive feedback loop that accelerates the clotting process.

Completion

Using the terms below, complete the following statements. Use each term only once.

thrombocytopenia	leukopenia	hypovolemic
hemolysis	differential	leukocytosis
fibrin	bilirubin	hemoglobinuria
fractionated	lipoproteins	hematocrit
metalloproteins	hematopoiesis	hematocytoblasts
hemoglobin		

1. When the components of whole blood are separated, they are said to be _____.

2. When clinicians refer to low blood volumes, they use the term _____.

3. The insoluble fibers that provide the basic framework for a blood clot are called _____.

4. Transport globulins that bind metal ions are called _____.

5. Globulins involved in lipid transport are called _____.

6. The percentage of whole blood occupied by cellular elements is the _____.

7. The presence of hemoglobin in urine is called _____.

8. The rupturing of red blood cells is called _____.

9. When heme is stripped of its iron, it is converted to _____.

10. An inadequate number of white blood cells is called _____.

11. Excessive numbers of white blood cells refers to a condition called _____.

12. A low blood platelet count refers to a condition called _____.

13. A blood cell count that determines the numbers and kinds of leukocytes is called a _____.

14. Formed elements in the blood are produced by the process of _____.

15. The stem cells that produce all of the blood cells are called _____.

16. The most common blood test used to determine if a person is anemic is _____.

Short Essay

Briefly answer the following questions in the spaces provided below.

1. List the eight primary functions of blood.

2. What are the five major components of plasma?

3. What three components make up the formed elements of blood?

4. What are the three primary classes of plasma proteins?

5. Why are mature red blood cells incapable of aerobic respiration, protein synthesis, and mitosis?

6. List the three kinds of *granular* WBCs and the two kinds of *agranular* WBCs.

7. What are the three primary functions of platelets?

8. List the five phases in the clotting response and summarize the results of each occurrence.

9. What two primary effects does erythropoietin have on the process of erythropoiesis?

LEVEL 3: CRITICAL THINKING AND CLINICAL APPLICATIONS

Using principles and concepts learned in Chapter 19, answer the following questions. Write your answers on a separate sheet of paper.

1. A man weighs 154 pounds and his hematocrit is 45 percent. The weight of blood is approximately 7 percent of his total body weight.

   a. What is his total blood volume in liters?

   b. What is his total cell volume in liters?

   c. What is his total plasma volume in liters?

   d. What percent of his blood constitutes the plasma volume?

   e. What percent of his blood comprises the formed elements?
      Note: 1 kilogram = 2.2 pounds

2. Considering the following conditions, what effect would each have on an individual's hematocrit (percentage of erythrocytes to the total blood volume)?

   a. an increase in viscosity of the blood

   b. an increased dehydration in the body

   c. problems with RBC production

   d. a result of internal bleeding

3. If it is necessary to increase the blood volume temporarily over a period of hours due to blood loss, and to buy time for lab work to determine a person's blood type, why would it be desirable to use a commercial plasma expander instead of donated plasma?

4. A person with type B blood has been involved in an accident and excessive bleeding necessitates a blood transfusion. Due to an error by a careless laboratory technician, the person is given type A blood. Explain what will happen.

5. Suppose you are a young woman who has Rh-negative blood. You are intent on marrying a young man who has Rh-positive blood. During your first pregnancy your gynecologist suggests that you be given RhoGAM injections. What is the purpose of the injections, and how does the RhoGAM affect your immune system?

6. Why is the life span of RBCs of humans and other mammals relatively short—normally less than 120 days?

7. Your doctor has recommended that you take a vitamin supplement of your choice. You look for a supplement that lists all the vitamins; however, you cannot find one that has vitamin K.

   a. Why?

   b. What are some dietary sources of vitamin K?

   c. What is the function of vitamin K in the body?

The Heart

OVERVIEW

The heart is one of nature's most efficient and durable pumps. Throughout life, it beats 60 to 80 times per minute, supplying oxygen and other essential nutrients to every cell in the body and removing waste for elimination from the lungs or the kidneys. Every day approximately 2,100 gallons of blood are pumped through a vast network of about 60,000 miles of blood vessels. The "double pump" design of this muscular organ facilitates the continuous flow of blood through a closed system that consists of a pulmonary circuit to the lungs and a systemic circuit serving all other regions of the body. A system of valves in the heart maintains the "one-way" direction of blood flow through the heart. Nutrient and oxygen supplies are delivered to the heart's tissues by coronary arteries that encircle the heart like a crown. Normal cardiac rhythm is maintained by the heart's electrical system, originating in a group of specialized pacemaker cells located in the right atrium.

Chapter 20 examines the structural features of the heart, cardiac physiology, cardiac dynamics, and the cardiovascular system.

LEVEL 1: REVIEWING FACTS AND TERMS

Review of Learning Outcomes

After completing this chapter, you should be able to do the following:

OUTCOME 20-1   Describe the anatomy of the heart, including vascular supply and pericardium structure, and trace the flow of blood through the heart, identifying the major blood vessels, chambers, and heart valves.

OUTCOME 20-2   Explain the events of an action potential in cardiac muscle, indicate the importance of calcium ions to the contractile process, describe the conducting system of the heart, and identify the electrical events associated with a normal electrocardiogram.

OUTCOME 20-3   Explain the events of the cardiac cycle, including atrial and ventricular systole and diastole, and relate the heart sounds to specific events in the cycle.

OUTCOME 20-4   Define cardiac output, describe the factors that influence heart rate and stroke volume, and explain how adjustments in stroke volume and cardiac output are coordinated at different levels of physical activity.

Multiple Choice

Place the letter corresponding to the best answer in the space provided.

OUTCOME 20-1 _____ 1. The blood vessels in the cardiovascular system are subdivided into the
   a. lymphatic and blood circuits.
   b. dorsal aorta and inferior-superior vena cava.
   c. cardiac and vascular circuits.
   d. pulmonary and systemic circuits.

OUTCOME 20-1 _____ 2. Blood is carried away from the heart by
   a. veins.
   b. capillaries.
   c. arteries.
   d. venules.

OUTCOME 20-1 _____ 3. The left atrium receives blood from the pulmonary circuit and empties it into the
   a. left ventricle.
   b. right atrium.
   c. right ventricle.
   d. conus arteriosus.

OUTCOME 20-1 _____ 4. The "double pump" function of the heart includes the right side, which serves as the _____ circuit pump, while the left side serves as the _____ pump.
   a. systemic; pulmonary
   b. pulmonary; hepatic portal
   c. hepatic portal; cardiac
   d. pulmonary; systemic

OUTCOME 20-1 _____ 5. The major difference between the left and right ventricles relative to their role in heart function is
   a. the L.V. pumps blood through the short, low-resistance pulmonary circuit.
   b. the R.V. pumps blood through the low-resistance systemic circulation.
   c. the L.V. pumps blood through the high-resistance systemic circulation.
   d. The R.V. pumps blood through the short, high-resistance pulmonary circuit.

OUTCOME 20-1 _____ 6. The great cardiac vein drains blood from the heart muscle to the
   a. right atrium.
   b. left ventricle.
   c. left atrium.
   d. right ventricle.

OUTCOME 20-1 _____ 7. The visceral pericardium, or epicardium, covers the
   a. inner surface of the heart.
   b. outer surface of the heart.
   c. vessels in the mediastinum.
   d. endothelial lining of the heart.

OUTCOME 20-1 _____ 8. The valves of the heart are covered by a squamous epithelium, the
    a. endocardium.
    b. epicardium.
    c. visceral pericardium.
    d. parietal pericardium.

OUTCOME 20-1 _____ 9. The three distinct layers of the heart wall include the
    a. epicardium, myocardium, and endocardium.
    b. skeletal, cardiac, and smooth.
    c. visceral, parietal, and fibrous.
    d. arteries, veins, and capillaries.

OUTCOME 20-1 _____ 10. Atrioventricular valves prevent backflow of blood into the
    _____; semilunar valves prevent backflow into the
    _____.
    a. atria; ventricles
    b. lungs; systemic circulation
    c. ventricles; atria
    d. capillaries; lungs

OUTCOME 20-1 _____ 11. Blood flows from the left atrium into the left ventricle through the
    _____ valve.
    a. bicuspid
    b. L. atrioventricular
    c. mitral
    d. a, b, and c are correct

OUTCOME 20-1 _____ 12. When deoxygenated blood leaves the right ventricle through a semilunar valve, it is forced into the
    a. pulmonary veins.
    b. aortic arch.
    c. pulmonary arteries.
    d. lung capillaries.

OUTCOME 20-1 _____ 13. Blood from systemic circulation is returned to the right atrium by the
    a. superior and inferior vena cava.
    b. pulmonary veins.
    c. pulmonary arteries.
    d. brachiocephalic veins.

OUTCOME 20-1 _____ 14. Oxygenated blood from the systemic arteries flows into
    a. peripheral tissue capillaries.
    b. systemic veins.
    c. the right atrium.
    d. the left atrium.

OUTCOME 20-1 _____ 15. The lung capillaries receive deoxygenated blood from the
    a. pulmonary veins.
    b. pulmonary arteries.
    c. aorta.
    d. superior and inferior vena cava.

OUTCOME 20-1   _____ 16. One of the important differences between skeletal muscle tissue and cardiac muscle tissue is that cardiac muscle tissue is

    a. striated voluntary muscle.

    b. multinucleated.

    c. comprised of unusually large cells.

    d. striated involuntary muscle.

OUTCOME 20-1   _____ 17. Cardiac muscle tissue

    a. will not contract unless stimulated by nerves.

    b. does not require nerve activity to stimulate a contraction.

    c. is under voluntary control.

    d. a, b, and c are correct.

OUTCOME 20-1   _____ 18. The primary difference(s) that characterize cardiac muscle cells when comparing them to skeletal muscle fibers is (are)

    a. small size.

    b. a single, centrally located nucleus.

    c. the presence of intercalated discs.

    d. all of the above.

OUTCOME 20-1   _____ 19. Blood from coronary circulation is returned to the right atrium of the heart via

    a. anastomoses.

    b. the circumflex branch.

    c. the coronary sinus.

    d. the anterior interventricular branch.

OUTCOME 20-1   _____ 20. The right coronary artery supplies blood to

    a. the right atrium.

    b. portions of the conducting system of the heart.

    c. portions of the right and left ventricles.

    d. a, b, and c are correct.

OUTCOME 20-2   _____ 21. The correct sequential path of a normal action potential in the heart is:

    a. SA node → AV bundle → AV node → Purkinje fibers.

    b. AV node → SA node → AV bundle → AV bundle.

    c. SA node → AV node → AV bundle → bundle branches → Purkinje fibers.

    d. SA node → AV node → bundle branches → AV bundle → Purkinje fibers.

OUTCOME 20-2   _____ 22. If the papillary muscles fail to contract, the

    a. atria will not pump blood.

    b. semilunar valves will not open.

    c. AV valves will not close properly.

    d. ventricles will not pump blood.

OUTCOME 20-2 _____ 23. The sinoatrial node acts as the pacemaker of the heart because these cells are

    a. located in the wall of the left atrium.

    b. the only cells in the heart that can conduct an impulse.

    c. the only cells in the heart innervated by the autonomic nervous system.

    d. the ones that depolarize and reach threshold first.

OUTCOME 20-2 _____ 24. After the SA node is depolarized and the impulse spreads through the atria, there is a slight delay before the impulse spreads to the ventricles. The importance of this delay is that it allows

    a. the atria to finish contracting.

    b. the ventricles to repolarize.

    c. a greater venous return.

    d. nothing; there is no reason for the delay.

OUTCOME 20-2 _____ 25. If each heart muscle cell contracted at its own individual rate, the condition would resemble

    a. heart flutter.

    b. bradycardia.

    c. fibrillation.

    d. myocardial infarction.

OUTCOME 20-2 _____ 26. The P wave of a normal electrocardiogram indicates

    a. atrial repolarization.

    b. atrial depolarization.

    c. ventricular repolarization.

    d. ventricular depolarization.

OUTCOME 20-2 _____ 27. The QRS complex of the ECG appears as the

    a. atria depolarize.

    b. atria repolarize.

    c. ventricles repolarize.

    d. ventricles depolarize.

OUTCOME 20-2 _____ 28. ECGs are useful in detecting and diagnosing abnormal patterns of cardiac activity called

    a. myocardial infarctions.

    b. cardiac arrhythmias.

    c. excitation–contraction coupling.

    d. autorhythmicity.

OUTCOME 20-2 _____ 29. An excessively large QRS complex often indicates that the

    a. mass of the heart muscle has decreased.

    b. heart has become enlarged.

    c. ventricular repolarization has slowed down.

    d. cardiac energy reserves are low.

OUTCOME 20-3 _____ 30. The events between the start of one heartbeat and the start of the next are called the
a. nodal rhythm.
b. property of automaticity.
c. cardiac cycle.
d. prepotential depolarization.

OUTCOME 20-3 _____ 31. The "lubb-dupp" sounds of the heart have practical clinical value because they provide information concerning the
a. cardiac output.
a. heart rate.
c. action and efficiency of the AV and semilunar valves.
d. stroke volume.

OUTCOME 20-3 _____ 32. When a chamber of the heart fills with blood and prepares for the start of the next beat, the chamber is in
a. systole.
b. ventricular ejection.
c. diastole.
d. isovolumetric contraction.

OUTCOME 20-3 _____ 33. At the start of atrial systole, the ventricles are filled to around _____ of capacity.
a. 10 percent
b. 30 percent
c. 50 percent
d. 70 percent

OUTCOME 20-4 _____ 34. The amount of blood ejected by the left ventricle per minute is the
a. stroke volume.
b. cardiac output.
c. end-diastolic volume.
d. end-systolic volume.

OUTCOME 20-4 _____ 35. The amount of blood pumped out of each ventricle during a single beat is the
a. stroke volume.
b. EDV.
c. cardiac output.
d. ESV.

OUTCOME 20-4 _____ 36. Under normal circumstances, the factors responsible for making delicate adjustments to the heart rate as circulatory demands change are
a. nerve and muscular activity.
b. cardiac output and stroke volume.
c. autonomic activity and circulatory hormones.
d. a, b, and c are correct.

OUTCOME 20-4 _____ 37. The cardiac centers in the medulla oblongata monitor baroreceptors and chemoreceptors innervated by the

    a. trochlear N IV and trigeminal N V.

    b. accessory N XI and hypoglossal N XII.

    c. glossopharyngeal N IX and vagus N X.

    d. facial N VII and vestibulocochlear N VIII.

OUTCOME 20-4 _____ 38. The atrial reflex (Bainbridge reflex) involves adjustments in heart rate stimulated by an increase in

    a. the influx of Ca^{2+}.

    b. venous return.

    c. acetylcholine released by parasympathetic neurons.

    d. autonomic tone.

OUTCOME 20-4 _____ 39. The difference between the end-diastolic volume (EDV) and the end-systolic volume (ESV) is the

    a. stroke volume.

    b. cardiac output.

    c. cardiac reserve.

    d. preload and afterload.

OUTCOME 20-4 _____ 40. The Frank-Starling principle describes the observed relationship that

    a. decreasing the EDV results in an increase in the stroke volume.

    b. increasing the EDV results in an increase in the stroke volume.

    c. increasing the EDV results in a decrease in the stroke volume.

    d. the EDV does not affect the stroke volume.

OUTCOME 20-4 _____ 41. Parasympathetic stimulation from the vagus nerve results in

    a. an increased heart rate.

    b. more forceful ventricular contractions.

    c. a decrease in heart rate.

    d. no effect on the heart rate.

OUTCOME 20-4 _____ 42. Cutting the vagus nerve

    a. decreases the heart rate.

    b. causes the heart to stop beating.

    c. increases the heart rate.

    d. does not affect the heart rate.

OUTCOME 20-4 _____ 43. If the heart rate is 78 bpm and the stroke volume is 80 m*l* per beat, the cardiac output will be

    a. 6240 m*l*/min.

    b. 158 m*l*/min.

    c. 1580 m*l*/min.

    d. 6420 m*l*/min.

Completion

Using the terms below, complete the following statements. Use each term only once.

preload	SA node	coronary sinus	chemoreceptors
intercalated discs	medulla oblongata	myocardium	cardiac skeleton
atrial reflex	repolarization	ESV	automaticity
pulmonary	atria	epicardium	veins
electrocardiogram	nodal cells	right ventricle	exchange vessels
pulmonary veins	auscultation	left ventricle	endocardium
pericardium	deoxygenated	stroke volume	

OUTCOME 20-1   1. Because of their thin walls and permeability capabilities, the capillaries are called _____.

OUTCOME 20-1   2. The afferent vessels that return blood to the heart are the _____.

OUTCOME 20-1   3. The right atrium receives blood from the systemic circuit and passes it to the _____.

OUTCOME 20-1   4. The chambers in the heart with thin walls that are highly distensible are the _____.

OUTCOME 20-1   5. The internal connective tissue network of the heart is called the _____.

OUTCOME 20-1   6. Each cardiac muscle fiber contacts several others at specialized sites known as _____.

OUTCOME 20-1   7. The pericardial tissue that covers and adheres closely to the outer surface of the heart is called the _____.

OUTCOME 20-1   8. The serous membrane lining the pericardial cavity is called the _____.

OUTCOME 20-1   9. The right side of the heart contains blood that is _____.

OUTCOME 20-1   10. When blood leaves the left atrium, it passes through an atrioventricular valve into the _____.

OUTCOME 20-1   11. Oxygenated blood is returned to the left atrium via the _____.

OUTCOME 20-1   12. The only arteries in the body that carry deoxygenated blood are the _____ arteries.

OUTCOME 20-1   13. The muscular wall of the heart that forms both atria and ventricles is the _____.

OUTCOME 20-1   14. The heart valves on the inner surface of the heart are covered by the _____.

OUTCOME 20-1   15. Blood returns to the heart from coronary circulation via the _____.

OUTCOME 20-2   16. When slow potassium channels begin opening in cardiac muscle, the result is a period of _____.

OUTCOME 20-2   17. The individual units responsible for establishing the rate of cardiac contraction are the _____.

OUTCOME 20-2   18. The action of cardiac muscle tissue contracting on its own in the absence of neural stimulation is called _____.

OUTCOME 20-2   19. A recording of electrical events that occur in the heart constitutes a(n) _____.

OUTCOME 20-3   20. Listening to sounds in the chest to determine the condition of the heart and the lungs is called _____.

OUTCOME 20-4   21. An increase in heart rate in response to stretching the right atrial wall is the _____.

OUTCOME 20-4   22. Blood CO_2, pH, and O_2 levels are monitored by _____.

OUTCOME 20-4   23. The autonomic headquarters for cardiac control is located in the _____.

OUTCOME 20-4   24. The _____ is the primary source of the impulses that establish the heart rate.

OUTCOME 20-4   25. The amount of blood that remains in the ventricles at the end of ventricular systole is the _____.

OUTCOME 20-4   26. The degree of muscle-cell stretching during ventricular diastole is called the _____.

OUTCOME 20-4   27. The amount of blood ejected by a ventricle during a single beat is the _____.

Matching

Match the terms in column B with the terms in column A. Use letters for answers in the spaces provided. Use each term only once.

Part I

		Column A	Column B
OUTCOME 20-1	_____	1. arteries	A. blood to systemic arteries
OUTCOME 20-1	_____	2. fossa ovalis	B. activates contraction process
OUTCOME 20-1	_____	3. carditis	C. muscular wall of the heart
OUTCOME 20-1	_____	4. serous membrane	D. sodium ion permeability
OUTCOME 20-1	_____	5. tricuspid valve	E. atrial septum structure
OUTCOME 20-1	_____	6. aortic valve	F. right atrioventricular
OUTCOME 20-1	_____	7. trabeculae carneae	G. interconnections between blood vessels
OUTCOME 20-1	_____	8. aorta	H. inflammation of the heart
OUTCOME 20-1	_____	9. myocardium	I. pericardium
OUTCOME 20-1	_____	10. anastomoses	J. muscular ridges
OUTCOME 20-2	_____	11. rapid depolarization	K. efferent vessels
OUTCOME 20-2	_____	12. calcium ions	L. semilunar

Part II

		Column A	Column B
OUTCOME 20-2	_____	13. cardiac pacemaker	M. Bainbridge reflex
OUTCOME 20-2	_____	14. bundle of His	N. AV bundle
OUTCOME 20-2	_____	15. P wave	O. monitor blood pressure
OUTCOME 20-3	_____	16. "lubb" sound	P. increases heart rate
OUTCOME 20-3	_____	17. "dupp" sound	Q. medulla oblongata
OUTCOME 20-4	_____	18. atrial reflex	R. AV valves close
OUTCOME 20-4	_____	19. baroreceptors	S. ventricular diastole
OUTCOME 20-4	_____	20. sympathetic activity	T. semilunar valve closes
OUTCOME 20-4	_____	21. fills atria	U. atrial depolarization
OUTCOME 20-4	_____	22. filling time	V. $SV \times HR$
OUTCOME 20-4	_____	23. cardiac output	W. venous return
OUTCOME 20-4	_____	24. cardiac centers	X. SA node

Drawing/Illustration Labeling

Identify each numbered structure by labeling the following figures. Place your answers in the spaces provided.

OUTCOME 20-1   **FIGURE 20-1**   Anatomy of the Heart (Frontal Section through the Heart)

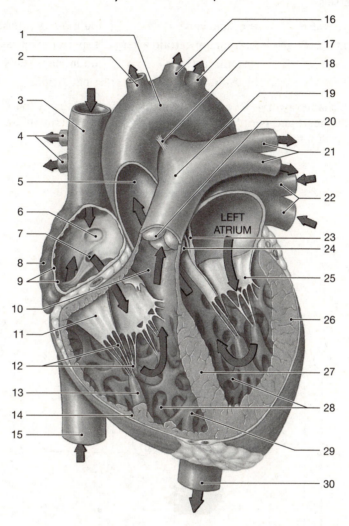

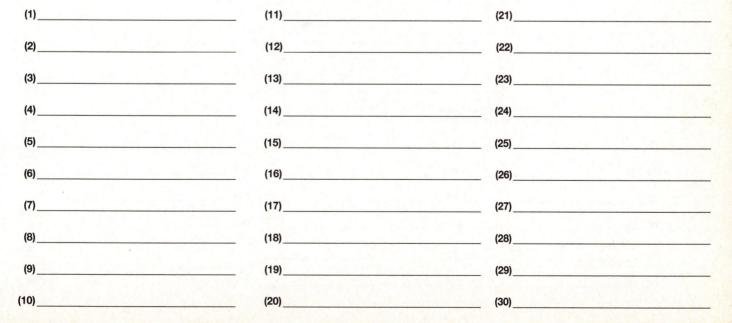

(1)_____

(2)_____

(3)_____

(4)_____

(5)_____

(6)_____

(7)_____

(8)_____

(9)_____

(10)_____

(11)_____

(12)_____

(13)_____

(14)_____

(15)_____

(16)_____

(17)_____

(18)_____

(19)_____

(20)_____

(21)_____

(22)_____

(23)_____

(24)_____

(25)_____

(26)_____

(27)_____

(28)_____

(29)_____

(30)_____

LEVEL 2: REVIEWING CONCEPTS

Chapter Overview

Using the terms below, fill in the blanks to correctly complete the chapter overview of the heart and peripheral blood vessels.

aortic semilunar valve	tricuspid valve	pulmonary arteries	left ventricle
pulmonary semilunar valve	L. common carotid artery	superior vena cava	systemic veins
inferior vena cava	systemic arteries	pulmonary veins	aorta
left atrium	bicuspid valve	right atrium	right ventricle

To complete the chapter overview you will use the circulatory pathway to follow a drop of blood as it goes through the heart and peripheral blood vessels. Identify each structure at the sequential, circled, numbered locations—1 through 16.

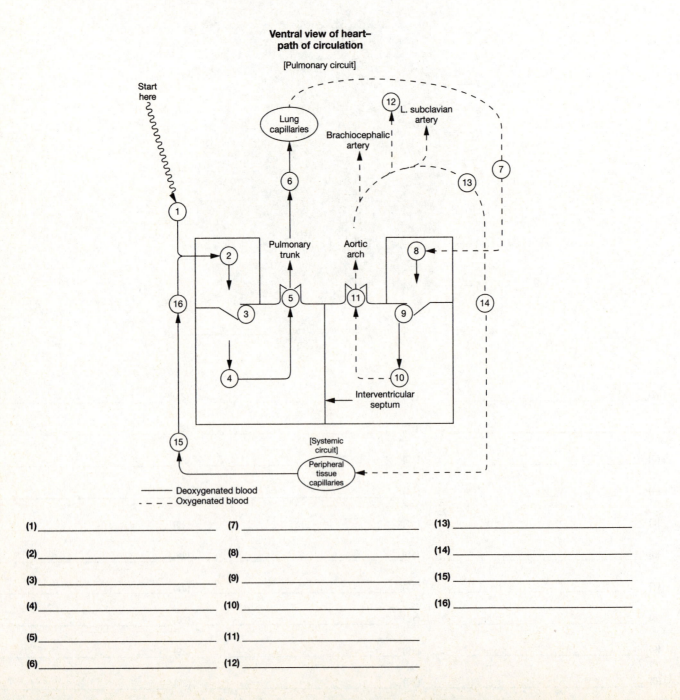

**Ventral view of heart–
path of circulation**

[Pulmonary circuit]

Start here

Lung capillaries

12 L. subclavian artery

Brachiocephalic artery

6

13

7

1

Pulmonary trunk

Aortic arch

8

2

5

11

16

3

9

14

4

10

Interventricular septum

15

[Systemic circuit]

Peripheral tissue capillaries

——— Deoxygenated blood
- - - - Oxygenated blood

(1) _____

(2) _____

(3) _____

(4) _____

(5) _____

(6) _____

(7) _____

(8) _____

(9) _____

(10) _____

(11) _____

(12) _____

(13) _____

(14) _____

(15) _____

(16) _____

Concept Map I

Using the following terms, fill in the circled, numbered, blank spaces to correctly complete the concept map. Use each term only once.

epicardium

pacemaker cells

endocardium

aortic

tricuspid

blood from atria

two semilunar

two atria

deoxygenated blood

oxygenated blood

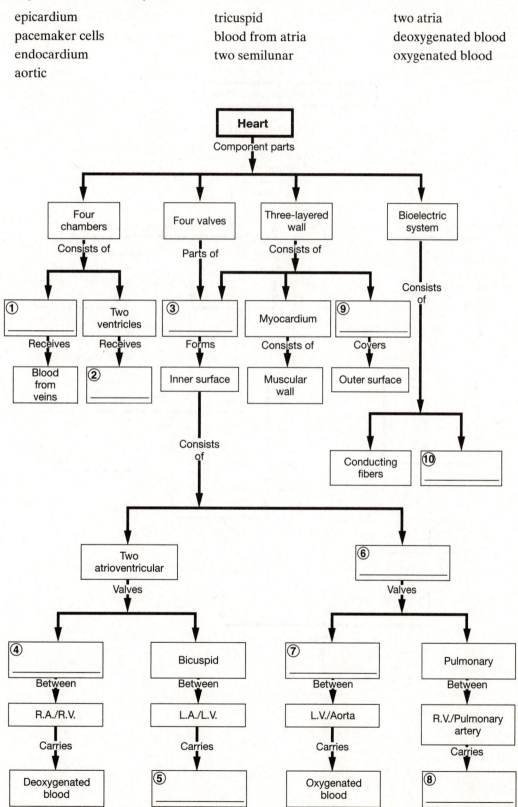

Concept Map II

Using the following terms, fill in the circled, numbered, blank spaces to correctly complete the concept map. Use each term only once.

afterload   hormones   autonomic innervation
preload   cardiac output   end-systolic volume
heart rate   filling time

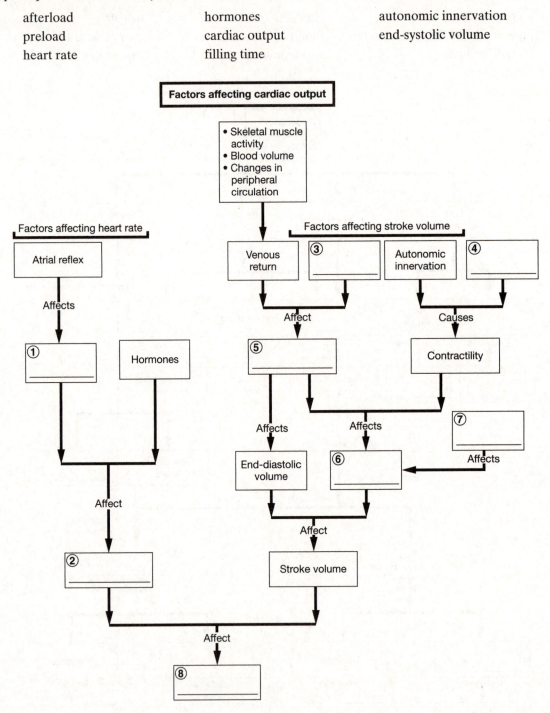

Concept Map III

Using the following terms, fill in the circled, numbered, blank spaces to correctly complete the concept map. Use each term only once.

increasing blood pressure,
    decreasing blood pressure
increasing epinephrine,
    norepinephrine

increasing heart rate
decreasing CO_2
increasing CO_2

increasing acetylcholine
chemoreceptors
decreasing heart rate

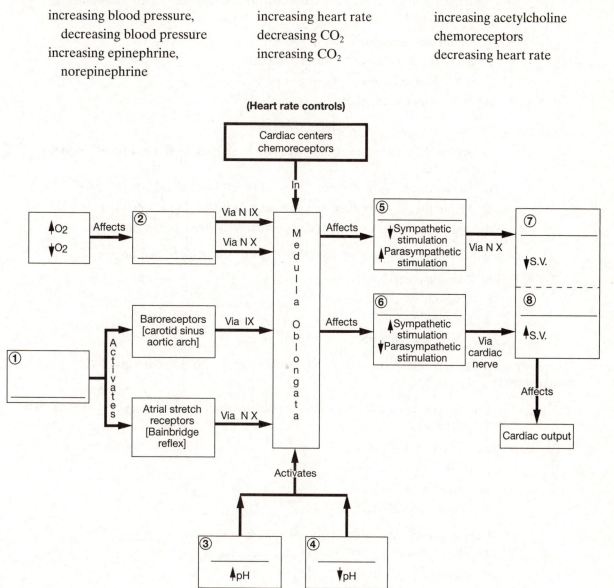

Multiple Choice

Place the letter corresponding to the best answer in the space provided.

_____ 1. Assuming anatomic position, the *best* way to describe the *specific* location of the heart in the body is

    a. within the mediastinum of the thorax.

    b. in the region of the fifth intercostal space.

    c. just behind the lungs.

    d. in the center of the chest.

_____ 2. The function of the chordae tendineae is to

    a. anchor the semilunar valve flaps and prevent backward flow of blood into the ventricles.

    b. anchor the AV valve flaps and prevent backflow of blood into the atria.

    c. anchor the AV valve flaps and prevent backflow of blood into the ventricle.

    d. anchor the aortic valve flaps and prevent backflow into the ventricles.

_____ 3. Which one of the following would *not* show up on an electrocardiogram?

    a. Abnormal heart block

    b. Murmurs

    c. Heart block

    d. Bundle branch block

[To answer questions 4–11 refer to the graph on page 389, which records the events of the cardiac cycle.]

_____ 4. During ventricular diastole, when the pressure in the left ventricle rises above that in the left atrium,

    a. the left AV valve closes.

    b. the left AV valve opens.

    c. the aortic valve closes.

    d. all the valves close.

_____ 5. During ventricular systole, the blood volume in the atria is _____, and the volume in the ventricle is _____.

    a. decreasing; increasing

    b. increasing; decreasing

    c. increasing; increasing

    d. decreasing; decreasing

_____ 6. During *most* of ventricular diastole, the

    a. pressure in the L. atrium is slightly lower than the pressure in the L. ventricle.

    b. pressure in the L. ventricle reaches 120 mm Hg while the pressure in the L. atrium reaches 30 mm Hg.

    c. pressure in the L. ventricle is slightly lower than the pressure in the L. atrium.

    d. pressures are the same in the L. ventricle and the L. atrium.

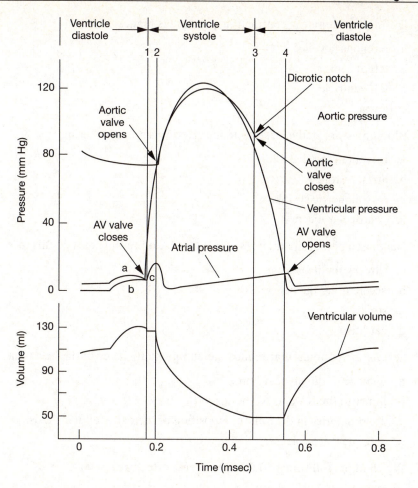

7. When the pressure within the L. ventricle becomes greater than the pressure within the aorta, the

   a. pulmonary semilunar valve is forced open.

   b. aortic semilunar valve closes.

   c. pulmonary semilunar valve closes.

   d. aortic semilunar valve is forced open.

8. The volume of blood in the L. ventricle is at its lowest when the

   a. ventricular pressure is 40 mm Hg.

   b. AV valve opens.

   c. atrial pressure is 30 mm Hg.

   d. AV valve closes.

9. The dicrotic notch indicates a brief rise in the aortic pressure. The rise in pressure is due to

   a. the closure of the semilunar valve.

   b. a decrease in the ventricular pressure.

   c. the opening of the AV valve.

   d. an increase in atrial pressure.

_____ 10. During isovolumetric contraction, pressure is highest in the

   a. pulmonary veins.

   b. left atrium.

   c. left ventricle.

   d. aorta.

_____ 11. Blood pressure in the large systemic arteries is greatest during

   a. isovolumetric diastole.

   b. atrial systole.

   c. isovolumetric systole.

   d. ventricular ejection.

_____ 12. Decreased parasympathetic (vagus) stimulation to the heart results in

   a. slowing the heart rate.

   b. accelerating the heart rate.

   c. stenosis.

   d. carditis.

_____ 13. Serious arrhythmias that reduce the pumping efficiency of the heart may indicate

   a. damage to the myocardium.

   b. injury to the SA and AV nodes.

   c. abnormalities in the ionic composition of the extracellular fluids.

   d. a, b, and c are correct.

_____ 14. Which of the following would *not* increase afterload?

   a. Constriction of peripheral blood vessels

   b. Arteriosclerosis in peripheral blood vessels

   c. Vasodilation of peripheral vessels

   d. Increased blood pressure

_____ 15. Ventricular diastole indicates that

   a. both atria are filling.

   b. the ventricles are "resting."

   c. the AV valves are closed.

   d. presystolic pressures are high.

_____ 16. Which of the following does *not* control the movement of blood through the heart?

   a. Opening and closing of the valves

   b. Contraction of the myocardium

   c. Size of the atria and ventricles

   d. Relaxation of the myocardium

_____ 17. The ANS can make very delicate adjustments in cardiovascular function to meet demands of other systems through

   a. heart rate and stroke volume.

   b. the difference between the end-diastolic volume and end-systolic volume.

   c. filling time and venous return.

   d. dual innervation and adjustments in autonomic tone.

_____ 18. Normally, the only electrical connection between the atria and the ventricles is the

     a. AV bundle.

     b. Purkinje fibers.

     c. anastomoses.

     d. trabeculae carnae.

_____ 19. If the SA node or internodal pathways are damaged, the

     a. ectopic pacemaker assumes command.

     b. Purkinje fibers take command.

     c. AV node assumes command.

     d. heart stops beating.

_____ 20. Tetanic contractions cannot occur in a normal cardiac muscle cell, regardless of the frequency or intensity of stimulation, because

     a. the refractory period ends before peak tension develops.

     b. summation is not possible.

     c. the action potential is relatively brief.

     d. twitches can summate and tetanus cannot occur.

Completion

Using the terms below, complete the following statements. Use each term only once.

angina pectoris	endocardium	auricle
cardiac reserve	anastomoses	systemic
pectinate muscles	myocardium	pacemaker
trabeculae carneae	bradycardia	afterload
filling time	tachycardia	infarct

1. Interconnections between vessels are referred to as _____.

2. An improperly functioning bicuspid (mitral) valve would affect _____ circulation.

3. The comb-like muscle ridges in the atria are called _____.

4. The muscular ridges in the internal surface of the ventricles are called _____.

5. The inner surface of the heart that is continuous with the endothelium of the attached blood vessels is the _____.

6. The muscular wall of the heart that contains cardiac muscle tissue, associated connective tissue, blood vessels, and nerves is the _____.

7. The term used to indicate a heart rate that is slower than the normal rate is _____.

8. The term used to indicate a heart rate that is faster than the normal rate is _____.

9. The spontaneous depolarization in the heart takes place in specialized _____ cells.

10. The duration of ventricular diastole is the _____.

11. The difference between resting and maximum cardiac output is the _____.

12. The amount of tension the contracting ventricle must produce to force open the semilunar valve and eject blood is the _____.

13. The expanded ear-like extension of the atrium is called a(n) _____.

14. Temporary heart insufficiency and ischemia causing severe chest pain is referred to as _____.

15. Tissue degeneration of the heart due to coronary circulation blockage creates a nonfunctional area known as a(n) _____.

Short Essay

Briefly answer the following questions in the spaces provided below.

1. If the heart beat rate (HR) is 80 bpm and the stroke volume (SV) is 75 m*l*, what is the cardiac output (CO) (*l*/min)?

2. If the end-systolic volume (ESV) is 60 m*l* and the end-diastolic volume (EDV) is 140 m*l*, what is the stroke volume (SV)?

3. If the cardiac output (CO) is 5 *l*/min and the heart rate (HR) is 100 bpm, what is the stroke volume (SV)?

4. What is the difference between the visceral pericardium and the parietal pericardium?

5. What are the chordae tendineae, papillary muscles, trabeculae carneae, and pectinate muscles? Where are they located?

6. What three distinct layers comprise the tissues of the heart wall?

7. What are the seven important functions of the connective tissues and the fibrous skeleton of the heart?

8. Because of specialized sites known as intercalated discs, cardiac muscle is a functional syncytium. What does this statement mean?

9. Beginning with the SA node, trace the pathway of an action potential through the conducting network of the heart. (Use arrows to indicate direction.)

10. What two common heart rate problems result from abnormal pacemaker function? Explain what each term means.

11. What two important factors have a direct effect on the heart rate?

LEVEL 3: CRITICAL THINKING AND CLINICAL APPLICATIONS

Using principles and concepts learned in Chapter 20, answer the following questions. Write your answers on a separate sheet of paper.

1. Why does a person who exhibits tachycardia eventually experience the loss of consciousness?

2. What effect does a drug have that blocks the calcium channels in cardiac muscle cells? What is the direct effect on the stroke volume?

3. A simple and effective method of cardiac assessment is auscultation (i.e., listening to the heart). What information does "heart sound" give to a physician or clinician who uses this method for analysis and diagnosis?

4. After a session of strenuous exercise, Joe experiences ischemia, chest constriction, and pain that radiates from the sternal area to the arms, back, and neck. He is taken to a hospital emergency room, where he is diagnosed with angina pectoris. What is this condition, and how may it typically be controlled?

5. Describe the significance of the P wave, the QRS complex, and the T wave as they relate to assessing the results of an ECG.

6. Explain the importance of exercise and the effect that it has on preload (i.e., the degree of stretching experienced by ventricular muscle cells during ventricular diastole).

7. Your anatomy and physiology instructor is lecturing on the importance of coronary circulation to the overall functional efficiency of the heart. Part of this efficiency is due to the presence of arterial anastomoses. What are arterial anastomoses and why are they important in coronary circulation?

8. The right and left atria of the heart look alike and perform similar functional demands. The right and left ventricles are very different, structurally and functionally. Why are the atrial similarities and the differences in the ventricles significant in the roles they play in the functional activities of the heart?

Blood Vessels and Circulation

OVERVIEW

Blood vessels form a closed system of tubes that transport blood and allow exchange of gases, nutrients, and wastes between the blood and the body cells. The general plan of the cardiovascular system includes the heart, arteries that carry blood away from the heart, capillaries that allow exchange of substances between blood and the cells, and veins that carry blood back to the heart. The pulmonary circuit carries deoxygenated blood from the right side of the heart to the lung capillaries, where the blood is oxygenated. The systemic circuit carries the oxygenated blood from the left side of the heart to the systemic capillaries in all parts of the body. The circular design of the system results in deoxygenated blood returning to the right atrium to continue the cycle of circulation.

Blood vessels in the muscles, the skin, the cerebral circulation, and the hepatic portal circulation are specifically adapted to serve the functions of organs and tissues in these important special areas of cardiovascular activity.

Chapter 21 describes the structure and functions of the blood vessels, the dynamics of the circulatory process, cardiovascular regulation, and patterns of cardiovascular response.

LEVEL 1: REVIEWING FACTS AND TERMS

Review of Learning Outcomes

After completing this chapter, you should be able to do the following:

OUTCOME 21-1   Distinguish among the types of blood vessels based on their structure and function, and describe how and where fluid and dissolved materials enter and leave the cardiovascular system.

OUTCOME 21-2   Explain the mechanisms that regulate blood flow through vessels, describe the factors that influence blood pressure, and discuss the mechanisms that regulate movement of fluids between capillaries and interstitial spaces.

OUTCOME 21-3   Describe the control mechanisms that regulate blood flow and pressure in tissues, and explain how the activities of the cardiac, vasomotor, and respiratory centers are coordinated to control blood flow through the tissues.

OUTCOME 21-4   Explain the cardiovascular system's homeostatic response to exercise and hemorrhaging, and identify the principal blood vessels and the functional characteristics of the special circulation to the brain, heart, and lungs.

OUTCOME 21-5   Describe the three general functional patterns seen in the pulmonary and systemic circuits of the cardiovascular system.

OUTCOME 21-6   Identify the major arteries and veins of the pulmonary circuit.

OUTCOME 21-7   Identify the major arteries and veins of the systemic circuit.

OUTCOME 21-8   Identify the differences between fetal and adult circulation patterns, and describe the changes in the patterns of blood flow that occur at birth.

OUTCOME 21-9   Discuss the effects of aging on the cardiovascular system, and give examples of interactions between the cardiovascular system and other organ systems.

Multiple Choice

Place the letter corresponding to the best answer in the space provided.

OUTCOME 21-1   _____   1. The layer of vascular tissue that consists of an endothelial lining and an underlying layer of connective tissue dominated by elastic fibers is the

    a. tunica intima.

    b. tunica media.

    c. tunica externa.

    d. tunica adventitia.

OUTCOME 21-1   _____   2. Smooth muscle fibers in arteries and veins are found in the

    a. endothelial lining.

    b. tunica externa.

    c. tunica intima.

    d. tunica media.

OUTCOME 21-1   _____   3. One of the major characteristics of the arteries supplying peripheral tissues is that they are

    a. elastic.

    b. muscular.

    c. rigid.

    d. a, b, and c are correct.

OUTCOME 21-1   _____   4. The only blood vessels whose walls permit exchange between the blood and the surrounding interstitial fluids are

    a. arterioles.

    b. venules.

    c. capillaries.

    d. a, b, and c are correct.

OUTCOME 21-1   _____   5. One of the primary characteristics of continuous capillaries is that they prevent

    a. the loss of blood cells and plasma proteins.

    b. the diffusion of water into the interstitial fluid.

    c. the bulk transport of small solutes into the interstitium.

    d. lipid-soluble materials from entering the interstitial fluid.

OUTCOME 21-1   _____   6. The unidirectional flow of blood in venules and medium-sized veins is maintained by

    a. the muscular walls of the veins.

    b. pressure from the left ventricle.

    c. arterial pressure.

    d. the presence of valves.

OUTCOME 21-2   _____   7. The specialized arteries that are able to tolerate the pressure shock produced each time ventricular systole occurs and blood leaves the heart are
     a. muscular arteries.
     b. elastic arteries.
     c. arterioles.
     d. fenestrated arteries.

OUTCOME 21-2   _____   8. Of the following blood vessels, the greatest drop in blood pressure occurs in the
     a. veins.
     b. capillaries.
     c. venules.
     d. arterioles.

OUTCOME 21-2   _____   9. If the systolic pressure is 120 mm Hg and the diastolic pressure is 90 mm Hg, the mean arterial pressure (MAP) is
     a. 30 mm Hg.
     b. 210 mm Hg.
     c. 100 mm Hg.
     d. 80 mm Hg.

OUTCOME 21-2   _____   10. The distinctive sounds of Korotkoff heard when taking the blood pressure are produced by
     a. turbulences as blood flows past the constricted portion of the artery.
     b. the contraction and relaxation of the ventricles.
     c. the opening and closing of the atrioventricular valves.
     d. a, b, and c are correct.

OUTCOME 21-2   _____   11. The phenomenon that helps maintain blood flow along the arterial network while the left ventricle is in diastole is
     a. peripheral resistance.
     b. hydrostatic pressure.
     c. vascular resistance.
     d. elastic rebound.

OUTCOME 21-2   _____   12. The most important factor in vascular resistance is
     a. a combination of neural and hormonal mechanisms.
     b. differences in the length of the blood vessels.
     c. friction between blood and the vessel walls.
     d. the diameter of the arterioles.

OUTCOME 21-2   _____   13. Two factors that assist the low venous pressures in propelling blood toward the heart are
     a. arterial pressure and pulse pressure.
     b. elastic rebound and mean arterial pressure.
     c. systolic pressure and diastolic pressure.
     d. muscular compression of peripheral veins and the respiratory pump.

OUTCOME 21-2   _____ 14. From the following selections, choose the answer that correctly
identifies all the factors that would increase blood pressure.
(*Note:* CO = cardiac output; SV = stroke volume; VR = venous return;
PR = peripheral resistance; BV = blood volume.)

a. Increasing CO, increasing SV, decreasing VR, decreasing PR,
increasing BV

b. Increasing CO, increasing SV, increasing VR, increasing PR,
increasing BV

c. Increasing CO, increasing SV, decreasing VR, increasing PR,
decreasing BV

d. Increasing CO, decreasing SV, increasing VR, decreasing PR,
increasing BV

OUTCOME 21-2   _____ 15. The two major factors affecting blood flow rates are

a. diameter and length of blood vessels.

b. pressure and resistance.

c. neural and hormonal control mechanisms.

d. turbulence and viscosity.

OUTCOME 21-2   _____ 16. For circulation to occur, the circulatory pressure must be sufficient
to overcome the

a. total peripheral resistance.

b. capillary hydrostatic pressure.

c. venous pressure.

d. mean arterial pressure.

OUTCOME 21-3   _____ 17. Atrial natriuretic peptide (ANP) reduces blood volume and pressure by

a. blocking release of ADH.

b. stimulating peripheral vasodilation.

c. increased water loss through the kidneys.

d. a, b, and c are correct.

OUTCOME 21-3   _____ 18. The circulatory-regulatory mechanisms that can assist in short-term
and long-term adjustments are

a. autoregulatory.

b. endocrine responses.

c. neural mechanisms.

d. tissue perfusions.

OUTCOME 21-3   _____ 19. The regulatory mechanism(s) that cause immediate, localized
homeostatic adjustments is (are)

a. autoregulation.

b. endocrine regulations.

c. neural regulation.

d. all of the above.

OUTCOME 21-3   _____ 20. The central regulation of cardiac output primarily involves the
activities of the

a. somatic nervous system.

b. autonomic nervous system.

c. central nervous system.

d. a, b, and c are correct.

OUTCOME 21-3 _____ 21. An increase in cardiac output normally occurs during
   a. widespread sympathetic stimulation.
   b. widespread parasympathetic stimulation.
   c. the process of vasomotion.
   d. stimulation of the vasomotor center.

OUTCOME 21-3 _____ 22. Stimulation of the vasomotor center in the medulla causes
   _____, and inhibition of the vasomotor center causes
   _____.
   a. vasodilation; vasoconstriction
   b. increasing diameter of arteriole; decreasing diameter of arteriole
   c. hyperemia; ischemia
   d. vasoconstriction; vasodilation

OUTCOME 21-3 _____ 23. Hormonal regulation by ADH, epinephrine, angiotensin II,
   and norepinephrine results in
   a. increasing peripheral vasodilation.
   b. decreasing peripheral vasoconstriction.
   c. increasing peripheral vasoconstriction.
   d. a, b, and c are correct.

OUTCOME 21-4 _____ 24. The three primary interrelated changes that occur as exercise begins are
   a. decreasing vasodilation, increasing venous return, and increasing
      cardiac output.
   b. increasing vasodilation, decreasing venous return, and increasing
      cardiac output.
   c. increasing vasodilation, increasing venous return, and increasing
      cardiac output.
   d. decreasing vasodilation, decreasing venous return, and decreasing
      cardiac output.

OUTCOME 21-4 _____ 25. The only area of the body where the blood supply is unaffected while
   exercising at maximum levels is the
   a. hepatic portal circulation.
   b. pulmonary circulation.
   c. brain.
   d. peripheral circulation.

OUTCOME 21-5 _____ 26. Other than near the heart, the peripheral distributions of arteries
   and veins on the left and right sides are
   a. completely different.
   b. large vessels that connect to the atria and ventricles.
   c. generally identical on both sides.
   d. different on the left but the same on the right.

OUTCOME 21-5 _____ 27. As the external iliac artery leaves the body trunk and enters the lower
   limb, it becomes the
   a. lower tibial artery.
   b. femoral artery.
   c. anterior fibular artery.
   d. lower abdominal aorta.

OUTCOME 21-5 _____ 28. The link between adjacent arteries or veins that reduces the impact of a temporary or permanent occlusion of a single blood vessel is a(n)

    a. arteriole.

    b. venule.

    c. anastomosis.

    d. cardiovascular bridge.

OUTCOME 21-6 _____ 29. The four large blood vessels, two from each lung, that empty into the left atrium, completing the pulmonary circuit, are the

    a. venae cavae.

    b. pulmonary arteries.

    c. pulmonary veins.

    d. subclavian veins.

OUTCOME 21-6 _____ 30. The blood vessels that provide blood to capillary networks that surround the alveoli in the lungs are

    a. pulmonary arterioles.

    b. fenestrated capillaries.

    c. left and right pulmonary arteries.

    d. left and right pulmonary veins.

OUTCOME 21-7 _____ 31. The three elastic arteries that originate along the aortic arch and deliver blood to the head, neck, shoulders, and arms are the

    a. axillary, R. common carotid, and right subclavian.

    b. R. dorsoscapular, R. thoracic, and R. vertebral.

    c. R. axillary, R. brachial, and L. internal carotid.

    d. brachiocephalic, L. common carotid, and left subclavian.

OUTCOME 21-7 _____ 32. The large blood vessel that collects most of the venous blood from organs below the diaphragm is the

    a. superior vena cava.

    b. inferior vena cava.

    c. hepatic portal vein.

    d. superior mesenteric vein.

OUTCOME 21-7 _____ 33. The three blood vessels that provide blood to all of the digestive organs in the abdominopelvic cavity are the

    a. thoracic aorta, abdominal aorta, and superior phrenic artery.

    b. intercostal, esophageal, and bronchial arteries.

    c. celiac trunk and the superior and inferior mesenteric arteries.

    d. adrenal, renal, and lumbar arteries.

OUTCOME 21-7 _____ 34. The diaphragm divides the descending aorta into

    a. brachiocephalic and carotid arteries.

    b. right and left common iliac arteries.

    c. superior phrenic and adrenal arteries.

    d. superior thoracic aorta and inferior abdominal aorta.

OUTCOME 21-7 _____ 35. The three branches that originate from the celiac trunk are the

    a. L. gastric, splenic, and common hepatic arteries.

    b. celiac trunk and superior-inferior mesenteric arteries.

    c. bronchial, pericardial, and esophageal arteries.

    d. mediastinal, intercostal, and phrenic arteries.

OUTCOME 21-7 _____ 36. Except for the cardiac veins, all of the body's systemic veins drain into either the

    a. superior or inferior vena cava.

    b. internal or external jugular veins.

    c. sigmoid or cavernous sinuses.

    d. inferior or superior mesenteric veins.

OUTCOME 21-7 _____ 37. Blood from the tissues and organs of the head, neck, chest, shoulders, and upper limbs is delivered to the

    a. inferior vena cava.

    b. cavernous sinus.

    c. conus venosus.

    d. superior vena cava.

OUTCOME 21-7 _____ 38. Blood from the lower limbs, the pelvis, and the lower abdomen is delivered to the

    a. hepatic veins.

    b. hepatic portal system.

    c. external iliac veins.

    d. great saphenous veins.

OUTCOME 21-7 _____ 39. Blood leaving the capillaries supplied by the celiac trunk and superior and inferior mesenteric arteries flows into the

    a. hepatic portal system.

    b. inferior vena cava.

    c. superior vena cava.

    d. abdominal aorta.

OUTCOME 21-8 _____ 40. The nutritional and respiratory needs of a fetus are provided by

    a. the stomach and lungs of the fetus.

    b. the stomach and lungs of the mother.

    c. diffusion across the placenta.

    d. a, b, and c are correct.

OUTCOME 21-8 _____ 41. In early fetal life, the foramen ovale allows blood to flow freely from the

    a. right ventricle to the left ventricle.

    b. right atrium to the left atrium.

    c. right atrium to the left ventricle.

    d. left atrium to the right ventricle.

OUTCOME 21-8 _____ 42. In the adult, the ductus arteriosus persists as a fibrous cord called the

    a. fossa ovalis.

    b. ligamentum arteriosum.

    c. ductus venosus.

    d. umbilicus.

OUTCOME 21-8 _____ 43. A few seconds after birth, rising O_2 levels stimulate the constriction of the ductus arteriosus, isolating the
      a. ligamentum arteriosum and fossa ovalis.
      b. ductus venosus and ligamentum arteriosum.
      c. superior and inferior vena cava.
      d. pulmonary and aortic trunks.

OUTCOME 21-9 _____ 44. The primary effect of a decrease in the hematocrit of elderly individuals is
      a. thrombus formation in the blood vessels.
      b. a lowering of the oxygen-carrying capacity of the blood.
      c. a reduction in the maximum cardiac output.
      d. damage to ventricular cardiac muscle fibers.

OUTCOME 21-9 _____ 45. In the heart, age-related progressive atherosclerosis causes
      a. changes in activities of conducting cells.
      b. a reduction in the elasticity of the fibrous skeleton.
      c. restricted coronary circulation.
      d. a reduction in maximum cardiac input.

OUTCOME 21-9 _____ 46. The systems responsible for modifying heart rate and regulating blood pressure are the
      a. respiratory and lymphoid systems.
      b. muscular and urinary systems.
      c. nervous and endocrine systems.
      d. digestive and skeletal systems.

Completion

Using the terms below, complete the following statements. Use each term only once.

foramen ovale	turbulence	venules	metarteriole
angiogenesis	right atrium	fenestrated	hepatic portal vein
autoregulation	alveoli	viscosity	precapillary sphincter
respiratory pump	vasomotion	vasodilators	femoral artery
pulse pressure	shock	anastomoses	atrioventricular septal
arterioles	vasoconstriction	sphygmomanometer	integumentary
arteriosclerosis	stroke	circulatory pressure	

OUTCOME 21-1   1. The smallest vessels of the arterial system are the _____.

OUTCOME 21-1   2. Blood flowing out of the capillary complex first enters small _____.

OUTCOME 21-1   3. Capillaries that contain "windows" or pores that penetrate the endothelial lining are _____ capillaries.

OUTCOME 21-2   4. The entrance to each capillary is guarded by a band of smooth muscle, the _____.

OUTCOME 21-2   5. The total peripheral resistance of the cardiovascular system reflects a combination of vascular resistance, turbulence, and _____.

OUTCOME 21-2   6. The difference between the systolic and diastolic pressures is the _____.

OUTCOME 21-2   7. The instrument used to determine blood pressure is called a _____.

OUTCOME 21-2   8. The pressure difference between the base of the ascending aorta and the entrance to the right atrium is the _____.

OUTCOME 21-2   9. The passageway in a capillary bed that possesses smooth muscle capable of changing its diameter is called a(n) _____.

OUTCOME 21-2   10. The cycling of contraction and relaxation of smooth muscles that changes blood flow through capillary beds is called _____.

OUTCOME 21-2   11. Formation of new blood vessels under the direction of vascular endothelial growth factor (VEGF) is called _____.

OUTCOME 21-2   12. The phenomenon in which high flow rates, irregular surfaces, and sudden changes in vessel diameter upset the smooth flow of blood, creating eddies and swirls, is called _____.

OUTCOME 21-3   13. The regulation of blood flow at the tissue level is called _____.

OUTCOME 21-3   14. Decreased tissue oxygen levels or increased CO_2 levels are examples of local _____.

OUTCOME 21-3   15. Stimulation of the vasomotor center in the medulla causes _____.

OUTCOME 21-3   16. A rise in the respiratory rate accelerates venous return through the action of the _____.

OUTCOME 21-4   17. An acute circulatory crisis marked by low blood pressure and inadequate peripheral blood flow is referred to as _____.

OUTCOME 21-4   18. When an interruption of the vascular supply to a portion of the brain occurs, it is called a(n) _____.

OUTCOME 21-5   19. The structures between adjacent arteries or veins that reduce the impact of a temporary blockage of a single blood vessel are called _____.

OUTCOME 21-6   20. In the lungs, gas exchange happens in the _____.

OUTCOME 21-7   21. The _____ receives blood from the splenic, inferior mesenteric, and superior mesenteric veins.

OUTCOME 21-7   22. As the external iliac artery leaves the trunk and enters the lower limb, it becomes the _____.

OUTCOME 21-7   23. The systemic circuit, which at any moment contains about 84 percent of the total blood volume, begins at the left ventricle and ends at the _____.

OUTCOME 21-8   24. During fetal life, blood can flow freely from the right atrium to the left atrium due to the presence of the _____.

OUTCOME 21-8   25. In a(n) _____ defect, both the atria and the ventricles are incompletely separated.

OUTCOME 21-9   26. Most of the age-related changes in the blood vessels are related to _____.

OUTCOME 21-9   27. The system in which stimulation of mast cells produces localized changes in blood flow and capillary permeability is the _____ system.

Matching

Match the terms in column B with the terms in column A. Use letters for answers in the spaces provided. Use each term only once.

Part I

		Column A	Column B
OUTCOME 21-1	_____	1. tunica externa	A. detect changes in pressure
OUTCOME 21-1	_____	2. vasa vasorum	B. capillary bed
OUTCOME 21-1	_____	3. sinusoids	C. carotid bodies
OUTCOME 21-1	_____	4. metarteriole	D. minimum blood pressure
OUTCOME 21-2	_____	5. respiratory pump	E. fenestrated capillaries
OUTCOME 21-2	_____	6. systolic pressure	F. connected by tight junction
OUTCOME 21-2	_____	7. diastolic pressure	G. "vessels of vessels"
OUTCOME 21-2	_____	8. continuous capillaries	H. capillaries > arteries
OUTCOME 21-2	_____	9. total cross-sectional area	I. connective tissue sheath
OUTCOME 21-2	_____	10. mean arterial pressure	J. "peak" blood pressure
OUTCOME 21-3	_____	11. baroreceptors	K. vasodilation
OUTCOME 21-3	_____	12. atrial natriuretic peptide	L. single value for blood pressure
OUTCOME 21-3	_____	13. chemoreceptors	M. venous return

Part II

		Column A	Column B
OUTCOME 21-3	_____	14. medulla	N. pulmonary circuit
OUTCOME 21-4	_____	15. atrial reflex	O. carry deoxygenated blood
OUTCOME 21-4	_____	16. rise in body temperature	P. foramen ovale
OUTCOME 21-4	_____	17. constant even in exercise	Q. carry oxygenated blood
OUTCOME 21-5	_____	18. blood between heart and lungs	R. loss of elasticity
			S. ruptured aneurysm
OUTCOME 21-5	_____	19. blood to cells and tissues	T. increasing venous return
OUTCOME 21-6	_____	20. pulmonary arteries	U. internal jugular vein
OUTCOME 21-6	_____	21. pulmonary veins	V. ligamentum arteriosum
OUTCOME 21-7	_____	22. carries O_2 blood to head	W. systemic circuit
OUTCOME 21-7	_____	23. blood from brain	X. increased blood flow to skin
OUTCOME 21-8	_____	24. blood flow to placenta	Y. umbilical arteries
OUTCOME 21-8	_____	25. interatrial opening	Z. age-related change
OUTCOME 21-8	_____	26. ductus arteriosus	AA. blood flow to brain
OUTCOME 21-9	_____	27. decreased hematocrit	BB. carotid arteries
OUTCOME 21-9	_____	28. massive blood loss	CC. vasomotor center
OUTCOME 21-9	_____	29. aging cardiac skeleton	

Drawing/Illustration Labeling

Identify each numbered structure by labeling the following figures. Place your answers in the spaces provided.

OUTCOME 21-6 **FIGURE 21-1** An Overview of the Major Systemic Arteries

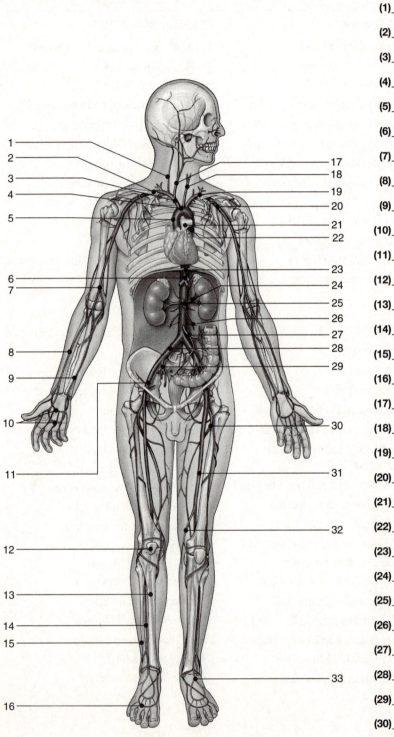

(1)_____

(2)_____

(3)_____

(4)_____

(5)_____

(6)_____

(7)_____

(8)_____

(9)_____

(10)_____

(11)_____

(12)_____

(13)_____

(14)_____

(15)_____

(16)_____

(17)_____

(18)_____

(19)_____

(20)_____

(21)_____

(22)_____

(23)_____

(24)_____

(25)_____

(26)_____

(27)_____

(28)_____

(29)_____

(30)_____

(31)_____

(32)_____

(33)_____

OUTCOME 21-6 **FIGURE 21-2** An Overview of the Major Systemic Veins

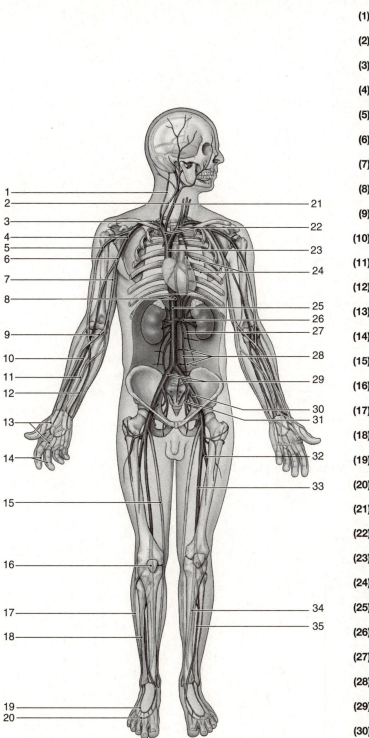

Veins

(1)_____

(2)_____

(3)_____

(4)_____

(5)_____

(6)_____

(7)_____

(8)_____

(9)_____

(10)_____

(11)_____

(12)_____

(13)_____

(14)_____

(15)_____

(16)_____

(17)_____

(18)_____

(19)_____

(20)_____

(21)_____

(22)_____

(23)_____

(24)_____

(25)_____

(26)_____

(27)_____

(28)_____

(29)_____

(30)_____

(31)_____

(32)_____

(33)_____

(34)_____

(35)_____

OUTCOME 21-6   **FIGURE 21-3**   Major Arteries of the Head and Neck

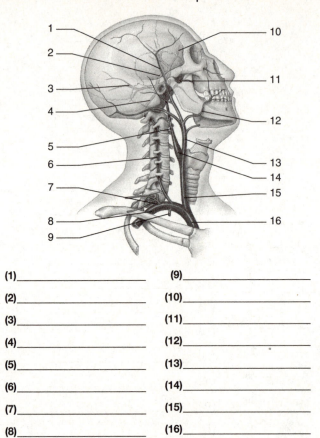

(1)_____	(9)_____
(2)_____	(10)_____
(3)_____	(11)_____
(4)_____	(12)_____
(5)_____	(13)_____
(6)_____	(14)_____
(7)_____	(15)_____
(8)_____	(16)_____

OUTCOME 21-6   **FIGURE 21-4**   Major Veins of the Head and Neck

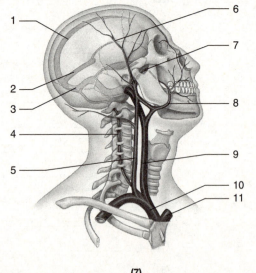

(1)_____	(7)_____
(2)_____	(8)_____
(3)_____	(9)_____
(4)_____	(10)_____
(5)_____	(11)_____
(6)_____	

LEVEL 2: REVIEWING CONCEPTS

Chapter Overview

Using the terms below, fill in the blanks to correctly complete the overview of the pathway of circulation through the aorta and its major branching arteries. Use each term only once. The objective is to identify the regions and major arteries that branch off the aorta from its superior position in the thorax to its inferior location in the abdominopelvic region.

renal	lumbar	abdominal aorta	L. common carotid
aortic valve	brachiocephalic	aortic arch	mediastinum
celiac	intercostal	common iliacs	superior phrenic
gonadal	adrenal	L. subclavian	inferior mesenteric
thoracic aorta	descending aorta	superior mesenteric	inferior phrenic

The major arteries and veins comprise the systemic circuit, which at any moment contains about 84 percent of the blood volume. The circuit begins at the left ventricle and ends at the right atrium. Oxygenated blood leaves the left ventricle as the ventricle contracts through the (1) _____ into the (2) _____, which curves across the superior surface of the heart, connecting the ascending aorta with the caudally directed (3) _____. The first artery branching off the aortic arch, the (4) _____, serves as a passageway for blood to the arteries that serve the arms, neck, and head. Continuing downward, the (5) _____ and the (6) _____ arteries arise separately from the aortic arch supplying arterial blood to vessels in the region that includes the neck, head, shoulders, and arms. Progressing more inferiorly, the (7) _____ is located within the (8) _____, providing blood to the (9) _____ arteries, which carry blood to the spinal cord and the body wall. Proceeding caudally, near the diaphragm, a pair of (10) _____ arteries delivers blood to the muscular diaphragm that separates the thoracic and abdominopelvic cavities. Leaving the thoracic aorta, the circulatory pathway continues into the (11) _____, which descends posteriorly to the peritoneal cavity. Just below the diaphragm, the (12) _____ arteries, which carry blood to the diaphragm, precede the (13) _____ trunk supplying blood to arterial branches that circulate blood to the stomach, liver, spleen, and pancreas. Further downward, the (14) _____ arteries serve the adrenal glands, and the (15)_____ arteries supplying the kidneys arise along the posteriolateral surface of the abdominal aorta behind the peritoneal lining. In the same region, the (16)_____ arteries arise on the anterior surface of the mesenteries. A pair of (17) _____ arteries provides blood to the testicles in the male and the ovaries in the female. They are located slightly superior to the (18) _____ artery, which branches into the connective tissues of the mesenteries supplying blood to the last third of the large intestine, colon, and rectum. In the final phase of aortic circulation, prior to the bifurcation of the aorta, small (19) _____ arteries extend on the posterior surface of the aorta, serving the spinal cord and the abdominal wall. Near the level of L 4, the abdominal aorta divides to form a pair of muscular arteries, the (20) _____, which are the major blood supply to the pelvis and the legs.

Concept Map I

Using the following terms, fill in the circled, numbered, blank spaces to correctly complete the concept map. Use each term only once.

pulmonary arteries   veins and venules   systemic circuit

arteries and arterioles   pulmonary veins

The cardiovascular system

(CO_2)   Lung capillaries   (O_2)

④ _Pulmonary_ Artery

Pulmonary circuit

① _Pulmonary_ veins

Pulmonary trunk

Heart

Vena cava

Aortic arch

⑤ _Systemic_ Circuit

Aorta

③ Veins and Venules

② _arteries_ and arterioles

(CO_2)   Systemic capillaries   (O_2)

—— Oxygenated blood

- - - - - Deoxygenated blood

Concept Map II

Using the following terms, fill in the circled, numbered, blank spaces to correctly complete the concept map. Use each term only once.

erythropoietin   increasing plasma volume   adrenal cortex
ADH (vasopressin)   epinephrine, norepinephrine   kidneys
increasing fluid   increasing blood pressure   atrial natriuretic peptide (ANP)

Endocrine system and cardiovascular regulation

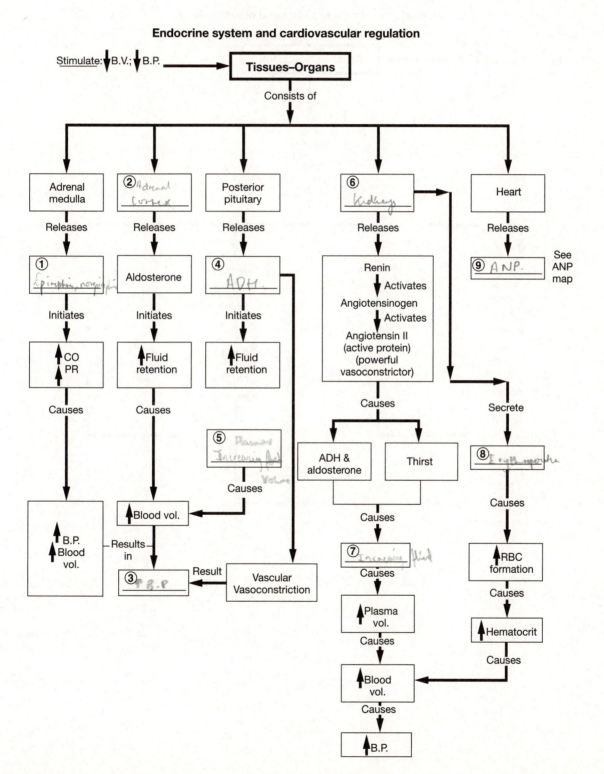

Concept Map III

Using the following terms, fill in the circled, numbered, blank spaces to correctly complete the concept map. Use each term only once.

decreasing blood pressure   increasing H_2O loss by kidneys
decreasing H_2O intake   increasing blood flow (*l*/min.)

ANP effects on blood volume and blood pressure

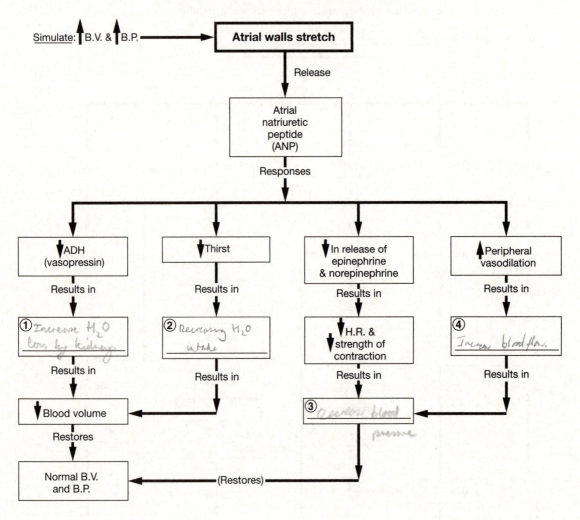

Simulate: ↑B.V. & ↑B.P. → **Atrial walls stretch**

Release

Atrial
natriuretic
peptide
(ANP)

Responses

| ↓ADH (vasopressin) | ↓Thirst | ↓In release of epinephrine & norepinephrine | ↑Peripheral vasodilation |

Results in

① *Increase H₂O loss by kidneys*   ② *Recucoring H₂O intake*   ↓H.R. & strength of contraction   ④ *Increase blood flow.*

Results in

↓Blood volume ← ← ③ *Decrease blood pressure* ←

Restores

Normal B.V. and B.P. ← (Restores) ←

Concept Map IV

Using the following terms, fill in the circled, numbered, blank spaces to correctly complete the concept map. Use each term only once.

L. subclavian artery   celiac trunk   R. gonadal artery
L. common iliac artery   superior mesenteric artery   brachiocephalic artery
ascending aorta   thoracic aorta

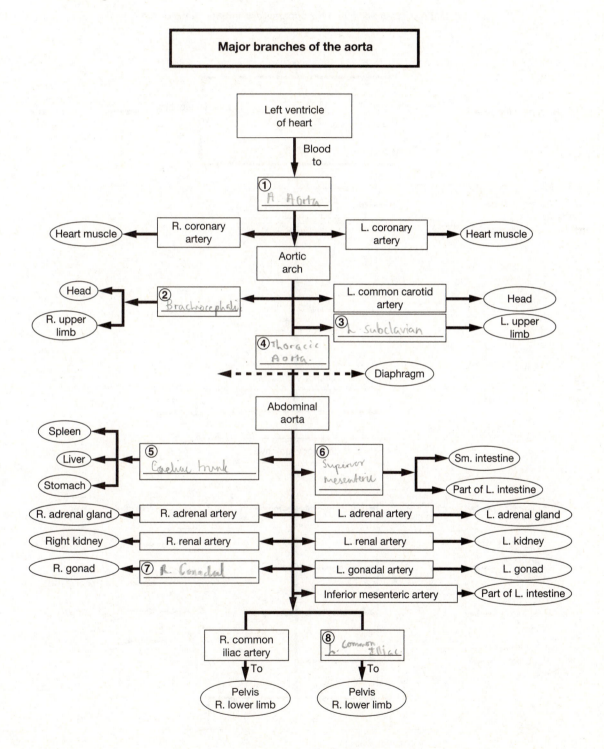

Major branches of the aorta

Left ventricle of heart

Blood to

① *A. Aorta*

Heart muscle ← R. coronary artery ← → L. coronary artery → Heart muscle

Aortic arch

Head ← ② *Brachiocephalic* ← → L. common carotid artery → Head

R. upper limb ←

③ *L. subclavian* → L. upper limb

④ *Thoracic Aorta.*

← → Diaphragm

Abdominal aorta

Spleen ←
Liver ← ⑤ *Caeliac trunk* ← → ⑥ *Superior mesenteric* → Sm. intestine
Stomach ← → Part of L. intestine

R. adrenal gland ← R. adrenal artery ← → L. adrenal artery → L. adrenal gland

Right kidney ← R. renal artery ← → L. renal artery → L. kidney

R. gonad ← ⑦ *R. Gonadal* ← → L. gonadal artery → L. gonad

Inferior mesenteric artery → Part of L. intestine

R. common iliac artery   ⑧ *L. Common Iliac.*

To ↓   To ↓

Pelvis R. lower limb   Pelvis R. lower limb

Concept Map V

Using the following terms, fill in the circled, numbered, blank spaces to correctly complete the concept map. (Note the direction of the arrows on this map.) Use each term only once.

azygos vein   L. common iliac vein   L. renal vein
L. hepatic veins   superior vena cava   R. adrenal vein

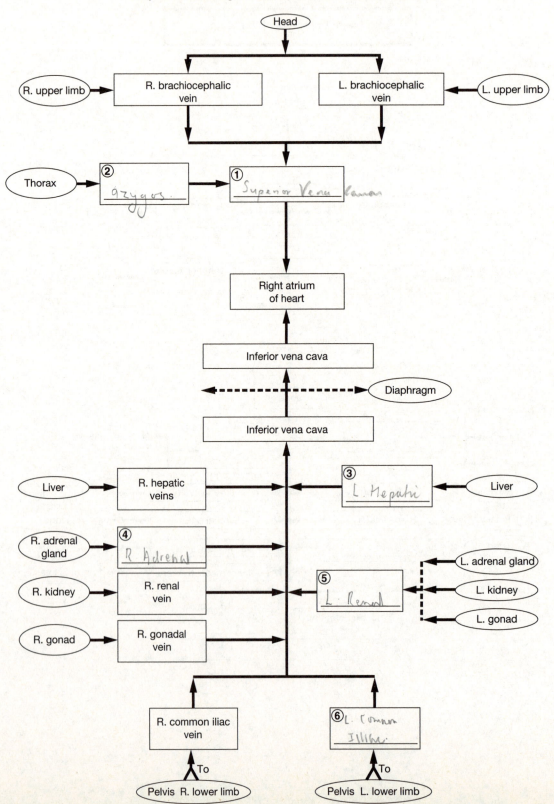

Major veins draining into the superior and inferior venae cavae

1. *Superior Vena cava*
2. *azygos*
3. *L. Hepatic*
4. *R Adrenal*
5. *L. Renal*
6. *L. Common Iliac*

Multiple Choice

Place the letter corresponding to the best answer in the space provided.

_____   1. In traveling from the heart to the peripheral capillaries, blood passes through

      a. arteries, arterioles, and venules.

      b. elastic arteries, muscular arteries, and arterioles.

      c. veins, venules, and arterioles.

      d. a, b, and c are correct.

_____   2. In general terms, blood flow (F) is directly proportional to

      a. resistance.

      b. increased viscosity.

      c. peripheral resistance.

      d. pressure.

_____   3. The goal of cardiovascular regulation is

      a. to equalize the stroke volume with the cardiac output.

      b. to increase pressure and reduce resistance.

      c. to equalize the blood flow with the pressure differences.

      d. maintenance of adequate blood flow through peripheral tissues and organs.

_____   4. Along the length of a typical capillary, blood pressure gradually falls from about

      a. 120 to 80 mm Hg.

      b. 75 to 50 mm Hg.

      c. 35 to 18 mm Hg.

      d. 15 to 5 mm Hg.

_____   5. In which of the following organs would you find *fenestrated* capillaries?

      a. Filtration areas of the kidneys

      b. Absorptive areas of the intestine

      c. Endocrine glands

      d. a, b, and c are correct

_____   6. The *average* pressure in arteries is approximately

      a. 120 mm Hg.

      b. 100 mm Hg.

      c. 140 mm Hg.

      d. 80 mm Hg.

_____   7. The effective pressure in the venous system is roughly

      a. 2 mm Hg.

      b. 10 mm Hg.

      c. 18 mm Hg.

      d. 80 mm Hg.

8. Net hydrostatic pressure forces water _____ a capillary; net osmotic pressure forces water _____ a capillary.

   a. out of; into
   b. into; out of
   c. out of; out of
   d. into; into

9. Of the following selections, the condition that would have the *greatest* influence on the level of peripheral resistance is doubling the

   a. length of a blood vessel.
   b. viscosity of the blood.
   c. diameter of a blood vessel.
   d. turbulence of the blood.

10. The relationship F = P/R means that flow is

    a. inversely proportional to resistance.
    b. proportional to the pressure gradient.
    c. inversely proportional to pressure, and directly proportional to resistance.
    d. directly proportional to the pressure gradient, and inversely proportional to resistance.

11. To increase blood flow to an adjacent capillary, the local controls that operate are

    a. decreasing O_2, increasing CO_2, and decreasing pH.
    b. increasing O_2, decreasing CO_2, and decreasing pH.
    c. increasing O_2, increasing CO_2, and increasing pH.
    d. decreasing O_2, decreasing CO_2, and decreasing pH.

12. The adrenergic fibers innervating arterioles are _____ fibers that release _____ and cause _____.

    a. parasympathetic; norepinephrine; vasodilation
    b. sympathetic; epinephrine; vasodilation
    c. sympathetic; norepinephrine; vasoconstriction
    d. parasympathetic; epinephrine; vasoconstriction

13. Two arteries formed by the bifurcation of the brachiocephalic artery are the

    a. aorta and subclavian.
    b. common iliac and common carotid.
    c. jugular and carotid.
    d. common carotid and subclavian.

14. The artery that serves the posterior thigh is the

    a. deep femoral.
    b. common iliac.
    c. internal iliac.
    d. celiac.

_____ 15. A major difference between the arterial and venous systems is that

      a. arteries are usually more superficial than veins.

      b. veins are not found in the abdominal cavity.

      c. there is dual venous drainage in the limbs.

      d. compared to arteries, veins are usually branched.

_____ 16. The large vein that drains the thorax is the

      a. superior vena cava.

      b. internal jugular.

      c. vertebral vein.

      d. azygos vein.

_____ 17. The veins that drain the head, neck, and upper extremities are the

      a. jugulars.

      b. brachiocephalics.

      c. subclavian.

      d. azygos.

_____ 18. The veins that drain venous blood from the legs and the pelvis are the

      a. posterior tibials.

      b. femorals.

      c. great saphenous.

      d. common iliacs.

_____ 19. The vein that drains the knee region of the body is the

      a. femoral.

      b. popliteal.

      c. external iliac.

      d. great saphenous.

_____ 20. The large artery that serves the brain is the

      a. internal carotid.

      b. external carotid.

      c. subclavian.

      d. cephalic.

_____ 21. The artery that links the subclavian and brachial arteries is the

      a. brachiocephalic.

      b. vertebral.

      c. cephalic.

      d. axillary.

_____ 22. The three arterial branches of the celiac trunk are the

      a. splenic, pancreatic, and mesenteric.

      b. phrenic, intercostal, and adrenolumbar.

      c. L. gastric, splenic, and common hepatic.

      d. brachial, ulnar, and radial.

_____ 23. The artery that supplies most of the small intestine and the first half of the large intestine is the

     a. adrenal artery.

     b. superior mesenteric.

     c. hepatic.

     d. inferior mesenteric.

_____ 24. The artery that supplies the pelvic organs is the

     a. internal iliac artery.

     b. external iliac artery.

     c. common iliac.

     d. femoral artery.

_____ 25. The branch(es) of the popliteal artery is (are) the

     a. great saphenous artery.

     b. anterior and posterior tibial arteries.

     c. peroneal artery.

     d. femoral and deep femoral arteries.

Completion

Using the terms below, complete the following statements. Use each term only once.

radial	precapillary sphincters	aorta
thrombus	great saphenous	cerebral arterial circle
edema	embolus	brachial
lumen	elastic rebound	venous return
veins	endothelium	recall of fluids

1. The anastomosis that encircles the infundibulum of the pituitary gland is the _____.

2. When the arteries absorb part of the energy provided by ventricular systole and give it back during ventricular diastole, the phenomenon is called _____.

3. When the blood volume increases at the expense of interstitial fluids, the process is referred to as _____.

4. The abnormal accumulation of interstitial fluid is called _____.

5. Rhythmic alterations in blood flow in capillary beds are controlled by _____.

6. A stationary blood clot within a blood vessel is called a(n) _____.

7. The lining of the lumen of blood vessels is comprised of a tissue layer called the _____.

8. Blockage of a pulmonary artery often caused by a detached blood clot is called a(n) _____.

9. The blood vessels that contain valves are the _____.

10. Changes in thoracic pressure during breathing and the action of skeletal muscles serve to aid in _____.

11. The largest artery in the body is the _____.

12. The artery generally auscultated to determine the blood pressure in the arm is the _____ artery.

13. The artery generally used to take the pulse at the wrist is the _____ artery.

14. The longest vein in the body is the _____.

15. The central cavity through which blood flows in a blood vessel is referred to as the _____.

Short Essay

Briefly answer the following questions in the spaces provided below.

1. What three distinct layers comprise the histological composition of typical arteries and veins?

2. List the types of blood vessels in the cardiovascular circuit and briefly describe their anatomical associations (use arrows to show this relationship).

3. a. What are sinusoids?

   b. Where are they found?

   c. Why are they important functionally?

4. Relative to gaseous exchange, what is the primary difference between the pulmonary circuit and the systemic circuit?

5. a. What are the three primary sources of peripheral resistance?

   b. Which one can be adjusted by the nervous or endocrine system to regulate blood flow?

6. Symbolically summarize the relationship among blood pressure (BP), peripheral resistance (PR), and blood flow (F). State what the formula means.

7. Explain what is meant by: BP = 120 mm Hg/80 mm Hg.

8. What is the mean arterial pressure (MAP) if the systolic pressure is 110 mm Hg and the diastolic pressure is 80 mm Hg?

9. What are the four important functions of continual movement of fluid from the plasma into tissues and lymphatic vessels?

10. What are the three primary factors that influence blood pressure and blood flow?

11. What three major baroreceptor populations enable the cardiovascular system to respond to alterations in blood pressure?

12. What hormones are responsible for long-term and short-term regulation of cardiovascular performance?

13. What are the clinical signs and symptoms of age-related changes in the blood?

14. How does arteriosclerosis affect blood vessels?

LEVEL 3: CRITICAL THINKING AND CLINICAL APPLICATIONS

Using principles and concepts learned in Chapter 21, answer the following questions. Write your answers on a separate sheet of paper.

1. Trace the circulatory pathway a drop of blood would take if it begins in the L. ventricle, travels down the left arm, and returns to the R. atrium. (Use arrows to indicate direction of flow.)

2. Suppose you are lying down and quickly rise to a standing position. What causes dizziness or a loss of consciousness to occur as a result of standing up quickly?

3. While pedaling an exercycle, Mary periodically takes her pulse by applying pressure to the carotid artery in the upper neck. Why might this method give her an erroneous pulse rate?

4. What cardiovascular changes occur during exercising?

The Lymphatic System and Immunity

OVERVIEW

The lymphatic system includes lymphoid organs and tissues, a network of lymphatic vessels that contain a fluid called lymph, and a dominant population of individual cells, the lymphocytes. The cells, tissues, organs, and vessels containing lymph perform three major functions:

1. The lymphoid organs and tissues serve as operating sites for the phagocytes and cells of the immune system that provide the body with protection from pathogens.
2. The lymphatic vessels containing lymph help to maintain blood volume in the cardiovascular system and absorb fats and other substances from the digestive tract.
3. The lymphocytes are the "defensive specialists" that protect the body from pathogenic microorganisms, foreign tissue cells, and diseased or infected cells in the body that pose a threat to the normal cell population.

Chapter 22 provides exercises focusing on topics that include the organization of the lymphatic system, the body's defense mechanisms, patterns of immune response, and interactions between the lymphatic system and other physiological systems.

LEVEL 1: REVIEWING FACTS AND TERMS

Review of Learning Outcomes

After completing this chapter, you should be able to do the following:

OUTCOME 22-1   Distinguish between innate (nonspecific) and adaptive (specific) defense, and explain the role of lymphocytes in the immune response.

OUTCOME 22-2   Identify the major components of the lymphatic system, describe the structure and functions of each component, and discuss the importance of lymphocytes.

OUTCOME 22-3   List the body's innate (nonspecific) defenses, and describe the components, mechanisms, and functions of each.

OUTCOME 22-4   Define adaptive (specific) defenses, identify the forms and properties of immunity, and distinguish between cell-mediated (cellular) immunity and antibody-mediated (humoral) immunity.

OUTCOME 22-5   Discuss the types of T cells and their roles in the immune response, and describe the mechanisms of T cell activation and differentiation.

OUTCOME 22-6   Discuss the mechanisms of B cell activation and differentiation, describe the structure and function of antibodies, and explain the primary and secondary responses to antigen exposure.

OUTCOME 22-7   Describe the development of immunological competence, list and explain examples of immune disorders and allergies, and discuss the effects of stress on immune function.

OUTCOME 22-8   Describe the effects of aging on the lymphatic system and the immune response.

OUTCOME 22-9   Give examples of interactions between the lymphatic system and other organ systems we have studied so far and explain how the nervous and endocrine systems influence the immune response.

Multiple Choice

Place the letter corresponding to the best answer in the space provided.

OUTCOME 22-1   _____   1. The primary responsibility(-ies) of the lymphocytes in the lymphatic system is (are) to respond to the presence of
   a. invading pathogens.
   b. abnormal body cells.
   c. foreign particles.
   d. a, b, and c are correct.

OUTCOME 22-1   _____   2. The anatomical barriers and defense mechanisms that cannot distinguish one potential threat from another are called
   a. the immune response.
   b. adaptive (specific) defenses.
   c. innate (nonspecific) defenses.
   d. abnormal nontoxicity.

OUTCOME 22-2   _____   3. The *major components* of the lymphatic system include
   a. lymph nodes, lymph, and lymphocytes.
   b. spleen, thymus, and tonsils.
   c. thoracic duct, R. lymphatic duct, and lymph nodes.
   d. lymphatic vessels, lymph, and lymphoid organs.

OUTCOME 22-2   _____   4. Lymphoid *organs* found in the lymphatic system include
   a. thoracic duct, R. lymphatic duct, and lymph nodes.
   b. lymphatic vessels, tonsils, and lymph nodes.
   c. spleen, thymus, and lymph nodes.
   d. a, b, and c are correct.

OUTCOME 22-2   _____   5. The *primary* function of the lymphatic system is
   a. transporting nutrients and oxygen to tissues.
   b. removal of carbon dioxide and waste products from tissues.
   c. regulation of temperature, fluid, electrolytes, and pH balance.
   d. production, maintenance, and distribution of lymphocytes.

OUTCOME 22-2   _____   6. Lymphocytes that assist in the regulation and coordination of the immune response are
   a. plasma cells.
   b. helper T and suppressor T cells.
   c. B cells.
   d. NK and B cells.

OUTCOME 22-2   _____   7. Normal lymphocyte populations are maintained through lymphopoiesis in the

      a. red bone marrow and lymphatic tissues.

      b. lymph in the lymphoid tissues.

      c. blood and the lymph.

      d. spleen and liver.

OUTCOME 22-2   _____   8. The largest collection of lymphoid tissue in the body is contained within the

      a. adult spleen.

      b. thymus gland.

      c. tonsils.

      d. lymphoid nodules.

OUTCOME 22-2   _____   9. The reticular epithelial cells in the cortex of the thymus maintain the blood–thymus barrier and secrete the hormones that

      a. form distinctive structures known as Hassall's corpuscles.

      b. cause the T cells to leave circulation via blood vessels.

      c. stimulate stem cell divisions and T-cell differentiation.

      d. cause the T cells to enter the circulation via blood vessels.

OUTCOME 22-3   _____   10. Of the following selections, the one that includes only innate (nonspecific) defenses is:

      a. T- and B-cell activation, complement, inflammation, phagocytosis.

      b. hair, skin, mucous membranes, antibodies.

      c. hair, skin, complement, inflammation, phagocytosis.

      d. antigens, antibodies, complement, macrophages.

OUTCOME 22-3   _____   11. The protective categories that prevent the approach of, deny entrance to, or limit the spread of microorganisms or other environmental hazards are called

      a. specific defenses.

      b. nonspecific defenses.

      c. specific immunity.

      d. immunological surveillance.

OUTCOME 22-3   _____   12. NK (natural killer) cells sensitive to the presence of abnormal cell membranes are primarily involved with

      a. defenses against specific threats.

      b. complex and time-consuming defense mechanisms.

      c. phagocytic activity for defense.

      d. immunological surveillance.

OUTCOME 22-3   _____   13. A physical barrier such as the epithelial covering of the skin provides effective immunity due to its makeup, which includes

      a. multiple layers.

      b. a keratin coating.

      c. a network of desmosomes that lock adjacent cells together.

      d. a, b, and c are correct.

OUTCOME 22-3 _____ 14. The "first line" of cellular defense against pathogenic invasion is
    a. interferon.
    b. pathogens.
    c. phagocytes.
    d. the complement system.

OUTCOME 22-3 _____ 15. Redness, swelling, heat, and pain are signs and symptoms associated with
    a. immunological surveillance.
    b. the inflammatory response.
    c. the complement system.
    d. phagocytosis.

OUTCOME 22-4 _____ 16. The four general characteristics of adaptive defenses include
    a. specificity, versatility, memory, and tolerance.
    b. innate, active, acquired, and passive.
    c. accessibility, recognition, compatibility, and immunity.
    d. a, b, and c are correct.

OUTCOME 22-4 _____ 17. The two major ways that the body "carries out" the immune response are
    a. phagocytosis and the inflammatory response.
    b. immunological surveillance and fever.
    c. direct attack by T cells and attack by circulating antibodies.
    d. physical barriers and the complement system.

OUTCOME 22-4 _____ 18. An *adaptive (specific)* defense mechanism is always activated by
    a. an antigen.
    b. an antibody.
    c. inflammation.
    d. fever.

OUTCOME 22-4 _____ 19. The type of immunity that develops as a result of natural exposure to an antigen in the environment is
    a. induced acquired immunity.
    b. naturally innate immunity.
    c. naturally acquired active immunity.
    d. naturally acquired passive immunity.

OUTCOME 22-4 _____ 20. A vaccine is an example of
    a. naturally acquired passive immunity.
    b. naturally acquired active passive immunity.
    c. artificially induced passive immunity.
    d. artificially induced active immunity.

OUTCOME 22-4 _____ 21. When an antigen appears, the immune response begins with the
    a. presence of immunoglobulins in body fluids.
    b. release of endogenous pyrogens.
    c. activation of the complement system.
    d. activation of specific T cells and B cells.

OUTCOME 22-4 _____ 22. If immune tolerance malfunctions in an individual, his or her B cells might begin to

a. manufacture antibodies against "self" cells and tissues.

b. activate cytotoxic T killer cells.

c. secrete lymphotoxins to destroy foreign antigens.

d. recall memory T cells to initiate the proper response.

OUTCOME 22-5 _____ 23. T-cell activation leads to the formation of cytotoxic T cells and memory T cells that provide

a. humoral immunity.

b. cellular immunity.

c. phagocytosis and immunological surveillance.

d. stimulation of inflammation and fever.

OUTCOME 22-5 _____ 24. Before an antigen can stimulate a lymphocyte, it must first be processed by a

a. macrophage.

b. NK cell.

c. cytotoxic T cell.

d. neutrophil.

OUTCOME 22-5 _____ 25. CD8 T cells are activated by exposure to antigens bound to

a. antigen-presenting cells (APCs).

b. Class I MHC proteins.

c. CD_4 receptor complex.

d. Class II MHC antigens.

OUTCOME 22-5 _____ 26. The T cells that limit the degree of immune system activation from a single stimulus are

a. memory T_C cells.

b. suppressor T cells.

c. cytotoxic T cells.

d. CD_4 T cells.

OUTCOME 22-6 _____ 27. Since each kind of B cell carries its own particular antibody molecule in its cell membrane, activation can only occur in the presence of a(n)

a. immunosuppressive drug.

b. cytotoxic T cell.

c. corresponding antigen.

d. CD_3 receptor complex.

OUTCOME 22-6 _____ 28. Activated B cells produce plasma cells that are specialized because they

a. synthesize and secrete antibodies.

b. produce helper T cells.

c. direct a physical and chemical attack.

d. a, b, and c are correct.

OUTCOME 22-6 _____ 29. An antibody is shaped like a(n)

    a. T.

    b. A.

    c. Y.

    d. B.

OUTCOME 22-6 _____ 30. Antigen binding changes antibody shape, exposing additional protein binding sites, which can lead to

    a. alteration in the cell membrane to increase phagocytosis.

    b. attraction of macrophages and neutrophils to the infected areas.

    c. activation of the complement system.

    d. precipitation.

OUTCOME 22-6 _____ 31. Antibodies may promote inflammation through the stimulation of

    a. basophils and mast cells.

    b. plasma cells and memory B cells.

    c. suppressor T cells.

    d. cytotoxic T cells.

OUTCOME 22-6 _____ 32. The antigenic determinant site is that portion of the antigen's exposed surface where

    a. the foreign "body" attacks.

    b. phagocytosis occurs.

    c. the antibody binds.

    d. the immune surveillance system is activated.

OUTCOME 22-6 _____ 33. In order for an antigenic molecule to be a complete antigen, it must

    a. be a large molecule.

    b. have at least two antigenic determinant sites.

    c. contain a hapten and a small organic molecule.

    d. be subject to antibody activity.

OUTCOME 22-7 _____ 34. The ability to demonstrate an immune response upon exposure to an antigen is called

    a. anaphylaxis.

    b. passive immunity.

    c. immunosuppression.

    d. immunological competence.

OUTCOME 22-7 _____ 35. Fetal antibody production is uncommon. Rather, the developing fetus has

    a. cell-mediated immunity.

    b. natural passive immunity.

    c. antibody-mediated immunity.

    d. endogenous pyrogens.

OUTCOME 22-7 _____ 36. When an immune response mistakenly targets normal body cells and tissues, the result is
   a. immune system failure.
   b. the development of an allergy.
   c. depression of the inflammatory response.
   d. an autoimmune disorder.

OUTCOME 22-7 _____ 37. Depression of the immune system due to chronic stress may cause
   a. depression of the inflammatory response.
   b. a reduction in the activities and numbers of phagocytes in peripheral tissues.
   c. the inhibition of interleukin secretion.
   d. a, b, and c are correct.

OUTCOME 22-7 _____ 38. The effect(s) of tumor necrosis factor (TNF) in the body is (are) to
   a. slow tumor growth and kill sensitive tumor cells.
   b. stimulate granulocyte production.
   c. increase T-cell sensitivity to interleukins.
   d. a, b, and c are correct.

OUTCOME 22-7 _____ 39. The major functions of interleukins in the immune system are to
   a. increase T-cell sensitivity to antigens exposed on macrophage membranes.
   b. stimulate B-cell activity, plasma-cell formation, and antibody production.
   c. enhance innate (nonspecific) defenses.
   d. a, b, and c are correct.

OUTCOME 22-8 _____ 40. With advancing age, B cells are less responsive, causing a
   a. decrease in antigen exposure.
   b. decreased antibody level after antigen exposure.
   c. deactivation of T-cell production.
   d. a, b, and c are correct.

OUTCOME 22-9 _____ 41. The lymphatic system influences nervous system activity by
   a. secreting endorphins and thymic hormones.
   b. returning tissue fluid to circulation.
   c. releasing cytokines that affect hypothalamic production of CRH and TRH.
   d. a, b, and c are correct.

Completion

Using the terms below, complete the following statements. Use each term only once.

lymphokines	innate	immunological competence
helper T	haptens	plasma cells
diapedesis	active	lymphatic capillaries
lacteals	passive	cell-mediated
precipitation	monokines	immunodeficiency disease
autoimmune disorder	neutralization	cytotoxic T
suppressor T	phagocytes	IgG
antibodies	immunity	lymphocytes
antigens	interferons	sensitization
costimulation	memory B cells	memory T
inflammation	thymic hormones	endocrine

OUTCOME 22-1   1. The ability to resist infection and disease through the activation of adaptive (specific) defenses constitutes _____.

OUTCOME 22-1   2. The cells that provide an adaptive (specific) defense known as the immune response are the _____.

OUTCOME 22-2   3. The special lymphatic vessels in the lining of the small intestines are the _____.

OUTCOME 22-2   4. Lymphocytes that attack foreign cells or body cells infected by viruses are called _____ cells.

OUTCOME 22-2   5. Plasma cells are responsible for the production and secretion of _____.

OUTCOME 22-2   6. The lymphatic system begins in the tissues as _____.

OUTCOME 22-3   7. Cells that represent the "first line" of cellular defense against pathogenic invasion are the _____.

OUTCOME 22-3   8. The process during which macrophages move through adjacent endothelial cells of capillary walls is called _____.

OUTCOME 22-3   9. The small proteins released by activated lymphocytes and macrophages and by tissue cells infected with viruses are called _____.

OUTCOME 22-4   10. An immunization where antibodies are administered to fight infection or prevent disease is _____.

OUTCOME 22-4   11. Cytotoxic T cells are responsible for the type of immunity referred to as _____.

OUTCOME 22-4   12. Immunity that is present at birth and has no relation to previous exposure to the pathogen involved is _____.

OUTCOME 22-4   13. Immunity that appears following exposure to an antigen as a consequence of the immune response is referred to as _____.

OUTCOME 22-5   14. The types of cells that inhibit the responses of other T cells and B cells are called _____ cells.

OUTCOME 22-5   15. Before B cells can respond to an antigen, they must be activated by _____ cells.

OUTCOME 22-5   16. The vital secondary binding process that eventually confirms the "OK to activate" signal is called _____.

OUTCOME 22-5   17. When an inactive cytotoxic T cell is activated and divides, it produces active T cells and _____.

OUTCOME 22-6   18. When binding occurs and the B cell prepares to undergo activation, the process is called _____.

OUTCOME 22-6   19. When an activated B cell divides, it ultimately produces daughter cells that differentiate into plasma cells and _____.

OUTCOME 22-6   20. Antibodies are produced and secreted by _____.

OUTCOME 22-6   21. When a direct attack by an antibody covers an antigen, its effect is _____.

OUTCOME 22-6   22. The formation of insoluble immune complexes is called _____.

OUTCOME 22-6   23. Small organic molecules that are not antigens by themselves are called _____.

OUTCOME 22-6   24. The ability to demonstrate an immune response upon exposure to an antigen is called _____.

OUTCOME 22-7   25. The only antibodies that cross the placenta from the maternal bloodstream are _____ antibodies.

OUTCOME 22-7   26. When the immune system fails to develop normally or the immune response is blocked in some way, the condition is termed _____.

OUTCOME 22-7   27. When the immune response inappropriately targets normal body cells and tissues, the result is a(n) _____.

OUTCOME 22-7   28. When glucocorticoids reduce the permeability of capillaries, they reduce the possibility of _____.

OUTCOME 22-7   29. Chemical messengers secreted by lymphocytes are called _____.

OUTCOME 22-7   30. Chemical messengers released by active macrophages are called _____.

OUTCOME 22-8   31. As a person advances in age, his or her T cells become less responsive to _____.

OUTCOME 22-8   32. A decrease in cytotoxic T cells may be associated with a reduction in circulating levels of _____.

OUTCOME 22-9   33. The system that secretes glucocorticoids with anti-inflammatory effects is the _____ system.

Matching

Match the terms in column B with the terms in column A. Use letters for answers in the spaces provided.
Use each term only once.

Part I

		Column A	Column B
OUTCOME 22-1	_____	1. adaptive immunity	A. thymus medullary cells
OUTCOME 22-2	_____	2. Peyer's patches	B. found on helper T cells
OUTCOME 22-2	_____	3. Hassall's corpuscles	C. active and passive
OUTCOME 22-3	_____	4. macrophages	D. chemical messengers
OUTCOME 22-3	_____	5. microphages	E. present at birth
OUTCOME 22-3	_____	6. mast cells	F. releases histamine
OUTCOME 22-3	_____	7. interferons	G. immune response
OUTCOME 22-4	_____	8. acquired immunity	H. monocytes
OUTCOME 22-4	_____	9. innate immunity	I. lymph nodules in small intestine
OUTCOME 22-4	_____	10. B cells	J. transfer of antibodies
OUTCOME 22-4	_____	11. passive immunity	K. neutrophils and eosinophils
OUTCOME 22-5	_____	12. CD 4 markers	L. antibody-mediated immunity

Part II

		Column A	Column B
OUTCOME 22-5	_____	13. cytotoxic T cells	M. resistance to viral infections
OUTCOME 22-5	_____	14. T lymphocytes	N. rapid allergic reaction
OUTCOME 22-5	_____	15. activated helper T cells	O. type III allergies
OUTCOME 22-6	_____	16. sensitization	P. slow tumor growth
OUTCOME 22-6	_____	17. antibody	Q. lyse cells directly
OUTCOME 22-6	_____	18. agglutination	R. exposure to an antigen
OUTCOME 22-6	_____	19. opsonization	S. microglia present antigens
OUTCOME 22-6	_____	20. active immunity	T. coating with antibodies
OUTCOME 22-6	_____	21. IgA antibodies	U. enhance nonspecific defenses
OUTCOME 22-7	_____	22. immediate hypersensitivity	V. decrease in immunity
OUTCOME 22-7	_____	23. immune complex disorder	W. two pairs of polypeptide chains
OUTCOME 22-7	_____	24. glucocorticoids	X. found in glandular secretions
OUTCOME 22-7	_____	25. interleukins	Y. secrete cytokines
OUTCOME 22-7	_____	26. interferons	Z. involution of thymus
OUTCOME 22-7	_____	27. TNFs	AA. anti-inflammatory effect
OUTCOME 22-8	_____	28. decreased thymic hormones	BB. antigens brought into cell
OUTCOME 22-8	_____	29. advancing age	CC. interaction between antigen and antibody
OUTCOME 22-9	_____	30. nervous system	DD. cell-mediated immunity

Drawing/Illustration Labeling

Identify each numbered structure in the following figures. Place your answers in the space provided.

OUTCOME 22-2   **FIGURE 22-1**   The Lymphoid Organs and Lymphatic Vessels

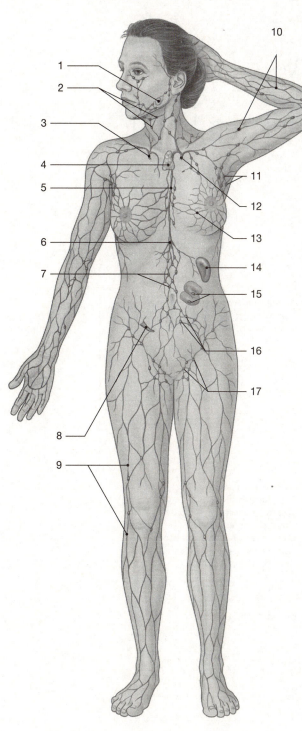

(1)_____

(2)_____

(3)_____

(4)_____

(5)_____

(6)_____

(7)_____

(8)_____

(9)_____

(10)_____

(11)_____

(12)_____

(13)_____

(14)_____

(15)_____

(16)_____

(17)_____

OUTCOME 22-3 **FIGURE 22-2** Innate (Nonspecific) Defenses

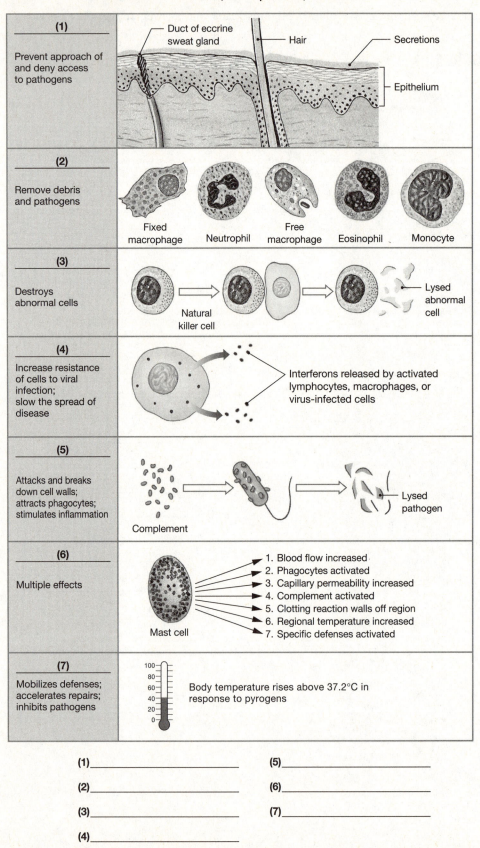

(1) _____

Prevent approach of and deny access to pathogens

Duct of eccrine sweat gland — Hair — Secretions — Epithelium

(2) _____

Remove debris and pathogens

Fixed macrophage   Neutrophil   Free macrophage   Eosinophil   Monocyte

(3) _____

Destroys abnormal cells

Natural killer cell   Lysed abnormal cell

(4) _____

Increase resistance of cells to viral infection; slow the spread of disease

Interferons released by activated lymphocytes, macrophages, or virus-infected cells

(5) _____

Attacks and breaks down cell walls; attracts phagocytes; stimulates inflammation

Complement   Lysed pathogen

(6) _____

Multiple effects

Mast cell
1. Blood flow increased
2. Phagocytes activated
3. Capillary permeability increased
4. Complement activated
5. Clotting reaction walls off region
6. Regional temperature increased
7. Specific defenses activated

(7) _____

Mobilizes defenses; accelerates repairs; inhibits pathogens

Body temperature rises above 37.2°C in response to pyrogens

(1)_____   (5)_____

(2)_____   (6)_____

(3)_____   (7)_____

(4)_____

LEVEL 2: REVIEWING CONCEPTS

Chapter Overview

Using the terms below, fill in the blanks to complete the chapter overview of the body's defense against viruses.

killer T cells	viruses	B cells
antibodies	helper T cells	macrophages
natural killer cells	suppressor T cells	memory T and B cells

At each site (1–9) complete the event that is taking place by identifying the type of cells or proteins participating in the adaptive (specific) immune response. Use each term only once.

The body's department of viral defense

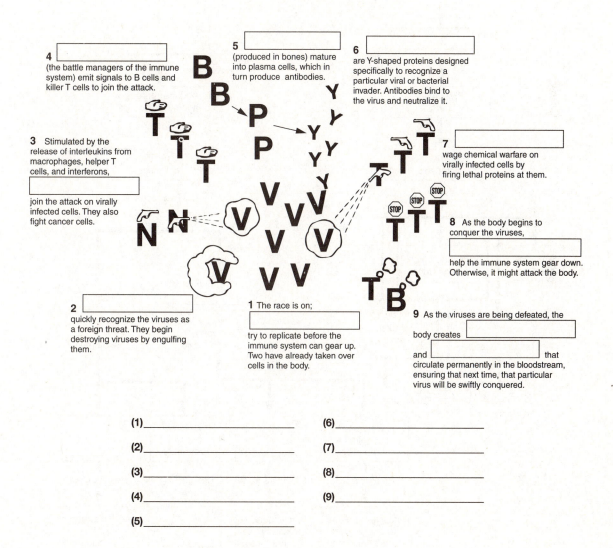

4 ☐
(the battle managers of the immune system) emit signals to B cells and killer T cells to join the attack.

5 ☐
(produced in bones) mature into plasma cells, which in turn produce antibodies.

6 ☐
are Y-shaped proteins designed specifically to recognize a particular viral or bacterial invader. Antibodies bind to the virus and neutralize it.

3 Stimulated by the release of interleukins from macrophages, helper T cells, and interferons, ☐ join the attack on virally infected cells. They also fight cancer cells.

7 ☐
wage chemical warfare on virally infected cells by firing lethal proteins at them.

8 As the body begins to conquer the viruses, ☐ help the immune system gear down. Otherwise, it might attack the body.

2 ☐
quickly recognize the viruses as a foreign threat. They begin destroying viruses by engulfing them.

1 The race is on; ☐ try to replicate before the immune system can gear up. Two have already taken over cells in the body.

9 As the viruses are being defeated, the body creates ☐ and ☐ that circulate permanently in the bloodstream, ensuring that next time, that particular virus will be swiftly conquered.

(1)_____

(2)_____

(3)_____

(4)_____

(5)_____

(6)_____

(7)_____

(8)_____

(9)_____

Concept Map I

Using the following terms, fill in the circled, numbered, blank spaces to correctly complete the concept map. Use each term only once.

active immunization   innate immunity   active

transfer of antibodies via   inflammation   phagocytic cells

placenta   passive immunization   adaptive immunity

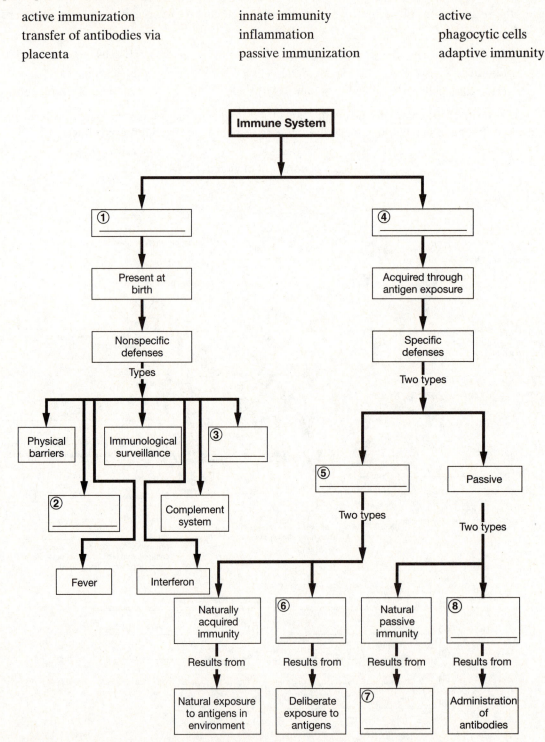

Concept Map II

Using the following terms, fill in the circled, numbered, blank spaces to correctly complete the concept map. Use each term only once.

increasing vascular permeability	tissue damage tissue repaired	phagocytosis of bacteria bacteria not destroyed

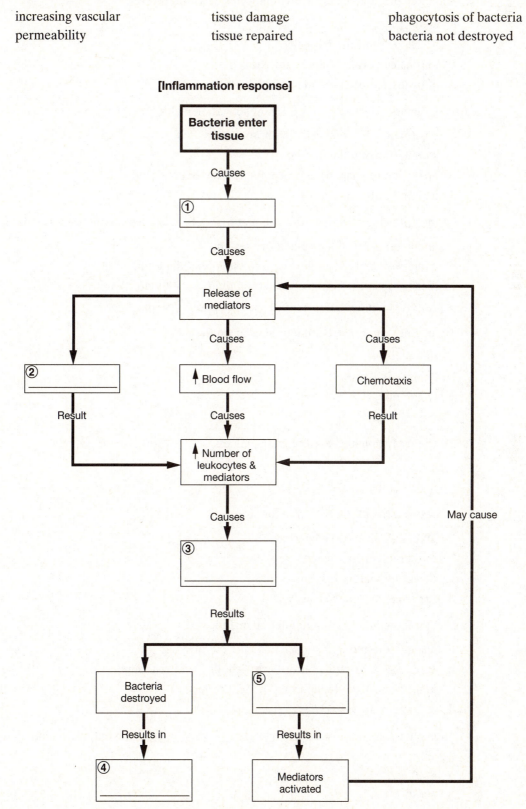

[Inflammation response]

Bacteria enter tissue

Causes

① _____

Causes

Release of mediators

Causes — ② _____ — Result

Causes — ↑ Blood flow

Causes — Chemotaxis — Result

↑ Number of leukocytes & mediators

Causes

③ _____

Results

Bacteria destroyed — Results in — ④ _____

⑤ _____ — Results in — Mediators activated

May cause

Multiple Choice

Place the letter corresponding to the best answer in the space provided.

_____ 1. The three different classes of lymphocytes in the blood are

    a. cytotoxic cells, helper cells, and suppressor cells.

    b. T cells, B cells, and NK cells.

    c. plasma cells, B cells, and cytotoxic cells.

    d. antigens, antibodies, and immunoglobulins.

_____ 2. The primary effect(s) of complement activation include

    a. destruction of target cell membranes.

    b. stimulation of inflammation.

    c. attraction of phagocytes and enhancement of phagocytosis.

    d. a, b, and c are correct.

_____ 3. Of the following selections, the one that *best* defines the lymphatic system is that it is

    a. an integral part of the circulatory system.

    b. a one-way route from the blood to the interstitial fluid.

    c. a one-way route from the interstitial fluid to the blood.

    d. closely related to and a part of the circulatory system.

_____ 4. Tissue fluid enters the lymphatic system via the

    a. thoracic duct.

    b. lymph capillaries.

    c. lymph nodes.

    d. bloodstream.

_____ 5. The larger lymphatic vessels contain valves.

    a. True

    b. False

_____ 6. A localized tissue response to injury is

    a. immunological surveillance.

    b. the complement system.

    c. the inflammatory response.

    d. the release of interferons.

_____ 7. Chemical mediators of inflammation include

    a. histamine, kinins, prostaglandins, and leukotrienes.

    b. epinephrine, norepinephrine, acetylcholine, and histamine.

    c. kinins, opsonins, epinephrine, and leukotrienes.

    d. a, b, and c are correct.

_____ 8. T lymphocytes comprise approximately _____ percent of circulating lymphocytes.

    a. 5–10

    b. 20–30

    c. 40–50

    d. 70–80

_____ 9. B lymphocytes differentiate into

    a. cytotoxic and suppressor cells.

    b. helper and suppressor cells.

    c. memory and helper cells.

    d. memory and plasma cells.

_____ 10. _____ cells may activate B cells, whereas _____ cells inhibit the activity of B cells.

    a. Memory; plasma

    b. Macrophages; microphages

    c. Memory; cytotoxic

    d. Helper T; suppressor T

_____ 11. The primary response of T-cell differentiation in cell-mediated immunity is the production of _____ cells.

    a. helper T

    b. suppressor T

    c. cytotoxic T

    d. memory

_____ 12. The vaccination of antigenic materials into the body is called

    a. naturally acquired active immunity.

    b. artificially induced active immunity.

    c. naturally acquired passive immunity.

    d. artificially induced passive immunity.

_____ 13. In passive immunity, _____ are transferred from another source.

    a. antibodies

    b. antigens

    c. T and B cells

    d. lymphocytes

_____ 14. The lymphatic function of the white pulp of the spleen is

    a. phagocytosis of abnormal blood cell components.

    b. release of splenic hormones into lymphatic vessels.

    c. initiation of immune responses by B cells and T cells.

    d. to degrade foreign proteins and toxins released by bacteria.

_____ 15. A person with type AB blood has

    a. anti-A and anti-B antibodies.

    b. only anti-O antibodies.

    c. no antigens.

    d. neither anti-A nor anti-B antibodies.

_____ 16. The antibodies produced and secreted by B lymphocytes are soluble proteins called

    a. lymphokines.

    b. agglutinins.

    c. immunoglobulins.

    d. leukotrienes.

_____ 17. The genes found in a region called the *major histocompatibility complex* code for

    a. immunoglobulins (IgG).

    b. human leukocyte antigens (HLAs).

    c. autoantibodies.

    d. alpha and gamma interferons.

_____ 18. Memory B cells do not differentiate into plasma cells unless they are

    a. initially subjected to a specific antigen.

    b. stimulated by active immunization.

    c. exposed to the same antigen a second time.

    d. stimulated by passive immunization.

_____ 19. The three-dimensional "fit" between the variable segments of the antibody molecule and the corresponding antigenic determinant site is referred to as the

    a. immunodeficiency complex.

    b. antibody–antigen complex.

    c. protein–complement complex.

    d. a, b, and c are correct.

_____ 20. One of the primary nonspecific effects that glucocorticoids have on the immune response is

    a. inhibition of interleukin secretion.

    b. increased release of T and B cells.

    c. decreased activity of cytotoxic T cells.

    d. depression of the inflammatory response.

Completion

Using the terms below, complete the following statements. Use each term only once.

cytokines	T cells	costimulation	mast
pyrogens	Kupffer cells	interferon	IgM
NK cells	properdin	helper T	IgG
Langerhans cells	opsonization	microglia	

1. Approximately 80 percent of circulating lymphocytes are classified as _____.

2. Fixed macrophages inside the CNS are called _____.

3. Macrophages in and around the liver sinusoids are referred to as _____.

4. Macrophages that reside within the epithelia of the skin and digestive tract are _____.

5. The vital secondary binding process that essentially confirms the initial T-cell activation is called _____.

6. Hormones released by tissue cells to coordinate local immune activities are classified as _____.

7. The complement factor involved in the alternative pathway in the complement system is _____.

8. Cells that play a pivotal role in initiating the inflammatory process are called _____ cells.

9. Circulating proteins that can reset the body's "thermostat" and cause a rise in body temperature are referred to as _____.

10. Enhanced phagocytosis resulting from the effects of antibody action on the antigen surface is called _____.

11. Immunological surveillance involves specific lymphocytes called _____.

12. The protein that appears to be effective in the treatment of some cancers is _____.

13. Antibodies that comprise approximately 80 percent of all antibodies in the body are _____.

14. Antibodies that occur naturally in blood plasma and are used to determine an individual's blood type are _____.

15. The human immunodeficiency virus (HIV) attacks _____ cells in humans.

Short Essay

Briefly answer the following questions in the spaces provided below.

1. What are the three primary organizational components of the lymphatic system?

2. What three major functions are performed by the lymphatic system?

3. What are the three different *classes* of lymphocytes found in the blood, and where does each class originate?

4. What three kinds of T cells comprise 80 percent of the circulating lymphocyte population? What is each type basically responsible for?

5. What is the primary function of activated B cells, and what are they ultimately responsible for?

6. What three lymphoid organs are important in the lymphatic system?

7. What are the six defenses that provide the body with a defensive capability known as *innate (nonspecific) immunity*?

8. What are the primary differences between the "recognition" mechanisms of NK (natural killer) cells and T and B cells?

9. What four primary effects result from complement activation?

10. What are the four general characteristics of immunity?

11. What is the primary difference between active and passive immunity?

12. What two primary mechanisms are involved in order for the body to reach its goal of the immune response?

13. What seven possible processes could operate when the antibody–antigen complex is found to eliminate the threat posed by the antigen?

14. What five different classes of antibodies (immunoglobulins) are found in body fluids?

15. Identify the classes of cytokines important to the immune response.

16. What is an autoimmune disorder, and what happens in the body when it occurs?

LEVEL 3: CRITICAL THINKING AND CLINICAL APPLICATIONS

Using principles and concepts learned in Chapter 22, answer the following questions. Write your answers on a separate sheet of paper.

1. Organ transplants are common surgical procedures in a number of hospitals throughout the United States. What happens if the donor and recipient are not compatible?
2. Using arrows, draw a schematic model to illustrate your understanding of the body's defense response against bacteria.
3. Using arrows, draw a schematic model to illustrate your understanding of the body's defense response against viruses.
4. Why are NK cells effective in fighting viral infections?
5. We usually associate a fever with illness or disease. In what ways may a fever be beneficial?
6. Stress of some kind is an everyday reality in the life of most human beings throughout the world. How does stress affect the effectiveness of the immune response?

The Respiratory System

OVERVIEW

The human body can survive without food for several weeks, without water for several days, but if the oxygen supply is cut off for more than several minutes, the possibility of death is imminent. Most cells in the body require a continuous supply of oxygen and must continuously get rid of carbon dioxide, a major waste product.

The respiratory system delivers oxygen from air to the blood and removes carbon dioxide, the gaseous waste product of cellular metabolism. It also plays an important role in regulating the pH of the body fluids.

The respiratory system consists of the lungs (which contain air sacs with moist membranes), a number of accessory structures, and a system of passageways that conduct inspired air from the atmosphere to the membranous sacs of the lungs, and expired air from the sacs in the lungs back to the atmosphere.

Chapter 23 includes exercises that examine the functional anatomy and organization of the respiratory system, respiratory physiology, the control of respiration, and respiratory interactions with other systems.

LEVEL 1: REVIEWING FACTS AND TERMS

Review of Learning Outcomes

After completing this chapter, you should be able to do the following:

OUTCOME 23-1   Describe the primary functions of the respiratory system, and explain how the delicate respiratory exchange surfaces are protected from pathogens, debris, and other hazards.

OUTCOME 23-2   Identify the organs of the upper respiratory system, and describe their functions.

OUTCOME 23-3   Describe the structure of the larynx, and discuss its roles in normal breathing and in the production of sound.

OUTCOME 23-4   Discuss the structure of the extrapulmonary airways.

OUTCOME 23-5   Describe the superficial anatomy of the lungs, the structure of a pulmonary lobule, and the functional anatomy of alveoli.

OUTCOME 23-6   Define and compare the processes of external respiration and internal respiration.

OUTCOME 23-7   Summarize the physical principles governing the movement of air into the lungs, and describe the origins and actions of the muscles responsible for respiratory movements.

OUTCOME 23-8   Summarize the physical principles governing the diffusion of gases into and out of the blood and body tissue.

OUTCOME 23-9   Describe the structure and function of hemoglobin, and the transport of oxygen and carbon dioxide in the blood.

OUTCOME 23-10   List the factors that influence respiration rate, and discuss reflex respiratory activity and the brain center involved in the control of respiration.

OUTCOME 23-11   Describe age-related changes in the respiratory system.

OUTCOME 23-12   Give examples of interactions between the respiratory system and other organ systems studied so far.

Multiple Choice

Place the letter corresponding to the best answer in the space provided.

OUTCOME 23-1   _____   1. The primary function(s) of the respiratory system is (are)

    a. to move air to and from the exchange surfaces of the lungs.

    b. to provide an area for gas exchange between air and circulating blood.

    c. to protect respiratory surfaces from dehydration and environmental variations.

    d. a, b, and c are correct

OUTCOME 23-1   _____   2. The conversion of angiotensin I to angiotensin II in the lung capillaries indirectly assists in the regulation of

    a. nonverbal auditory communication.

    b. blood volume and blood pressure.

    c. sound production.

    d. all of the above.

OUTCOME 23-1   _____   3. The air-filled pockets within the lungs where all gas exchange between air and blood occurs are the

    a. bronchioles.

    b. pharyngeal pouches.

    c. alveoli.

    d. lamina propria.

OUTCOME 23-1   _____   4. The "patrol force" of the alveolar epithelium involved with phagocytosis consists primarily of alveolar

    a. NK cells.

    b. cytotoxic cells.

    c. macrophages.

    d. plasma cells.

OUTCOME 23-2   _____   5. The respiratory system consists of structures that

    a. provide an extensive surface area for gas exchange between air and circulating blood.

    b. permit vocalization and production of sound.

    c. move air to and from the exchange surfaces of the lungs along the respiratory passageways.

    d. a, b, and c are correct.

OUTCOME 23-3 _____ 6. The entry of liquids or solid food into the respiratory passageways during swallowing is prevented by the
    a. glottis folding back over the epiglottis.
    b. expansion of the cricoid cartilages.
    c. epiglottis folding down over the glottis.
    d. depression of the larynx.

OUTCOME 23-4 _____ 7. Structures in the trachea that prevent its collapse or overexpansion as pressures change in the respiratory system are the
    a. O-ringed tracheal cartilages.
    b. C-shaped tracheal cartilages.
    c. irregular circular bones.
    d. S-shaped tracheal bones.

OUTCOME 23-4 _____ 8. The trachea allows for the passage of large masses of food through the esophagus due to
    a. incomplete esophageal cartilages.
    b. the elasticity of the ringed tracheal cartilages.
    c. distortion of the posterior tracheal wall.
    d. a, b, and c are correct.

OUTCOME 23-4 _____ 9. The function of the hilum along the medial surface of the lung is to
    a. mark the line of separation between the two bronchi.
    b. provide access to pulmonary vessels and nerves.
    c. prevent foreign objects from entering the trachea.
    d. a, b, and c are correct.

OUTCOME 23-5 _____ 10. Pulmonary surfactant is a phospholipid secretion produced by alveolar cells to
    a. increase the surface area of the alveoli.
    b. reduce the cohesive force of H_2O molecules and lower surface tension.
    c. increase the cohesive force of air molecules and raise surface tension.
    d. reduce the attractive forces of O_2 molecules and increase surface tension.

OUTCOME 23-5 _____ 11. Dilation and relaxation of the bronchioles is possible because the walls of bronchioles contain
    a. smooth muscle tissue regulated by the ANS.
    b. C-shaped cartilaginous rings.
    c. cuboidal epithelial cells containing cilia.
    d. bands of skeletal muscles in the submucosa.

OUTCOME 23-5 _____ 12. Structural features that make the lungs highly pliable and capable of tolerating great changes in volume are
    a. the air-filled passageways and C-shaped ringed cartilages.
    b. the bronchi, bronchioles, and the pleura.
    c. the parenchyma, lobules, and the costal surfaces.
    d. the elastic fibers in the trabeculae, the septa, and the pleurae.

OUTCOME 23-5   _____ 13. After passing through the trachea, the most complete pathway a molecule of inspired air would take to reach an alveolus is:

    a. primary bronchus → secondary bronchus → bronchioles → respiratory bronchioles → terminal bronchioles → alveolus.

    b. primary bronchus → bronchioles → terminal bronchioles → respiratory bronchioles → alveolus.

    c. primary bronchus → secondary bronchus → bronchioles → terminal bronchioles → respiratory bronchioles → alveolus.

    d. bronchi → secondary bronchus → bronchioles → terminal bronchioles → alveolus.

OUTCOME 23-5   _____ 14. The serous membrane in contact with the lung is the

    a. visceral pleura.

    b. parietal pleura.

    c. pulmonary mediastinum.

    d. pulmonary mesentery.

OUTCOME 23-6   _____ 15. The diffusion of gases between interstitial fluid and cytoplasm is

    a. internal respiration.

    b. external respiration.

    c. alveolar ventilation.

    d. pulmonary ventilation.

OUTCOME 23-6   _____ 16. Breathing, which involves the physical movement of air into and out of the lungs, is

    a. internal respiration.

    b. pulmonary ventilation.

    c. gas diffusion.

    d. alveolar ventilation.

OUTCOME 23-6   _____ 17. The process that prevents the buildup of carbon dioxide in the alveoli and ensures a continuous supply of oxygen that keeps pace with absorption by the bloodstream is

    a. external respiration.

    b. internal respiration.

    c. pulmonary ventilation.

    d. alveolar ventilation.

OUTCOME 23-6   _____ 18. The absorption of oxygen and the release of carbon dioxide by cells is

    a. internal respiration.

    b. lung compliance.

    c. the respiratory rate.

    d. external respiration.

OUTCOME 23-7   _____ 19. A single respiratory cycle consists of

    a. active and passive breathing.

    b. inspiration and expiration.

    c. costal and shallow breathing.

    d. a, b, and c are correct.

OUTCOME 23-7 _____ 20. Air enters the respiratory passageways when the pressure inside the lungs is lower than the _____ pressure.

    a. blood

    b. abdominal

    c. arterial

    d. atmospheric

OUTCOME 23-7 _____ 21. The movement of air into and out of the lungs is primarily dependent on

    a. pressure differences between the air in the atmosphere and air in the lungs.

    b. pressure differences between the air in the atmosphere and the anatomic dead space.

    c. pressure differences between the air in the atmosphere and individual cells.

    d. a, b, and c are correct.

OUTCOME 23-7 _____ 22. During inspiration, there will be an increase in the volume of the thoracic cavity and a(n)

    a. decreasing lung volume, increasing intrapulmonary pressure.

    b. decreasing lung volume, decreasing intrapulmonary pressure.

    c. increasing lung volume, decreasing intrapulmonary pressure.

    d. increasing lung volume, increasing intrapulmonary pressure.

OUTCOME 23-7 _____ 23. During expiration, the diaphragm

    a. contracts and progresses inferiorly toward the stomach.

    b. contracts and the dome rises into the thoracic cage.

    c. relaxes and the dome rises into the thoracic cage.

    d. relaxes and progresses inferiorly toward the stomach.

OUTCOME 23-7 _____ 24. Stiffening and reduction in chest movement effectively limit the

    a. normal partial pressure of CO_2.

    b. respiratory minute volume.

    c. partial pressure of O_2.

    d. rate of hemoglobin saturation.

OUTCOME 23-7 _____ 25. During expiration, there is a(n)

    a. decrease in intrapulmonary pressure.

    b. increase in intrapulmonary pressure.

    c. increase in atmospheric pressure.

    d. increase in the volume of the lungs.

OUTCOME 23-8 _____ 26. A lack of surfactant secretion onto alveolar surfaces causes the alveoli to

    a. burst.

    b. open.

    c. collapse.

    d. retain their form and shape.

OUTCOME 23-8   _____   27. If there is a P_{O_2} of 104 mm Hg and a P_{CO_2} of 40 mm Hg in the alveoli, and a P_{O_2} of 40 mm Hg and a P_{CO_2} of 45 mm Hg within the pulmonary blood, there will be a net diffusion of

    a. CO_2 into the blood from the alveoli; O_2 from the blood into the alveoli.

    b. O_2 and CO_2 into the blood from the alveoli.

    c. O_2 and CO_2 from the blood into the alveoli.

    d. O_2 into the blood from the alveoli; CO_2 from the blood into the alveoli.

OUTCOME 23-8   _____   28. When the partial pressure difference is greater across the respiratory membrane, the rate of gas diffusion is

    a. not affected.

    b. faster.

    c. slower.

    d. gradually decreased.

OUTCOME 23-8   _____   29. The arrangement that improves the efficiency of pulmonary ventilation and pulmonary circulation occurs when

    a. the surface area of the respiratory membrane is increased.

    b. the gases are liquid soluble.

    c. blood flow and air flow are coordinated.

    d. distances involved in gas exchange are small.

OUTCOME 23-8   _____   30. If the partial pressure of oxygen is lower in the pulmonary capillaries than in the alveolus, then

    a. O_2 will diffuse out of the alveolus into the pulmonary capillary.

    b. O_2 will diffuse out of the pulmonary capillary into the alveolus.

    c. there will be no net diffusion of oxygen.

    d. O_2 will diffuse from the area of lesser concentration to the area of greater concentration.

OUTCOME 23-8   _____   31. Blood entering the systemic circuit normally has a P_{CO_2} of 40 mm Hg, while peripheral tissues have a P_{CO_2} of 45 mm Hg; therefore

    a. CO_2 diffuses out of the blood.

    b. there is no net diffusion of CO_2.

    c. CO_2 diffuses into the interstitium.

    d. CO_2 diffuses into the blood.

OUTCOME 23-9   _____   32. Each molecule of hemoglobin has the capacity to carry _____ molecules of oxygen (O_2).

    a. six

    b. eight

    c. four

    d. two

OUTCOME 23-9   _____   33. What percentage of total oxygen (O_2) is carried within red blood cells chemically bound to hemoglobin?

    a. 5 percent

    b. 68 percent

    c. 98 percent

    d. 100 percent

OUTCOME 23-9 _____ 34. Factors that cause a decrease in hemoglobin saturation at a given P_{O_2} are

    a. increasing P_{O_2}, decreasing CO_2, and increasing temperature.

    b. increasing 2, 3-bisphosphoglycerate (BPG), increasing temperature, and decreasing pH.

    c. decreasing BPG, increasing pH, and increasing CO_2.

    d. decreasing temperature, decreasing CO_2, and decreasing P_{O_2}.

OUTCOME 23-9 _____ 35. Each hemoglobin molecule consists of

    a. four globular protein subunits, each containing one heme unit.

    b. one globular protein subunit containing four heme units.

    c. four globular protein subunits and one heme unit.

    d. one heme unit surrounded by four globular protein subunits.

OUTCOME 23-9 _____ 36. When hemoglobin binds with molecules of oxygen, the end product is

    a. carbaminohemoglobin.

    b. carbon dioxide.

    c. oxyhemoglobin.

    d. carbon monoxide.

OUTCOME 23-9 _____ 37. Carbon dioxide is transported in the blood by

    a. conversion to a molecule of carbonic acid.

    b. binding to the protein part of the hemoglobin molecule.

    c. dissolving in plasma.

    d. a, b, and c are correct.

OUTCOME 23-10 _____ 38. If the rate and depth of respiration exceed the demands for oxygen delivery and carbon dioxide removal, the condition is called

    a. hypoventilation.

    b. apnea.

    c. hypoxia.

    d. hyperventilation.

OUTCOME 23-10 _____ 39. Under normal conditions, the greatest effect on the respiratory centers is initiated by

    a. decreases in P_{O_2}.

    b. increases and decreases in P_{O_2} and P_{CO_2}.

    c. increases and decreases in P_{CO_2}.

    d. increases in P_{O_2}.

OUTCOME 23-10 _____ 40. Emotional states that initiate sympathetic activation in the ANS cause

    a. bronchoconstriction and decrease the respiratory rate.

    b. bronchodilation and decrease the respiratory rate.

    c. bronchodilation and increase the respiratory rate.

    d. bronchoconstriction and increase the respiratory rate.

OUTCOME 23-10 _____ 41. The initiation of inspiration originates with discharge of inspiratory neurons in the

    a. diaphragm.

    b. medulla.

    c. pons.

    d. lungs.

OUTCOME 23-10 _____ 42. Examples of protective reflexes that operate when you are exposed to toxic vapors, chemical irritants, or mechanical stimulation of the respiratory tract include

    a. sneezing, coughing, and laryngeal spasms.

    b. hypoventilation and hyperventilation.

    c. pain, changes in body temperature, and changes in lung volume.

    d. hypercapnia, hypoxia, and epitaxis.

OUTCOME 23-10 _____ 43. As the volume of the lungs increases during times of forced breathing, the

    a. inspiratory center is stimulated, the expiration center inhibited.

    b. inspiratory center is inhibited, the expiratory center stimulated.

    c. inspiratory and expiratory centers are stimulated.

    d. inspiratory and expiratory centers are inhibited.

OUTCOME 23-11 _____ 44. With increasing age, elastic tissue deterioration and stiffening and reduction in chest movement effectively limit

    a. chemoreceptor stimulation.

    b. baroreceptor stimulation.

    c. the respiratory minute volume.

    d. respiratory center activity.

OUTCOME 23-12 _____ 45. The nervous system interacts with the respiratory system by

    a. eliminating organic wastes generated by cells or the respiratory system.

    b. monitoring respiratory volume, blood gas levels, and blood and CSF pH.

    c. increasing hemoglobin's affinity for oxygen.

    d. mobilizing specific defenses when infection occurs.

Completion

Using the terms below, complete the following statements. Use each term only once.

chemoreceptor	expiration	external respiration
mucus escalator	external nares	baroreceptors
inspiration	Bohr effect	surfactant
internal respiration	respiratory bronchioles	larynx
apneustic	temperature	bronchioles
skeletal muscles	residual volume	simple squamous cells
equilibrium	intrapulmonary pressure	diaphragm
iron ion	sympathetic	alveolar ventilation
gas diffusion	sound waves	cardiovascular
glottis	carbaminohemoglobin	emphysema
apnea		

OUTCOME 23-1   1. The movement of air out of the lungs is referred to as _____.

OUTCOME 23-1   2. The movement of air into the lungs is called _____.

OUTCOME 23-1   3. The phospholipid secretion that coats the alveolar epithelium and keeps the alveoli from collapsing is called _____.

OUTCOME 23-1   4. The mechanism in the respiratory tree that carries debris and dust particles toward the pharynx and an acid bath in the stomach is the _____.

OUTCOME 23-2   5. Air normally enters the respiratory system via the paired _____.

OUTCOME 23-2   6. Inspired air leaves the pharynx by passing through a narrow opening called the _____.

OUTCOME 23-3   7. The glottis is surrounded and protected by the _____.

OUTCOME 23-3   8. Air passing through the glottis vibrates the vocal cords and produces _____.

OUTCOME 23-4   9. The components of the bronchial tree dominated by smooth muscle tissue are the _____.

OUTCOME 23-4   10. ANS control of the normal diameter of the trachea is under the control of the _____ division.

OUTCOME 23-5   11. Within a pulmonary lobule, the terminal bronchioles branch to form several _____.

OUTCOME 23-5   12. The alveolar epithelium consists primarily of unusually thin and delicate _____.

OUTCOME 23-6   13. The diffusion of gases between the alveoli and the alveolar capillaries is called _____.

OUTCOME 23-6   14. The absorption of oxygen and the release of carbon dioxide by cells is called _____.

OUTCOME 23-6   15. External respiration occurs across respiratory membranes between alveolar air spaces and alveolar capillaries by _____.

OUTCOME 23-6   16. The primary function of pulmonary ventilation is to maintain adequate _____.

OUTCOME 23-7   17. During inhalation, the lungs expand, and there is a drop in the _____.

OUTCOME 23-7   18. The changes of volume in the lungs occur through the contraction of the _____.

OUTCOME 23-7   19. The most important muscles involved in inhalation are the external intercostal muscles and the _____.

OUTCOME 23-8   20. The rate of diffusion of gases varies according to factors such as the size of the concentration gradient and the _____.

OUTCOME 23-8   21. When the total number of gas molecules in solution remains constant, the solution is in _____.

OUTCOME 23-8   22. The amount of air that remains in the lungs after a maximal exhalation is the _____.

OUTCOME 23-9   23. In the hemoglobin molecule, the oxygen molecules specifically bind to the _____.

OUTCOME 23-9   24. The effect of pH on the hemoglobin saturation curve is called the _____.

OUTCOME 23-9   25. CO_2 molecules bound to exposed amino groups of the globin portions of hemoglobin molecules form _____.

OUTCOME 23-10   26. When respiration is suspended due to exposure to toxic vapors or chemical irritants, reflex activity may result in _____.

OUTCOME 23-10   27. Changes in blood pressure modify activities of the respiratory centers due to _____ in the aortic or carotid sinuses.

OUTCOME 23-10   28. Reflexes that respond to changes in the P_{O_2}, P_{CO_2}, or pH of the blood and/or cerebrospinal fluid are _____ reflexes.

OUTCOME 23-10   29. The _____ center in the pons helps increase the intensity of inhalation during quiet breathing.

OUTCOME 23-11   30. A chronic progressive condition characterized by shortness of breath and an inability to tolerate physical exertion is _____.

OUTCOME 23-12   31. The respiratory system has extensive and physiological connections to the _____ system.

Matching

Match the terms in column B with the terms in column A. Use letters for answers in the spaces provided. Use each term only once.

Part I

		Column A	Column B
OUTCOME 23-1	_____	1. upper respiratory tract	A. nostrils
OUTCOME 23-1	_____	2. lower respiratory tract	B. includes lungs
OUTCOME 23-2	_____	3. external nares	C. phonation
OUTCOME 23-3	_____	4. vocal cords	D. surfactant cells
OUTCOME 23-3	_____	5. epiglottis	E. delivers air to larynx
OUTCOME 23-4	_____	6. trachea	F. pulmonary ventilation
OUTCOME 23-5	_____	7. pneumocytes type II	G. dust cells
OUTCOME 23-5	_____	8. alveolar macrophages	H. windpipe
OUTCOME 23-6	_____	9. air movement	I. external respiration
OUTCOME 23-6	_____	10. anoxia	J. elevation of the ribs
OUTCOME 23-6	_____	11. gas diffusion	K. pressure/volume relationship
OUTCOME 23-7	_____	12. Boyle's law	L. tissue death
OUTCOME 23-7	_____	13. external intercostals	M. covering of glottis

Part II

		Column A	Column B
OUTCOME 23-7	_____	14. expandability of lungs	N. four protein subunits
OUTCOME 23-8	_____	15. P_{O_2} decreases	O. 40 mm Hg
OUTCOME 23-8	_____	16. partial pressure	P. surface tension
OUTCOME 23-8	_____	17. surfactant layer	Q. bronchodilation
OUTCOME 23-8	_____	18. alveolar air P_{O_2}	R. elastic tissue deterioration
OUTCOME 23-8	_____	19. alveolar air P_{CO_2}	S. promote passive or active exhalation
OUTCOME 23-9	_____	20. heme unit	T. O_2 released by hemoglobin
OUTCOME 23-9	_____	21. hemoglobin molecule	U. stimulation to the DRG
OUTCOME 23-9	_____	22. CO_2 transport	V. 100 mm Hg
OUTCOME 23-10	_____	23. increase in P_{CO_2}	W. mobilization of specific defenses
OUTCOME 23-10	_____	24. decrease in P_{CO_2}	X. lung compliance
OUTCOME 23-10	_____	25. apneustic center	Y. bronchoconstriction
OUTCOME 23-10	_____	26. pneumotaxic center	Z. single gas in a mixture
OUTCOME 23-11	_____	27. lowers vital capacity	AA. bicarbonate ion
OUTCOME 23-12	_____	28. lymphatic system	BB. iron ion

Drawing/Illustration Labeling

Identify each numbered structure in the following figures. Place your answers in the spaces provided.

OUTCOME 23-2 **FIGURE 23-1** Structures of the Upper Respiratory System

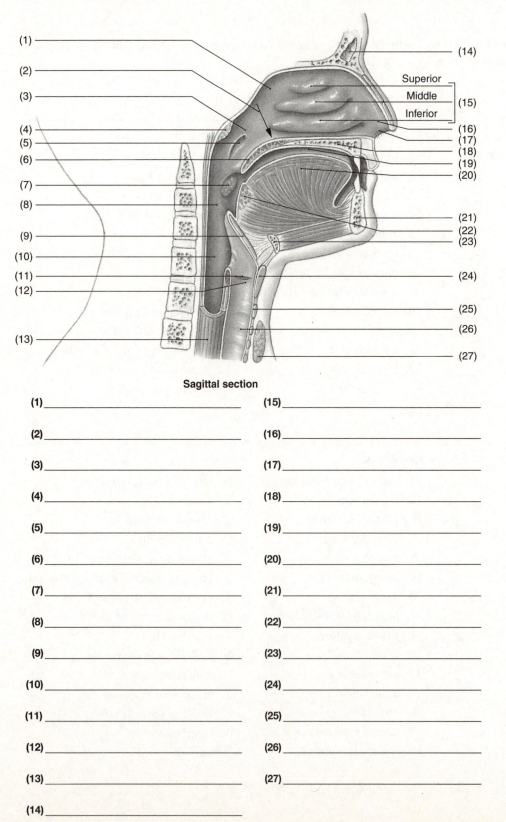

Sagittal section

(1) _____

(2) _____

(3) _____

(4) _____

(5) _____

(6) _____

(7) _____

(8) _____

(9) _____

(10) _____

(11) _____

(12) _____

(13) _____

(14) _____

(15) _____

(16) _____

(17) _____

(18) _____

(19) _____

(20) _____

(21) _____

(22) _____

(23) _____

(24) _____

(25) _____

(26) _____

(27) _____

OUTCOME 23-4, 23-5 **FIGURE 23-2** Bronchi and Lobules of the Lungs

(1)_____

(2)_____

(3)_____

(4)_____

(5)_____

(6)_____

(7)_____

(8)_____

(9)_____

(10)_____

Bronchopulmonary
segment

LEVEL 2: REVIEWING CONCEPTS

Chapter Overview

Using the terms below, fill in the blanks to complete the chapter overview identifying the air passageways through the respiratory system.

glottis	epiglottis	larynx	nasal conchae
nasopharynx	bronchioles	soft palate	vocal folds
trachea	carbon dioxide	vocal cords	nasal vestibule
nose	hard palate	primary bronchi	smooth muscle
oxygen	external nares	pharynx (use twice)	pulmonary capillaries
cilia	alveoli	nasal cavity	

The (1) _____ is the primary passageway for air entering the respiratory system. Air normally enters through the paired (2) _____ that communicate with the (3) _____. The reception area, called the (4) _____, contains coarse hairs that trap airborne particles and prevent them from entering the respiratory tract. To pass through the reception area to the internal nares, air tends to flow between adjacent (5) _____, projecting toward the nasal septum. As air bounces off these projections, it is warmed and humidified, which induces turbulence, causing airborne particles to come into contact with the mucus covering the surfaces of these lobular structures. A bony (6) _____ separates the oral and nasal cavities. A fleshy (7) _____ extends behind the hard palate, marking the boundary line between the (8) _____ and the superior portion of the (9) _____, a chamber with the wall coated by mucus secretions, and shared by the digestive and respiratory systems. Inhaled air leaves the chamber and enters the (10) _____ through a narrow opening called the (11) _____. The elongate, cartilaginous (12) _____ projects superior to the narrowing opening and forms a "lid" over it. This region also contains the (13) _____, which are involved with the production of sound, and are commonly known as the (14) _____. After the air passes through this region, it is conveyed into the (15) _____, a tough, flexible tube that contains 15 to 20 C-shaped cartilaginous rings that serve to stiffen the windpipe walls, protect the airway, and prevent its collapse or overexpansion as pressures change in the respiratory system. This air conduit contains finger-like cellular extensions, the (16) _____, which may cause a countercurrent that serves to move mucus toward the (17) _____. The tough, flexible windpipe branches within the mediastinum, giving rise to the right and left (18) _____, which also have cartilaginous C-shaped supporting rings. These tube-like branches provide the lungs with air. As air passes into the lungs, the passageways get narrower and are subject to constriction caused by (19) _____ tissue in the submucosal lining of the smaller tube-like (20) _____. After passing through these finer conducting branches, the air flows into the "sac-like" (21) _____, where the atmosphere is dry and the gaseous exchange of (22) _____ and (23) _____ occurs between the air inside the small spheres and the blood in the (24) _____. Movement of air into and out of the alveoli prevents the buildup of carbon dioxide in the alveoli, and ensures a continuous supply of oxygen that keeps pace with absorption by the bloodstream.

Concept Map I

Using the following terms, fill in the circled, numbered, blank spaces to correctly complete the concept map. Use each term only once.

pulmonary veins   respiratory bronchioles   trachea

left ventricle   right atrium   pharynx

pulmonary arteries   systemic capillaries   alveoli

secondary bronchi

External–internal respiration
(Air flow and O_2–CO_2 transport)

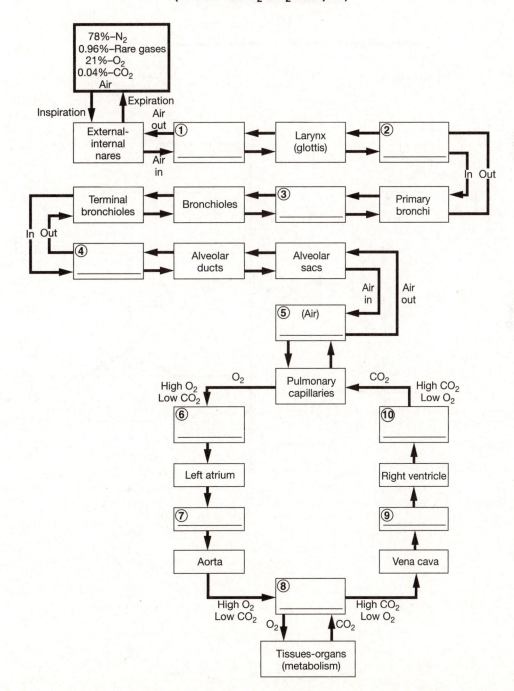

Concept Map II

Using the following terms, fill in the circled, numbered, blank spaces to correctly complete the concept map. Use each term only once.

tertiary bronchi	trachea	left primary bronchus
laryngopharynx	alveoli	vocal folds
respiratory bronchioles	cricoid	larynx
nasal sinuses	nose	

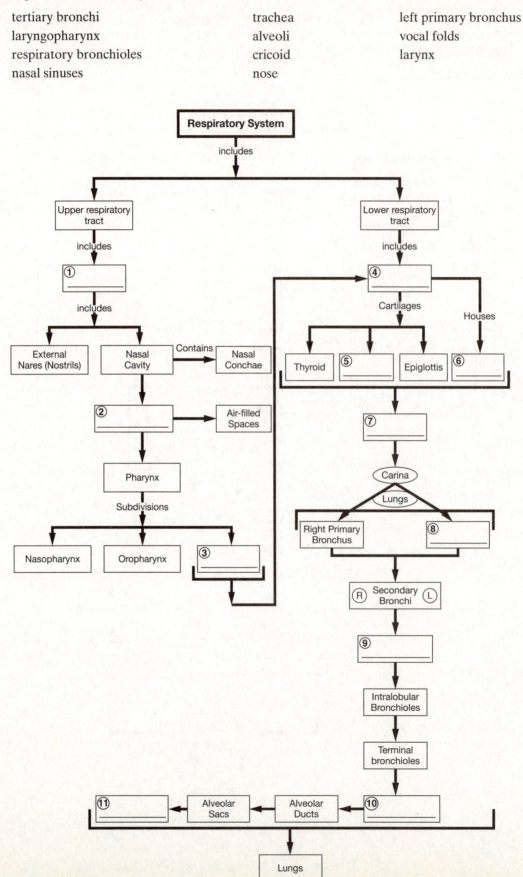

Multiple Choice

Place the letter corresponding to the best answer in the space provided.

_____ 1. The paranasal sinuses include

    a. occipital, parietal, temporal, and mandibular.

    b. ethmoid, parietal, sphenoid, and temporal.

    c. frontal, sphenoid, ethmoid, and maxillary.

    d. mandibular, maxillary, frontal, and temporal.

_____ 2. A rise in arterial P_{CO_2} elevates cerebrospinal fluid (CSF) carbon dioxide levels and stimulates the chemoreceptive neurons of the medulla to produce

    a. hypercapnia.

    b. hyperventilation.

    c. hypoventilation.

    d. eupnea.

_____ 3. The primary function of pulmonary ventilation is to maintain adequate

    a. air in the anatomic dead space.

    b. pulmonary surfactant.

    c. vital capacity.

    d. alveolar ventilation.

_____ 4. The purpose of the fluid in the pleural cavity is to

    a. provide a medium for the exchange of O_2 and CO_2.

    b. reduce friction between the parietal and visceral pleura.

    c. allow for the exchange of electrolytes during respiratory movements.

    d. provide lubrication for diaphragmatic constriction.

_____ 5. When the diaphragm and external intercostal muscles contract,

    a. intrapulmonary pressure increases.

    b. intrapleural pressure decreases.

    c. expiration occurs.

    d. the volume of the lungs decreases.

_____ 6. If a person is stabbed in the chest and the *thoracic wall* is punctured but the lung is not penetrated, the

    a. lung will probably collapse.

    b. intrapleural pressure becomes subatmospheric.

    c. intrapulmonary pressure becomes subatmospheric.

    d. intra-alveolar pressure becomes subatmospheric.

_____ 7. The *most important* factor determining airway resistance is

    a. interactions between flowing gas molecules.

    b. the length of the airway.

    c. the thickness of the wall of the airway.

    d. the airway radius.

_____ 8. The sympathetic division of the ANS causes _____ of airway smooth muscle; therefore, resistance is _____.

    a. relaxation; decreased

    b. relaxation; increased

    c. constriction; increased

    d. constriction; decreased

_____ 9. The substance often administered during an asthmatic attack to decrease resistance via airway dilation is

    a. histamine.

    b. epinephrine.

    c. norepinephrine.

    d. angiotensin.

_____ 10. Decreased amounts of CO_2 concentrations in the bronchioles cause a(n)

    a. increase in bronchiolar dilation.

    b. decrease in bronchiolar constriction.

    c. increase in bronchiolar constriction.

    d. none of these.

_____ 11. Matching air flow and blood flow in the right proportion at each alveolus improves

    a. pulmonary surfactant action.

    b. respiratory reflex action.

    c. pulmonary circulation.

    d. external respiration.

_____ 12. If a person is breathing 15 times a minute and has a tidal volume of 500 m*l*, the *total* minute respiratory volume is

    a. 7,500 min/m*l*.

    b. 515 m*l*.

    c. 7,500 m*l*.

    d. 5,150 m*l*.

_____ 13. The *residual volume* is the volume of air

    a. at the end of normal respiration if expiratory muscles are actively contracted.

    b. that remains in lungs after maximal expiration.

    c. that can be inspired over and above resting tidal volume.

    d. inspired and expired during one breath.

_____ 14. The maximum amount of air moved in and out during a single respiratory cycle is the

    a. tidal volume.

    b. vital capacity.

    c. alveolar ventilation.

    d. residual volume.

_____ 15. If a person is breathing 12 times per minute, the tidal volume is 350 m*l*, and the volume in the anatomic dead space is 150 m*l*, what is the alveolar ventilation rate?

   a. 0 m*l*/min
   b. 5250 m*l*/min
   c. 6000 m*l*/min
   d. 2400 m*l*/min

_____ 16. The most effective means of increasing alveolar ventilation is to

   a. increase rapid shallow breathing.
   b. breathe normally.
   c. breathe slowly and deeply.
   d. increase total pulmonary circulation.

_____ 17. When a person does not produce enough surfactant and becomes exhausted by the effort required to keep inflating and deflating the lungs, the condition is called

   a. hypoxia.
   b. atelectasis.
   c. respiratory distress syndrome.
   d. eupnea.

_____ 18. The partial pressure of O_2 in the atmosphere at sea level is

   a. 104 mm Hg.
   b. 200 mm Hg.
   c. 760 mm Hg.
   d. 160 mm Hg.

_____ 19. Which of the following is *not* a process involved with internal respiration?

   a. Bicarbonate ions are formed in the red blood cells.
   b. Hemoglobin binds more oxygen.
   c. Carbon dioxide diffuses from the interstitial spaces to the blood.
   d. Oxygen diffuses from the blood to the interstitial spaces.

_____ 20. Movement of air into and out of the lungs is accomplished by the process of _____, while all movement of gases across membranes is by _____.

   a. endocytosis; exocytosis
   b. diffusion; osmosis
   c. bulk flow; passive diffusion
   d. pinocytosis; diffusion

_____ 21. The correct sequential transport of O_2 from the tissue capillaries to O_2 consumption in cells is:

   a. lung, alveoli, plasma, erythrocytes, cells.
   b. erythrocytes, plasma, interstitial fluid, cells.
   c. plasma, erythrocytes, alveoli, cells.
   d. erythrocytes, interstitial fluid, plasma, cells.

_____ 22. It is important that free H+ resulting from dissociation of H_2CO_3 combine with hemoglobin to reduce the possibility of

    a. CO_2 escaping from the RBC.

    b. recombining with H_2O.

    c. an acidic condition within the blood.

    d. maintaining a constant pH in the blood.

_____ 23. Intracellular bicarbonate ions are exchanged for extracellular chloride ions, resulting in a mass movement of chloride ions into the

    a. red blood cells.

    b. interstitial fluid.

    c. plasma.

    d. alveoli.

_____ 24. A respiratory disorder characterized by fluid leakage into the alveoli or swelling and constriction of the respiratory bronchioles is

    a. anoxia.

    b. atelectasis.

    c. pneumonia.

    d. asthma.

_____ 25. In the chronic, progressive condition of emphysema,

    a. respiratory bronchioles and alveoli are functionally eliminated.

    b. the accessory respiratory muscles become inactive, decreasing respiration.

    c. the intrapleural pressure decreases, causing the lungs to deflate.

    d. compliance is decreased due to the loss of supporting tissues resulting from alveolar damage.

Completion

Using the terms below, complete the following statements. Use each term only once.

hypercapnia	thyroid cartilage	atelectasis
hilum	intrapleural	parenchyma
pneumothorax	phonation	epiglottis
alveolar ventilation	carina	anatomic dead space
laryngopharynx	eupnea	oropharynx
anoxia	lamina propria	Henry's law
Dalton's law	vestibule	hypoxia

1. The layer of connective tissue between the respiratory epithelium and the underlying bones or cartilages is the _____.

2. The portion of the nasal cavity containing epithelium with coarse hairs is the _____.

3. The chamber shared by the digestive and respiratory systems is the _____.

4. The portion of the pharynx lying between the hyoid and entrance to the esophagus is the _____.

5. The elastic cartilage that prevents the entry of liquids or solid food into the respiratory passageway is the _____.

6. The "Adam's apple" is the prominent ridge on the anterior surface of the _____.

7. Sound production at the larynx is called _____.

8. The groove along the medial surface of a lung that provides access for the pulmonary vessels and nerves is called the _____.

9. Quiet breathing in which inhalation involves muscular contractions and exhalation is a passive process called _____.

10. A decline in oxygen content causes affected tissues to suffer from _____.

11. The effects of cerebrovascular accidents and myocardial infarctions are the result of localized _____.

12. A collapsed lung is referred to as a(n) _____.

13. An injury to the chest wall that permits air to enter the intrapleural space is called a(n) _____.

14. The pressure measured in the slender space between the parietal and visceral pleura is the _____ pressure.

15. An increase in the P_{CO_2} of arterial blood is called _____.

16. The ridge that marks the line of separation between the two primary bronchi is the _____.

17. The substance of each lung is referred to as the _____.

18. The volume of air in the conducting passages is known as the _____.

19. The amount of air reaching the alveoli each minute is the _____.

20. The relationship among pressure, solubility, and the number of gas molecules in solution is known as _____.

21. The principle that states, "Each of the gases that contributes to the total pressure is in proportion to its relative abundance" is known as _____.

Short Essay

Briefly answer the following questions in the spaces provided below.

1. What are the five primary functions of the respiratory system?

2. How is the inhaled air warmed and humidified in the nasal cavities, and why is this type of environmental condition necessary?

3. What is surfactant, and what is its function in the respiratory membrane?

4. What protective mechanisms comprise the respiratory defense system?

5. What is the difference among external respiration, internal respiration, and cellular respiration?

6. What is the relationship between the volume and the pressure of a gas?

7. What do the terms *pneumothorax* and *atelectasis* refer to, and what is the relationship between the two terms?

8. What is vital capacity and how is it measured?

9. What is alveolar ventilation and how is it calculated?

10. What are the three forms of CO_2 transport in the blood? What is the percent of each form?

11. What regulatory sources modify the activities of the respiratory centers to alter the patterns of respiration?

12. What is the relationship between rise in arterial P_{CO_2} and hyperventilation?

LEVEL 3: CRITICAL THINKING AND CLINICAL APPLICATIONS

Using principles and concepts learned in Chapter 23, answer the following questions. Write your answers on a separate sheet of paper.

1. I. M. Good decides to run outside all winter long in spite of the cold weather. How does running in cold weather affect the respiratory passageways and the lungs?

2. U. R. Hurt is a 68-year-old man with arteriosclerosis. His condition has promoted the development of a pulmonary embolism that has obstructed blood flow to the right upper lobe of the lung. Eventually he experiences heart failure. Why?

3. C. U. Later's son is the victim of a tracheal defect that resulted in the formation of complete cartilaginous rings instead of the normal C-shaped tracheal rings. After eating a meal he experiences difficulty when swallowing. Why?

4. M. I. Right has been a two-pack-a-day smoker for the last 20 years. Recently, he has experienced difficulty breathing. After a series of pulmonary tests his doctor informs Mr. Right that he has emphysema. What is the relationship between his condition and his breathing difficulties?

5. On a recent flight to a vacation destination, the passengers on the flight experienced a sudden loss in cabin pressure. Many of them immediately got sick. Why?

The Digestive System

OVERVIEW

The digestive system is the "food processor" in the human body. The system works at two levels—one mechanical, the other chemical. The digestive functions are a series of integrated steps that include: (1) ingestion, (2) mechanical processing, (3) digestion, (4) secretion, (5) absorption, and (6) excretion (defecation).

By the time you are 65 years old, you will have consumed more than 70,000 meals and disposed of 50 tons of food, most of which has to be converted into chemical forms that cells can utilize for their metabolic functions.

After ingestion and mastication, the food is propelled through the muscular gastrointestinal tract (GI), where it is chemically split into smaller molecular substances. The products of the digestive process are then absorbed from the GI tract into the circulating blood and transported wherever they are needed in the body or, if undigested, are eliminated from the body. As the food is processed through the GI tract, accessory organs such as the salivary glands, liver, and pancreas assist in the "breakdown" process.

Chapter 24 includes exercises organized to guide your study of the structure and function of the digestive tract and accessory organs, hormonal and nervous system influences on the digestive process, and the chemical and mechanical events that occur during absorption.

LEVEL 1: REVIEWING FACTS AND TERMS

Review of Learning Outcomes

After completing this chapter, you should be able to do the following:

OUTCOME 24-1　　Identify the organs of the digestive system, list their major functions, describe the functional histology of the digestive tract, and outline the mechanisms that regulate digestion.

OUTCOME 24-2　　Discuss the anatomy of the oral cavity, and list the functions of its major structures and regions.

OUTCOME 24-3　　Describe the structure and functions of the pharynx.

OUTCOME 24-4　　Describe the structure and functions of the esophagus.

OUTCOME 24-5　　Describe the anatomy of the stomach, including its histological features, and discuss its roles in digestion and absorption.

OUTCOME 24-6　　Describe the anatomical and histological characteristics of the small intestine, explain the functions and regulation of intestinal secretions, and describe the structure, functions, and regulation of the accessory digestive organs.

OUTCOME 24-7   Describe the gross and histological structure of the large intestine, including its regional specializations and role in nutrient absorption.

OUTCOME 24-8   List the nutrients required by the body, describe the chemical events responsible for the digestion of organic nutrients, and describe the mechanisms involved in the absorption of organic and inorganic nutrients.

OUTCOME 24-9   Summarize the effects of aging on the digestive system.

OUTCOME 24-10   Give examples of interactions between the digestive system and other organ systems studied so far.

Multiple Choice

Place the letter corresponding to the best answer in the space provided.

OUTCOME 24-1   _____   1. Which one of the following organs is *not* a part of the digestive system?
   a. Liver
   b. Gallbladder
   c. Spleen
   d. Pancreas

OUTCOME 24-1   _____   2. Of the following selections, the one that contains *only* accessory structures is
   a. pharynx, esophagus, small and large intestine.
   b. oral cavity, stomach, pancreas, and liver.
   c. salivary glands, pancreas, liver, gallbladder.
   d. tongue, teeth, stomach, small and large intestine.

OUTCOME 24-1   _____   3. The active process that occurs when materials enter the digestive tract via the mouth is
   a. secretion.
   b. ingestion.
   c. absorption.
   d. excretion.

OUTCOME 24-1   _____   4. The lining of the digestive tract plays a defensive role by protecting surrounding tissues against
   a. corrosive effects of digestive acids and enzymes.
   b. mechanical stresses.
   c. pathogenic organisms swallowed with food.
   d. a, b, and c are correct.

OUTCOME 24-1   _____   5. Sympathetic stimulation of the muscularis externa promotes
   a. muscular inhibition and relaxation.
   b. increased muscular tone and activity.
   c. muscular contraction and increased excitation.
   d. increased digestive and gastric motility.

OUTCOME 24-1   _____   6. The mucous-producing, unicellular glands found in the mucosal epithelium of the stomach and small and large intestine are
   a. enteroendocrine cells.
   b. parietal cells.
   c. chief cells.
   d. mucous (goblet) cells.

OUTCOME 24-1   _____   7. Which of the layers of the digestive tube is most responsible
                              for peristalsis along the esophagus?
                              a. Lamina propria
                              b. Muscularis externa
                              c. Submucosa
                              d. Mucosal epithelium

OUTCOME 24-1   _____   8. The muscular layers involved with peristalsis are described as
                              a. skeletal.
                              b. sigmoidal and segmented.
                              c. circular and longitudinal.
                              d. a, b, and c are correct.

OUTCOME 24-1   _____   9. Swirling, mixing, and churning motions of the digestive tract provide
                              a. action of acids, enzymes, and buffers.
                              b. mechanical processing after ingestion.
                              c. chemical breakdown of food into small fragments.
                              d. a, b, and c are correct.

OUTCOME 24-1   _____   10. Powerful peristaltic contractions moving the contents of the colon
                              toward the rectum are called
                              a. segmentation.
                              b. the defecation reflex.
                              c. pendular movements.
                              d. mass movements.

OUTCOME 24-1   _____   11. Accelerated secretions by the salivary glands, resulting in the production
                              of watery saliva containing abundant enzymes, are promoted by
                              a. sympathetic stimulation.
                              b. parasympathetic stimulation.
                              c. the gastroenteric reflex.
                              d. excessive secretion of salivary amylase.

OUTCOME 24-2   _____   12. The submandibular glands produce saliva, which is
                              a. primarily a mucous secretion.
                              b. primarily a serous secretion.
                              c. both mucus and serous.
                              d. neither mucus nor serous.

OUTCOME 24-2   _____   13. The three pairs of salivary glands that secrete into the oral cavity are the
                              a. pharyngeal, palatoglossal, and palatopharyngeal.
                              b. lingual, labial, and frenulum.
                              c. uvular, ankyloglossal, and hypoglossal.
                              d. parotid, sublingual, and submandibular.

OUTCOME 24-2   _____   14. Crushing, mashing, and grinding of food are accomplished by the
                              action of the
                              a. incisors.
                              b. bicuspids.
                              c. cuspids.
                              d. molars.

OUTCOME 24-2   _____   15. The three phases of deglutition are
    a. parotid, sublingual, and submandibular.
    b. pharyngeal, palatopharyngeal, and stylopharyngeal.
    c. buccal, pharyngeal, and esophageal.
    d. palatal, lingual, and mesial.

OUTCOME 24-3   _____   16. On its way to the esophagus, food normally passes through the
    a. esophageal sphincter.
    b. oropharynx and laryngopharynx.
    c. pharyngeal sphincter.
    d. a, b, and c are correct.

OUTCOME 24-3   _____   17. The pharyngeal muscles that push the food bolus toward the esophagus are the
    a. pharyngeal constrictor muscles.
    b. palatopharyngeus muscles.
    c. palatal muscles.
    d. stylopharyngeus muscles.

OUTCOME 24-4   _____   18. The esophageal glands that produce a mucous secretion that reduces friction between the bolus and the esophageal lining are located in the
    a. mucosa.
    b. muscularis mucosa.
    c. serosa.
    d. submucosa.

OUTCOME 24-4   _____   19. The primary function of the esophagus is to
    a. convey solid foods and liquids from the mouth to the pharynx.
    b. convey solid foods and liquids to the stomach.
    c. convey solid foods and liquids from the stomach to the small intestine.
    d. process solid foods and liquids for digestion and absorption.

OUTCOME 24-4   _____   20. The inferior end of the esophagus normally remains in a state of active contraction that
    a. prevents air from entering into the esophagus.
    b. allows food to move rapidly into the stomach.
    c. prevents the backflow of materials from the stomach into the esophagus.
    d. initiates the movement of food from the stomach into the small intestine.

OUTCOME 24-5   _____   21. The hormone gastrin
    a. is produced in response to sympathetic stimulation.
    b. is secreted by the pancreatic islets.
    c. increases the activity of parietal and chief cells.
    d. inhibits the activity of the muscularis externa of the stomach.

OUTCOME 24-5   _____   22. The release of chyme from the stomach into the duodenum is regulated by the
a. pyloric sphincter.
b. lower esophageal sphincter.
c. pyloric antrum.
d. pyloric canal.

OUTCOME 24-5   _____   23. Which of the following inhibit(s) gastric contractions?
a. secretin.
b. somatostatin.
c. distention of duodenal stretch receptors.
d. a, b, and c are correct.

OUTCOME 24-5   _____   24. Which of the following is secreted by the stomach?
a. Galactase
b. Somatostatin
c. Ptyalin
d. Secretin

OUTCOME 24-6   _____   25. The three divisions of the small intestine are
a. cephalic, gastric, and intestinal.
b. buccal, pharyngeal, and esophageal.
c. duodenum, jejunum, and ileum.
d. fundus, body, and pylorus.

OUTCOME 24-6   _____   26. The myenteric plexus (Auerbach) of the intestinal tract is found
a. within the mucosa.
b. within the submucosa.
c. within the circular muscle layer.
d. between the circular and longitudinal muscle layers.

OUTCOME 24-6   _____   27. An enzyme *not* found in pancreatic juice is
a. lipase.
b. amylase.
c. nuclease.
d. GIP.

OUTCOME 24-6   _____   28. What happens to salivary amylase after it is swallowed?
a. It is compacted and becomes part of the feces.
b. It is absorbed and resecreted by the salivary glands.
c. It is absorbed in the duodenum and broken down into amino acids in the liver.
d. It is digested and absorbed in the small intestine.

OUTCOME 24-6   _____   29. The three phases of gastric function are
a. parotid, sublingual, and submandibular.
b. duodenal, jejunal, and iliocecal.
c. cephalic, gastric, and intestinal.
d. buccal, pharyngeal, and esophageal.

OUTCOME 24-6 _____ 30. Bile is stored and modified in the
   a. enterohepatic circulation.
   b. gallbladder.
   c. pancreas.
   d. duodenum.

OUTCOME 24-6 _____ 31. The primary function(s) of the liver is (are)
   a. metabolic regulation.
   b. hematological regulation.
   c. bile production.
   d. a, b, and c are correct.

OUTCOME 24-6 _____ 32. The hormone that promotes the emptying of the gallbladder
   and of pancreatic juice containing enzymes is
   a. secretin.
   b. gastrin.
   c. enterogastrone.
   d. cholecystokinin.

OUTCOME 24-7 _____ 33. Undigested food residues are moved through the *large intestine* in the
   following sequence:
   a. cecum, colon, and rectum.
   b. colon, cecum, and rectum.
   c. ileum, colon, and rectum.
   d. duodenum, jejunum, and ileum.

OUTCOME 24-7 _____ 34. The longitudinal ribbon of smooth muscle visible on the outer surfaces
   of the colon just beneath the serosa is (are) the
   a. haustra.
   b. taenia coli.
   c. epiploic appendages.
   d. vermiform appendix.

OUTCOME 24-7 _____ 35. Undigested food residues are moved through the colon
   in the following sequence:
   a. ascending, descending, transverse, sigmoid.
   b. descending, transverse, ascending, sigmoid.
   c. ascending, transverse, descending, sigmoid.
   d. descending, ascending, transverse, sigmoid.

OUTCOME 24-7 _____ 36. Material arriving from the ileum first enters an expanded pouch
   called the
   a. appendix.
   b. cecum.
   c. haustra.
   d. ileocecal valve.

OUTCOME 24-7 _____ 37. In addition to storage of fecal material and absorption of some vitamins, an important function of the large intestine is

    a. reabsorption of water.

    b. to produce bilirubin, which gives feces its brown coloration.

    c. production of intestinal gas.

    d. a, b, and c are correct.

OUTCOME 24-7 _____ 38. The vitamins liberated by bacterial action and absorbed in the large intestine are

    a. A, D, E, and K.

    b. thiamin, riboflavin, and niacin.

    c. C, D, and E.

    d. biotin, pantothenic acid, and vitamin K.

OUTCOME 24-8 _____ 39. The organic nutrients that provide energy for the human body are

    a. vitamins, minerals, and water.

    b. carbohydrates, lipids, and proteins.

    c. electrolytes, nucleic acids, and vitamins.

    d. a, b, and c are correct.

OUTCOME 24-8 _____ 40. The nutrients that can be absorbed without preliminary processing but may involve special transport mechanisms are

    a. nucleic acids, minerals, and enzymes.

    b. sucrose, amino acids, and fatty acids.

    c. water, electrolytes, and vitamins.

    d. lactose, fructose, and galactose.

OUTCOME 24-8 _____ 41. The enzyme lactase, which digests lactose to glucose and galactose, is synthesized by

    a. the stomach.

    b. the pancreas.

    c. epithelial cells lining the small intestine.

    d. Brunner's glands.

OUTCOME 24-8 _____ 42. Hydrochloric acid in the stomach functions primarily to

    a. facilitate protein digestion.

    b. facilitate lipid digestion.

    c. facilitate carbohydrate digestion.

    d. hydrolyze peptide bonds.

OUTCOME 24-8 _____ 43. A molecule absorbed into the lacteals of the lymphatic system within the walls of the small intestine is

    a. glucose.

    b. a lipid.

    c. vitamin B_{12}.

    d. an amino acid.

OUTCOME 24-8 _____ 44. The intestinal epithelium absorbs monosaccharides by

    a. carrier micelles and chylomicrons.

    b. net osmosis and diffusion.

    c. facilitated diffusion and cotransport mechanisms.

    d. active transport and net osmosis.

OUTCOME 24-8 _____ 45. When two fluids are separated by a selectively permeable membrane, water tends to flow into the solution that has the

    a. lower concentration of solutes.

    b. same osmolarity on each side of the membrane.

    c. higher concentration of solutes.

    d. proper osmotic equilibrium.

OUTCOME 24-9 _____ 46. In the elderly, the decline in olfactory and gustatory sensitivities with age can lead to

    a. a decrease in stem cell division.

    b. increased colon and stomach cancers.

    c. dietary changes that affect the entire body.

    d. decreased motility and peristaltic contractions.

OUTCOME 24-10 _____ 47. The system that interacts with the digestive tract to control hunger, satiation, and feeding behaviors is the

    a. endocrine system.

    b. nervous system.

    c. lymphoid system.

    d. cardiovascular system.

Completion

Using the terms below, complete the following statements. Use each term only once.

tongue	duodenum	lipase	chyme
nasopharynx	kidneys	acini	peristalsis
enteroendocrine	colon	vitamin K	haustra
carbohydrates	gastrin	heartburn	pepsin
Kupffer cells	vitamin B_{12}	integumentary	digestion
uvula	plicae	cardiac sphincter	exocytosis
lamina propria	esophagus	pacesetter	olfactory
cholecystokinin	mucins	cuspids	

OUTCOME 24-1   1. The digestive tube between the pharynx and the stomach is the _____.

OUTCOME 24-1   2. The chemical breakdown of food into small organic fragments suitable for absorption by the digestive epithelium refers to _____.

OUTCOME 24-1   3. The cells of the digestive epithelium scattered among the columnar cells that secrete hormones that coordinate the activities of the digestive tract and the accessory glands are called _____ cells.

OUTCOME 24-1   4. The layer of loose connective tissue that contains blood vessels, nerve endings, and lymph vessels is the _____.

OUTCOME 24-1   5. Permanent transverse folds in the lining of the digestive tract that increase the surface area available for absorption are called _____.

OUTCOME 24-1   6. The smooth muscle along the digestive tract has rhythmic cycles of activity due to the presence of _____ cells.

OUTCOME 24-1   7. The muscularis externa propels materials from one portion of the digestive tract to another by waves of muscular contractions referred to as _____.

OUTCOME 24-2   8. Saliva originating in the submandibular and sublingual salivary glands contains large quantities of glycoproteins called _____.

OUTCOME 24-2   9. The conical teeth used for tearing or slashing are the _____.

OUTCOME 24-2   10. Salivary amylase in the oral cavity begins the digestive process by acting on _____.

OUTCOME 24-2   11. The structure in the oral cavity that prevents food from entering the pharynx prematurely is the _____.

OUTCOME 24-2   12. Mechanical processing by compression, abrasion, and distortion is one of the functions of the _____.

OUTCOME 24-3   13. The region of the pharynx through which food does not normally pass is the _____.

OUTCOME 24-4   14. The circular muscle layer at the inferior end of the esophagus is called the _____.

OUTCOME 24-5   15. Neural stimulation and chemoreceptors in the mucosa trigger the release of _____.

OUTCOME 24-5   16. The agitation of ingested materials with the gastric juices of the stomach produces a viscous, soupy mixture called _____.

OUTCOME 24-6   17. The portion of the small intestine that receives chyme from the stomach and exocrine secretions from the pancreas and liver is the _____.

OUTCOME 24-6   18. The two intestinal hormones that inhibit gastric secretion to some degree are secretin and _____.

OUTCOME 24-6   19. The sinusoids of the liver contain large numbers of phagocytic cells called _____.

OUTCOME 24-6   20. The blind pockets lined by simple cuboidal epithelium that produce pancreatic juice are called _____.

OUTCOME 24-7   21. The pouches that form the wall of the colon, permitting considerable distention and elongation, are called _____.

OUTCOME 24-7   22. The section of the large intestine that bears taenia coli, and sacs of fat, is the _____.

OUTCOME 24-7   23. When significant amounts of ammonia and smaller amounts of other toxins cross the colonic epithelium, they are processed by the liver and ultimately excreted as relatively nontoxic compounds at the _____.

OUTCOME 24-8   24. A vitamin that is produced in the colon and is absorbed with other lipids released through bacterial action is _____.

OUTCOME 24-8   25. The pancreatic enzyme that hydrolyzes triglycerides into monoglycerides and fatty acids is pancreatic _____.

OUTCOME 24-8   26. Secreted initially as a proenzyme by stomach chief cells, the enzyme that digests protein into smaller peptide chains is _____.

OUTCOME 24-8   27. The nutrient that must be bound to intrinsic factor before being absorbed across the intestinal epithelium via active transport is _____.

OUTCOME 24-8   28. Interstitial cells secrete chylomicrons into interstitial fluid by _____.

OUTCOME 24-9   29. Weakening of muscular sphincters in the elderly can lead to esophageal reflux, and the elderly frequently experience _____.

OUTCOME 24-9   30. Dietary changes associated with age usually result from a decline in gustatory and _____ sensitivities.

OUTCOME 24-10   31. The system that provides vitamin D_3 needed for the absorption of calcium and phosphorus in the digestive tract is the _____ system.

Matching

Match the terms in column B with the terms in column A. Use letters for answers in the spaces provided. Use each term only once.

Part I

		Column A	Column B
OUTCOME 24-1	_____	1. sheets of serous membrane	A. roof of oral cavity
OUTCOME 24-1	_____	2. accessory organ	B. deglutition
OUTCOME 24-1	_____	3. lamina propria	C. elevates larynx
OUTCOME 24-1	_____	4. swallowing	D. mesenteries
OUTCOME 24-1	_____	5. D cells	E. peptide hormone
OUTCOME 24-1	_____	6. secretin	F. esophageal hiatus
OUTCOME 24-2	_____	7. gingivae	G. release somatostatin
OUTCOME 24-2	_____	8. hard and soft palate	H. liver
OUTCOME 24-2	_____	9. incisors	I. crush, mush, grind
OUTCOME 24-2	_____	10. bicuspids	J. gums
OUTCOME 24-3	_____	11. stylopharyngeus	K. loose connective tissue
OUTCOME 24-4	_____	12. opening in diaphragm	L. clipping or cutting

Part II

		Column A	Column B
OUTCOME 24-5	_____	13. chief cell	M. bands of smooth muscles
OUTCOME 24-5	_____	14. parietal cells	N. secrete intrinsic factor
OUTCOME 24-6	_____	15. villi	O. provides vitamin D_3
OUTCOME 24-6	_____	16. duodenal glands	P. produce mucus
OUTCOME 24-6	_____	17. liver cells	Q. facilitated diffusion
OUTCOME 24-7	_____	18. taenia coli	R. gradual loss of teeth
OUTCOME 24-7	_____	19. sigmoid	S. secrete pepsinogen
OUTCOME 24-7	_____	20. large intestine	T. starch breakdown
OUTCOME 24-8	_____	21. B vitamins and C	U. digestive epithelial damage by acids
OUTCOME 24-8	_____	22. pancreatic alpha-amylase	V. vitamin K absorption
OUTCOME 24-8	_____	23. transport mechanism	W. small intestine
OUTCOME 24-9	_____	24. gingivitis	X. colon
OUTCOME 24-9	_____	25. peptic ulcers	Y. hepatocytes
OUTCOME 24-10	_____	26. integumentary system	Z. water soluble

Drawing/Illustration Labeling

Identify each numbered structure in the following figures. Place your answers in the spaces provided.

OUTCOME 24-1 **FIGURE 24-1** The Components of the Digestive System

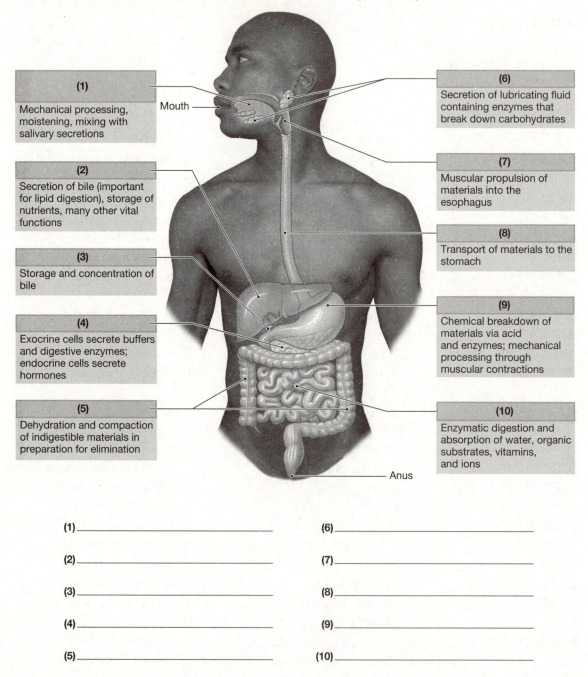

(1)

Mechanical processing, moistening, mixing with salivary secretions

(2)

Secretion of bile (important for lipid digestion), storage of nutrients, many other vital functions

(3)

Storage and concentration of bile

(4)

Exocrine cells secrete buffers and digestive enzymes; endocrine cells secrete hormones

(5)

Dehydration and compaction of indigestible materials in preparation for elimination

(6)

Secretion of lubricating fluid containing enzymes that break down carbohydrates

(7)

Muscular propulsion of materials into the esophagus

(8)

Transport of materials to the stomach

(9)

Chemical breakdown of materials via acid and enzymes; mechanical processing through muscular contractions

(10)

Enzymatic digestion and absorption of water, organic substrates, vitamins, and ions

Mouth

Anus

(1)_____

(2)_____

(3)_____

(4)_____

(5)_____

(6)_____

(7)_____

(8)_____

(9)_____

(10)_____

OUTCOME 24-1   **FIGURE 24-2**   The Structure of the Digestive Tract

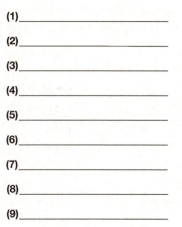
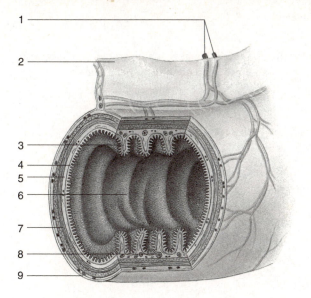

(1)_____

(2)_____

(3)_____

(4)_____

(5)_____

(6)_____

(7)_____

(8)_____

(9)_____

OUTCOME 24-2   **FIGURE 24-3**   Structure of a Typical Tooth—Sectional View

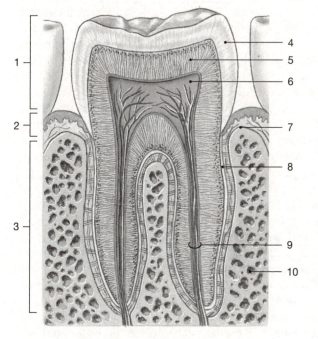

(1)_____

(2)_____

(3)_____

(4)_____

(5)_____

(6)_____

(7)_____

(8)_____

(9)_____

(10)_____

LEVEL 2: REVIEWING CONCEPTS

Chapter Overview

Using the terms below, fill in the blanks to complete the chapter overview, which traces the pathway of food as it enters the digestive tract and passes along its length. Use each term only once.

proteins	lubrication	swallowing
palates	anus	oral cavity
uvula	peristalsis	absorption
pancreas	gingivae (gums)	chyme
stomach	pyloric canal	salivary amylase
teeth	complex carbohydrates	hydrochloric
mucus	tongue	mechanical processing
esophagus	duodenum	lower esophageal sphincter
salivary glands	gastric juices	rectum
feces	mastication	rugae
pharynx	pepsins	defecation

The digestive tract begins at the (1) _____. Food preparation, which includes analysis, (2) _____, digestion by salivary amylase, and (3) _____, occurs in this region of the food processing system. Noticeable features in this region include pink ridges, the (4) _____ that surround the base of the glistening white (5) _____ used in handling and (6) _____ of the food. The roof of the cavity is formed by the hard and soft (7) _____, while the floor contains the (8) _____. Three pairs of (9) _____ secrete saliva, which contains the enzyme (10) _____, which initiates the digestion of (11) _____. After food preparation and the start of the digestive process, the food bolus passes into the (12) _____, an anatomical space that serves as a common passageway for solid food, liquids, and air. The region that contains structures involved in the food movement process is characterized by the presence of the hanging, dangling process, the (13) _____, supported by the posterior part of the soft palate. Additional movement is provided by powerful constrictor muscles involved in (14) _____, causing propulsion into the elongated, flattened tube, the (15) _____. Alternating wave-like contractions called (16) _____ facilitate passage to a region that looks like a ring of muscle tissue, the (17) _____, which serves as the entrance to the (18) _____. In this enlarged chamber the food bolus is "agitated" because of glandular secretions called (19) _____ that produce a viscous, soupy mixture referred to as (20) _____. The walls of the chamber look "wrinkled" due to the presence of folds called (21) _____, which are covered with a slippery, slimy (22) _____ coat that offers protection from erosion and other potential damage. In this region, (23) _____ acid and enzymes called (24) _____ are involved in the preparation and partial digestion of (25) _____. Following the digestive activity in this enlarged chamber, the chyme is conveyed through the (26) _____ into the (27) _____, the proximal segment of the small intestine. This portion is the "mixing bowl" that includes chyme and receives digestive secretions from the liver and the (28) _____. The small intestine plays the key role in the digestion and (29) _____ of nutrients. The digestive process continues into the large intestine, where the reabsorption of water and the compaction of the intestinal contents into (30) _____ occurs. Other activities include the absorption of important vitamins liberated by bacterial action and the storage of fecal matter prior to (31) _____. The (32) _____, an expandable organ for the temporary storage of feces, forms the last segment of the digestive tract. The (33) _____ is the exit of the anal canal.

Concept Map I

Using the following terms, fill in the circled, numbered, blank spaces to correctly complete the concept map. Use each term only once.

stomach   digestive tract movements   hormones
pancreas   hydrochloric acid   large intestine
saliva   intestinal mucosa   bile

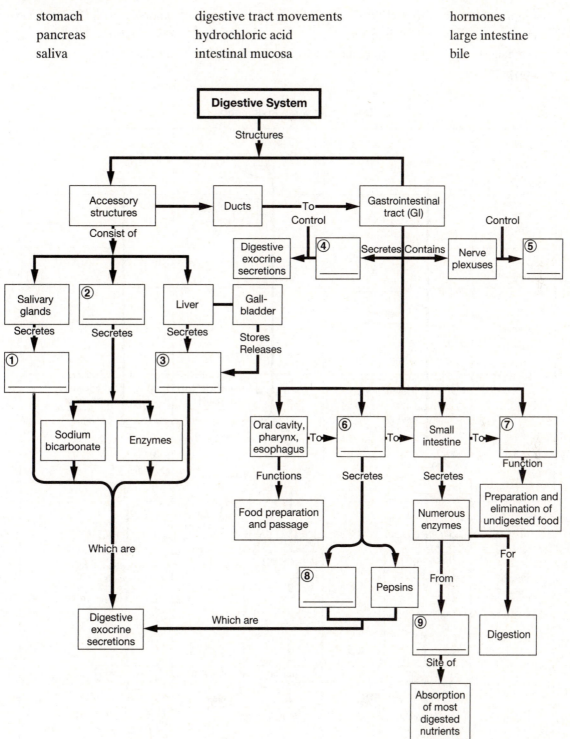

Concept Map II

Using the following terms, fill in the circled, numbered, blank spaces to correctly complete the concept map. Use each term only once.

simple sugars	triglycerides	complex sugars and starches
lacteal	small intestine	disaccharides, trisaccharides
esophagus	monoglycerides, fatty acids in micelles	amino acids
polypeptides		

Chemical events in digestion

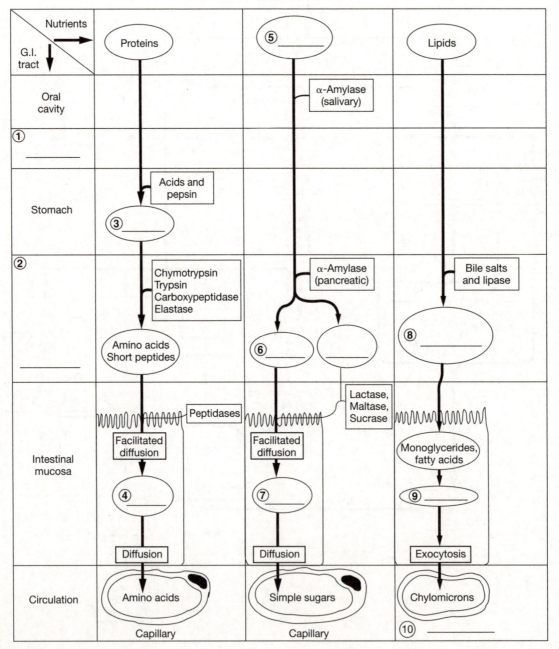

Concept Map III

Using the following terms, fill in the circled, numbered, blank spaces to correctly complete the concept map. Use each term only once.

acid production   material in jejunum   nutrient utilization by tissues
GIP   VIP   bile in gallbladder
insulin

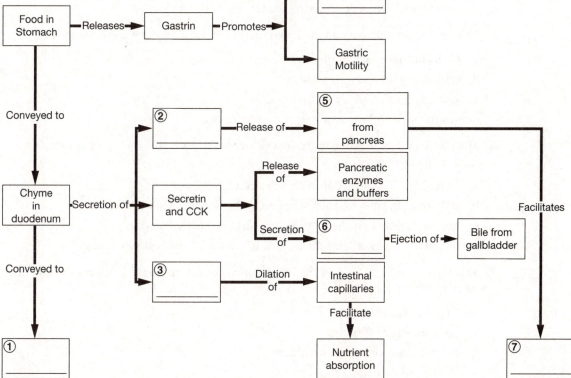

Action of Major Digestive Tract Hormones

Multiple Choice

Place the letter corresponding to the best answer in the space provided.

_____ 1. The final enzymatic steps in the digestion of carbohydrates are accomplished by

    a. brush border enzymes of the microvilli.

    b. enzymes secreted by the stomach.

    c. enzymes secreted by the pancreas.

    d. the action of bile by the gallbladder.

_____ 2. Many visceral smooth muscle networks show rhythmic cycles of activity in the absence of neural stimulation due to the presence of

    a. direct contact with motor neurons carrying impulses to the CNS.

    b. an action potential generated and conducted over the sarcolemma.

    c. the single motor units that contract independently of each other.

    d. pacesetter cells that spontaneously depolarize and trigger contraction of entire muscular sheets.

_____ 3. A drop in pH below 4.5 in the duodenum stimulates the secretion of

    a. cholecystokinin.

    b. secretin.

    c. gastrin.

    d. GIP.

_____ 4. The stomach is divided into the following regions:

    a. cardia, rugae, pylorus.

    b. greater omentum, lesser omentum, gastric body.

    c. cardia, fundus, body, pylorus.

    d. parietal cells, chief cells, gastral glands.

_____ 5. The two factors that play an important part in controlling gastric emptying are

    a. CNS and ANS regulation.

    b. release of HCl and gastric juice.

    c. stomach distention and gastrin release.

    d. sympathetic and parasympathetic stimulation.

_____ 6. The plicae of the intestinal mucosa, which bears the intestinal villi, are structural features that provide for

    a. increased total surface area for absorption.

    b. stabilizing the mesenteries attached to the dorsal body wall.

    c. gastric contractions that churn and swirl the gastric contents.

    d. initiating enterogastric reflexes that accelerate the digestive process.

_____ 7. The enteroendocrine cells of the intestinal glands are responsible for producing the intestinal hormones

    a. gastrin and pepsinogen.

    b. biliverdin and bilirubin.

    c. enterokinase and aminopeptidase.

    d. cholecystokinin and secretin.

_____ 8. Most intestinal absorption occurs in the

    a. duodenum.

    b. ileocecum.

    c. ileum.

    d. jejunum.

_____ 9. The primary function(s) of the intestinal juice is (are) to

    a. moisten the chyme.

    b. neutralize acid.

    c. dissolve digestive enzymes and products of digestion.

    d. a, b, and c are correct.

_____ 10. An immediate increase in the rates of glandular secretion and peristaltic activity along the entire small intestine is a result of the

    a. presence of intestinal juice.

    b. gastroenteric reflex.

    c. gastroileal reflex.

    d. action of hormonal and CNS controls.

_____ 11. One of the primary effects of secretin is to cause a(n)

    a. decrease in duodenal submucosal secretions.

    b. increase in release of bile from the gallbladder into the duodenum.

    c. increase in secretion of buffers by the pancreas.

    d. increase in gastric motility and secretory rates.

_____ 12. The peptide hormone that causes the release of insulin from the pancreatic islets is

    a. GIP.

    b. CCK.

    c. VIP.

    d. GTP.

_____ 13. The large intestine can be divided into three major parts, the

    a. duodenum, jejunum, and ileum.

    b. cecum, colon, and rectum.

    c. haustra, taenia coli, and appendix.

    d. right colic flexure, left colic flexure, and sigmoid flexure.

_____ 14. The muscular sphincter that guards the entrance between the ileum and the cecum is the

    a. pyloric sphincter.

    b. gastrointestinal sphincter.

    c. ileocecal valve.

    d. taenia coli.

_____ 15. The contractions that force fecal material into the rectum and produce the urge to defecate are called _____ movements.

    a. primary

    b. mass

    c. secondary

    d. bowel

_____ 16. A large meal containing small amounts of protein, large amounts of carbohydrates, wine, and after-dinner coffee will leave your stomach very quickly because

    a. carbohydrate and protein initiate neural responses.

    b. the arrival of chyme in the duodenum stimulates secretion of hormones.

    c. chyme initiates the triggering of the enterogastric reflex.

    d. alcohol and caffeine stimulate gastric secretion and motility.

_____ 17. The external anal sphincter is under voluntary control.

    a. True

    b. False

_____ 18. The two positive feedback loops involved in the defecation reflex are

    a. internal and external sphincter muscles.

    b. the anorectal canal and the rectal columns.

    c. stretch receptors in rectal walls, and the sacral parasympathetic system.

    d. mass movements and peristaltic contractions.

_____ 19. When you see, smell, taste, or think of food the _____ phase of gastric secretion begins.

    a. gastric

    b. cephalic

    c. intestinal

    d. gastrointestinal

_____ 20. Triglycerides coated with proteins create a complex known as a

    a. micelle.

    b. co-transport.

    c. glycerolproteinase.

    d. chylomicron.

Completion

Using the terms below, complete the following statements. Use each term only once.

Brunner's glands	duodenum	lacteals
bicuspids	submucosal plexus	ileum
mesenteries	parietal peritoneum	adventitia
segmentation	ankyloglossia	Peyer's patches
enterocrinin	incisors	alkaline tide

1. The submucosal region that contains sensory nerve cells, parasympathetic ganglia, and sympathetic postganglionic fibers is referred to as the _____.

2. The serosa that lines the inner surfaces of the body wall is called the _____.

3. Double sheets of peritoneal membrane that connect the parietal peritoneum with the visceral peritoneum are called _____.

4. The fibrous sheath formed by a dense network of collagen fibers that firmly attaches the digestive tract to adjacent structures is called a(n) _____.

5. Movements that churn and fragment digested materials in the small intestine are termed _____.

6. Restrictive movements of the tongue that cause abnormal eating or speaking are a result of a condition known as _____.

7. Blade-shaped teeth found at the front of the mouth used for clipping or cutting are called _____.

8. The teeth used for crushing, mashing, and grinding are the molars and the premolars called _____.

9. The sudden rush of bicarbonate ions from gastric parietal cells into the bloodstream is referred to as the _____.

10. The portion of the small intestine that receives chyme from the stomach and exocrine secretions from the pancreas and the liver is the _____.

11. The most inferior part of the small intestine is the _____.

12. Terminal lymphatics contained within each villus are called _____.

13. The submucosal glands in the duodenum that secrete mucus, providing protection and buffering action, are _____.

14. Aggregations of lymphatic nodules forming lymphatic centers in the ileum are called _____.

15. The hormone that stimulates the duodenal glands when acid chyme enters the small intestine is _____.

Short Essay

Briefly answer the following questions in the spaces provided below.

1. What six integrated steps comprise the digestive functions?

2. What six major layers show histological features reflecting specializations throughout the digestive tract? (List the layers in correct order from the inside to the outside.)

3. Briefly describe the difference between a peristaltic movement of food and movement by segmentation.

4. What three pairs of salivary glands secrete saliva into the oral cavity?

5. What four types of teeth are found in the oral cavity and what is the function of each type?

6. Starting at the superior end of the digestive tract and proceeding inferiorly, what six sphincters control movement of materials through the tract?

7. What are the similarities and differences between parietal cells and chief cells in the wall of the stomach?

8. a. What three phases are involved in the regulation of gastric function?

   b. What regulatory mechanism(s) dominate(s) each phase?

9. What are the most important hormones that regulate intestinal activity?

10. What are the three principal functions of the large intestine?

11. What three basic categories describe the functions of the liver?

12. What two distinct functions are performed by the pancreas?

LEVEL 3: CRITICAL THINKING AND CLINICAL APPLICATIONS

Using principles and concepts learned in Chapter 24, answer the following questions. Write your answers on a separate sheet of paper.

1. Charlotte complains of recurring abdominal pains, usually 2 to 3 hours after eating. Additional eating seems to relieve the pain. After weeks of digestive stress, she agrees to see her doctor, who diagnoses her condition as a peptic ulcer. How would you explain this diagnosis to the patient?

2. How does a decrease in smooth muscle tone affect the digestive processes and promote constipation and the production of hemorrhoids in the elderly?

3. O. B. Good likes his four to six cups of caffeinated coffee daily, and he relieves his thirst by drinking a few alcoholic beverages during the day. He constantly complains of heartburn. What is happening in his GI tract to cause the "burning" sensation?

4. Jerry is an alcoholic who has developed cirrhosis of the liver. He has been told by his physician that this condition is a serious threat to his life. Why?

5. A 14-year-old girl complains of gassiness, rumbling sounds, cramps, and diarrhea after drinking a glass of milk. At first she assumes indigestion; however, every time she consumes milk or milk products the symptoms reappear. What is her problem? What strategies will be necessary for her to cope effectively with the problem?

CHAPTER 25

Metabolism and Energetics

OVERVIEW

The processes of metabolism and energetics are intriguing to most students of anatomy and physiology because of the mystique associated with the "fate of absorbed nutrients." This mysterious phrase raises questions such as:

- What happens to the food we eat?
- What determines the "fate" of the food and the way it will be used?
- How are the nutrients from the food utilized by the body?

All cells require energy for anabolic processes provided by breakdown mechanisms involving catabolic processes. Provisions for these cellular activities are made by the food we eat; these activities provide energy, promote tissue growth and repair, and serve as regulatory mechanisms for the body's metabolic machinery.

The exercises in Chapter 25 provide for study and review of cellular metabolism; metabolic interactions including the absorptive and postabsorptive state; diet and nutrition; and bioenergetics. Much of the material you have learned in previous chapters will be of value as you "see" the integrated forces of the digestive and cardiovascular system exert a coordinated effort to initiate and advance the work of metabolism.

LEVEL 1: REVIEWING FACTS AND TERMS

Review of Learning Outcomes

After completing this chapter, you should be able to do the following:

OUTCOME 25-1   Define energetics and metabolism, and explain why cells must synthesize new organic components.

OUTCOME 25-2   Describe the basic steps in glycolysis, the citric acid cycle, and the electron transport system, and summarize the energy yields of glycolysis and cellular respiration.

OUTCOME 25-3   Describe the pathways involved in lipid metabolism, and summarize the mechanisms of lipid transport and distribution.

OUTCOME 25-4   Summarize the main processes of protein metabolism, and discuss the use of proteins as an energy source.

OUTCOME 25-5   Differentiate between the absorptive and postabsorptive metabolic states, and summarize the characteristics of each.

OUTCOME 25-6   Explain what constitutes a balanced diet and why such a diet is important.

OUTCOME 25-7   Define metabolic rate, discuss the factors involved in determining an individual's BMR, and discuss the homeostatic mechanisms that maintain a constant body temperature.

Multiple Choice

Place the letter corresponding to the best answer in the space provided.

OUTCOME 25-1   _____   1. The term *metabolism* refers to all of the

    a. synthesis of new organic molecules.

    b. energy released during the breakdown of organic substances.

    c. chemical reactions that occur in the body.

    d. a, b, and c are correct.

OUTCOME 25-1   _____   2. The source that cells rely on to provide energy and to create new intracellular components is the

    a. nutrient pool.

    b. mitochondria.

    c. electron transport system.

    d. citric acid cycle.

OUTCOME 25-1   _____   3. The periodic breakdown and replacement of the organic components of a cell is called

    a. catabolism.

    b. metabolic turnover.

    c. anaerobic glycolysis.

    d. oxidative phosphorylation.

OUTCOME 25-1   _____   4. The process that breaks down organic substrates, releasing energy that can be used to synthesize ATP or other high-energy compounds, is

    a. metabolism.

    b. anabolism.

    c. catabolism.

    d. oxidation.

OUTCOME 25-2   _____   5. Acetyl-CoA cannot be used to make glucose because the

    a. catabolism of glucose ends with the formation of acetyl-CoA.

    b. catabolic pathways in the cells are all organic substances.

    c. glucose is used exclusively to provide energy for cells.

    d. decarboxylation step between pyruvic acid and acetyl-CoA cannot be reversed.

OUTCOME 25-2   _____   6. In glycolysis, six-carbon glucose molecules are broken down into 2 three-carbon molecules of

    a. pyruvic acid.

    b. acetyl-CoA.

    c. citric acid.

    d. oxaloacetic acid.

OUTCOME 25-2   _____   7. The first step in a sequence of enzymatic reactions in the citric acid cycle is the formation of

     a. acetyl-CoA.

     b. citric acid.

     c. oxaloacetic acid.

     d. pyruvic acid.

OUTCOME 25-2   _____   8. For each glucose molecule converted to 2 pyruvates, the anaerobic reaction sequence in glycolysis provides a net gain of

     a. 2 ATP for the cell.

     b. 4 ATP for the cell.

     c. 36 ATP for the cell.

     d. 38 ATP for the cell.

OUTCOME 25-2   _____   9. The complete catabolism (anaerobic + aerobic) of a glucose molecule yields a net of _____ molecules of ATP.

     a. 4

     b. 32

     c. 36

     d. 24

OUTCOME 25-3   _____   10. Lipids cannot provide large amounts of adenosine triphosphate (ATP) in a short period of time because

     a. lipid reserves are difficult to mobilize.

     b. lipids are insoluble and it is difficult for water-soluble enzymes to reach them.

     c. most lipids are processed in mitochondria, and mitochondrial activity is limited by the availability of oxygen.

     d. a, b, and c are correct.

OUTCOME 25-3   _____   11. Lipogenesis is an effective lipid metabolic pathway because lipids, as well as carbohydrates and amino acids, can be converted into

     a. linoleic acid.

     b. acetyl-CoA.

     c. chylomicrons.

     d. all of the above.

OUTCOME 25-3   _____   12. Most lipids circulate through the bloodstream as

     a. free fatty acids.

     b. lipoproteins.

     c. cholesterol.

     d. triglycerides.

OUTCOME 25-3   _____   13. Lipids are needed throughout the body because all cells require lipids to

     a. provide the basic service of energy for cells.

     b. maintain their cell membranes.

     c. provide for tissue growth and repair.

     d. maintain optimum temperature in cells.

OUTCOME 25-4 _____ 14. Protein catabolism is an impractical primary energy source because
   a. proteins are more difficult to break apart than are carbohydrates or lipids.
   b. one of the by-products of protein catabolism, ammonium ions, is toxic to cells.
   c. proteins are important structural and functional cellular components.
   d. a, b, and c are correct.

OUTCOME 25-4 _____ 15. The first step in amino acid catabolism is the removal of the
   a. carboxyl group.
   b. amino group.
   c. keto acid.
   d. hydrogen from the central carbon.

OUTCOME 25-5 _____ 16. An important energy reserve that can be broken down when a cell cannot obtain enough glucose from interstitial fluid is
   a. pyruvic acid.
   b. glycogen.
   c. amino acids.
   d. ATP.

OUTCOME 25-5 _____ 17. The metabolic pattern characterized by nutrient absorption, anabolism, and insulin dominance is the
   a. absorptive state.
   b. postabsorptive state.

OUTCOME 25-5 _____ 18. The metabolic pattern characterized by catabolism of internal energy reserves, elevated levels of ketone bodies in the blood, and a focus on maintaining homeostatic blood glucose levels is the
   a. absorptive state.
   b. postabsorptive state.

OUTCOME 25-6 _____ 19. A balanced diet contains all of the ingredients necessary to
   a. prevent starvation.
   b. prevent life-threatening illnesses.
   c. prevent deficiency diseases.
   d. maintain homeostasis.

OUTCOME 25-6 _____ 20. Foods that provide all essential amino acids in sufficient quantities are called
   a. incomplete proteins.
   b. minerals.
   c. complete proteins.
   d. vitamins.

OUTCOME 25-6 _____ 21. Inorganic ions released through dissociation of electrolytes and involved in a number of physiological functions are known as
   a. minerals.
   b. porphyrins.
   c. vitamins.
   d. N compounds.

OUTCOME 25-6   _____   22. Minerals, vitamins, and water are classified as *essential* nutrients because
    a. they are used by the body in large quantities.
    b. the body cannot synthesize the nutrients in sufficient quantities.
    c. they are the major providers of calories for the body.
    d. a, b, and c are correct.

OUTCOME 25-6   _____   23. Examples of trace minerals found in small quantities in the body include
    a. sodium, potassium, chloride, and calcium.
    b. phosphorus, magnesium, calcium, and iron.
    c. iron, zinc, copper, and manganese.
    d. phosphorus, zinc, copper, and potassium.

OUTCOME 25-6   _____   24. Hypervitaminosis involving water-soluble vitamins is relatively uncommon because
    a. excessive amounts are stored in adipose tissue.
    b. excessive amounts are readily excreted in the urine.
    c. the excess amount is stored in the bones.
    d. excesses are readily absorbed into skeletal muscle tissue.

OUTCOME 25-7   _____   25. To examine the metabolic rate of an individual, results may be expressed as
    a. calories per hour.
    b. calories per day.
    c. calories per unit of body weight per day.
    d. a, b, and c are correct.

OUTCOME 25-7   _____   26. An individual's basal metabolic rate ideally represents
    a. the minimum resting energy expenditure of an awake, alert person.
    b. genetic differences among ethnic groups.
    c. the amounts of circulating hormone levels in the body.
    d. a measurement of the daily energy expenditures for a given individual.

OUTCOME 25-7   _____   27. If your daily intake exceeds your total energy demands, you will store the excess energy primarily as
    a. triglycerides in adipose tissue.
    b. glycogen in the liver.
    c. lipoproteins in the liver.
    d. glucose in the bloodstream.

OUTCOME 25-7   _____   28. The four processes involved in heat exchange with the environment are
    a. physiological, behavioral, generational, and acclimatization.
    b. radiation, conduction, convection, and evaporation.
    c. sensible, insensible, heat loss, and heat gain.
    d. thermogenesis, dynamic action, pyrexia, and thermalphasic.

OUTCOME 25-7   _____   29. The primary mechanisms for increasing heat loss in the body include
    a. sensible and insensible.
    b. acclimatization and pyrexia.
    c. physiological mechanisms and behavioral modifications.
    d. vasomotor and respiratory.

Completion

Using the terms below, complete the following statements. Use each term only once.

glucagon	malnutrition	deamination
minerals	thermoregulation	urea
liver	beta oxidation	avitaminosis
pyrexia	adipose	nutrition
insulin	calorie	glycogen
transamination	triglycerides	anabolism
glycolysis	free fatty acids	oxidative phosphorylation
hypervitaminosis		

OUTCOME 25-1   1. The synthesis of new organic components that involves the formation of new chemical bonds is _____.

OUTCOME 25-2   2. The major pathway of carbohydrate metabolism is _____.

OUTCOME 25-2   3. The process that produces over 90 percent of the ATP used by our cells is _____.

OUTCOME 25-3   4. The most abundant lipids in the body are the _____.

OUTCOME 25-3   5. Fatty acid molecules are broken down into two-carbon fragments by means of a sequence of reactions known as _____.

OUTCOME 25-3   6. The lipids that can diffuse easily across cell membranes are the _____.

OUTCOME 25-4   7. The process that prepares an amino acid for breakdown in the citric acid cycle is _____.

OUTCOME 25-4   8. As a result of deamination, the relatively harmless water-soluble compound excreted in the urine is _____.

OUTCOME 25-4   9. The process that attaches the amino group of an amino acid to a keto acid is called _____.

OUTCOME 25-5   10. In the liver and in skeletal muscle, glucose molecules are stored as _____.

OUTCOME 25-5   11. The focal point for metabolic regulation and control in the body is the _____.

OUTCOME 25-5   12. _____ tissue stores lipids.

OUTCOME 25-5   13. During the absorptive state, increased glucose uptake and utilization is affected by the hormone _____.

OUTCOME 25-5   14. During the postabsorptive state, glycogenolysis is promoted by the hormone _____.

OUTCOME 25-6   15. The absorption of nutrients from food is called _____.

OUTCOME 25-6   16. An unhealthy state resulting from inadequate intake of one or more nutrients that becomes life-threatening as the deficiencies accumulate is called _____.

OUTCOME 25-6   17. Inorganic ions released through the dissociation of electrolytes are _____.

OUTCOME 25-6   18. The deficiency disease resulting from a vitamin deficiency is referred to as _____.

OUTCOME 25-6   19. The intake of vitamins that exceeds the ability to store, utilize, or excrete a particular vitamin is called _____.

OUTCOME 25-7   20. The amount of energy required to raise the temperature of 1 gram of water 1 degree centigrade is the _____.

OUTCOME 25-7   21. The homeostatic process that keeps body temperatures within acceptable limits regardless of environmental conditions is called _____.

OUTCOME 25-7   22. Any elevated body temperature is called _____.

Matching

Match the terms in column B with the terms in column A. Use letters for answers in the spaces provided. Use each term only once.

Part I

		Column A	Column B
OUTCOME 25-1	_____	1. catabolism	A. fat catabolism
OUTCOME 25-1	_____	2. anabolism	B. substrate-level phosphorylation
OUTCOME 25-2	_____	3. glycolysis	C. urea formation
OUTCOME 25-2	_____	4. GTP from citric acid cycle	D. lipoproteins
OUTCOME 25-3	_____	5. lipolysis	E. formation of new chemical bonds
OUTCOME 25-3	_____	6. lipogenesis	F. energy release
OUTCOME 25-3	_____	7. lipid transport	G. lipid synthesis
OUTCOME 25-3	_____	8. good cholesterol	H. lipid breakdown
OUTCOME 25-3	_____	9. ketone bodies	I. anaerobic reaction sequence
OUTCOME 25-4	_____	10. amino acid catabolism	J. removal of amino group
OUTCOME 25-4	_____	11. deamination	K. HDLs

Part II

		Column A	Column B
OUTCOME 25-5	_____	12. insulin	L. glycogen reserves
OUTCOME 25-5	_____	13. gluconeogenesis	M. B complex and C
OUTCOME 25-5	_____	14. glycogenesis	N. A, D, E, K
OUTCOME 25-5	_____	15. skeletal muscle	O. absorptive state hormone
OUTCOME 25-6	_____	16. fat-soluble vitamins	P. increasing muscle tone
OUTCOME 25-6	_____	17. water-soluble vitamins	Q. glucose synthesis
OUTCOME 25-7	_____	18. basal metabolic rate	R. release of hormones; increasing metabolism
OUTCOME 25-7	_____	19. shivering thermogenesis	S. glycogen formation
OUTCOME 25-7	_____	20. non-shivering thermogenesis	T. resting energy expenditure

LEVEL 2: REVIEWING CONCEPTS

Chapter Overview

Using the terms below, fill in the blanks to complete the chapter overview of metabolism and energetics. Use each term only once.

two	carbohydrates	nutrient pool
three	catabolism	homeostasis
four	anaerobic	anabolism
eight	glycolysis	ETS (electron
twenty-four	aerobic	transport system)
thirty-six	pyruvic acid	GTP
metabolism	cytosol	FADH$_2$
NAD	cellular metabolism	

(1) _____ refers to all the chemical reactions that occur in the body. Chemical reactions within cells, collectively known as (2) _____, provide the energy needed to maintain (3) _____ and to perform essential functions. The breakdown of organic substrates is called (4) _____. The ATP produced by mitochondria provides energy to support (5) _____, the synthesis of new organic molecules. The (6) _____ is the source of the substrates for both the breakdown and synthesis of organic substrates. Most cells generate ATP and other high-energy compounds by breaking down (7) _____, especially glucose. Although most ATP production occurs inside mitochondria, the first steps take place in the (8) _____. The process of glycolysis does not require oxygen, thus the reactions are said to be (9) _____. The mitochondrial activity responsible for ATP production is called (10) _____ metabolism, or cellular respiration. (11) _____ is the breakdown of glucose to (12) _____ through a series of enzymatic steps. There is a net gain of (13) _____ ATP molecules for each glucose molecule converted to two pyruvic acid molecules. In addition, two molecules of the coenzyme (14) _____ are converted to NADH. Once transferred to mitochondria, both pyruvic acid and NADH yield more energy. Each of the two NADH molecules produced in glycolysis provides another two ATP molecules. Each of the (15) _____ NADH molecules produced in the mitochondria, yields (16) _____ ATP molecules for a total of (17) _____. Another two ATP molecules are gained from each of the two (18) _____ molecules generated in the mitochondria, yielding a total of (19) _____ ATP. The citric acid cycle generates an additional two ATP molecules in the form of (20) _____. NAD and FADH$_2$ are coenzymes that provide electrons to the (21) _____. The shuffling from the citric acid cycle to the ETS yields 28 molecules of ATP. Summing up, for each glucose molecule processed, the cell gains (22) _____ molecules of ATP: 2 from glycolysis, 4 from NADH generated in glycolysis, 2 from the citric acid cycle (by means of GTP), and 28 from the ETS.

Concept Map I

Using the following terms, fill in the circled, numbered, blank spaces to correctly complete the concept map. Use each term only once.

vitamins	metabolic regulators	vegetables
meat	tissue growth and repair	9 cal/gram
proteins	carbohydrates	

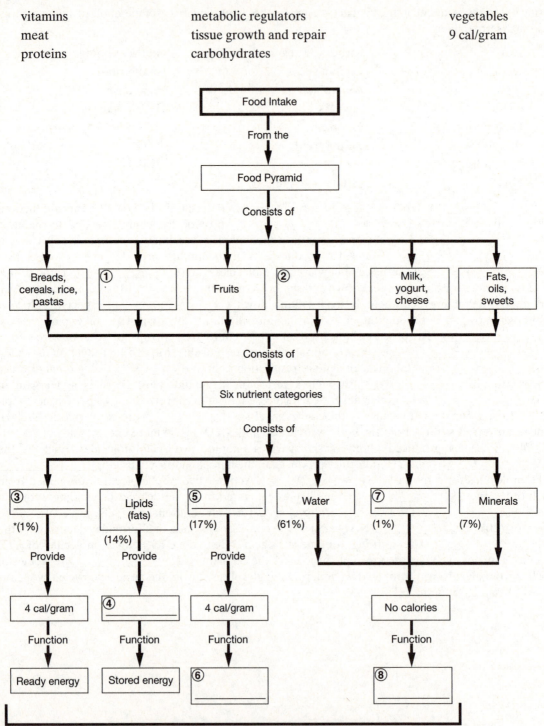

All nutrient categories are necessary – Interdependent.

*% of nutrients found in the "ideal" 25-year-old male weighing 65 kg (143 lb)

Note: For the latest information about food groups, visit www.choosemyplate.gov

Concept Map II

Using the following terms, fill in the circled, numbered, blank spaces to correctly complete the concept map. Use each term only once.

gluconeogenesis   lipogenesis   lipolysis
amino acids   beta oxidation   glycolysis

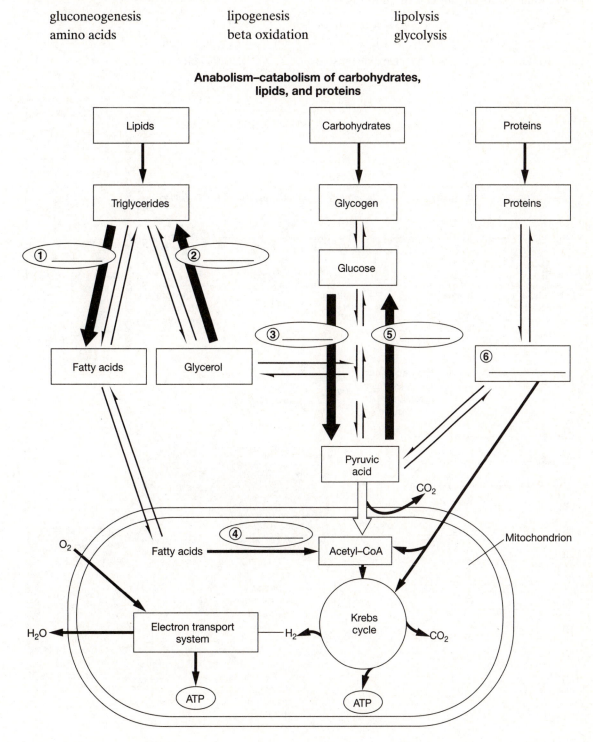

Anabolism–catabolism of carbohydrates, lipids, and proteins

Concept Map III

Using the following terms, fill in the numbered blank spaces to correctly complete the concept map.

Note: Some terms are duplicated on this map

HDL	skeletal muscle	IDL
adipose cells	VLDL	chylomicrons
LDL	cardiac muscle	liver cells

Intestinal tract → Lacteals → Thoracic duct → Circulatory system → Capillaries

Intestinal mucosa

① _____ Lacteal

Thoracic duct

Lipoprotein lipase in endothelium

② _____
③ _____
④ _____
⑤ _____

Fatty acids and monoglycerides released for use by tissue cells

Lipoprotein lipase

⑥   ⑦

LIVER CELLS

⑧

Cholesterol extracted

Excess cholesterol excreted in bile

⑩ High cholesterol

⑪ Low cholesterol

Cholesterol absorption by HDL

⑨ PERIPHERAL CELLS

Lysosomal breakdown

Used in synthesis of membranes, hormones, other materials

Cholesterol released

(1)_____

(2)_____

(3)_____

(4)_____

(5)_____

(6)_____

(7)_____

(8)_____

(9)_____

(10)_____

(11)_____

Concept Map IV

Using the following terms, fill in the circled, numbered, blank spaces to correctly complete the concept map. Use each term only once.

glycogen → glucose   fat cells   liver   glucagon
epinephrine   protein   pancreas

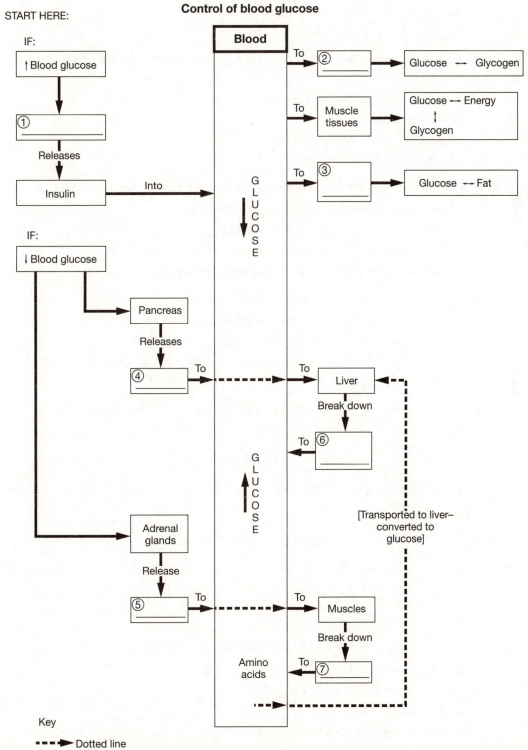

Control of blood glucose

Multiple Choice

Place the letter corresponding to the best answer in the space provided.

_____   1. The major metabolic pathways that provide most of the ATP used by typical cells are

    a. anabolism and catabolism.

    b. gluconeogenesis and glycogenesis.

    c. glycolysis and aerobic respiration.

    d. lipolysis and lipogenesis.

_____   2. A cell with excess carbohydrates, lipids, and amino acids will break down carbohydrates to

    a. provide tissue growth and repair.

    b. obtain energy.

    c. provide metabolic regulation.

    d. a, b, and c are correct.

_____   3. Fatty acids and many amino acids cannot be converted to glucose because

    a. their catabolic pathways produce acetyl-CoA.

    b. they are not used for energy production.

    c. other organic molecules cannot be converted to glucose.

    d. glucose is necessary to start the metabolic processes.

_____   4. Lipogenesis can use almost any organic substrate because

    a. lipid molecules contain carbon, hydrogen, and oxygen.

    b. triglycerides are the most abundant lipids in the body.

    c. lipids can be converted and channeled directly into the Krebs cycle.

    d. lipids, amino acids, and carbohydrates can be converted to acetyl-CoA.

_____   5. When more nitrogen is absorbed than excreted, an individual is said to be in a state of

    a. protein sparing.

    b. negative nitrogen balance.

    c. positive nitrogen balance.

    d. ketoacidosis.

_____   6. The catabolism of _lipids_ results in the release of

    a. 4.18 C/g.

    b. 4.32 C/g.

    c. 9.46 C/g.

    d. a, b, and c are incorrect.

_____   7. The catabolism of _carbohydrates_ and _proteins_ results in the release of

    a. equal amounts of C/g.

    b. 4.18 C/g and 4.32 C/g, respectively.

    c. 9.46 C/g.

    d. 4.32 C/g and 4.18 C/g, respectively.

_____ 8. The process that produces more than 90 percent of the ATP used by body cells is

    a. glycolysis.

    b. the citric acid cycle.

    c. oxidative phosphorylation.

    d. the electron transport system.

_____ 9. Examples of essential fatty acids that cannot be synthesized by the body but must be included in the diet are

    a. cholesterol, glycerol, and butyric acid.

    b. leucine, lysine, and phenylalanine acids.

    c. linoleic and linolenic acids.

    d. LDL, IDL, and HDL.

_____ 10. Most lipids circulate through the bloodstream as

    a. lipoproteins.

    b. saturated fats.

    c. unsaturated fats.

    d. polyunsaturated fats.

_____ 11. The primary function of very-low-density lipoproteins (VLDLs) is

    a. transporting cholesterol from peripheral tissues back to the liver.

    b. transporting cholesterol to peripheral tissues.

    c. carrying absorbed lipids from the intestinal tract to circulation.

    d. transporting triglycerides to peripheral tissues.

_____ 12. LDLs are absorbed by cells through the process of

    a. simple diffusion.

    b. receptor-mediated endocytosis.

    c. active transport.

    d. facilitated diffusion.

_____ 13. In the beta-oxidation of an 18-carbon fatty acid molecule, the cell gains

    a. 36 ATP.

    b. 38 ATP.

    c. 78 ATP.

    d. 144 ATP.

_____ 14. Most of the lipids absorbed by the digestive tract are immediately transferred to the

    a. liver.

    b. red blood cells.

    c. venous circulation via the left lymphatic duct.

    d. hepatocytes for storage.

_____ 15. After a meal, the absorptive state continues for about

    a. 2 hours.

    b. 4 hours.

    c. 12 hours.

    d. 24 hours.

_____ 16. When the amount of nitrogen you absorb from the diet is equal to the amount you lose in urine and feces, you are in

    a. positive nitrogen balance.

    b. negative nitrogen balance.

    c. dynamic equilibrium.

    d. nitrogen balance.

_____ 17. An important energy source during periods of starvation, when glucose supplies are limited, is

    a. glycerol.

    b. lipoproteins.

    c. cholesterol.

    d. free fatty acids.

_____ 18. Hypervitaminosis usually involves one of the fat-soluble vitamins because the excess

    a. cannot be absorbed from the digestive tract.

    b. is retained and stored in the body.

    c. is not metabolized properly.

    d. is usually converted to fat.

_____ 19. Milk and eggs are complete proteins because they contain

    a. more protein than fat.

    b. the recommended intake for vitamin B_{12}.

    c. all the essential amino acids in sufficient quantities.

    d. all the essential fatty acids and amino acids.

_____ 20. A nitrogen compound important in energy storage in muscle tissue is

    a. hemoglobin.

    b. cAMP.

    c. glycoprotein.

    d. creatine.

Completion

Using the terms below, complete the following statements. Use each term only once.

convection	nutrient pool	lipoproteins
HDLs	transamination	chylomicrons
evaporation	sensible perspiration	deamination
LDLs	insensible water loss	calorie
acclimatization	metabolic rate	BMR

1. All of the anabolic and catabolic pathways in the cell utilize a collection of organic substances known as the _____.

2. Complexes that contain large insoluble glycerides and cholesterol and a coating dominated by phospholipids and protein are called _____.

3. The largest lipoproteins in the body that are produced by the intestinal epithelial cells are the _____.

4. Lipoproteins that deliver cholesterol to peripheral tissues are _____.

5. Lipoproteins that transport excess cholesterol from peripheral tissues back to the liver are _____.

6. The amino group of an amino acid may be transferred to a keto acid by the process of _____.

7. The removal of an amino group in a reaction that generates an ammonium ion is called _____.

8. The four basic processes involved in heat exchange with the environment are radiation, conduction, convection, and _____.

9. The amount of energy required to raise the temperature of 1 gram of water 1 degree centigrade defines a _____.

10. The value of the sum total of the anabolic and catabolic processes occurring in the body is the _____.

11. The minimum resting energy expenditure of an awake, alert person is the individual's _____.

12. The result of conductive heat loss to the air that overlies the surface of the body is called _____.

13. Water losses that occur via evaporation from alveolar surfaces and the surface of the skin are referred to as _____.

14. Loss of water from the sweat glands is responsible for _____.

15. Making physiological adjustment to a particular environment over time is referred to as _____.

Short Essay

Briefly answer the following questions in the spaces provided below.

1. List examples of essential fatty acids (EFAs) that cannot be synthesized by the body.

2. List the five different classes of lipoproteins and give the relative proportions of lipids and proteins.

3. What is the difference between *transamination* and *deamination*?

4. What is the difference between an essential amino acid (EAA) and a nonessential amino acid?

5. What four factors make protein catabolism an impractical source of quick energy?

6. From a metabolic standpoint, what five distinctive components function in the human body?

7. What is the primary difference between the absorptive and postabsorptive states?

8. What are the basic food groups needed in sufficient quantity and quality to provide the basis for a balanced diet?

9. What is the meaning of the term *nitrogen balance?*

10. Even though minerals do not contain calories, why are they important in good nutrition?

11. a. List the fat-soluble vitamins.

    b. List the water-soluble vitamins.

12. What four basic processes are involved in heat exchange with the environment?

13. What two primary general mechanisms are responsible for increasing heat loss in the body?

LEVEL 3: CRITICAL THINKING AND CLINICAL APPLICATIONS

Using principles and concepts learned in Chapter 25, answer the following questions. Write your answers on a separate sheet of paper.

1. Greg S., a 32-year-old male, received a blood test assessment that reported the following values: *total cholesterol*—240 mg/d*l* (high normal); *HDL cholesterol*—21 mg/d*l* (below normal); *triglycerides*—125 mg/d*l* (normal). If the minimal normal value for HDLs is 35 mg/d*l*, what are the physiological implications of his test results?

2. Charlene S. expresses a desire to lose weight. She decides to go on a diet, selecting food from the basic food groups; however, she limits her total caloric intake to approximately 300–400 calories daily. Within a few weeks Charlene begins to experience symptoms marked by a fruity, sweetish breath odor, excessive thirst, weakness, and at times nausea accompanied by vomiting. What are the physiological implications of this low-calorie diet?

3. Ron S. offers to cook supper for the family. He decides to bake potatoes to make home fries. He puts three tablespoons of fat in a skillet to fry the cooked potatoes. How many calories does he add to the potatoes by frying them? (1 tbsp = 13 grams of fat)

4. Steve S., age 24, eats a meal that contains 40 grams of protein, 50 grams of fat, and 69 grams of carbohydrates.
   a. How many calories does Steve consume at this meal?
   b. What are his percentages of total calories from carbohydrate, fat, and protein intake?

5. Renee R. has enrolled in a weight-control course at a local community college. During the first class the instructor presents information about calorie counting and exercise. What is the significance of calorie counting and exercise in a weight-control program?

CHAPTER 26

The Urinary System

OVERVIEW

As you have learned in previous chapters, the body's cells break down the food we eat, release energy from it, and produce chemical by-products that collect in the bloodstream. The kidneys, playing a crucial homeostatic role, cleanse the blood of these waste products and excess water and ions by forming urine, which is then transferred to the urinary bladder and eventually eliminated by urination. The kidneys receive more blood from the heart than any other organ of the body. They handle 1½ quarts of blood every minute through a complex blood circulation housed in the typical adult kidney, which is about 10 cm (4 in.) long, 5.5 cm (2.2 in.) wide, 3 cm (1.2 in.) thick, and weighs about 150 g (5.25 oz).

In addition to the kidneys, which produce the urine, the urinary system consists of the ureters, urinary bladder, and urethra, components that are responsible for transport, storage, and conduction of the urine to the exterior.

The activities for Chapter 26 focus on the structural and functional organization of the urinary system, the regulatory mechanisms that control urine formation and modification, and urine transport, storage, and elimination.

LEVEL 1: REVIEWING FACTS AND TERMS

Review of Learning Outcomes

After completing this chapter, you should be able to do the following:

OUTCOME 26-1   Identify the components of the urinary system, and describe the functions it performs.

OUTCOME 26-2   Describe the location and structural features of the kidneys, identify major blood vessels associated with each kidney, trace the path of blood flow through a kidney, describe the structure of a nephron, and identify the functions of each region of the nephron and collecting system.

OUTCOME 26-3   Describe the basic processes responsible for urine production.

OUTCOME 26-4   Describe the factors that influence glomerular filtration pressure and the rate of filtrate formation.

OUTCOME 26-5   Identify the types and functions of transport mechanisms found along each segment of the nephron, explain the role of countercurrent multiplication, describe hormonal influence on the volume and concentration of urine, and describe the characteristics of a normal urine sample.

OUTCOME 26-6   Describe the structures and functions of the ureters, urinary bladder, and urethra; discuss the voluntary and involuntary regulation of urination; and describe the micturition reflex.

OUTCOME 26-7   Describe the effects of aging on the urinary system.

OUTCOME 26-8   Give examples of interactions between the urinary system and organ systems studied so far.

Multiple Choice

Place the letter corresponding to the best answer in the space provided.

OUTCOME 26-1 _____ 1. Urine leaving the kidneys travels along the following sequential pathway to the exterior.

a. Ureters, urinary bladder, urethra

b. Urethra, urinary bladder, ureters

c. Urinary bladder, ureters, urethra

d. Urinary bladder, urethra, ureters

OUTCOME 26-1 _____ 2. Which organ or structure does *not* belong to the urinary system?

a. Urethra

b. Gallbladder

c. Kidneys

d. Ureters

OUTCOME 26-2 _____ 3. Seen in section, the kidney is divided into

a. renal columns and renal pelvises.

b. an outer cortex and an inner medulla.

c. major and minor calyces.

d. a renal tubule and renal corpuscle.

OUTCOME 26-2 _____ 4. The basic functional unit in the kidney is the

a. glomerulus.

b. nephron loop.

c. Glomerular capsule.

d. nephron.

OUTCOME 26-2 _____ 5. The three concentric layers of connective tissue that protect and anchor the kidneys are the

a. hilus, renal sinus, and renal corpuscle.

b. cortex, medulla, and papillae.

c. fibrous capsule, perinephritic fat capsule, and renal fascia.

d. major calyces, minor calyces, and renal pyramids.

OUTCOME 26-2 _____ 6. Blood supply to the proximal and distal convoluted tubules of the nephron is provided by the

a. peritubular capillaries.

b. afferent arterioles.

c. segmental veins.

d. interlobular veins.

OUTCOME 26-2 _____ 7. In a nephron, the long tubular passageway through which the filtrate passes includes the

a. collecting tubule, collecting duct, and papillary duct.

b. renal corpuscle, renal tubule, and renal pelvis.

c. proximal and distal convoluted tubules and nephron loop.

d. nephron loop, collecting duct, and papillary duct.

OUTCOME 26-2 _____ 8. The primary site(s) of regulating water, sodium, and potassium ion loss in the nephron is (are) the
   a. distal convoluted tubule.
   (b.) nephron loop and collecting duct.
   c. proximal convoluted tubule.
   d. glomerulus.

OUTCOME 26-2 _____ 9. The primary site for secretion of substances into the filtrate is the
   a. renal corpuscle.
   b. nephron loop.
   (c.) distal convoluted tubule.
   d. proximal convoluted tubule.

OUTCOME 26-3 _____ 10. The initial factor that determines if urine production occurs is
   a. secretion.
   b. absorption.
   c. sympathetic activation.
   (d.) filtration.

OUTCOME 26-3 _____ 11. The three processes involved in urine formation are
   a. diffusion, osmosis, and filtration.
   b. cotransport, counter transport, and facilitated diffusion.
   c. regulation, elimination, and micturition.
   (d.) filtration, reabsorption, and secretion.

OUTCOME 26-3 _____ 12. The filtration of plasma that generates approximately 180 liters/day of filtrate occurs in the
   a. nephron loop.
   b. proximal convoluted tubule.
   (c.) renal corpuscle.
   d. distal convoluted tubule.

OUTCOME 26-3 _____ 13. Approximately 60–70 percent of the water is reabsorbed in the
   a. renal corpuscle.
   (b.) proximal convoluted tubule.
   c. distal convoluted tubule.
   d. collecting duct.

OUTCOME 26-3 _____ 14. The portion of the renal tubule that is under aldosterone stimulation is the
   a. proximal convoluted tubule.
   b. nephron loop.
   c. vasa recta.
   (d.) distal convoluted tubule.

OUTCOME 26-3 _____ 15. The proximal convoluted tubule (PCT) and the distal convoluted tubule (DCT) are separated by a U-shaped tube called the
   (a.) nephron loop.
   b. glomerulus.
   c. vasa recta.
   d. macula densa.

OUTCOME 26-3 _____ 16. The location and orientation of the carrier proteins determine whether a particular substance is reabsorbed or secreted in

    a. active transport.

    b. cotransport.

    c. countertransport.

    (d.) all of the above.

OUTCOME 26-4 _____ 17. Dilation of the afferent arteriole, contraction of mesengial cells, and constriction of the efferent arteriole causes

    (a.) elevation of glomerular blood pressure.

    b. a decrease in glomerular blood pressure.

    c. a decrease in the glomerular filtration rate.

    d. an increase in the secretion of renin and erythropoietin.

OUTCOME 26-4 _____ 18. The glomerular filtration rate is regulated by

    a. autoregulation.

    b. hormonal regulation.

    c. autonomic regulation.

    (d.) a, b, and c are correct.

OUTCOME 26-4 _____ 19. The pressure that tends to draw water out of the filtrate and into the plasma, opposing filtration, is the

    (a.) blood colloid osmotic pressure (BCOP).

    b. glomerular hydrostatic pressure (GHP).

    c. capsular hydrostatic pressure (CHP).

    d. capsular colloid osmotic pressure (CCOP).

OUTCOME 26-5 _____ 20. The three primary waste products found in a representative urine sample are

    (a.) ions, metabolites, and nitrogenous wastes.

    b. urea, creatinine, and uric acid.

    c. glucose, lipids, and proteins.

    d. sodium, potassium, and chloride.

OUTCOME 26-5 _____ 21. The average pH for normal urine is about

    a. 5.0.

    (b.) 6.0.

    c. 7.0.

    d. 8.0.

OUTCOME 26-5 _____ 22. The mechanism(s) responsible for the reabsorption of organic molecules from the tubular fluid is (are)

    a. active transport.

    b. diffusion.

    (c.) cotransport and facilitated diffusion.

    d. countertransport.

OUTCOME 26-5 _____ 23. Countertransport resembles cotransport in all respects *except*

    a. calcium ions are exchanged for sodium ions.

    b. chloride ions are exchanged for bicarbonate ions.

    (c.) the two transported ions move in opposite directions.

    d. a, b, and c are correct.

OUTCOME 26-5   _____   24. The primary site of nutrient reabsorption in the nephron is the
   a. proximal convoluted tubule.
   b. distal convoluted tubule.
   c. nephron loop.
   d. renal corpuscle.

OUTCOME 26-5   _____   25. In countercurrent multiplication, the *countercurrent* refers to the fact
   that an exchange occurs between
   a. sodium ions and chloride ions.
   b. fluids moving in opposite directions.
   c. potassium and chloride ions.
   d. solute concentrations in the nephron loop.

OUTCOME 26-5   _____   26. The result of the countercurrent multiplication mechanism is
   a. decreased solute concentration in the descending limb of the
      nephron loop.
   b. decreased transport of sodium and chloride in the ascending limb
      of the nephron loop.
   c. increased solute concentration in the descending limb of the
      nephron loop.
   d. osmotic flow of water from peritubular fluid into the descending
      limb of the nephron loop.

OUTCOME 26-5   _____   27. When antidiuretic hormone levels rise, the distal convoluted tubule
   becomes
   a. less permeable to water; reabsorption of water decreases.
   b. more permeable to water; water reabsorption increases.
   c. less permeable to water; reabsorption of water increases.
   d. more permeable to water; water reabsorption decreases.

OUTCOME 26-5   _____   28. The result of the effect of aldosterone along the DCT, the collecting
   tubule, and the collecting duct is
   a. increased conservation of sodium ions and water.
   b. increased sodium ion excretion.
   c. decreased sodium ion reabsorption in the DCT.
   d. increased sodium ion and water excretion.

OUTCOME 26-6   _____   29. The openings of the urethra and the two ureters mark an area on the
   internal surface of the urinary bladder called the
   a. internal urethral sphincter.
   b. external urethral sphincter.
   c. trigone.
   d. renal sinus.

OUTCOME 26-6   _____   30. When urine leaves the kidney, it travels to the urinary bladder via the
   a. urethra.
   b. ureters.
   c. renal hilus.
   d. renal calyces.

OUTCOME 26-6 _____ 31. The expanded, funnel-shaped upper end of the ureter in the kidney is the
    a. renal pelvis.
    b. urethra.
    c. renal hilus.
    d. renal calyces.

OUTCOME 26-6 _____ 32. Contraction of the muscular bladder forces the urine out of the body through the
    a. ureter.
    b. urethra.
    c. nephron loop.
    d. collecting duct.

OUTCOME 26-6 _____ 33. During the micturition reflex, increased afferent fiber activity in the pelvic nerves facilitates
    a. parasympathetic motor neurons in the sacral spinal cord.
    b. sympathetic sensory neurons in the sacral spinal cord.
    c. the action of stretch receptors in the wall of the bladder.
    d. urine ejection due to internal and external sphincter contractions.

OUTCOME 26-6 _____ 34. Urine reaches the urinary bladder by the
    a. action of stretch receptors in the bladder wall.
    b. fluid pressures in the renal pelvis.
    c. peristaltic contractions of the ureters.
    d. sustained contractions and relaxation of the urinary bladder.

OUTCOME 26-7 _____ 35. In males, enlargement of the prostate gland compresses the urethra and restricts the flow of urine, causing
    a. incontinence.
    b. urinary retention.
    c. a stroke.
    d. Alzheimer's disease.

OUTCOME 26-8 _____ 36. Along with the urinary system, the other systems of the body that affect the composition of body fluids are the
    a. nervous, endocrine, and cardiovascular.
    b. lymphoid, cardiovascular, and respiratory.
    c. integumentary, respiratory, and digestive.
    d. muscular, digestive, and lymphoid.

OUTCOME 26-8 _____ 37. The system that monitors distention of the urinary bladder and controls urination is the
    a. nervous system.
    b. endocrine system.
    c. cardiovascular system.
    d. lymphatic system.

Completion

Using the terms below, complete the following statements. Use each term only once.

cerebral cortex	urethra	concentration
composition	rugae	parathyroid hormone
secretion	endocrine	nephron loop
glomerulus	filtrate	internal urethral sphincter
renal threshold	neck	countertransport
glomerular filtration rate (GFR)	antidiuretic hormone	ureters
	interlobar veins	nephrons
kidneys	glomerular hydrostatic	
countercurrent multiplication	glomerular filtration	
	micturition reflex	

OUTCOME 26-1   1. The excretory functions of the urinary system are performed by the ___kidneys___.

OUTCOME 26-1   2. Urine leaving the kidneys flows through paired tubes called ___ureters___.

OUTCOME 26-1   3. The structure that carries urine from the urinary bladder to the exterior of the body is the ___urethra___.

OUTCOME 26-2   4. The renal corpuscle contains a capillary knot referred to as the ___glomerulus___.

OUTCOME 26-2   5. Urine production begins in microscopic, tubular structures called ___nephrons___.

OUTCOME 26-2   6. In a mirror image of arterial distribution, the interlobular veins deliver blood to arcuate veins that empty into ___interlobar veins___.

OUTCOME 26-2   7. The outflow across the walls of the glomerulus produces a protein-free solution known as the ___filtrate nephro___.

OUTCOME 26-3   8. The creation of the concentration gradient in the medulla occurs in the ___nephron loop___.

OUTCOME 26-3   9. The primary method of elimination for some compounds, including many drugs, is the process of ___secretion___.

OUTCOME 26-3   10. The plasma concentration at which a specific component or ion will begin appearing in the urine is called the ___renal threshold___.

OUTCOME 26-4   11. The pressure that tends to drive water and solute molecules across the glomerular wall is the ___glomerular hydrostatic___ pressure.

OUTCOME 26-4   12. The vital first step essential to all other kidney functions is _____.

OUTCOME 26-5   13. The filtration, absorption, and secretion activities of the nephrons determine the urine's _____.

OUTCOME 26-5   14. The osmotic movement of water across the walls of the tubules and collecting ducts determines the urine's _____.

OUTCOME 26-5   15. The hormone that stimulates carrier proteins to reduce the urinary loss of calcium ions is ___parathyroid___.

OUTCOME 26-5   16. Secretion of H^+ in the DCT is coupled with Na^+ reabsorption via a(n) ___countertransport___ mechanism.

OUTCOME 26-5   17. The mechanism that operates efficiently to reabsorb solutes and water before the tubular fluid reaches the DCT and collecting system is ___countercurrent___.

OUTCOME 26-5   18. Passive reabsorption of water from urine in the collecting system is regulated by circulating levels of ___antidiuretic hormone___.

OUTCOME 26-6   19. The muscular ring that provides involuntary control over the discharge of urine from the urinary bladder is the _Internal Urethral Sphincter_

OUTCOME 26-6   20. In a relaxed condition, the epithelium of the urinary bladder forms a series of prominent folds called _____.

OUTCOME 26-6   21. The area surrounding the urethral opening of the urinary bladder is the _neck_.

OUTCOME 26-6   22. The process of urination is coordinated by the _cerebral_ _Micturition reflex._

OUTCOME 26-6   23. We become consciously aware of the fluid pressure in the urinary bladder because of sensations relayed to the _cerebral cortex_.

OUTCOME 26-7   24. Age-related cumulative damage to the filtration apparatus due to fewer glomeruli results in a reduction of the _____.

OUTCOME 26-8   25. The system that adjusts rates of fluid and electrolyte reabsorption in the kidneys is the _____ system.

Matching

Match the terms in column B with the terms in column A. Use letters for answers in the spaces provided. Use each term only once.

Part I

	Column A	Column B
OUTCOME 26-1 ___	1. micturition	A. blood to kidney
OUTCOME 26-1 ___	2. urinary bladder	B. ADH stimulation
OUTCOME 26-2 ___	3. fibrous capsule	C. blood colloid osmotic pressure
OUTCOME 26-2 ___	4. renal medulla	D. glucose in urine
OUTCOME 26-2 ___	5. renal artery	E. active transport, for example
OUTCOME 26-2 ___	6. renal vein	F. urine storage
OUTCOME 26-2 ___	7. glomerular epithelium	G. outer connective tissue layer
OUTCOME 26-3 ___	8. collecting system	H. blood from kidney
OUTCOME 26-3 ___	9. carrier-mediated	I. elimination of urine
OUTCOME 26-4 ___	10. opposes filtration	J. podocytes
OUTCOME 26-5 ___	11. glycosuria	K. renal pyramids

Part II

	Column A	Column B
OUTCOME 26-5 ___	12. countertransport	L. voluntary control
OUTCOME 26-5 ___	13. cotransport processes	M. urge to urinate
OUTCOME 26-5 ___	14. nephron loop	N. sodium—hydrogen ion exchange
OUTCOME 26-5 ___	15. aldosterone	O. detrusor muscle
OUTCOME 26-6 ___	16. urinary bladder	P. incontinence
OUTCOME 26-6 ___	17. involuntary urination	Q. ion pump—Na+ channels
OUTCOME 26-6 ___	18. internal sphincter	R. countercurrent multiplication
OUTCOME 26-6 ___	19. external sphincter	S. involuntary control
OUTCOME 26-6 ___	20. 200 ml of urine in bladder	T. role of sodium ion
OUTCOME 26-7 ___	21. reduction in GFR	U. sphincter controls urination
OUTCOME 26-8 ___	22. muscular system	V. fewer glomeruli

Drawing/Illustration Labeling

Identify each numbered structure in the following figures. Place your answers in the spaces provided.

OUTCOME 26-1 **FIGURE 26-1** Anterior View of the Urinary System

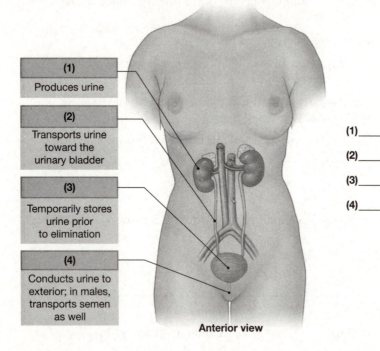

(1)
Produces urine

(2)
Transports urine toward the urinary bladder

(3)
Temporarily stores urine prior to elimination

(4)
Conducts urine to exterior; in males, transports semen as well

Anterior view

(1)_____

(2)_____

(3)_____

(4)_____

OUTCOME 26-2 **FIGURE 26-2** Sectional View of the Left Kidney

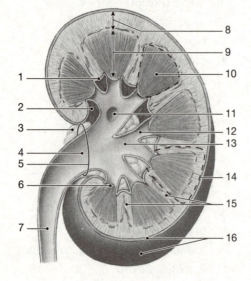

(1) INNER LAYER OF FIBROUS CAPSULE RENAL PYRAMID

(2) RENAL SINUS

(3) Adipose Tissue

(4) HILUM

(5) RENAL PELVIS

(6) RENAL PAPILLA

(7) Ureter

(8) CORTEX

(9) MEDULLA

(10) RENAL PYRAMID

(11) connection to minor

(12) minor calyx

(13) major calyx

(14) Renal lobe

(15) Renal Column

(16) outer Fibrous capsule

OUTCOME 26-3   **FIGURE 26-3**   Functional Anatomy of a Representative Nephron

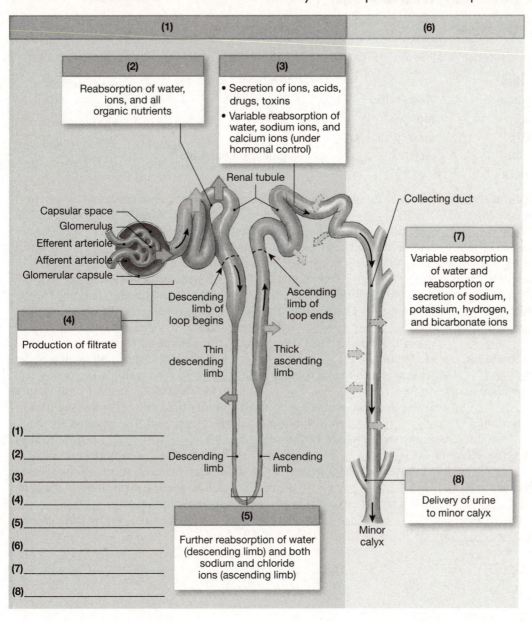

(2) Reabsorption of water, ions, and all organic nutrients

(3)
• Secretion of ions, acids, drugs, toxins
• Variable reabsorption of water, sodium ions, and calcium ions (under hormonal control)

Renal tubule

Collecting duct

Capsular space
Glomerulus
Efferent arteriole
Afferent arteriole
Glomerular capsule

(7) Variable reabsorption of water and reabsorption or secretion of sodium, potassium, hydrogen, and bicarbonate ions

(4) Production of filtrate

Descending limb of loop begins

Ascending limb of loop ends

Thin descending limb

Thick ascending limb

(1)_____

(2)_____

Descending limb

Ascending limb

(3)_____

(4)_____

(5)_____

(5) Further reabsorption of water (descending limb) and both sodium and chloride ions (ascending limb)

(6)_____

(7)_____

(8)_____

(8) Delivery of urine to minor calyx

Minor calyx

LEVEL 2: REVIEWING CONCEPTS

Chapter Overview

Using the terms below, fill in the blanks to complete the chapter overview of the urinary system.

urethra	urine	isotonic	peritubular
organic	glomerulus	water	nephron loop
out	tubular	ions	aldosterone
ADH	urinary bladder	proximal	concentrating
filtrate	distal	glomerular capsule	collecting
ureters			

Each nephron consists of a renal tubule, a long tubular passageway; a renal corpuscle, a spherical structure consisting of (1) _glomerulus capsule_ ; a cup-shaped chamber; and a capillary network known as the (2) _glomerulus_ , which consists of a tuft of capillaries. The renal corpuscle is the site where the process of filtration occurs. Filtration produces a solution, known as a (3) _filtrate_ , which is similar to blood plasma. As the solution leaves the renal corpuscle it enters the first segment of therenal tubule, the (4) _proximal_ convoluted tubule (PCT), identifiable because of its close proximity to the vascular pole of the glomerulus. As the protein-free filtrate travels along the tubule, this tubular fluid gradually changes its composition. In this first tubular segment, the active removal of (5) _ions_ and (6) _organic_ substrates produces a continuous osmotic flow of water (7) _out_ of the tubular fluid. This reduces the volume of filtrate but keeps the solutions inside and outside the tubule (8) _isotonic_ . In the PCT and descending limb of the (9) _nephron loop_ , water moves into the surrounding (10) _peritubular_ fluids, leaving a small volume of highly concentrated (11) _tubular_ fluid. This reduction occurs by obligatory (12) _water_ reabsorption. The somewhat "refined" tubular fluid enters the last segment of the nephron, the (13) _distal_ convoluted tubule, which is specifically adapted for reabsorption of sodium ions due to the influence of the hormone (14) _Aldosterone_ , and in its most distal part, the effect of (15) _ADH_ , causing an osmotic flow of water that assists in (16) _concentrating_ the filtrate. Final adjustments to the sodium ion concentration and the volume of the fluid are made in the (17) _collecting_ ducts, which deliver the waste products in the form of (18) _urine_ to the renal pelvis. It will then be conducted into the (19) _Ureters_ on its way to the (20) _bladder_ for storage until it leaves the body through the (21) _urethra_ .

Concept Map I

Using the following terms, fill in the circled, numbered, blank spaces to correctly complete the concept map. Use each term only once.

ureter   renal sinus   glomerulus
urinary bladder   renal tubules   papillary duct
major calyces   renal (inner) medulla proximal convoluted tubule
nephrons

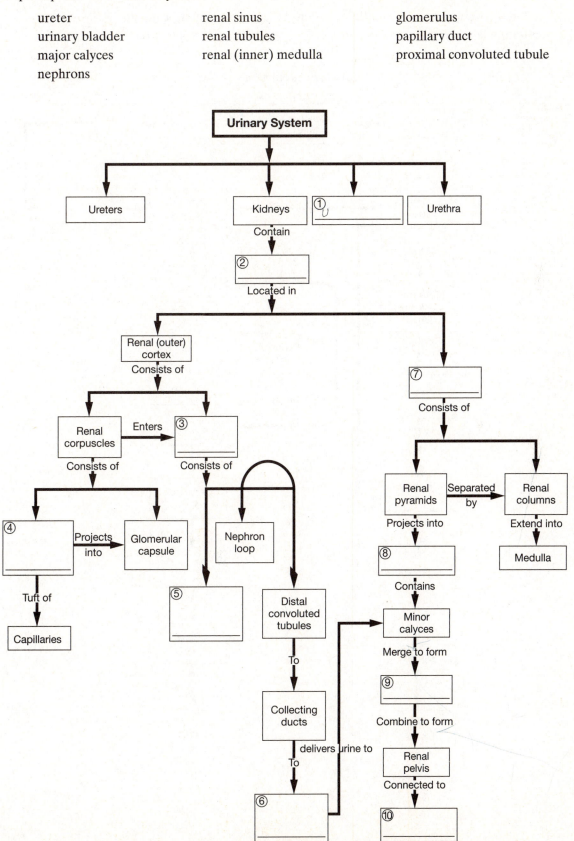

Concept Map II

Using the following terms, fill in the circled, numbered, blank spaces to correctly complete the concept map. Use each term only once.

efferent arteriole   afferent arteriole   interlobar vein
interlobular vein   renal artery   arcuate artery

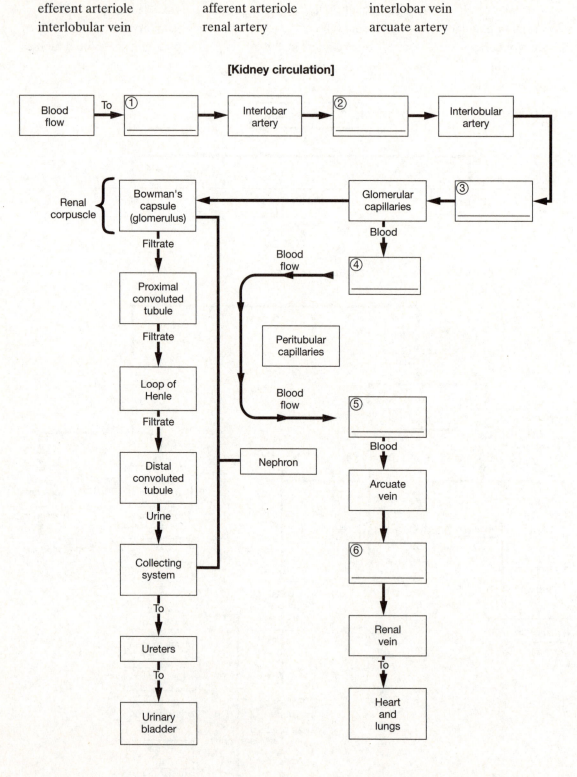

[Kidney circulation]

Concept Map III

Using the following terms, fill in the circled, numbered, blank spaces to correctly complete the concept map. Use each term only once.

adrenal cortex $\qquad$ $\downarrow Na^+$ excretion $\qquad$ $\downarrow H_2O$ excretion

liver $\qquad$ $\downarrow$ plasma volume $\qquad$ Na^+ reabsorption

angiotensin I $\qquad$ rennin

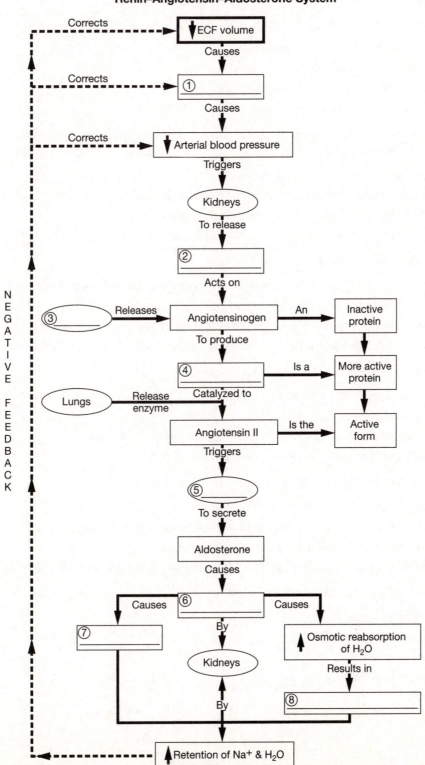

Renin–Angiotensin–Aldosterone System

Corrects ----→ ↓ECF volume

Causes

Corrects ----→ ① _____

Causes

Corrects ----→ ↓ Arterial blood pressure

Triggers

Kidneys

To release

② _____

Acts on

③ _____ Releases → Angiotensinogen — An → Inactive protein

To produce

④ _____ Is a → More active protein

Lungs — Release enzyme → Catalyzed to

Angiotensin II — Is the → Active form

Triggers

⑤ _____

To secrete

Aldosterone

Causes

Causes — ⑥ _____ — Causes

⑦ _____ By ↑ Osmotic reabsorption of H_2O

Kidneys Results in

⑧ _____

By

↑ Retention of Na^+ & H_2O

NEGATIVE FEEDBACK

Multiple Choice

Place the letter corresponding to the best answer in the space provided.

_____ 1. The vital function(s) performed by the nephrons in the kidneys is (are)

    a. production of filtrate.

    b. reabsorption of organic substrates.

    c. reabsorption of water and ions.

    d. a, b, and c are correct.

_____ 2. The renal corpuscle consists of the

    a. renal columns and renal pyramids.

    b. major and minor calyces.

    c. glomerular capsule and the glomerulus.

    d. cortex and the medulla.

_____ 3. The filtration process within the renal corpuscle involves passage across three physical barriers, the

    a. podocytes, pedicels, and slit pores.

    b. capillary endothelium, dense layer, visceral epithelium.

    c. capsular space, tubular pole, and macula densa.

    d. collecting tubules, collecting ducts, and papillary ducts.

_____ 4. The thin segments in the nephron loop are

    a. relatively impermeable to water, but freely permeable to ions and other solutes.

    b. freely permeable to water, ions, and other solutes.

    c. relatively impermeable to water, ions, and other solutes.

    d. freely permeable to water, but relatively impermeable to ions and other solutes.

_____ 5. The thick segments in the nephron loop contain

    a. ADH-regulated permeability.

    b. an aldosterone-regulated pump.

    c. transport mechanisms that pump materials out of the filtrate.

    d. diffusion mechanisms for getting rid of excess water.

_____ 6. The collecting system in the kidney is responsible for

    a. active secretion and reabsorption of sodium ions.

    b. absorption of nutrients, plasma proteins, and ions from the filtrate.

    c. creation of the medullary concentration gradient.

    d. making final adjustments to the sodium ion concentration and volume of urine.

_____ 7. Blood arrives at the renal corpuscle by way of a(n)

    a. efferent arteriole.

    b. afferent arteriole.

    c. renal artery.

    d. arcuate arteries.

_____ 8. Sympathetic innervation into the kidney is responsible for

    a. decreasing the GFR and slowing the production of filtrate.

    b. stimulation of renin release.

    c. altering the GFR by changing the regional pattern of blood circulation.

    d. a, b, and c are correct.

_____ 9. When plasma glucose concentrations are higher than the renal threshold, glucose concentrations in the filtrate exceed the tubular maximum (T_m) and

    a. glucose is transported across the membrane by countertransport.

    b. the glucose is filtered out at the glomerulus.

    c. glucose appears in the urine.

    d. the individual has eaten excessive amounts of sweets.

_____ 10. The outward pressure forcing water and solute molecules across the glomerulus wall is the _____ pressure.

    a. capsular hydrostatic

    b. glomerular hydrostatic

    c. blood osmotic

    d. filtration

_____ 11. The opposing forces of the filtration pressure at the glomerulus are the

    a. glomerular hydrostatic pressure and blood osmotic pressure.

    b. capsular hydrostatic pressure and glomerular hydrostatic pressure.

    c. blood pressure and glomerular filtration rate.

    d. net hydrostatic pressure and colloid osmotic pressure.

_____ 12. The amount of filtrate produced in the kidneys each minute is the

    a. glomerular filtration rate.

    b. glomerular hydrostatic pressure.

    c. capsular hydrostatic pressure.

    d. osmotic gradient.

_____ 13. Inadequate ADH secretion results in the inability to reclaim the water entering the filtrate, causing

    a. glycosuria.

    b. dehydration.

    c. anuria.

    d. dysuria.

_____ 14. Under normal circumstances, virtually all the glucose, amino acids, and other nutrients are reabsorbed before the filtrate leaves the

    a. distal convoluted tubule.

    b. glomerulus.

    c. collecting ducts.

    d. proximal convoluted tubule.

_____ 15. Aldosterone stimulates ion pumps along the distal convoluted tubule (DCT), the collecting tubule, and the collecting duct, causing a(n)

    a. increase in the number of sodium ions lost in the urine.

    b. decrease in the concentration of the filtrate.

    c. reduction in the number of sodium ions lost in the urine.

    d. countercurrent multiplication.

_____ 16. The high osmotic concentrations found in the kidney medulla are primarily due to

    a. the presence of sodium ions, chloride ions, and urea.

    b. the presence of excessive amounts of water.

    c. hydrogen and ammonium ions.

    d. a, b, and c are correct.

_____ 17. The substances that influence the glomerular filtration rate (GFR) by regulating blood pressure and volume are

    a. PTH, epinephrine, and oxytocin.

    b. insulin, glucagon, and glucocorticoids.

    c. renin, ANP, and ADH.

    d. a, b, and c are correct.

_____ 18. Angiotensin II is a potent hormone that

    a. causes constriction of the efferent arteriole at the nephron.

    b. triggers the release of ADH in the CNS.

    c. stimulates secretion of aldosterone by the adrenal cortex and epinephrine by the adrenal medulla.

    d. a, b, and c are correct.

_____ 19. Sympathetic innervation of the afferent arterioles causes a(n)

    a. decrease in GFR and decrease of filtrate production.

    b. increase in GFR and an increase in filtrate production.

    c. decrease in GFR and an increase in filtrate production.

    d. increase in GFR and a decrease of filtrate production.

_____ 20. During periods of strenuous exercise, sympathetic activation causes the blood flow to

    a. decrease to the skin and skeletal muscles, but increase to the kidneys.

    b. cause an increase in GFR.

    c. increase to the skin and skeletal muscles, but decrease to the kidneys.

    d. be shunted toward the kidneys.

Completion

Using the terms below, complete the following statements. Use each term only once.

aldosterone	vasa recta	angiotensin II
filtration	diabetes insipidus	osmotic gradient
macula densa	renin	glomerular filtration rate
secretion	transport maximum	cortical
retroperitoneal	reabsorption	glomerular filtration

1. The kidneys lie between the muscles of the dorsal body wall and the peritoneal lining in a position referred to as _____.

2. The amount of filtrate produced in the kidneys each minute is the _____.

3. The process responsible for concentrating the filtrate as it travels through the collecting system toward the renal pelvis is the _____.

4. When all available carrier proteins are occupied at one time, the saturation concentration is called the _____.

5. The region in the distal convoluted tubule (DCT) where the tubule cells adjacent to the tubular pole are taller and contain clustered nuclei is called the _____.

6. Approximately 85 percent of the nephrons in a kidney are called _____ nephrons.

7. A capillary that accompanies the nephron loop into the medulla is termed the _____.

8. When hydrostatic pressure forces water across a membrane, the process is referred to as _____.

9. The removal of water and solute molecules from the filtrate is termed _____.

10. Transport of solutes across the tubular epithelium and into the filtrate is _____.

11. The vital first step essential to all kidney function is _____.

12. The ion pump and the sodium ion channels are controlled by the hormone _____.

13. When a person produces large quantities of dilute urine, the condition is termed _____.

14. The hormone that causes a brief but powerful vasoconstriction in peripheral capillary beds is _____.

15. Sympathetic activation stimulates the juxtaglomerular apparatus to release _____.

Short Essay

Briefly answer the following questions in the spaces provided below.

1. What are six essential functions of the urinary system?

2. Trace the pathway of a drop of urine from the time it leaves the kidney until it is urinated from the body. (Use arrows to show direction.)

3. What three concentric layers of connective tissue protect and anchor the kidneys?

4. Trace the pathway of a drop of filtrate from the time it goes through the filtration membrane in the glomerulus until it enters the collecting tubules as concentrated urine. (Use arrows to indicate direction of flow.)

5. What vital functions of the kidneys are performed by the nephrons?

6. a. What three physical barriers are involved with passage of materials during the filtration process?

   b. Cite one structural characteristic of each barrier.

7. What two substances are secreted by the juxtaglomerular apparatus?

8. What effects result from sympathetic innervation into the kidneys?

9. What are the three processes involved in urine formation?

10. Write a formula to show the following relationship and define each component of the formula: The *filtration pressure* (FP) at the glomerulus is the difference between the glomerular hydrostatic pressure and the opposing capsular and osmotic pressures.

11. Why does hydrogen ion secretion accelerate during starvation?

12. What three basic concepts describe the mechanism of countercurrent multiplication?

13. What three control mechanisms are involved with regulation of the glomerular filtration rate (GFR)?

14. What four hormones affect urine production? Describe the role of each one.

LEVEL 3: CRITICAL THINKING AND CLINICAL APPLICATIONS

Using principles and concepts learned in Chapter 26, answer the following questions. Write your answers on a separate sheet of paper.

1. David spends an evening with the boys at the local tavern. During the course of the night he makes numerous trips to the restroom to urinate. How does the "diuretic effect" of alcohol affect the urinary system?

2. Lori has had a series of laboratory tests, including a CBC, lipid profile series, and a urinalysis. The urinalysis revealed the presence of an abnormal amount of plasma proteins and white blood cells. (a) What is your diagnosis? (b) What effect does her condition have on her urine output?

3. On an anatomy and physiology examination your instructor asks the following question:

   What is the *principal function* associated with each of the following components of the nephron?

   a. renal corpuscle

   b. proximal convoluted tubule

   c. distal convoluted tubule

   d. nephron loop and collecting system

4. a. Given the following information, determine the net filtration pressure (NFP) at the glomerulus:

glomerular hydrostatic pressure	GHP	65 mm Hg
capsular hydrostatic pressure	CsHP	18 mm Hg
blood colloid osmotic pressure	BCOP	32 mm Hg

   b. How does the net filtration pressure affect the GFR?

   c. What are the implications of a net filtration pressure (NFP) of 15 mm Hg?

5. Cindy F. is a marathon runner who, from time to time, experiences acute kidney dysfunction during the event. Laboratory tests have shown elevated serum concentrations of K^+, lowered serum concentrations of Na^+, and a decrease in the GFR. At one time she experienced renal failure. Explain why these symptoms and conditions may occur while running a marathon.

Fluid, Electrolyte, and Acid–Base Balance

OVERVIEW

In many of the previous chapters we examined the various organ systems, focusing on the structural components and the functional activities necessary to *support* life. This chapter details the roles the organ systems play in *maintaining* life in the billions of individual cells in the body. Early in the study of anatomy and physiology we established that the cell is the basic unit of structure and function of all living things, no matter how complex the organism is as a whole. The last few chapters concentrated on the necessity of meeting the nutrient needs of cells and getting rid of waste products that might interfere with normal cell functions. To meet the nutrient needs of cells, substances must move in; and to get rid of wastes, materials must move out. The result is a constant movement of materials into and out of the cells.

Both the external environment (interstitial fluid) and the internal environment of the cell (intracellular fluid) comprise an "exchange system" operating within controlled and changing surroundings, producing a dynamic equilibrium by which homeostasis is maintained.

Chapter 27 considers the mechanics and dynamics of fluid balance, electrolyte balance, acid–base balance, and the interrelationships of functional patterns that operate in the body to support and maintain a constant state of equilibrium.

LEVEL 1: REVIEWING FACTS AND TERMS

Review of Learning Outcomes

After completing this chapter, you should be able to do the following:

OUTCOME 27-1   Explain what is meant by the terms *fluid balance*, *electrolyte balance*, and *acid–base balance*, and discuss their importance for homeostasis.

OUTCOME 27-2   Compare the composition of intracellular and extracellular fluids, explain the basic concepts involved in the regulation of fluids and electrolytes, and identify the hormones that play important roles in fluid and electrolyte regulation.

OUTCOME 27-3   Describe the movement of fluid within the ECF, between the ECF and the ICF, and between the ECF and the environment.

OUTCOME 27-4   Discuss the mechanisms by which sodium, potassium, calcium, and chloride ion concentrations are regulated to maintain electrolyte balance.

OUTCOME 27-5   Explain the buffering systems that balance the pH of the intracellular and extracellular fluids, and describe the compensatory mechanisms involved in the maintenance of acid–base balance.

OUTCOME 27-6   Identify the most frequent disturbances of acid–base balance, and explain how the body responds when the pH of body fluids varies outside normal limits.

OUTCOME 27-7   Describe the effects of aging on fluid, electrolyte, and acid–base balance.

Multiple Choice

Place the letter corresponding to the best answer in the space provided.

OUTCOME 27-1   _____   1. When the amount of water you gain each day is equal to the amount you lose to the environment, you are in
   a. electrolyte balance.
   b. fluid balance.
   c. acid–base balance.
   d. dynamic equilibrium.

OUTCOME 27-1   _____   2. When the production of hydrogen ions in your body is precisely offset by their loss, you are in
   a. electrolyte balance.
   b. osmotic equilibrium.
   c. fluid balance.
   d. acid–base balance.

OUTCOME 27-1   _____   3. Electrolyte balance primarily involves balancing the rates of absorption across the digestive tract with rates of loss at the
   a. heart and lungs.
   b. kidneys and sweat glands.
   c. stomach and liver.
   d. pancreas and gallbladder.

OUTCOME 27-2   _____   4. Clinically, approximately two-thirds of the total body water content is
   a. extracellular fluid (ECF).
   b. intracellular fluid (ICF).
   c. tissue fluid.
   d. interstitial fluid (IF).

OUTCOME 27-2   _____   5. Extracellular fluids in the body consist of
   a. interstitial fluid, blood plasma, and lymph.
   b. cerebrospinal fluid, synovial fluid, and serous fluids.
   c. aqueous humor, perilymph, and endolymph.
   d. a, b, and c are correct.

OUTCOME 27-2   _____   6. The principal ions in the extracellular fluid (ECF) are
   a. sodium, chloride, and bicarbonate.
   b. potassium, magnesium, and phosphate.
   c. phosphate, sulfate, and magnesium.
   d. potassium, ammonium, and chloride.

OUTCOME 27-2 _____ 7. Physiological adjustments affecting fluid and electrolyte balance are mediated primarily by
   a. antidiuretic hormone (ADH).
   b. aldosterone.
   c. natriuretic peptides (ANP and BNP).
   d. a, b, and c are correct.

OUTCOME 27-2 _____ 8. The two important effects of increased release of ADH are
   a. increased rate of sodium absorption and decreased thirst.
   b. reduction of urinary water losses and stimulation of the thirst center.
   c. decrease in the plasma volume and elimination of the source of stimulation.
   d. decrease in plasma osmolarity and alteration of the composition of tissue fluid.

OUTCOME 27-2 _____ 9. Secretion of aldosterone occurs in response to
   a. a drop in plasma volume at the juxtaglomerular apparatus.
   b. a decline in filtrate osmotic concentration at the DCT.
   c. high potassium ion concentrations.
   d. a, b, and c are correct.

OUTCOME 27-2 _____ 10. Atrial natriuretic peptide hormone
   a. reduces thirst.
   b. blocks the release of ADH.
   c. blocks the release of aldosterone.
   d. a, b, and c are correct.

OUTCOME 27-2 _____ 11. The principal ions in the ICF are
   a. magnesium, phosphate, and sodium.
   b. sodium, potassium, and magnesium.
   c. potassium, magnesium, and phosphate.
   d. sodium, chloride, and bicarbonate.

OUTCOME 27-3 _____ 12. The force that tends to push water out of the plasma and into the interstitial fluid is the
   a. net hydrostatic pressure.
   b. colloid osmotic pressure.
   c. total peripheral resistance.
   d. mean arterial pressure.

OUTCOME 27-3 _____ 13. The exchange between plasma and interstitial fluid is determined by the relationship between the
   a. total peripheral resistance and mean arterial pressure.
   b. fluid balance and acid–base balance.
   c. net hydrostatic and net colloid osmotic pressures.
   d. none of the above.

OUTCOME 27-3   _____ 14. If the ECF is *hypertonic* with respect to the ICF, water will move
    a. from the ECF into the cell until osmotic equilibrium is restored.
    b. from the cells into the ECF until osmotic equilibrium is restored.
    c. in both directions until osmotic equilibrium is restored.
    d. in response to the pressure of carrier molecules.

OUTCOME 27-3   _____ 15. When water is lost but electrolytes are retained, the osmolarity of the ECF rises and osmosis then moves water
    a. out of the ECF and into the ICF until isotonicity is reached.
    b. back and forth between the ICF and the ECF.
    c. out of the ICF and into the ECF until isotonicity is reached.
    d. directly into the blood plasma until equilibrium is reached.

OUTCOME 27-3   _____ 16. When pure water is consumed, the extracellular fluid becomes
    a. hypotonic with respect to the ICF.
    b. hypertonic with respect to the ICF.
    c. isotonic with respect to the ICF.
    d. in equilibrium with the ICF.

OUTCOME 27-4   _____ 17. The concentration of potassium in the ECF is controlled by adjustments in the rate of active secretion
    a. in the proximal convoluted tubule of the nephron.
    b. in the nephron loop.
    c. along the distal convoluted tubule and collecting system of the nephron.
    d. along the collecting tubules.

OUTCOME 27-4   _____ 18. The activity that occurs in the body to maintain calcium homeostasis occurs primarily in the
    a. bone.
    b. digestive tract.
    c. kidneys.
    d. a, b, and c are correct.

OUTCOME 27-5   _____ 19. The hemoglobin buffer system helps prevent drastic alterations in pH when
    a. the plasma P_{CO_2} is rising or falling.
    b. there is an increase in hemoglobin production.
    c. there is a decrease in RBC production.
    d. the plasma P_{CO_2} is constant.

OUTCOME 27-5   _____ 20. The primary role of the carbonic acid–bicarbonate buffer system is to prevent pH changes caused by
    a. Na^+ movement.
    b. organic acid and fixed acids in the ECF.
    c. the rising and falling of the plasma P_{CO_2}.
    d. a, b, and c are correct.

OUTCOME 27-5 _____ 21. Pulmonary and renal mechanisms support the buffer systems by
  a. secreting or generating hydrogen ions.
  b. controlling the excretion of acids and bases.
  c. generating additional buffers when necessary.
  d. a, b, and c are correct.

OUTCOME 27-5 _____ 22. The lungs contribute to pH regulation by their effects on the
  a. hemoglobin buffer system.
  b. phosphate buffer system.
  c. carbonic acid–bicarbonate buffer system.
  d. protein buffer system.

OUTCOME 27-5 _____ 23. Increasing or decreasing the rate of respiration can have a profound effect on the buffering capacity of body fluids by
  a. lowering or raising the P_{O_2}.
  b. lowering or raising the P_{CO_2}.
  c. increasing the production of lactic acid.
  d. a, b, and c are correct.

OUTCOME 27-5 _____ 24. Examples of mechanisms involved in the renal response to acidosis include
  a. reabsorption of H^+ and secretion of HCO_3^-.
  b. reabsorption of H^+ and HCO_3^-.
  c. secretion of H^+ and reabsorption of HCO_3^-.
  d. secretion of H^+ and HCO_3^-.

OUTCOME 27-5 _____ 25. When carbon dioxide concentrations rise, additional hydrogen ions are produced and the pH
  a. goes up.
  b. goes down.
  c. remains the same.
  d. is not affected.

OUTCOME 27-6 _____ 26. Disorders that have the potential for disrupting pH balance in the body include
  a. emphysema and renal failure.
  b. neural damage and CNS disease.
  c. heart failure and hypotension.
  d. a, b, and c are correct.

OUTCOME 27-6 _____ 27. Respiratory alkalosis develops when
  a. respiratory activity raises plasma P_{CO_2} to above-normal levels.
  b. respiratory activity lowers plasma P_{CO_2} to below-normal levels.
  c. respiratory activity decreases plasma P_{CO_2} to below-normal levels.
  d. P_{CO_2} levels are not affected.

OUTCOME 27-6   _____ 28. The most frequent cause of metabolic acidosis is

    a. production of a large number of fixed or organic acids.

    b. a severe bicarbonate loss.

    c. an impaired ability to excrete hydrogen ions at the kidneys.

    d. generation of large quantities of ketone bodies.

OUTCOME 27-6   _____ 29. A mismatch between carbon dioxide generation in peripheral tissues and carbon dioxide excretion at the lungs is a

    a. metabolic acid–base disorder.

    b. condition known as ketoacidosis.

    c. respiratory acid–base disorder.

    d. severe bicarbonate loss.

OUTCOME 27-6   _____ 30. The major cause(s) of metabolic acidosis is (are)

    a. production of a large number of fixed or organic acids.

    b. impaired ability to excrete H^+ at the kidneys.

    c. a severe bicarbonate loss.

    d. a, b, and c are correct.

OUTCOME 27-6   _____ 31. The most important factor affecting the pH in body tissues is

    a. the P_{CO_2}.

    b. changes in K^+ concentrations.

    c. calcium ion concentrations in the ECF.

    d. a, b, and c are correct.

OUTCOME 27-7   _____ 32. As a result of the aging process, the ability to regulate pH through renal compensation declines due to

    a. a reduction in the number of functional nephrons.

    b. increased glomerular filtration.

    c. increased ability to concentrate urine.

    d. a reduction in the rate of insensible perspiration.

OUTCOME 27-7   _____ 33. The risk of respiratory acidosis in the elderly is increased due to

    a. a reduction in the number of nephrons.

    b. a reduction in vital capacity.

    c. increased insensible perspiration.

    d. a decrease in ADH and aldosterone sensitivity.

Completion

Using the terms below, complete the following statements. Use each term only once.

respiratory compensation	hypercapnia	hemoglobin	calcium
osmoreceptors	fluid	colloid osmotic pressure	edema
acidosis	hypotonic	skeletal mass	net hydrostatic pressure
aldosterone	hypertonic	alkalosis	
antidiuretic hormone	renal compensation	fluid shift	
buffers	kidneys	electrolyte	

OUTCOME 27-1   1. When there is neither a net gain nor a net loss of any ion in the body fluid, an _____ balance exists.

OUTCOME 27-1   2. When the amount of water gained each day is equal to the amount lost to the environment, a person is in _____ balance.

OUTCOME 27-2   3. Osmotic concentrations of the plasma are monitored by special cells in the hypothalamus called _____.

OUTCOME 27-2   4. The rate of sodium absorption along the DCT (distal convoluted tubule) and collecting system of the kidneys is regulated by the hormone _____.

OUTCOME 27-2   5. The hormone that stimulates water conservation at the kidneys and the thirst center to promote the drinking of fluids is _____.

OUTCOME 27-3   6. Water movement between the ECF and the ICF is termed a(n) _____.

OUTCOME 27-3   7. If the ECF becomes more concentrated with respect to the ICF, the ECF is _____.

OUTCOME 27-3   8. If the ECF becomes more dilute with respect to the ICF, the ECF is _____.

OUTCOME 27-3   9. The movement of abnormal amounts of water from plasma into interstitial fluid is called _____.

OUTCOME 27-3   10. The force that tends to push water out of the plasma and into the interstitial fluid is the _____.

OUTCOME 27-3   11. The force that tends to draw water out of the interstitial fluid and into the plasma is the net _____.

OUTCOME 27-4   12. The most important site of sodium ion regulation is the _____.

OUTCOME 27-4   13. Calcitonin from the C cells of the thyroid gland promotes a loss of _____.

OUTCOME 27-5   14. Dissolved compounds that can provide or remove hydrogen ions and thereby stabilize the pH of a solution are _____.

OUTCOME 27-5   15. The only intracellular buffer system that can have an immediate effect on the pH of the ECF is the _____ buffer system.

OUTCOME 27-5   16. A change in the respiratory rate that helps stabilize pH is called _____.

OUTCOME 27-5   17. A change in the rates of hydrogen ion and bicarbonate ion secretion or absorption in response to changes in plasma pH is called _____.

OUTCOME 27-6   18. When the pH in the body falls below 7.35, the condition is called _____.

OUTCOME 27-6   19. When the pH in the body increases above 7.45, the condition is called _____.

OUTCOME 27-6   20. A low plasma pH due to an elevated plasma P_{CO_2} is called _____.

OUTCOME 27-7   21. Many people over age 60 experience a net loss in their body mineral content due to decreased muscle mass and _____.

Matching

Match the terms in column B with the terms in column A. Use letters for answers in the spaces provided. Use each term only once.

Part I

		Column A	Column B
OUTCOME 27-1	_____	1. fluid balance	A. H_2O moves cells into ECF
OUTCOME 27-2	_____	2. tissue fluid	B. dominant cation—ECF
OUTCOME 27-2	_____	3. fluid compartments	C. concentration of dissolved solutes
OUTCOME 27-2	_____	4. potassium	D. dominant cation—ICF
OUTCOME 27-2	_____	5. sodium	E. ICF and ECF
OUTCOME 27-2	_____	6. key components of ECF	F. posterior pituitary gland
OUTCOME 27-2	_____	7. osmolarity	G. hyponatremia
OUTCOME 27-2	_____	8. cardiac muscle fiber	H. plasma, interstitial fluid
OUTCOME 27-2	_____	9. antidiuretic hormone	I. water gain = water loss
OUTCOME 27-3	_____	10. hypertonic plasma	J. atrial natriuretic peptide
OUTCOME 27-3	_____	11. overhydration	K. interstitial fluid

Part II

		Column A	Column B
OUTCOME 27-4	_____	12. ADH secretion decreases	L. buffers pH of ICF
OUTCOME 27-4	_____	13. increased atrial pressure	M. hyperventilation
OUTCOME 27-5	_____	14. phosphate buffer system	N. buffers pH of ECF
OUTCOME 27-5	_____	15. carbonic acid–bicarbonate buffer	O. respiratory acidosis
OUTCOME 27-5	_____	16. balancing H^+ gains/losses	P. water loss at kidneys increases
OUTCOME 27-5	_____	17. decreased P_{CO_2}	Q. release of ANP
OUTCOME 27-6	_____	18. starvation	R. pH increases
OUTCOME 27-6	_____	19. hypocapnia	S. ↓ renal compensation
OUTCOME 27-6	_____	20. hypercapnia	T. ketoacidosis
OUTCOME 27-7	_____	21. aging effect	U. maintenance of acid–base balance

Drawing/Illustration Labeling

Using the terms and figure below, fill in each numbered box to correctly complete the information regarding pH. Use each term only once.

alkalosis   pH 6.80   pH 7.80
acidosis   pH 7.35   pH 7.45

FIGURE 27-1   The pH Scale

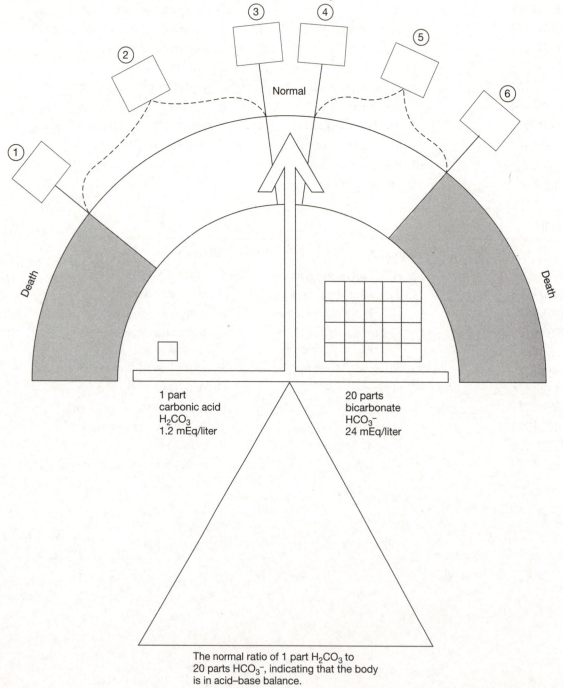

The normal ratio of 1 part H_2CO_3 to 20 parts HCO_3^-, indicating that the body is in acid–base balance.

LEVEL 2: REVIEWING CONCEPTS

Chapter Overview

Using the terms below, fill in the blanks to correctly complete the chapter overview of respiratory acid–base regulation. Use each term only once.

(A)
decreased
acidosis
H^+; HCO_3^-
increased
increased P_{CO_2}

(B)
P_{CO_2}
alkalosis
respiratory rate
reabsorption; secretion
decreased P_{CO_2}

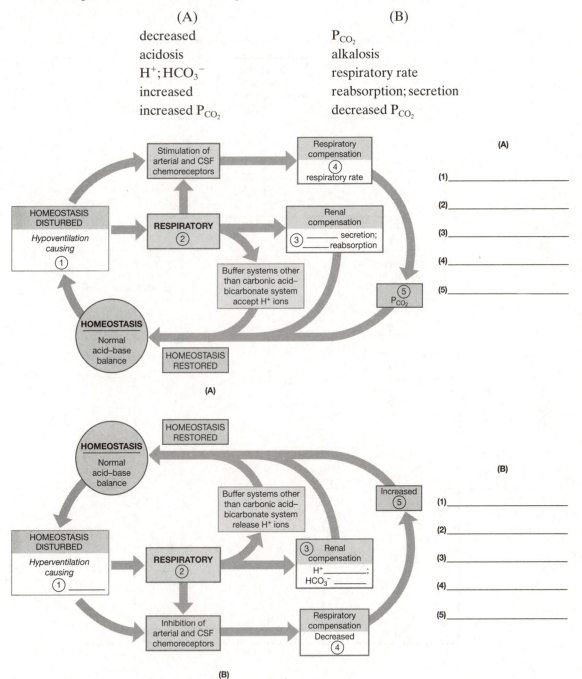

(A)
(1) _____
(2) _____
(3) _____
(4) _____
(5) _____

(B)
(1) _____
(2) _____
(3) _____
(4) _____
(5) _____

Concept Map I

Using the following terms, fill in the circled, numbered, blank spaces to correctly complete the concept map. Use each term only once.

↑ H_2O retention at kidneys   ↓ volume of body H_2O

aldosterone secretion by adrenal cortex

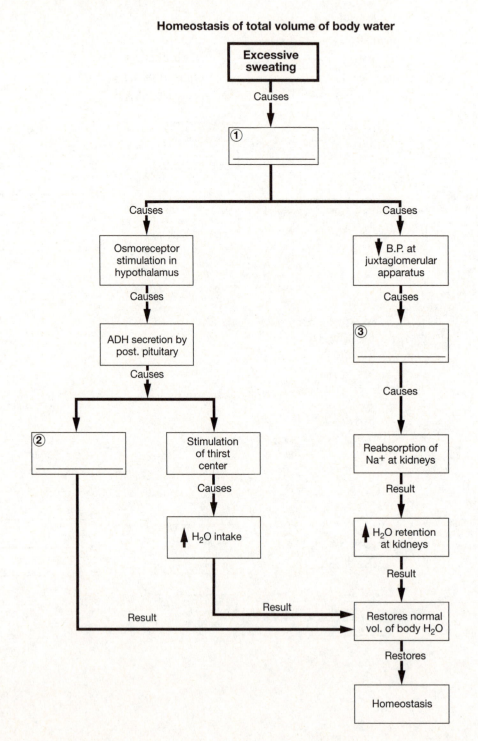

Concept Map II

Using the following terms, fill in the circled, numbered, blank spaces to correctly complete the concept map. Use each term only once.

↓ ECF volume

↑ ICF volume

↓ pH

ECF hypotonic to ICF

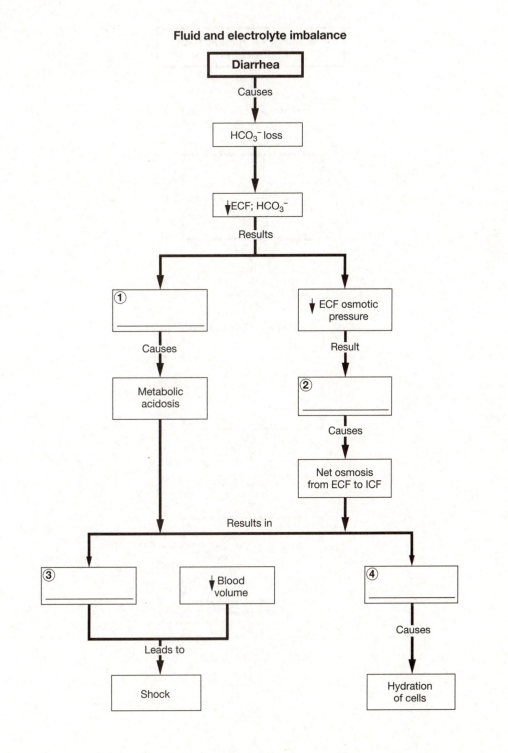

Fluid and electrolyte imbalance

Concept Map III

Using the following terms, fill in the circled, numbered, blank spaces to correctly complete the concept map. Use each term only once.

↑ blood pH   ↑ depth of breathing   ↑ blood CO₂
hyperventilation   normal blood pH

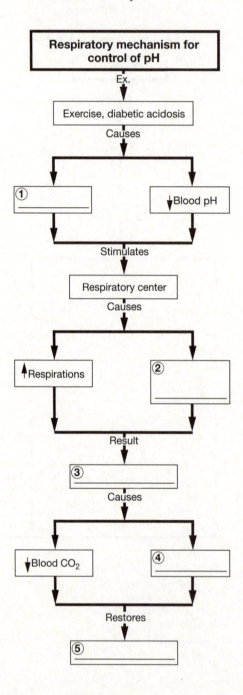

Concept Map IV

Using the following terms, fill in the circled, numbered, blank spaces to correctly complete the concept map. Use each term only once.

HCO_3^- ↑ blood pH   ↓ blood pH

```
          ┌─────────────────────────┐
          │  Urinary mechanisms for │
          │ maintaining homeostasis │
          │      of blood pH        │
          └─────────────────────────┘
                      │ Ex.
                      ▼
            ┌───────────────────┐
            │ Diabetic ketosis  │
            └───────────────────┘
                   Causes
                      ▼
            ┌───────────────────┐
            │ ① _____│
            └───────────────────┘
                  Stimulates
                      ▼
            ┌───────────────────┐
            │   Renal tubules   │
            └───────────────────┘
                      To
                      │
          Secrete          Reabsorb
             ▼                ▼
     ┌──────────────┐  ┌──────────────┐
     │ H⁺ and NH₃   │  │② _____│
     └──────────────┘  └──────────────┘
             │                │
              Causes
                 ▼
     ┌──────────────┐  ┌──────────────┐
     │③ _____ │  │ ↓ Urine pH   │
     └──────────────┘  └──────────────┘
            To
            ▼
     ┌──────────────┐
     │   Normal     │
     │   blood pH   │
     └──────────────┘
```

Concept Map V

Using the following terms, fill in the circled, numbered, blank spaces to correctly complete the concept map. Use each term only once.

↑ plasma volume   ↑ H₂O loss

↓ H₂O loss   ↓ B.P. at kidneys

↑ ANP release   ↑ aldosterone release

↓ ADH release   ↓ aldosterone release

Homeostasis Fluid volume regulation—sodium ion concentrations

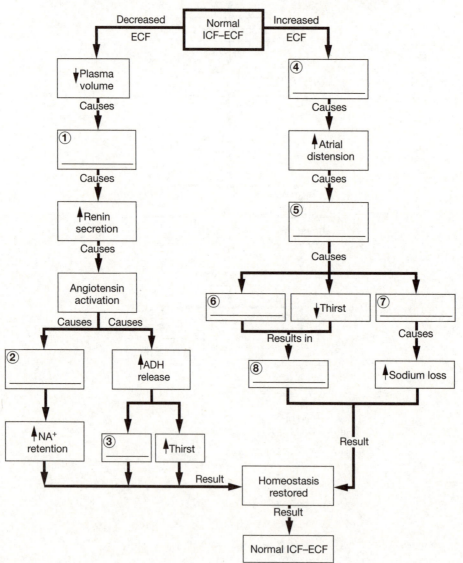

Multiple Choice

Place the letter corresponding to the best answer in the space provided.

_____ 1. All of the homeostatic mechanisms that monitor and adjust the composition of body fluids respond to changes in the

   a. intracellular fluid.
   b. extracellular fluid.
   c. regulatory hormones.
   d. fluid balance.

_____ 2. Important homeostatic adjustments occur in response to changes in

   a. cell receptors that respond to ICF volumes.
   b. hypothalamic osmoreceptors.
   c. hormone levels.
   d. plasma volume or osmolarity.

_____ 3. All water transport across cell membranes and epithelia occurs passively, in response to

   a. active transport and cotransport.
   b. countertransport and facilitated diffusion.
   c. osmotic gradients and hydrostatic pressure.
   d. cotransport and endocytosis.

_____ 4. Whenever the rate of sodium intake or output changes, there is a corresponding gain or loss of water that tends to

   a. keep the sodium concentration constant.
   b. increase the sodium concentration.
   c. decrease the sodium concentration.
   d. alter the sodium concentration.

_____ 5. Angiotensin II produces a coordinated elevation in the extracellular fluid volume by

   a. stimulating thirst.
   b. causing the release of ADH.
   c. triggering the secretion of aldosterone.
   d. a, b, and c are correct.

_____ 6. The rate of tubular secretion of potassium ions changes in response to

   a. alterations in the potassium ion concentration in the ECF.
   b. changes in pH.
   c. aldosterone levels.
   d. a, b, and c are correct.

_____ 7. The most important factor affecting the pH in body tissues is

   a. the protein buffer system.
   b. carbon dioxide concentration.
   c. the bicarbonate reserve.
   d. the presence of ammonium ions.

_____ 8. The body content of water or electrolytes will rise if

    a. losses exceed gains.

    b. intake is less than outflow.

    c. outflow exceeds intake.

    d. intake exceeds outflow.

_____ 9. When an individual loses body water, plasma volume

    a. decreases and electrolyte concentrations rise.

    b. increases and electrolyte concentrations decrease.

    c. increases and electrolyte concentrations increase.

    d. decreases and electrolyte concentrations decrease.

_____ 10. The most common problems with electrolyte balance are caused by

    a. shifts of the bicarbonate ion.

    b. an imbalance between sodium gains and losses.

    c. an imbalance between chloride gains and losses.

    d. a, b, and c are correct.

_____ 11. Sodium ions enter the ECF by crossing the digestive epithelium via

    a. endocytosis.

    b. active transport.

    c. diffusion and carrier-mediated transport.

    d. exocytosis.

_____ 12. Deviations outside of the normal pH range due to changes in hydrogen ion concentrations

    a. disrupt the stability of cell membranes.

    b. alter protein structure.

    c. change the activities of important enzymes.

    d. a, b, and c are correct.

_____ 13. When the P_{CO_2} increases and additional hydrogen ions and bicarbonate ions are released into the plasma, the pH

    a. goes up; ↑ alkalinity.

    b. goes down; ↑ acidity.

    c. goes up; ↓ acidity.

    d. is not affected.

_____ 14. Important examples of organic acids found in the body are

    a. sulfuric acid and phosphoric acid.

    b. hydrochloric acid and carbonic acid.

    c. lactic acid and ketone bodies.

    d. a, b, and c are correct.

_____ 15. In a protein buffer system, if the pH increases, a carboxyl group (COOH) of an amino acid dissociates and releases a

a. hydroxyl ion.

b. molecule of carbon monoxide.

c. hydrogen ion.

d. molecule of carbon dioxide.

_____ 16. Normal pH values are limited to the range of

a. 7.35 to 7.45.

b. 6.8 to 7.0.

c. 6.0 to 8.0.

d. 6.35 to 8.35.

_____ 17. The condition that results when the respiratory system cannot eliminate all the carbon dioxide generated by peripheral tissues is

a. respiratory alkalosis.

b. hypocapnia.

c. respiratory acidosis.

d. metabolic alkalosis.

_____ 18. When a pulmonary response cannot reverse respiratory acidosis, the kidneys respond by

a. increasing the reabsorption of hydrogen ions.

b. increasing the rate of hydrogen ion secretion into the filtrate.

c. decreasing the rate of hydrogen ion secretion into the filtrate.

d. increased loss of bicarbonate ions.

_____ 19. Chronic diarrhea causes a severe loss of bicarbonate ions, resulting in

a. respiratory acidosis.

b. respiratory alkalosis.

c. metabolic acidosis.

d. metabolic alkalosis.

_____ 20. Compensation for metabolic alkalosis involves

a. ↑ pulmonary ventilation; ↑ loss of bicarbonates in the urine.

b. ↑ pulmonary ventilation; ↓ loss of bicarbonates in the urine.

c. ↓ pulmonary ventilation; ↓ loss of bicarbonates in the urine.

d. ↓ pulmonary ventilation; ↑ loss of bicarbonates in the urine.

Completion

Using the terms below, complete the following statements. Use each term only once.

buffer system   alkaline tide   kidneys   angiotensin II

lactic acidosis   organic acids   ketoacidosis   respiratory acidosis

volatile acid   hypoventilation   fixed acids   hyperventilation

1. The most important sites of sodium ion regulation are the _____.

2. Renin release by kidney cells initiates a chain of events leading to the synthesis of _____.

3. An acid that can leave solution and enter the atmosphere is referred to as a(n) _____.

4. Acids that remain in body fluids until excreted at the kidneys are called _____.

5. Acid participants in or by-products of cellular metabolism are referred to as _____.

6. A combination of a weak acid and its dissociation products comprise a(n) _____.

7. When the respiratory system is unable to eliminate normal amounts of CO_2 generated by peripheral tissues, the result is the development of _____.

8. The usual cause of respiratory acidosis is _____.

9. Physical or psychological stresses or conscious effort may produce an increased respiratory rate referred to as _____.

10. Severe exercise or prolonged oxygen starvation of cells may develop into _____.

11. Generation of large quantities of ketone bodies during the postabsorptive state results in a condition known as _____.

12. An influx of large numbers of bicarbonate ions into the ECF due to secretion of HCl by the gastric mucosa is known as the _____.

Short Essay

Briefly answer the following questions in the spaces provided below.

1. What three different, interrelated types of homeostasis are involved in the maintenance of normal volume and composition in the ECF and the ICF?

2. What three primary hormones mediate physiological adjustments that affect fluid and electrolyte balance?

3. What two major effects does ADH have on maintaining homeostatic volumes of water in the body?

4. What three factors control the rate of tubular secretion of potassium ions along the DCT of the nephron?

5. What two primary steps are involved in the regulation of sodium ion concentrations?

6. Write the chemical equation to show how CO_2 interacts with H_2O in solution to form molecules of carbonic acid. Continue the equation to show the dissociation of carbonic acid molecules to produce hydrogen ions and bicarbonate ions.

7. a. What three *chemical* buffer systems represent the first line of defense against pH shift?

   b. What two *physiological* mechanisms represent the second line of defense against pH shift?

8. How do pulmonary and renal mechanisms support the chemical buffer systems?

9. What is the difference between hypercapnia and hypocapnia?

10. What are the three major causes of metabolic acidosis?

LEVEL 3: CRITICAL THINKING AND CLINICAL APPLICATIONS

Using principles and concepts learned in Chapter 27, answer the following questions. Write your answers on a separate sheet of paper. Some of the questions in this section will require the following information for your reference.

Normal arterial blood gas values:

pH:	7.35–7.45
P_{CO_2}:	35 to 45 mm Hg
P_{O_2}:	80 to 100 mm Hg
HCO_3^-:	22 to 26 mEq/liter

1. A comatose teenager is taken to the nearby hospital emergency room by the local rescue squad. His friends reported that he had taken some drug with a large quantity of alcohol. His arterial blood gas (ABG) studies reveal: pH = 7.17; P_{CO_2} = 73 mm Hg; HCO_3^- = 26 mEq/liter.

    Identify the teenager's condition and explain what the clinical values reveal.

2. A 62-year-old woman has been vomiting and experiencing anorexia for several days. After being admitted to the hospital, her ABG studies are reported as follows: pH = 7.65; P_{CO_2} = 52 mm Hg; HCO_3^- = 55 mEq/liter.

    Identify the woman's condition and explain what the clinical values reveal.

3. After analyzing the ABG values below, identify the condition in each one of the following four cases.

    a. pH = 7.30; P_{CO_2} = 37 mm Hg; HCO_3^- = 16 mEq/liter

    b. pH = 7.52; P_{CO_2} = 32 mm Hg; HCO_3^- = 25 mEq/liter

    c. pH = 7.36; P_{CO_2} = 67 mm Hg; HCO_3^- = 23 mEq/liter

    d. pH = 7.58; P_{CO_2} = 43 mm Hg; HCO_3^- = 42 mEq/liter

4. Why does maintaining fluid balance in older people require a higher water intake than in a normal, healthy adult under age 40?

CHAPTER 28

The Reproductive System

OVERVIEW

The structures and functions of the reproductive system are notably different from those of any other organ system in the human body. The other systems of the body are functional at birth or shortly thereafter; however, the reproductive system does not become functional until it is acted on by hormones during puberty.

Most of the other body systems function to support and maintain the individual, but the reproductive system is specialized to ensure survival not of the individual but of the species.

Even though major differences exist between the reproductive organs of the male and female, both are primarily concerned with propagation of the species and passing genetic material from one generation to another. In addition, the reproductive system produces hormones that stimulate the development of secondary sex characteristics.

This chapter provides a series of exercises that will assist you in reviewing and reinforcing your understanding of the anatomy and physiology of the male and female reproductive systems, the effects of male and female hormones, and changes that occur during the aging process.

LEVEL 1: REVIEWING FACTS AND TERMS

Review of Learning Outcomes

After completing this chapter, you should be able to do the following:

OUTCOME 28-1   List the basic components of the human reproductive system, and summarize the functions of each.

OUTCOME 28-2   Describe the components of the male reproductive system and the roles played by the reproductive tract and accessory glands in producing spermatozoa; specify the composition of semen; and summarize the hormonal mechanisms that regulate male reproductive functions.

OUTCOME 28-3   Describe the components of the female reproductive system and the ovarian roles in oogenesis; explain the complete ovarian and uterine cycles; outline the histology, anatomy, and functions of the vagina; and summarize all aspects of the female reproductive cycle.

OUTCOME 28-4   Discuss the physiology of sexual intercourse in males and females.

OUTCOME 28-5   Describe the reproductive system changes that occur with aging.

OUTCOME 28-6   Give examples of interactions between the reproductive system and each of the other organ systems.

Multiple Choice

Place the letter corresponding to the best answer in the space provided.

OUTCOME 28-1 _____ 1. Collectively, the functional male and female reproductive cells are called
    a. gonads.
    b. ova.
    c. sperm.
    d. gametes.

OUTCOME 28-1 _____ 2. The reproductive organs that produce gametes and hormones are the
    a. accessory glands.
    b. gonads.
    c. vagina and penis.
    d. a, b, and c are correct.

OUTCOME 28-2 _____ 3. In the male, the important function(s) of the epididymis is (are)
    a. to monitor and adjust the composition of the tubular fluid.
    b. to act as a recycling center for damaged spermatozoa.
    c. as the site of physical maturation of spermatozoa.
    d. a, b, and c are correct.

OUTCOME 28-2 _____ 4. Beginning inferior to the urinary bladder, sperm travel to the exterior through the urethral regions in the following order:
    a. membranous urethra, prostatic urethra, spongy urethra.
    b. prostatic urethra, spongy urethra, membranous urethra.
    c. prostatic urethra, membranous urethra, spongy urethra.
    d. spongy urethra, prostatic urethra, membranous urethra.

OUTCOME 28-2 _____ 5. The external genitalia of the male includes the
    a. scrotum and penis.
    b. urethra and bulbo-urethral glands.
    c. raphe and dartos.
    d. prepuce and glans.

OUTCOME 28-2 _____ 6. The three masses of erectile tissue that comprise the body of the penis are
    a. two cylindrical corpora cavernosa and a slender corpus spongiosum.
    b. two slender corpora spongiosa and a cylindrical corpus cavernosum.
    c. preputial glands, a corpus cavernosum, and a corpus spongiosum.
    d. two corpora cavernosa and a preputial gland.

OUTCOME 28-2 _____ 7. In the process of spermatogenesis, the developmental sequence includes
    a. spermatids, spermatozoa, spermatogonia, and spermatocytes.
    b. spermatogonia, spermatocytes, spermatids, and spermatozoa.
    c. spermatocytes, spermatogonia, spermatids, and spermatozoa.
    d. spermatogonia, spermatids, spermatocytes, and spermatozoa.

OUTCOME 28-2   _____   8. An individual spermatozoon completes its development in the seminiferous tubules and its physical maturation in the epididymis in approximately

    a. 2 weeks.

    b. 3 weeks.

    c. 5 weeks.

    d. 9 weeks.

OUTCOME 28-2   _____   9. In the male, sperm cells, before leaving the body, travel from the testes to the

    a. ductus deferens → epididymis → urethra → ejaculatory duct.

    b. ejaculatory duct → epididymis → ductus deferens → urethra.

    c. epididymis → ductus deferens → ejaculatory duct → urethra.

    d. epididymis → ejaculatory duct → ductus deferens → urethra.

OUTCOME 28-2   _____   10. The accessory organs in the male that secrete into the ejaculatory ducts and the urethra are the

    a. epididymis, seminal vesicles, and vas deferens.

    b. prostate gland, inguinal canals, and raphe.

    c. adrenal glands, bulbo-urethral glands, and seminal glands.

    d. seminal vesicles, prostate gland, and bulbo-urethral glands.

OUTCOME 28-2   _____   11. Semen, the volume of fluid called the ejaculate, contains

    a. spermatozoa, seminal plasmin, and enzymes.

    b. alkaline and acid secretions and sperm.

    c. mucus, sperm, and enzymes.

    d. spermatozoa, seminal fluid, and enzymes.

OUTCOME 28-2   _____   12. The correct average characteristics and composition of semen include

    a. vol., 2.5 m*l*; specific gravity, 1.028; pH, 7.19; sperm count, 20–100 million/m*l*.

    b. vol., 1.2 m*l*; specific gravity, 0.050; pH, 7.50; sperm count, 10–100 million/m*l*.

    c. vol., 4.5 m*l*; specific gravity, 2.010; pH, 6.90; sperm count, 5–10 million/m*l*.

    d. vol., 0.5 m*l*; specific gravity, 3.010; pH, 7.0; sperm count, 2–4 million/m*l*.

OUTCOME 28-2   _____   13. The hormone synthesized in the hypothalamus that initiates release of pituitary hormones is

    a. FSH (follicle-stimulating hormone).

    b. ICSH (interstitial-cell-stimulating hormone).

    c. LH (luteinizing hormone).

    d. GnRH (gonadotropin-releasing hormone).

OUTCOME 28-2   _____   14. The hormone that promotes spermatogenesis along the seminiferous tubules is

    a. TSH.

    b. FSH.

    c. GnRH.

    d. LH.

OUTCOME 28-3 _____ 15. The function of the uterus in the female is to
    a. ciliate the sperm into the uterine tube for possible fertilization.
    b. provide ovum transport via peristaltic contractions of the uterine wall.
    c. provide mechanical protection and nutritional support to the developing embryo.
    d. encounter spermatozoa during the first 12–24 hours of its passage.

OUTCOME 28-3 _____ 16. In the female, after leaving the ovaries, the ovum travels in the uterine tubes to the uterus via the
    a. ampulla → infundibulum → isthmus → intramural portion.
    b. infundibulum → ampulla → isthmus → intramural portion.
    c. isthmus → ampulla → infundibulum → intramural portion.
    d. intramural portion → ampulla → isthmus → infundibulum.

OUTCOME 28-3 _____ 17. Starting at the superior end, the uterus in the female is divided into
    a. body, isthmus, and cervix.
    b. isthmus, body, and cervix.
    c. body, cervix, and isthmus.
    d. cervix, body, and isthmus.

OUTCOME 28-3 _____ 18. Ovum transport in the uterine tubes involves a combination of
    a. flagellar locomotion and ciliary movement.
    b. active transport and ciliary movement.
    c. ciliary movement and peristaltic contractions.
    d. movement in uterine fluid and flagellar locomotion.

OUTCOME 28-3 _____ 19. The outer limits of the vulva are established by the
    a. vestibule and the labia minora.
    b. mons pubis and labia majora.
    c. lesser and greater vestibular glands.
    d. prepuce and vestibule.

OUTCOME 28-3 _____ 20. The ovarian cycle begins as activated follicles develop into
    a. ova.
    b. primary follicles.
    c. graafian follicles.
    d. primordial follicles.

OUTCOME 28-3 _____ 21. The process of oogenesis produces three nonfunctional polar bodies that eventually disintegrate and
    a. a primordial follicle.
    b. a granulosa cell.
    c. one functional ovum.
    d. a zona pellucida.

OUTCOME 28-3 _____ 22. The proper sequence that describes the ovarian cycle involves the formation of
    a. primary follicles, secondary follicles, tertiary follicles, ovulation, and formation and destruction of the corpus luteum.
    b. primary follicles, secondary follicles, tertiary follicles, corpus luteum, and ovulation.
    c. corpus luteum; primary, secondary, and tertiary follicles; and ovulation.
    d. primary and tertiary follicles, secondary follicles, ovulation, and formation and destruction of the corpus luteum.

OUTCOME 28-3 _____ 23. Under normal circumstances, in a 28-day cycle, ovulation occurs on _____, and the menses begins on _____.
    a. day 1; day 14
    b. day 28; day 14
    c. day 14; day 1
    d. day 6; day 14

OUTCOME 28-3 _____ 24. The reproductive function(s) of the vagina is (are)
    a. as a passageway for the elimination of menstrual fluids.
    b. to receive the penis during coitus.
    c. to form the lower portion of the birth canal during childbirth.
    d. a, b, and c are correct.

OUTCOME 28-3 _____ 25. The dominant hormone of the follicular phase is
    a. progesterone.
    b. follicle-stimulating hormone.
    c. luteinizing hormone.
    d. estradiol.

OUTCOME 28-3 _____ 26. The principal hormone of the luteal phase is
    a. human chorionic gonadotrophin.
    b. progesterone.
    c. estrogen.
    d. luteinizing hormone.

OUTCOME 28-3 _____ 27. High estrogen levels
    a. stimulate LH secretion.
    b. increase pituitary sensitivity to GnRH.
    c. increase the GnRH pulse frequency.
    d. a, b, and c are correct.

OUTCOME 28-4 _____ 28. Peristaltic contractions of the ampulla, pushing fluid and spermatozoa into the prostatic urethra, are called
    a. emission.
    b. ejaculation.
    c. detumescence.
    d. subsidence.

OUTCOME 28-4 _____ 29. In the male, erotic thoughts and stimulation of sensory nerves in the genital region lead to an increase in parasympathetic outflow over the pelvic nerves, which in turn leads to
    a. emission and ejaculation.
    b. erection of the penis.
    c. orgasm and detumescence.
    d. sexual dysfunction or impotence.

OUTCOME 28-4 _____ 30. Engorgement of the erectile tissues of the clitoris and increased secretion of the greater vestibular glands involve neural activity that includes
    a. somatic motor neurons.
    b. sympathetic activation.
    c. parasympathetic activation.
    d. a, b, and c are correct.

OUTCOME 28-5 _____ 31. Menopause is accompanied by a sharp and sustained rise in the production of _____ while circulating concentrations of _____ decline.
    a. estrogen and progesterone; GnRH, FSH, and LH
    b. GnRH, FSH, and LH; estrogen and progesterone
    c. LH and estrogen; progesterone, GnRH, and FSH
    d. FSH, LH, and progesterone; estrogen and GnRH

OUTCOME 28-5 _____ 32. In the male, between the ages of 50 and 60 circulating _____ levels begin to decline, coupled with increases in circulating levels of _____.
    a. FSH and LH; testosterone
    b. FSH; testosterone and LH
    c. testosterone; FSH and LH
    d. a, b, and c are correct

OUTCOME 28-6 _____ 33. The system(s) that control(s) sexual behavior and sexual function is (are) the
    a. endocrine system.
    b. integumentary system.
    c. nervous system.
    d. a, b, and c are correct.

Completion

Using the terms below, complete the following statements. Use each term only once.

menopause	fertilization	LH
ovaries	uterus	testes
placenta	cardiovascular	raphe
ovulation	inhibin	oogenesis
seminiferous tubules	graafian	ductus deferens
greater vestibular	gonads	fructose
spermiogenesis	prostate gland	spermatids
infundibulum	muscular	orgasm

OUTCOME 28-1   1. The fusion of a sperm contributed by the father and an egg, or ovum, from the mother is called _____.

OUTCOME 28-1   2. In the male and the female, the reproductive organs that produce gametes and hormones are the _____.

OUTCOME 28-2   3. Sperm production occurs within the slender, tightly coiled _____.

OUTCOME 28-2   4. The male gonads are the _____.

OUTCOME 28-2   5. The physical transformation of a spermatid to a spermatozoon is called _____.

OUTCOME 28-2   6. The process of meiosis in the male produces undifferentiated male gametes called _____.

OUTCOME 28-2   7. After passing along the tail of the epididymis, the spermatozoa arrive at the _____.

OUTCOME 28-2   8. The boundary between the two chambers in the scrotum is marked by the _____.

OUTCOME 28-2   9. Thirty percent of the secretions in a typical sample of seminal fluid is contributed by the _____.

OUTCOME 28-2   10. The primary energy source for the mobilization of sperm is _____.

OUTCOME 28-2   11. A peptide hormone that causes the secretion of androgens by the interstitial cells of the testes is _____.

OUTCOME 28-3   12. The female gonads are the _____.

OUTCOME 28-3   13. By the tenth day of the ovarian cycle, the mature tertiary follicle is also called the mature _____ follicle.

OUTCOME 28-3   14. Egg release from the ovary into the uterine tube is called _____.

OUTCOME 28-3   15. After ovulation, the ovum passes from the ovary into the expanded funnel called the _____.

OUTCOME 28-3   16. The structure that provides mechanical protection, nutritional support, and waste removal for the developing embryo is the _____.

OUTCOME 28-3   17. Ovum production, which occurs on a monthly basis as part of the ovarian cycle, is referred to as _____.

OUTCOME 28-3   18. The mucous glands that discharge secretions into the vestibule near the vaginal entrance are known as _____ glands.

OUTCOME 28-3   19. As secondary follicles develop, FSH levels decline due to the negative feedback effects of _____.

OUTCOME 28-3   20. Nutrients for embryonic and fetal development are provided by a temporary structure called the _____.

OUTCOME 28-4   21. Female _____ is accompanied by peristaltic contractions of the uterine and vaginal walls.

OUTCOME 28-5   22. In females, the time that ovulation and menstruation cease is referred to as _____.

OUTCOME 28-6   23. The system responsible for the ejection of semen from the male reproductive tract is the _____ system.

Matching

Match the terms in column B with the terms in column A. Use letters for answers in the spaces provided. Use each term only once.

Part I

		Column A	Column B
OUTCOME 28-1	_____	1. gametes	A. sperm production
OUTCOME 28-1	_____	2. external genitalia	B. produce alkaline secretion
OUTCOME 28-1	_____	3. FSH, LH, ICSH	C. anterior pituitary
OUTCOME 28-2	_____	4. nurse cells	D. hormone—male secondary sex characteristics
OUTCOME 28-2	_____	5. interstitial cells	E. androgen production
OUTCOME 28-2	_____	6. seminiferous tubules	F. perineal structures
OUTCOME 28-2	_____	7. puberty in male	G. blood–testis barrier
OUTCOME 28-2	_____	8. seminal glands	H. oocyte transport
OUTCOME 28-2	_____	9. seminalplasmin	I. spermatogenesis
OUTCOME 28-2	_____	10. testosterone	J. reproductive cells
OUTCOME 28-3	_____	11. uterine tubes	K. antibiotic enzyme in semen

Part II

		Column A	Column B
OUTCOME 28-3	_____	12. ovaries	L. follicular degeneration
OUTCOME 28-3	_____	13. puberty in female	M. endocrine structure
OUTCOME 28-3	_____	14. immature eggs	N. egg production
OUTCOME 28-3	_____	15. atresia	O. milk production
OUTCOME 28-3	_____	16. vaginal folds	P. menarche
OUTCOME 28-3	_____	17. corpus luteum	Q. parasympathetic activation
OUTCOME 28-3	_____	18. lactation	R. rugae
OUTCOME 28-4	_____	19. arousal	S. cessation of ovulation and menstruation
OUTCOME 28-4	_____	20. detumescence	T. oocytes
OUTCOME 28-5	_____	21. menopause	U. reduced estrogen concentrations
OUTCOME 28-5	_____	22. osteoporosis	V. subsidence of erection
OUTCOME 28-6	_____	23. endocrine system	W. oxytocin—smooth muscle contractions

Drawing/Illustration Labeling

Identify each numbered structure by labeling the following figures.

OUTCOME 28-2 **FIGURE 28-1** Male Reproductive Organs (Sagittal Section)

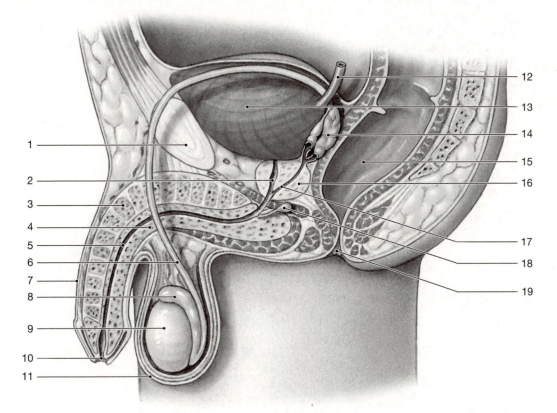

(1)_____   (11)_____

(2)_____   (12)_____

(3)_____   (13)_____

(4)_____   (14)_____

(5)_____   (15)_____

(6)_____   (16)_____

(7)_____   (17)_____

(8)_____   (18)_____

(9)_____   (19)_____

(10)_____

OUTCOME 28-2   **FIGURE 28-2**   The Testes and Epididymis

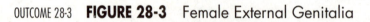

(1)_____

(2)_____

(3)_____

(4)_____

(5)_____

(6)_____

(7)_____

(8)_____

(9)_____

OUTCOME 28-3   **FIGURE 28-3**   Female External Genitalia

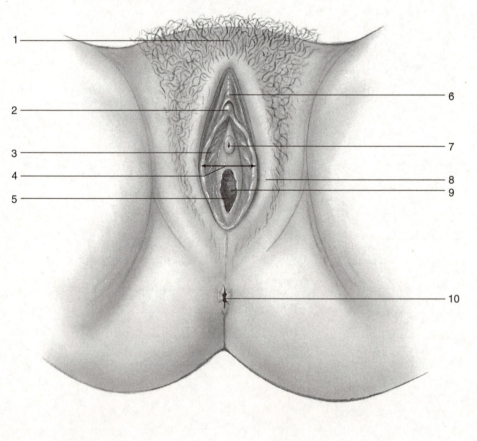

(1)_____   (6)_____

(2)_____   (7)_____

(3)_____   (8)_____

(4)_____   (9)_____

(5)_____   (10)_____

OUTCOME 28-3 **FIGURE 28-4** Female Reproductive System (Sagittal Section)

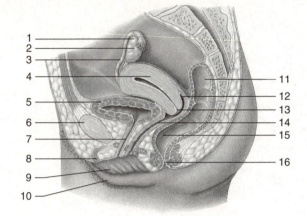

(1)_____	(9)_____
(2)_____	(10)_____
(3)_____	(11)_____
(4)_____	(12)_____
(5)_____	(13)_____
(6)_____	(14)_____
(7)_____	(15)_____
(8)_____	(16)_____

OUTCOME 28-3 **FIGURE 28-5** Female Reproductive Organs (Posterior View)

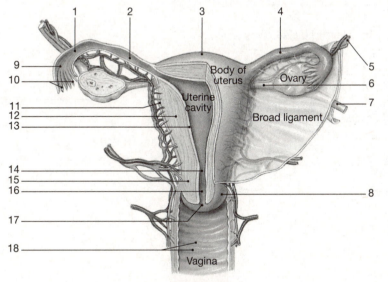

Posterior view

(1)_____	(7)_____	(13)_____
(2)_____	(8)_____	(14)_____
(3)_____	(9)_____	(15)_____
(4)_____	(10)_____	(16)_____
(5)_____	(11)_____	(17)_____
(6)_____	(12)_____	(18)_____

OUTCOME 28-2   **FIGURE 28-6**   Spermatogenesis

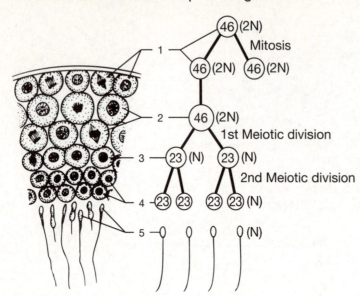

(1)_____

(2)_____

(3)_____

(4)_____

(5)_____

KEY:
N = haploid
2N = diploid

OUTCOME 28-3   **FIGURE 28-7**   Oogenesis

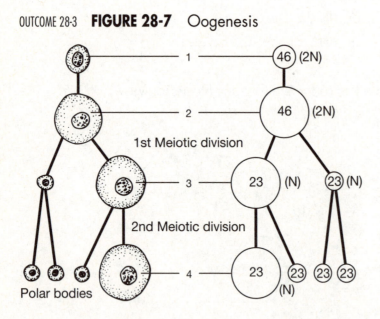

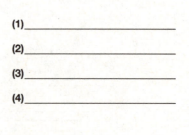

(1)_____

(2)_____

(3)_____

(4)_____

KEY:
N = haploid
2N = diploid

LEVEL 2: REVIEWING CONCEPTS

Chapter Overview

Using the terms below, identify the numbered locations on the reproductive system of the male. Use each term only once.

spongy urethra	external urethral orifice	· ejaculatory duct
ductus deferens	seminiferous tubules	urethra
body of epididymis	and rete testis	

For the chapter overview of the male reproductive system you will follow the pathway that sperm travel from the time they are formed in the testes until they leave the body of the male. Identify the locations along the numbered (1–7) route and record them in the spaces provided below.

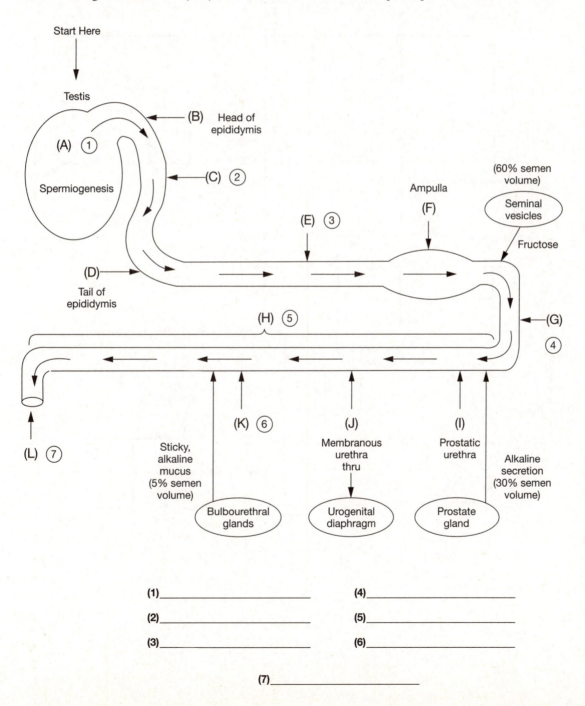

(1)_____   (4)_____

(2)_____   (5)_____

(3)_____   (6)_____

(7)_____

Concept Map I

Using the following terms, fill in the circled, numbered, blank spaces to correctly complete the concept map. Use each term only once.

urethra	seminiferous tubules	penis
produce testosterone	FSH	seminal vesicles
ductus deferens	bulbo-urethral glands	

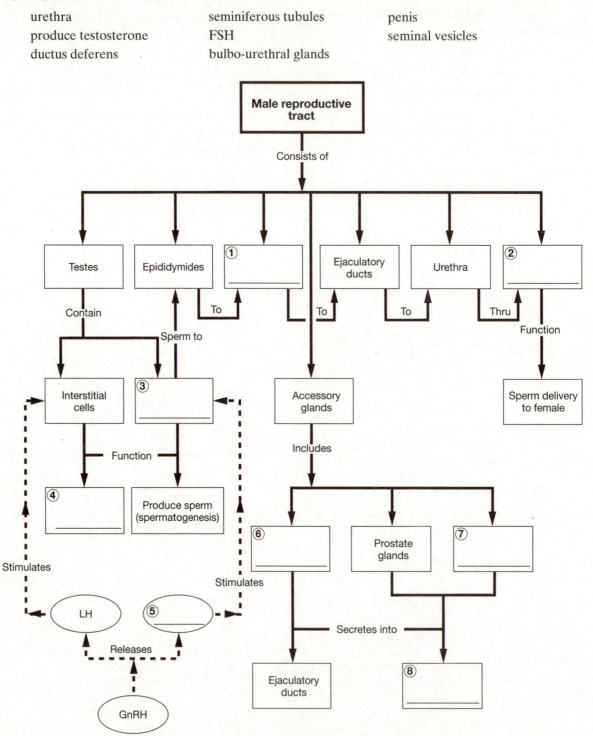

Concept Map II

Using the following terms, fill in the circled, numbered, blank spaces to correctly complete the concept map. Use each term only once.

external urethral orifice   body   crus
corpus spongiosum   prepuce   frenulum

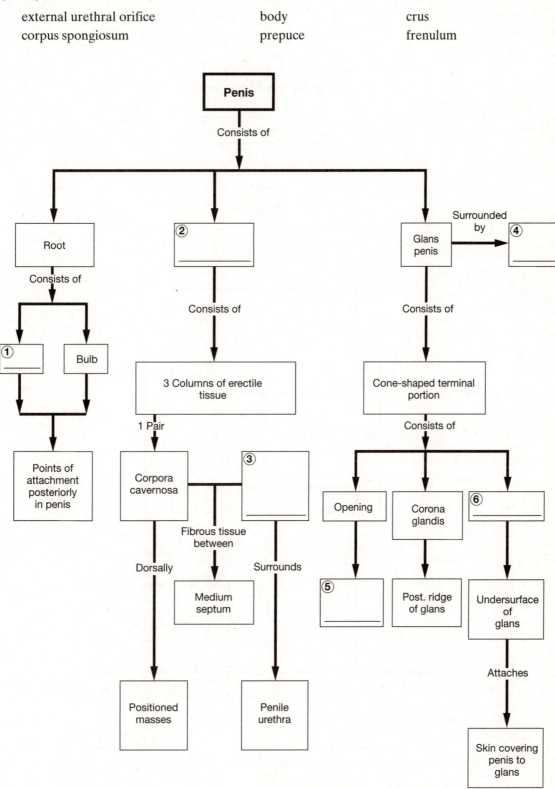

Concept Map III

Using the following terms, fill in the circled, numbered, blank spaces to correctly complete the concept map. Use each term only once.

inhibin	testes	male secondary sex characteristics
FSH	CNS	anterior pituitary
interstitial cells		

Male Hormone Regulation

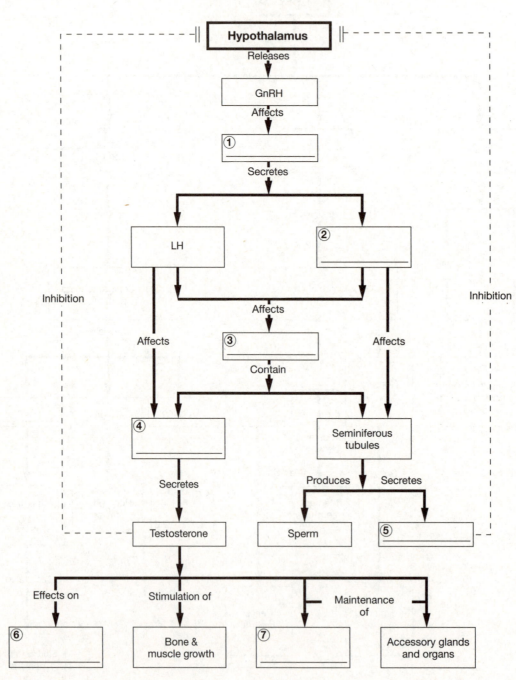

Concept Map IV

Using the following terms, fill in the circled, numbered, blank spaces to correctly complete the concept map. Use each term only once.

nutrients   supports fetal development   endometrium
vulva   granulosa and thecal cells   clitoris
vagina   uterine tubes   labia majora and minora
follicles

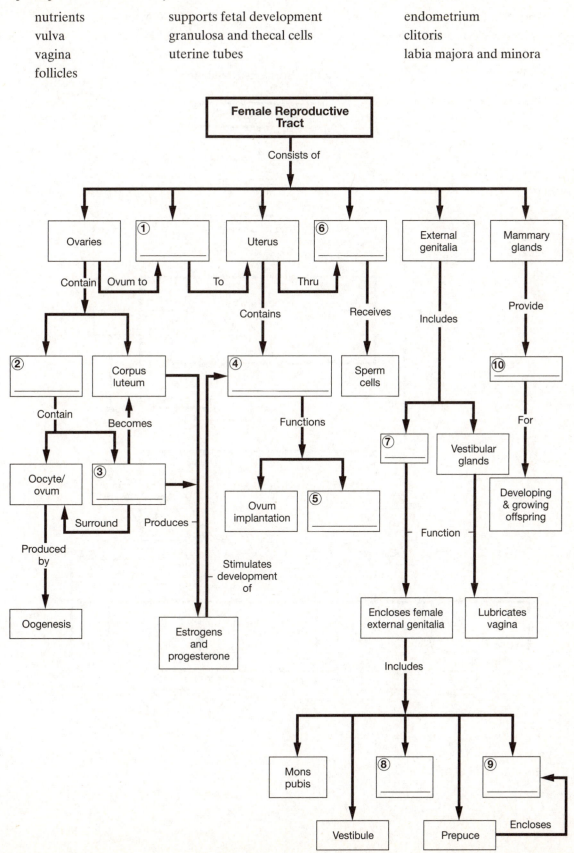

Concept Map V

Using the following terms, fill in the circled, numbered, blank spaces to correctly complete the concept map. Use each term only once.

bone and muscle growth   follicles   GnRH

accessory glands and organs   progesterone   LH

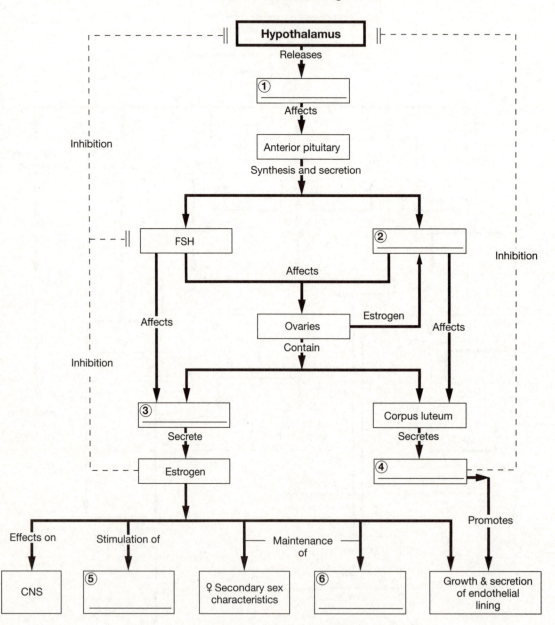

Female Hormone Regulation

Multiple Choice

Place the letter corresponding to the best answer in the space provided.

_____ 1. The part of the endometrium that undergoes cyclical changes in response to sexual hormone levels is the

    a. functional zone.

    b. basilar zone.

    c. serosa.

    d. muscular myometrium.

_____ 2. In a 28-day cycle, estrogen levels peak at

    a. day 1.

    b. day 7.

    c. day 14.

    d. day 28.

_____ 3. The rupture of the follicular wall and ovulation are caused by

    a. a sudden surge in LH (luteinizing hormone) concentration.

    b. a sudden surge in the secretion of estrogen.

    c. an increase in the production of progesterone.

    d. increased production and release of GnRH (gonadotrophin-releasing hormone).

_____ 4. The spermatic cord is a structure that includes the

    a. epididymis, ductus deferens, blood vessels, and nerves.

    b. vas deferens, prostate gland, blood vessels, and urethra.

    c. ductus deferens, blood vessels, nerves, and lymphatics.

    d. a, b, and c are correct.

_____ 5. The function(s) of the nurse cells is (are)

    a. maintenance of the blood–testis barrier.

    b. support of spermiogenesis.

    c. secretion of inhibin and androgen-binding protein.

    d. a, b, and c are correct.

_____ 6. The tail of the sperm has the unique distinction of being the

    a. only flagellum in the body that contains chromosomes.

    b. only flagellum in the human body.

    c. only flagellum in the body that contains mitochondria.

    d. only flagellum in the body that contains centrioles.

_____ 7. The function(s) of the prostate gland is (are) to

    a. produce acid and alkaline secretions.

    b. produce acidic secretions and a compound with antibiotic properties.

    c. secrete a thick, sticky, alkaline mucus for lubrication.

    d. secrete seminal fluid with a distinctive ionic and nutrient composition.

_____ 8. Powerful, rhythmic contractions in the ischiocavernosus and bulbocavernosus muscles of the pelvic floor result in

   a. erection.

   b. emission.

   c. ejaculation.

   d. a, b, and c are correct.

_____ 9. The process of erection involves complex neural processes that include

   a. increased sympathetic outflow over the pelvic nerves.

   b. increased parasympathetic outflow over the pelvic nerves.

   c. decreased parasympathetic outflow over the pelvic nerves.

   d. somatic motor neurons in the upper sacral segments of the spinal cord.

_____ 10. Impotence, a common male sexual dysfunction, is the

   a. inability to produce sufficient sperm for fertilization.

   b. term used to describe male infertility.

   c. inability of the male to ejaculate.

   d. inability to achieve or maintain an erection.

_____ 11. The ovaries, uterine tubes, and uterus are enclosed within an extensive mesentery known as the

   a. rectouterine pouch.

   b. suspensory ligament.

   c. ovarian ligament.

   d. broad ligament.

_____ 12. If fertilization is to occur, the ovum must encounter spermatozoa during the first _____ hours of its passage.

   a. 2–4

   b. 6–8

   c. 12–24

   d. 30–36

_____ 13. The three pairs of suspensory ligaments that stabilize the position of the uterus and limit its range of movement are the

   a. broad, ovarian, and suspensory.

   b. uterosacral, round, and cardinal.

   c. anteflexure, retroflexure, and tunica albuginea.

   d. endometrium, myometrium, and serosa.

_____ 14. The histological composition of the uterine wall consists of the

   a. body, isthmus, and cervix.

   b. mucosa, epithelial lining, and lamina propria.

   c. mucosa, functional zone, and basilar layer.

   d. endometrium, myometrium, and perimetrium.

_____ 15. The hormone that acts to reduce the rate of GnRH and FSH production by the anterior pituitary is

    a. inhibin.

    b. LH.

    c. ICSH.

    d. testosterone.

_____ 16. The three sequential stages of the menstrual cycle are

    a. menses, luteal phase, and postovulatory phase.

    b. menses, follicular phase, and preovulatory phase.

    c. menses, proliferative phase, and follicular phase.

    d. menses, proliferative phase, and secretory phase.

_____ 17. At the time of ovulation, basal body temperature

    a. declines sharply.

    b. increases slightly.

    c. stays the same.

    d. is one-half a degree Fahrenheit lower than normal.

_____ 18. The seminal vesicles

    a. store semen.

    b. secrete a fructose-rich fluid.

    c. conduct spermatozoa into the epididymis.

    d. secrete mucus.

_____ 19. A recently marketed drug is said to "remind the pituitary" to produce the gonadotropins, FSH, and LH. This drug might be useful as

    a. a male contraceptive.

    b. a female contraceptive.

    c. both a male and female contraceptive.

    d. a fertility drug.

Completion

Using the terms below, complete the following statements. Use each term only once.

cervical os	ampulla	inguinal canals
detumescence	androgens	smegma
raphe	prepuce	fimbriae
hymen	tunica albuginea	rete testis
zona pellucida	ejaculation	corona radiata
clitoris	acrosome	corpus luteum
cremaster	corpus albicans	menopause
menses	mesovarium	fornix

1. The interstitial cells are responsible for the production of male sex hormones, which are called _____.

2. The narrow canals linking the scrotal chambers with the peritoneal cavity are called the _____.

3. The scrotum is divided into two separate chambers, and the boundary between the two is marked by a raised thickening in the scrotal surface known as the _____.

4. The layer of skeletal muscle that contracts and tenses the scrotum, pulling the testes closer to the body, is the _____ muscle.

5. Semen is expelled from the body by a process called _____.

6. The maze of tubules to which the seminiferous tubules connect is known as the _____.

7. The tip of the sperm containing an enzyme that plays a role in fertilization is the _____.

8. The fold of skin that surrounds the glans of the penis is the _____.

9. Preputial glands in the skin of the neck and the inner surface of the prepuce secrete a waxy material known as _____.

10. Subsidence of erection mediated by the sympathetic nervous system is referred to as _____.

11. The time after the last menstrual cycle of the female is known as _____.

12. Microvilli are present in the *space* between the developing oocyte and the innermost follicular cells called the _____.

13. Follicular cells surrounding the oocyte prior to ovulation are known as the _____.

14. Degenerated follicular cells proliferate to create an endocrine structure known as the _____.

15. A knot of pale scar tissue produced by fibroblasts invading a degenerated corpus luteum is called a(n) _____.

16. The thickened fold of mesentery that supports and stabilizes the position of each ovary is the _____.

17. The thickened mesothelium that overlies a layer of dense connective tissue covering the exposed surfaces of each ovary is the _____.

18. The fingerlike projections on the infundibulum that extend into the pelvic cavity are called _____.

19. The uterine cavity opens into the vagina at the _____.

20. The shallow recess surrounding the cervical protrusion is known as the _____.

21. Prior to sexual activity, the thin epithelial fold that partially or completely blocks the entrance to the vagina is the _____.

22. The female equivalent of the penis derived from the same embryonic structure is the _____.

23. The period marked by the destruction and shedding of the functional zone of the endometrium is the _____.

24. Just before it reaches the prostate and seminal vesicles, the ductus deferens becomes enlarged, and the expanded portion is known as the _____.

Short Essay

Briefly answer the following questions in the spaces provided below.

1. What are the four important functions of the nurse cells in the testes?

2. What are the three important functions of the epididymis?

3. a. What three glands secrete their products into the male reproductive tract?

    b. What are the four primary functions of these glands?

4. What is the difference between seminal fluid and semen?

5. What is the difference between emission and ejaculation?

6. What are the five primary functions of testosterone in the male?

7. What are the three reproductive functions of the vagina?

8. What are the three phases of female sexual function, and what occurs in each phase?

9. Beginning with the formation of primary follicles, what are the steps involved in the ovarian cycle?

10. What are the five primary functions of the estrogens?

11. What are the three stages of the menstrual cycle?

LEVEL 3: CRITICAL THINKING AND CLINICAL APPLICATIONS

Using principles and concepts learned in Chapter 28, answer the following questions. Write your answers on a separate sheet of paper.

1. I. M. Hurt was struck in the abdomen with a baseball bat while playing in a young men's baseball league. As a result, his testes ascend into the abdominopelvic region quite frequently, causing sharp pains. He has been informed by his urologist that he is sterile due to his unfortunate accident.

    a. Why does his condition cause sterility?

    b. What primary factors are necessary for fertility in males?

2. A 19-year-old female is aware of differences between her body and those of her female classmates. Compared to most of her female peers, she has a low-pitched voice, growth of hair on her face, and much smaller breasts. Physiologically speaking, what might be happening on a hormonal level to produce these differences?

3. A contraceptive pill "tricks the brain" into thinking you are pregnant. What does this mean?

4. Sexually transmitted diseases in males do not result in inflammation of the peritoneum (peritonitis) as they sometimes do in females. Why?

Development and Inheritance

OVERVIEW

It is difficult to imagine that today there is a single cell that 40 weeks from now, if fertilized, will develop into a complex organism containing trillions of cells organized into tissues, organs, and organ systems — so it is the miracle of life!

The previous chapter considered the male and female reproductive tracts, where gametes develop, receive sustenance, and through which they are transported. This chapter will examine the details of pregnancy and parturition.

A new life begins in the tubes of these systems, and it is here that new genetic combinations, similar to the parents, yet different, are made and nourished until they emerge from the female tract to take up life on their own — at first highly dependent on extrinsic support but growing independent with physical maturation.

All of these events occur as a result of the complex, unified process of development — an orderly sequence of progressive changes that begin at fertilization and have profound effects on the individual for a lifetime.

Chapter 29 highlights the major aspects of development and development processes, regulatory mechanisms, and how developmental patterns can be modified for the good or ill of the individual. The chapter concludes by addressing the topic of inheritance.

LEVEL 1: REVIEWING FACTS AND TERMS

Review of Learning Outcomes

After completing this chapter, you should be able to do the following:

OUTCOME 29-1 Explain the relationship between differentiation and development, and specify the various stages of development.

OUTCOME 29-2 Describe the process of fertilization, and explain how developmental processes are regulated.

OUTCOME 29-3 List the three stages of prenatal development, and describe the major events of each.

OUTCOME 29-4 Explain how the three germ layers participate in the formation of extraembryonic membranes, and discuss the importance of the placenta as an endocrine organ.

OUTCOME 29-5   Describe the interplay between the maternal organ systems and the developing fetus, and discuss the structural and functional changes in the uterus during gestation.

OUTCOME 29-6   List and discuss the events that occur during labor and delivery.

OUTCOME 29-7   Identify the features and physiological changes of the postnatal stages of life.

OUTCOME 29-8   Relate basic principles of genetics to the inheritance of human traits.

Multiple Choice

Place the letter corresponding to the best answer in the space provided.

OUTCOME 29-1   _____   1. The creation of different types of cells during the processes of development is called
    a. fertilization.
    b. differentiation.
    c. maturity.
    d. implantation.

OUTCOME 29-1   _____   2. The gradual modification of anatomical structures and physiological characteristics from fertilization to maturity is called
    a. differentiation.
    b. conception.
    c. inheritance.
    d. development.

OUTCOME 29-1   _____   3. Fetal development begins at the start of the
    a. birth process.
    b. second month after fertilization.
    c. ninth week after fertilization.
    d. process of fertilization.

OUTCOME 29-1   _____   4. The stage of development that commences at birth and continues to maturity is the
    a. postnatal period.
    b. prenatal period.
    c. embryological period.
    d. childhood period.

OUTCOME 29-2   _____   5. Typically, fertilization occurs in the
    a. lower part of the uterine tube.
    b. junction between the ampulla and isthmus of the uterine tube.
    c. upper part of the uterus.
    d. antrum of a tertiary follicle.

OUTCOME 29-2   _____   6. Sterility in males may result from a sperm count of less than _____ million sperm/ml of semen.
    a. 20
    b. 40
    c. 60
    d. 100

OUTCOME 29-2   _____   7. Fertilization is completed with the
         a. formation of a gamete containing 23 chromosomes.
         b. formation of the male and female pronuclei.
         c. completion of the meiotic process.
         d. formation of a zygote containing 46 chromosomes.

OUTCOME 29-3   _____   8. Alterations in genetic activity during development occur as a result of
         a. the maturation of the sperm and ovum.
         b. the chromosome complement in the nucleus of the cell.
         c. differences in the cytoplasmic composition of individual cells.
         d. conception.

OUTCOME 29-3   _____   9. As development proceeds, some embryonic cells affect the differentiation patterns of other embryonic cells by
         a. initiating capacitation.
         b. releasing RNAs, polypeptides, and small proteins.
         c. secreting progesterone and hGC.
         d. producing the three germ layers.

OUTCOME 29-3   _____   10. Organs and organ systems complete most of their development by the end of the
         a. first trimester.
         b. second trimester.
         c. third trimester.
         d. time of placentation.

OUTCOME 29-4   _____   11. The most dangerous period in prenatal or postnatal life is the
         a. first trimester.
         b. second trimester.
         c. third trimester.
         d. expulsion stage.

OUTCOME 29-4   _____   12. The four general processes that occur during the first trimester are
         a. dilation, expulsion, placental, and labor.
         b. blastocyst, blastomere, morula, and trophoblast.
         c. cleavage, implantation, placentation, and embryogenesis.
         d. yolk sac, amnion, allantois, and chorion.

OUTCOME 29-4   _____   13. Germ-layer formation results from the process of
         a. embryogenesis.
         b. organogenesis.
         c. gastrulation.
         d. parturition.

OUTCOME 29-4   _____   14. The extraembryonic membranes that develop from the endoderm and mesoderm are
         a. amnion and chorion.
         b. yolk sac and allantois.
         c. allantois and chorion.
         d. yolk sac and amnion.

OUTCOME 29-4 _____ 15. The chorion develops from the
a. endoderm and mesoderm.
b. ectoderm and mesoderm.
c. trophoblast and endoderm.
d. mesoderm and trophoblast.

OUTCOME 29-4 _____ 16. Blood flows to and from the placenta via
a. paired umbilical veins and a single umbilical artery.
b. paired umbilical arteries and a single umbilical vein.
c. a single umbilical artery and a single umbilical vein.
d. two umbilical arteries and two umbilical veins.

OUTCOME 29-4 _____ 17. The hormone(s) produced by the placenta include(s)
a. human chorionic gonadotrophin hormone.
b. estrogen and progesterone.
c. relaxin and human placental lactogen.
d. a, b, and c are correct.

OUTCOME 29-4 _____ 18. Throughout embryonic and fetal development, metabolic wastes generated by the fetus are eliminated by transfer to the
a. maternal circulation.
b. amniotic fluid.
c. chorion.
d. allantois.

OUTCOME 29-4 _____ 19. The umbilical cord or umbilical stalk contains
a. the amnion, allantois, and chorion.
b. paired umbilical arteries and the amnion.
c. a single umbilical vein and the chorion.
d. the allantois, blood vessels, and the yolk stalk.

OUTCOME 29-5 _____ 20. During gestation, the mother's lungs deliver extra oxygen and remove excess carbon dioxide generated by the fetus, requiring
a. increased maternal respiratory rate and decreased tidal volume.
b. decreased maternal respiratory rate and increased tidal volume.
c. increased maternal respiratory rate and tidal volume.
d. decreased maternal respiratory rate and tidal volume.

OUTCOME 29-5 _____ 21. The thickest portion of the uterine wall that provides much of the force needed to move a fetus out of the uterus and into the vagina is the
a. endometrium.
b. perimetrium.
c. myometrium.
d. basilar zone.

OUTCOME 29-5 _____ 22. Prostaglandins in the endometrium
a. stimulate smooth muscle contractions.
b. cause the mammary glands to begin secretory activity.
c. cause an increase in the maternal blood volume.
d. initiate the process of organogenesis.

OUTCOME 29-5 _____ 23. During gestation, the primary major compensatory adjustment(s) is (are)

    a. increased respiratory rate and tidal volume.

    b. increased maternal requirements for nutrients.

    c. increased glomerular filtration rate.

    d. all of the above.

OUTCOME 29-6 _____ 24. The sequential stages of labor are

    a. dilation, expulsion, and placental.

    b. fertilization, cleavage, and implantation.

    c. fertilization, implantation, and placental.

    d. expulsion, placental, birth.

OUTCOME 29-6 _____ 25. When a woman's water breaks, it occurs late in the

    a. expulsion stage.

    b. placental state.

    c. dilation stage.

    d. neonatal period.

OUTCOME 29-7 _____ 26. The sequential stages that identify the features and functions associated with the human experience are

    a. neonatal, childhood, infancy, and maturity.

    b. neonatal, postnatal, childbirth, and adolescence.

    c. neonatal, infancy, childhood, adolescence, and maturity.

    d. prenatal, neonatal, postnatal, and infancy.

OUTCOME 29-7 _____ 27. The systems that were relatively nonfunctional during the fetus's prenatal period that must become functional at birth are the

    a. circulatory, muscular, and skeletal.

    b. integumentary, reproductive, and nervous.

    c. endocrine, nervous, and circulatory.

    d. respiratory, digestive, and excretory.

OUTCOME 29-8 _____ 28. The normal chromosome complement of a typical somatic, or body, cell is

    a. 23.

    b. N, or haploid.

    c. 46.

    d. 92.

OUTCOME 29-8 _____ 29. Gametes are different from ordinary somatic cells because

    a. they contain only half the normal number of chromosomes.

    b. they contain the full complement of chromosomes.

    c. the chromosome number doubles in gametes.

    d. gametes are diploid, or 2N.

OUTCOME 29-8 _____ 30. During gamete formation, meiosis splits the chromosome pairs, producing

    a. diploid gametes.

    b. haploid gametes.

    c. gametes with a full chromosome complement.

    d. duplicate gametes.

OUTCOME 29-8 _____ 31. The normal male genotype is _____, and the normal female genotype is _____.

    a. XX; XY

    b. X; Y

    c. XY; XX

    d. Y; X

OUTCOME 29-8 _____ 32. If an allele must be present on both the maternal and paternal chromosomes to affect the phenotype, the allele is said to be

    a. dominant.

    b. codominant.

    c. recessive.

    d. complementary.

OUTCOME 29-8 _____ 33. When dominant alleles on two genes interact to produce a phenotype different from that seen when one gene contains recessive alleles, the interaction is called

    a. complementary gene action.

    b. suppression.

    c. crossing over.

    d. codominance.

Completion

Using the terms below, complete the following statements. Use each term only once.

childhood	fertilization	second trimester	heterozygous
capacitation	first trimester	true labor	development
meiosis	yolk sac	parturition	human chorionic
expulsion	gametogenesis	placenta	gonadotropin
chorion	polyspermy	autosomal	embryological
homozygous	infancy	induction	development

OUTCOME 29-1 1. The gradual modification of anatomical structures during the period from fertilization to maturity is called _____.

OUTCOME 29-1 2. The events that occur during the first two months after fertilization comprise _____.

OUTCOME 29-2 3. One chromosome in each pair is contributed by the sperm and the other by the egg at _____.

OUTCOME 29-2 4. Sperm cannot fertilize an egg until they have undergone an activation in the vagina called _____.

OUTCOME 29-2 5. The process of fertilization by more than one sperm that produces a nonfunctional zygote is known as _____.

OUTCOME 29-3 6. As development proceeds, the chemical interplay among developing cells that can influence differentiation patterns is called _____.

OUTCOME 29-3 7. The time during prenatal development when the fetus begins to look distinctively human is referred to as the _____.

OUTCOME 29-3 8. The period of time during which the rudiments of all the major organ systems appear is referred to as the _____.

OUTCOME 29-4 9. The first of the extraembryonic membranes to appear, which ultimately becomes involved in blood formation, is the _____.

OUTCOME 29-4   10. The extraembryonic membrane formed from the mesoderm and trophoblast is the _____.

OUTCOME 29-4   11. The placental hormone present in blood or urine samples that provides a reliable indication of pregnancy is _____.

OUTCOME 29-5   12. The vital link between maternal and embryonic systems that support the fetus during development is the _____.

OUTCOME 29-5   13. When the biochemical and mechanical factors reach the point of no return in the uterus, it indicates the beginning of _____.

OUTCOME 29-6   14. The goal of labor is _____.

OUTCOME 29-6   15. The stage in which contractions reach maximum intensity is the _____ stage.

OUTCOME 29-7   16. The life stage characterized by events that occur prior to puberty is called _____.

OUTCOME 29-7   17. The life stage that follows the neonatal period and continues to two years of age is referred to as _____.

OUTCOME 29-8   18. The special form of cell division leading to the production of sperm or eggs is _____.

OUTCOME 29-8   19. The formation of gametes is called _____.

OUTCOME 29-8   20. Chromosomes with genes that affect mostly somatic characteristics are referred to as _____ chromosomes.

OUTCOME 29-8   21. If both chromosomes of a homologous pair carry the same allele of a particular gene, the individual is _____ for that trait.

OUTCOME 29-8   22. When an individual has two different alleles carrying different instructions, the individual is _____ for that trait.

Matching

Match the terms in column B with the terms in column A. Use letters for answers in the spaces provided. Use each term only once.

Part I

		Column A	Column B
OUTCOME 29-1	_____	1. conception	A. cellular chemical interplay
OUTCOME 29-2	_____	2. cell specialization	B. differentiation
OUTCOME 29-2	_____	3. fertilization	C. prenatal development
OUTCOME 29-2	_____	4. amphimixis	D. milk production
OUTCOME 29-2	_____	5. corona radiata	E. pronuclei fusion
OUTCOME 29-2	_____	6. induction	F. fetus looks human
OUTCOME 29-3	_____	7. acrosin	G. fertilization
OUTCOME 29-3	_____	8. gestation	H. zygote formation
OUTCOME 29-4	_____	9. second trimester	I. mesoderm and endoderm
OUTCOME 29-4	_____	10. allantois	J. mesoderm and trophoblast
OUTCOME 29-4	_____	11. amnion	K. proteolytic enzyme
OUTCOME 29-5	_____	12. human placental lactogen	L. thick envelope surrounding oocyte
OUTCOME 29-5	_____	13. the chorion	M. mesoderm and ectoderm

Part II

		Column A	Column B
OUTCOME 29-5	_____	14. prostaglandin	N. individual gene variation
OUTCOME 29-5	_____	15. relaxin	O. full set of genetic material
OUTCOME 29-6	_____	16. parturition	P. visible characteristics
OUTCOME 29-6	_____	17. true labor	Q. chromosomes and component genes
OUTCOME 29-7	_____	18. neonate	R. begins at puberty
OUTCOME 29-7	_____	19. adolescence	S. forcible expulsion of the fetus
OUTCOME 29-8	_____	20. genome	T. newborn infant
OUTCOME 29-8	_____	21. mutation	U. positive feedback
OUTCOME 29-8	_____	22. alleles	V. forms of a given gene
OUTCOME 29-8	_____	23. phenotype	W. softens symphysis pubis
OUTCOME 29-8	_____	24. genotype	X. stimulates smooth muscle contractions

Drawing/Illustration Labeling

Identify each numbered structure in the following figure. Place your answers in the spaces provided.

OUTCOME 29-1 **FIGURE 29-1** Major Forms of Inheritance

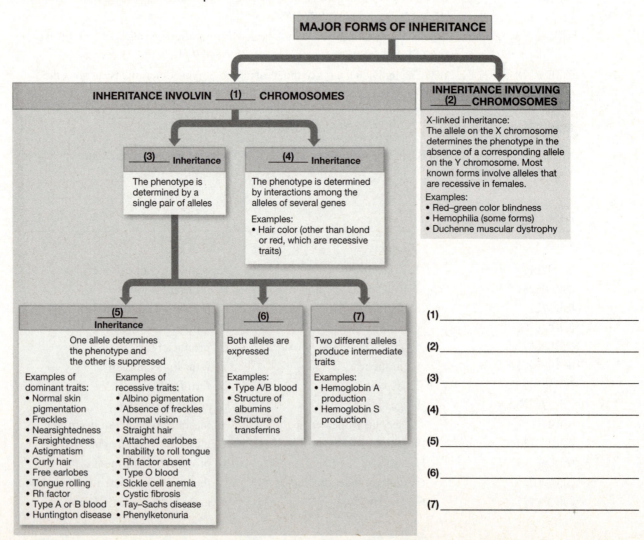

(1) _____

(2) _____

(3) _____

(4) _____

(5) _____

(6) _____

(7) _____

LEVEL 2: REVIEWING CONCEPTS

Chapter Overview

Using the terms below, identify the numbered locations on the pathway through the female reproductive tract.

early blastocyst	secondary oocyte	morula	2-cell stage
fertilization	implantation	zygote	8-cell stage

Follow the path the ovum takes as it develops in the ovary and travels into the uterine tube, where it is fertilized and undergoes cleavage until implantation takes place in the uterine wall. Your task is to identify the structures or processes at the numbered locations on the diagram. Record your answers in the spaces below the diagram.

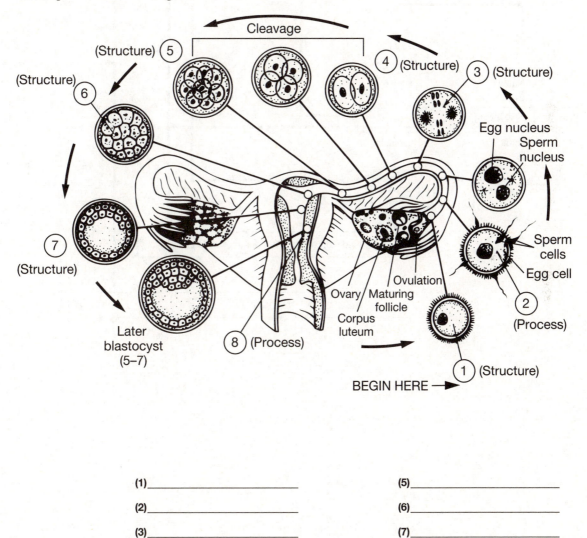

(1)_____   (5)_____

(2)_____   (6)_____

(3)_____   (7)_____

(4)_____   (8)_____

Concept Map I

Using the following terms, fill in the circled, numbered, blank spaces to correctly complete the concept map. Use each term only once.

fetal oxytocin release   parturition   placenta

maternal oxytocin   positive feedback   estrogen

distortion of myometrium

[Initiation of Labor and Delivery]

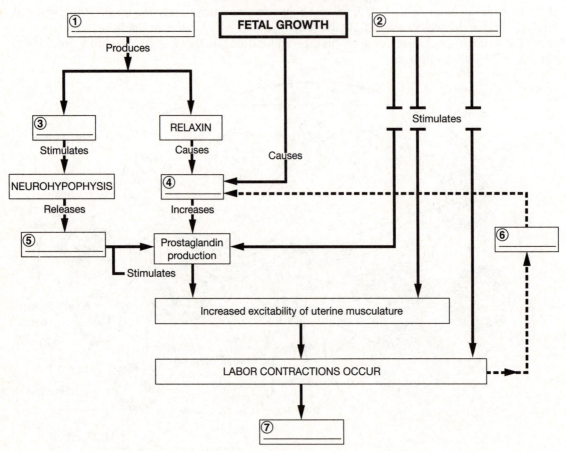

Concept Map II

Using the following terms, fill in the circled, numbered, blank spaces to correctly complete the concept map. Use each term only once.

placenta   ovary   anterior pituitary
relaxin   progesterone   mammary gland development

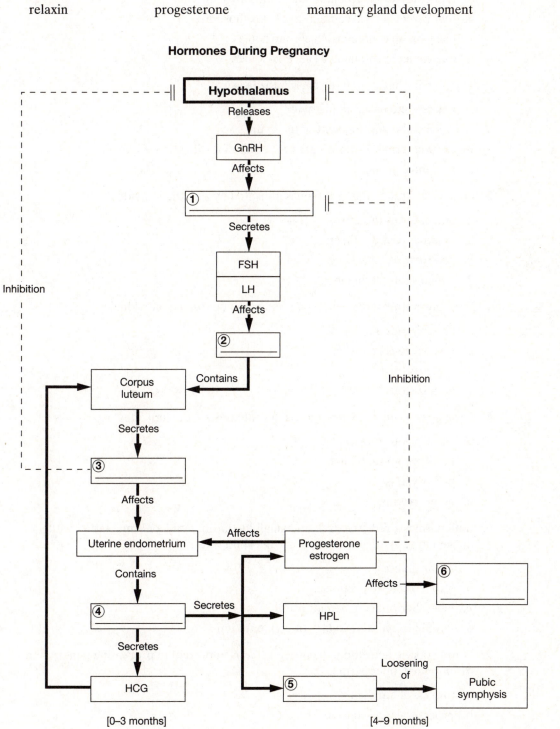

Hormones During Pregnancy

Multiple Choice

Place the letter corresponding to the best answer in the space provided.

_____ 1. In the male, the completion of metaphase II, anaphase II, and telophase II produces

    a. four gametes, each containing 46 chromosomes.

    b. four gametes, each containing 23 chromosomes

    c. one gamete containing 46 chromosomes.

    d. one gamete containing 23 chromosomes.

_____ 2. The primary function of the spermatozoa is to

    a. nourish and program the ovum.

    b. support the development of the ovum.

    c. carry paternal chromosomes to the site of fertilization.

    d. a, b, and c are correct.

_____ 3. For a given trait, if the genotype is indicated by AA, the individual is

    a. homozygous recessive.

    b. heterozygous dominant.

    c. heterozygous recessive.

    d. homozygous dominant.

_____ 4. For a given trait, if the genotype is indicated by Aa, the individual is

    a. homozygous dominant.

    b. homozygous recessive.

    c. heterozygous.

    d. homozygous.

_____ 5. For a given trait, if the genotype is indicated by aa, the individual is

    a. homozygous recessive.

    b. homozygous dominant.

    c. homozygous.

    d. heterozygous.

_____ 6. If albinism is a recessive trait and an albino mother and a normal father with the genotype AA have an offspring, the child will

    a. be an albino.

    b. have normal coloration.

    c. have abnormal pigmentation.

    d. have blue eyes due to lack of pigmentation.

_____ 7. During oocyte activation, the process that is important in preventing penetration by more than one sperm is the

    a. process of amphimixis.

    b. process of capacitation.

    c. process of monospermy.

    d. cortical reaction.

_____ 8. The sequential developmental stages that occur during cleavage are:

    a. blastocyst → blastomeres → trophoblast → morula.

    b. blastomeres → morula → blastocyst → trophoblast.

    c. yolk sac → amnion → allantois → chorion.

    d. amnion → allantois → chorion → yolk sac.

_____ 9. The zygote arrives in the uterine cavity as a

    a. morula.

    b. blastocyst.

    c. trophoblast.

    d. chorion.

_____ 10. During implantation, the inner cell mass of the blastocyst separates from the trophoblast, creating a fluid-filled chamber called the

    a. blastodisk.

    b. blastocoele.

    c. allantois.

    d. amniotic cavity.

_____ 11. The formation of extraembryonic membranes occurs in the correct sequential steps, which are

    a. blastomeres, morula, blastocyst, and trophoblast.

    b. amnion, allantois, chorion, and yolk sac.

    c. yolk sac, amnion, allantois, and chorion.

    d. blastocyst, trophoblast, amnion, and chorion.

_____ 12. In the event of "fraternal" or dizygotic twins,

    a. the blastomeres separate early during cleavage.

    b. the inner cell mass splits prior to gastrulation.

    c. two separate eggs are ovulated and fertilized.

    d. a, b, and c are correct.

_____ 13. The enzyme hyaluronidase is necessary to

    a. induce labor.

    b. permit fertilization.

    c. promote gamete formation.

    d. support the process of gastrulation.

_____ 14. The important and complex development event(s) that occur during the first trimester is (are)

    a. cleavage.

    b. implantation and placentation.

    c. embryogenesis.

    d. a, b, and c are correct.

_____ 15. Exchange between the embryonic and maternal circulations occurs by diffusion across the syncytial and cellular trophoblast layers via the

    a. umbilicus.

    b. allantois.

    c. chorionic blood vessels.

    d. yolk sac.

_____ 16. Karyotyping is the determination of

    a. an individual's chromosome complement.

    b. the life stages of an individual.

    c. the stages of prenatal development.

    d. an individual's stage of labor.

_____ 17. Colostrum, produced and secreted by the mammary glands, contains proteins that help the infant

    a. activate the digestive system to become fully functional.

    b. get adequate amounts of fat to help control body temperature.

    c. to initiate the milk let-down reflex.

    d. ward off infections until its own immune system becomes fully functional.

_____ 18. Implantation begins when the

    a. morula first reaches the uterus.

    b. surface of the blastocyst nearest the inner cell mass makes contact with the uterine lining.

    c. yolk sac is fully formed.

    d. cortical reaction has triggered the onset of embryogenesis.

_____ 19. Embryogenesis is the process that establishes the foundation for

    a. the formation of the blastocyst.

    b. implantation to occur.

    c. the formation of the placenta.

    d. all of the major organ systems.

_____ 20. An ectopic pregnancy refers to

    a. implantation occurring within the endometrium.

    b. implantation occurring somewhere other than within the uterus.

    c. the formation of a gestational neoplasm.

    d. the formation of extraembryonic membranes in the uterus.

Completion

Using the terms below, complete the following statements. Use each term only once.

cleavage	genetic recombination	synapsis
alleles	chromatids	spontaneous mutations
tetrad	X-linked	differentiation
inheritance	chorion	hyaluronidase
activation	amniocentesis	simple inheritance
corona radiata	genetics	polygenic inheritance

1. Transfer of genetically determined characteristics from generation to generation refers to _____.

2. The study of the mechanisms responsible for inheritance is called _____.

3. Attached duplicate, doubled chromosomes resulting from the first meiotic division are called _____.

4. When corresponding maternal and paternal chromosomes pair off, the event is known as _____.

5. Paired, duplicated chromosomes visible at the start of meiosis form a combination of four chromatids called a(n) _____.

6. _____ is a phenomenon that greatly increases the range of possible variations among gametes.

7. _____ are the result of random errors in DNA replication.

8. The collection and analysis of amniotic fluid and fetal cells is a process known as _____.

9. The alternative forms of any one gene are called _____.

10. Characteristics carried by genes on the X chromosome that affect somatic traits are termed _____.

11. Phenotypic characters determined by interactions between a single pair of alleles are known as _____.

12. Interactions between alleles on several genes involve _____.

13. When the oocyte leaves the ovary, it is surrounded by a layer of follicle cells, the _____.

14. The enzyme in the acrosomal cap of sperm that is used to break down the follicular cells surrounding the oocyte is _____.

15. The sequence of cell divisions that begins immediately after fertilization and ends at the first contact with the uterine wall is called _____.

16. The fetal contribution to the placenta is the _____.

17. The creation of different cell types during development is called _____.

18. Conditions inside the oocyte resulting from the sperm entering the ooplasm initiate _____.

Short Essay

Briefly answer the following questions in the spaces provided below.

1. What is the primary difference between simple inheritance and polygenic inheritance?

2. What does the term *capacitation* refer to?

3. What are the four general processes that occur during the first trimester?

4. What three germ layers result from the process of gastrulation?

5. a. What are the four extraembryonic membranes that are formed from the three germ layers?

   b. From which given layers does each membrane originate?

6. What are the major physiological adjustments necessary in the maternal systems to support the developing fetus?

7. What three major factors oppose the inhibitory effect of progesterone on the uterine smooth muscle?

8. What primary factors interact to produce labor contractions in the uterine wall?

9. What are the three stages of labor?

10. What are the identifiable life stages that comprise postnatal development of distinctive characteristics and abilities?

11. What three events interact to promote increased hormone production and sexual maturation at adolescence?

12. What four processes are involved with aging that influence the genetic programming of individual cells?

LEVEL 3: CRITICAL THINKING AND CLINICAL APPLICATIONS

Using principles and concepts learned in Chapter 29, answer the following questions. Write your answers on a separate sheet of paper.

1. A common form of color blindness is associated with the presence of a dominant or recessive gene on the X chromosome. Normal color vision is determined by the presence of a dominant gene (C), and color blindness results from the presence of the recessive gene (c). Suppose a heterozygous normal female marries a normal male. Is it possible for any of their children to be color blind? Show the possibilities by using a Punnett square.

2. Albinism (aa) is inherited as a homozygous recessive trait. If a homozygous-recessive mother and a heterozygous father decide to have children, what are the possibilities of their offspring inheriting albinism? Use a Punnett square to show the possibilities.

3. Tongue rolling is inherited as a dominant trait. Even though a mother and father are tongue rollers (T), show how it would be possible to bear children who do not have the ability to roll the tongue. Use a Punnett square to show the possibilities.

4. After two years of trying desperately to have children, Bob and Sharon are informed by their physician that Sharon is pregnant. As the time for the birth approaches, they are informed that the delivery will be a breech birth, a condition in which the legs or buttocks of the fetus enter the vaginal canal first. What are the risks associated with a breech birth and what alternatives might be required?

Chapter 1:
An Introduction to Anatomy and Physiology

LEVEL 1: REVIEWING FACTS AND TERMS

Multiple Choice

1. d	6. d	11. d	16. c	21. d
2. b	7. d	12. c	17. d	22. b
3. c	8. b	13. d	18. c	23. b
4. a	9. c	14. a	19. a	24. c
5. c	10. c	15. d	20. d	

Completion

1. homeostasis	10. organs	19. liver
2. procrastination	11. urinary	20. medial
3. body structures	12. digestive	21. distal
4. eponym	13. integumentary	22. transverse
5. histologist	14. regulation	23. pericardial
6. embryology	15. equilibrium	24. mediastinum
7. physiology	16. autoregulation	25. peritoneal
8. tissues	17. positive feedback	
9. molecules	18. endocrine	

Matching

1. C	5. G	9. H	13. O	17. P
2. A	6. J	10. F	14. L	18. N
3. B	7. D	11. M	15. K	
4. I	8. E	12. Q	16. R	

Drawing/Illustration Labeling

Figure 1-1   Planes of the Body

1. frontal (coronal)	2. transverse	3. sagittal

Figure 1-2   Human Body Orientation and Direction

1. superior	6. anterior (ventral)	11. inferior
2. cranial (cephalic)	7. superior	12. proximal
3. posterior (dorsal)	8. lateral	13. medial
4. caudal	9. proximal	14. distal
5. inferior	10. distal	

Figure 1-3   Anatomical Landmarks

1. cephalic (head, craniofacial)
2. oral
3. axillary
4. brachial
5. carpal
6. palmar
7. digits (phalanges)
8. patellar
9. crural
10. tarsal

11. digits (phalanges)
12. hallux
13. ocular or orbital
14. buccal
15. thoracic
16. abdominal
17. umbilical
18. pelvic (pelvis)
19. pubic (pubis)
20. femoral

21. acromial
22. olecranon
23. lumbar
24. gluteal
25. popliteal
26. sural
27. calcaneal
28. plantar
29. cephalic

Figure 1-4   Body Cavities

(a)

1. pleural cavity
2. pericardial cavity
3. thoracic cavity
4. diaphragm
5. peritoneal cavity
6. abdominal cavity
7. pelvic cavity
8. abdominopelvic cavity

(b)

9. visceral pericardium
10. pericardial cavity
11. parietal pericardium

(c)

12. pericardial cavity
13. pleural cavity
14. pleura
15. mediastinum
16. spinal cord

LEVEL 2: REVIEWING CONCEPTS

Chapter Overview

1. atoms
2. molecules
3. organelles
4. cells
5. tissues
6. organs
7. systems
8. organism

Concept Maps

I Anatomy

1. macroscopic anatomy
2. regional anatomy
3. structure of organ systems
4. surgical anatomy
5. embryology
6. cytology
7. tissues

II Physiology

1. cell functions
2. chemical interactions
3. organ physiology
4. cardiac physiology
5. specific organ systems
6. respiratory
7. pathological physiology
8. effects of diseases

III Ventral Body Cavity—Coelom

1. thoracic cavity
2. mediastinum
3. left lung
4. trachea, esophagus
5. heart
6. peritoneal cavity
7. pelvic cavity
8. digestive glands and organs
9. reproductive organs

IV Clinical Technology—Radiological Procedures

1. PET scan
2. computerized (axial) tomography
3. x-rays
4. MRI
5. magnetic field and radio waves
6. imaging technique
7. internal structures

Multiple Choice

1. a	4. c	7. c	10. b	13. b
2. b	5. a	8. c	11. d	14. a
3. d	6. b	9. d	12. a	

Completion

1. histologist
2. cardiovascular
3. lymphatic
4. extrinsic regulation
5. nervous
6. knee
7. appendicitis
8. elbow
9. negative feedback
10. sternum

Short Essay

1. a. In the anatomical position, the hands are at the sides with the palms facing forward, and the feet are together.

   b. A supine person is lying down with face up.

   c. A prone person is lying down with face down.

2. Molecular → cells → tissues → organs → organ systems → organism

3. Homeostatic regulation refers to adjustments in physiological systems that are responsible for the preservation of homeostasis (the steady state).

4. In negative feedback a variation outside of normal limits triggers an automatic response that corrects the situation. In positive feedback the initial stimulus produces a response that exaggerates the stimulus.

5. Ventral (anterior), dorsal (posterior), cranial (cephalic), caudal

6. A sagittal section separates right and left portions. A transverse or horizontal section separates superior and inferior portions of the body.

7. a. pelvic cavity b. ventral cavity c. thoracic cavity d. abdominopelvic cavity

8. a. Body cavities protect delicate organs from accidental shocks, and cushion them from the thumps and bumps that occur while walking, running, and jumping.

   b. Body cavities permit significant changes in the size and shape of internal organs.

LEVEL 3: CRITICAL THINKING AND CLINICAL APPLICATIONS

1. Since there is a lung in each compartment, if one lung is diseased or infected the other lung may remain functional. Also, if one lung is traumatized due to injury, the other one may be spared and function sufficiently to save the life of the injured person.

2.
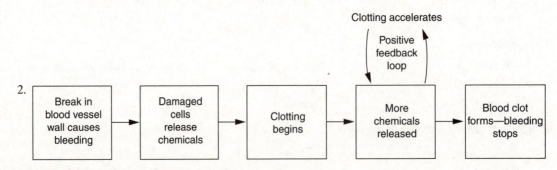

3. Pericardial, mediastinal, abdominal, pelvic

4. Extrinsic regulation. The nervous and endocrine systems can control or adjust the activities of many different systems simultaneously. This usually causes more extensive and potentially more effective adjustments in system activities.

5. When calcium levels are elevated, calcitonin is released. This hormone should bring about a decrease in blood calcium levels, thus decreasing the stimulus for its own release.

6. a. Urinary—elimination or conservation of water from the blood

   b. Digestive—absorption of water; loss of water in feces

   c. Integumentary—loss of water through perspiration

   d. Cardiovascular—distribution of water

7. At no. 1 on the graph there will be *increased blood flow to the skin and increased sweating* if the body temperature rises above 99°F. The *body surface cools* and the *temperature declines*. If the temperature falls below 98°F there is a *decrease in blood flow to the skin* and *shivering* occurs. These activities help to *conserve body heat* and the body temperature rises.

Chapter 2:
The Chemical Level of Organization

LEVEL 1: REVIEWING FACTS AND TERMS

Multiple Choice

1. d	6. c	11. b	16. c	20. a	24. b
2. b	7. d	12. d	17. b	21. c	25. c
3. c	8. c	13. d	18. a	22. b	26. a
4. b	9. a	14. b	19. c	23. d	27. d
5. a	10. c	15. c			

Completion

1. protons
2. mass number
3. covalent bonds
4. ionic bond
5. H_2
6. cation
7. Decomposition, synthesis
8. exergonic, endergonic
9. catalysts
10. organic
11. inorganic
12. solvent, solute
13. water
14. acidic
15. buffers
16. salt
17. carbonic acid
18. glucose
19. lipids
20. proteins
21. DNA
22. ATP
23. metabolic turnover

Matching

1. I	5. A	9. G	13. L	17. N
2. E	6. C	10. H	14. K	18. O
3. J	7. D	11. S	15. M	19. Q
4. B	8. F	12. P	16. R	

Drawing/Illustration Labeling

Figure 2-1   Diagram of an Atom

1. nucleus
2. electron shell
3. electron
4. protons and neutrons

Figure 2-2   Identification of Types of Bonds

1. nonpolar covalent bond
2. polar covalent bond
3. ionic bond

Figure 2-3   Identification of Organic Molecules

1. monosaccharide
2. disaccharide
3. polysaccharide
4. saturated fatty acid
5. polyunsaturated fatty acid
6. amino acid
7. cholesterol
8. DNA

LEVEL 2: REVIEWING CONCEPTS

Chapter Overview

1. electrons
2. electron clouds
3. nucleus
4. protons
5. neutrons
6. deuterium
7. isotope
8. oxygen
9. six
10. eight
11. double covalent bond
12. molecule
13. oxygen gas
14. polar covalent
15. negatively
16. positively
17. carbon dioxide
18. oxygen
19. water
20. salts
21. carbohydrates
22. lipids
23. proteins
24. nucleic acids
25. DNA
26. RNA
27. ATP
28. cells

Concept Maps

I Carbohydrates

1. monosaccharides
2. glucose
3. disaccharides
4. sucrose
5. complex carbohydrates
6. glycogen
7. plants

II Major Lipids

1. saturated
2. glycerides
3. Di-
4. glycerol + 3 fatty acids
5. local hormones
6. phospholipid
7. carbohydrates + diglyceride
8. steroids

III Proteins

1. amino acids
2. amino group
3. –COOH
4. variable group
5. structural proteins
6. keratin
7. enzymes
8. primary
9. alpha helix
10. globular proteins
11. quaternary

IV Nucleic Acids

1. deoxyribose nucleic acid
2. deoxyribose
3. purines
4. thymine
5. ribonucleic acid
6. N bases
7. ribose
8. pyrimidines
9. adenine

Multiple Choice

1. a
2. b
3. d
4. d
5. a
6. a
7. c
8. a
9. d
10. c
11. b
12. d
13. c
14. b
15. d
16. d
17. b
18. d
19. a
20. c
21. c
22. a
23. d
24. b

Completion

1. inert
2. mole
3. molecule
4. ions
5. ionic bond
6. molecular weight
7. salts
8. hydrophobic
9. alkaline
10. nucleic acids
11. isomers
12. hydrolysis
13. dehydration synthesis
14. saturated
15. peptide bond

Short Essay

1.

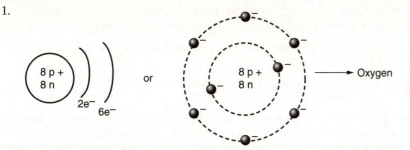

2. a. hydrolysis b. dehydration synthesis

3. Their outer energy levels contain the maximum number of electrons and they will not react with one another nor combine with atoms of other elements.

4. The oxygen atom has a much stronger attraction for the shared electrons than do the hydrogen atoms, so the electrons spend most of this time in the vicinity of the oxygen nucleus. Because the oxygen atom has two extra electrons part of the time, it develops a slight negative charge. The hydrogens develop a slight positive charge because their electrons are away part of the time.

5. $6 \times 12 = 72$

   $12 \times 1 = 12$

   $6 \times 16 = \underline{96}$

   MW $= 180$

6. (1) Solubility (efficient solvent), (2) reactivity (participates in many chemical reactions), (3) high heat capacity (absorbs and releases heat slowly), (4) ability to serve as a lubricant

7. *Carbohydrates,* ex. glucose; *lipids,* ex. steroids; *proteins,* ex. enzymes; *nucleic acid,* ex. DNA

8. In a saturated fatty acid, each carbon atom in the hydrocarbon tail has four single covalent bonds. If some of the carbon-to-carbon bonds are double covalent bonds, the fatty acid is unsaturated.

9. (1) adenine nucleotide,

   (2) thymine nucleotide,

   (3) cytosine nucleotide,

   (4) guanine nucleotide

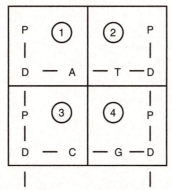

(Model)

10. Adenine, ribose, 3 phosphates

LEVEL 3: CRITICAL THINKING AND CLINICAL APPLICATIONS

1. $AB + CD \rightarrow AD + CB$ or

   $AB + CD \rightarrow AC + BD$

2. Baking soda is sodium bicarbonate ($NaHCO_3$). In solution, sodium bicarbonate reversibly dissociates into a sodium ion ($Na+$) and a bicarbonate ion (HCO_3-) [$NaHCO \leftrightarrow NA+ + HCO_3-$]. The bicarbonate ion will remove an excess hydrogen ion ($H+$) from the solution and form a weak acid, carbonic acid [$H+ + HCO_3^- \leftrightarrow H_2CO_3$], and the net effect is a less acidic (higher pH) stomach.

3. $C_6H_{12}O_6 + C_6H_{12}O_6 \leftrightarrow C_{12}H_{22}O_{11} + H_2O$

4. Carbohydrates are an immediate source of energy and can be metabolized quickly. Those not immediately metabolized may be stored as glycogen or excess may be stored as fat.

5. If a person exhales large amounts of CO_2, the equilibrium will shift to the left, and the level of H^+ in the blood will decrease, causing the pH to rise.

6. The heat of boiling breaks bonds in the protein's tertiary structure, quaternary structure, or both, resulting in structural change known as denaturation.

7. The RDA for fat intake per day is 30 percent of the total caloric intake. Fats are necessary for good health. They are involved in many body functions and serve as an energy reserve, provide insulation, cushion delicate organs, are essential structural components of all cells, and form membranes within and between cells that prevent the diffusion of solutes.

8. If the amount of enzyme is decreased at the second step, it would slow down the remaining steps of the pathway because less substrate would be available for the next two steps. This would cause a decrease in the amount of product.

Chapter 3:
The Cellular Level of Organization

LEVEL 1: REVIEWING FACTS AND TERMS

Multiple Choice

1. d	7. d	13. b	18. b	23. d	28. c
2. c	8. c	14. c	19. c	24. a	29. d
3. a	9. a	15. b	20. a	25. d	30. b
4. b	10. b	16. a	21. a	26. b	31. a
5. c	11. c	17. b	22. b	27. b	32. c
6. a	12. b				

Completion

1. phospholipids
2. cytoskeleton
3. fixed ribosomes
4. nuclear pores
5. nucleolus
6. gene
7. translation
8. ribosomes
9. osmosis
10. endocytosis
11. transmembrane
12. interphase
13. anaphase
14. MPF
15. cancer
16. differentiation

Matching

1. I	6. B	11. N	15. S	19. Q
2. H	7. C	12. R	16. M	20. U
3. J	8. D	13. T	17. L	21. O
4. A	9. F	14. K	18. P	
5. G	10. E			

Drawing/Illustration Labeling

Figure 3-1   Anatomy of a Model Cell

1. microvilli
2. secretory vesicles
3. cytosol
4. lysosome
5. centrosome
6. centriole
7. chromatin

8. nucleoplasm
9. nucleolus
10. nuclear envelope
11. cytoskeleton
12. plasma membrane
13. Golgi apparatus
14. proteasomes

15. mitochondrion
16. peroxisome
17. nuclear pores
18. smooth endoplasmic reticulum
19. rough endoplasmic reticulum
20. fixed ribosome
21. free ribosome

Figure 3-2   The Plasma Membrane

1. glycolipids or glycocalyx
2. integral glycoproteins

3. cholesterol
4. peripheral protein

5. integral protein with channel

Figure 3-3   The Stages of Mitosis

1. interphase
2. early prophase

3. late prophase
4. metaphase

5. anaphase
6. telophase

LEVEL 2: REVIEWING CONCEPTS

Chapter Overview

1. extracellular fluid
2. cell membrane
3. cilia
4. phospholipids
5. integral
6. peripheral
7. channels
8. cytosol
9. ions
10. proteins
11. waste

12. cytoskeleton
13. organelles
14. protein synthesis
15. mitochondria
16. cristae
17. matrix
18. metabolic enzymes
19. ATP
20. nucleus
21. nuclear envelope
22. nuclear pores

23. nucleoli
24. ribosomes
25. nucleoplasm
26. chromosomes
27. Golgi apparatus
28. saccules
29. lysosomes
30. digestive enzymes
31. autolysis

Concept Maps

I Typical Animal Cell

1. lipid bilayer
2. proteins
3. organelles

4. membranous
5. nucleolus
6. lysosomes

7. centrioles
8. ribosomes
9. fluid component

II Membrane Permeability

1. diffusion
2. molecular size
3. channel mediated
4. net diffusion of water

5. active transport
6. specificity
7. vesicular transport

8. pinocytosis
9. "cell eating"
10. exocytosis

III Cell Cycle

1. somatic cells
2. G_1 phase
3. G_2 phase

4. DNA replication
5. mitosis
6. metaphase

7. telophase
8. cytokinesis

Multiple Choice

1. d	7. c	12. a	17. d	22. c
2. b	8. a	13. d	18. b	23. b
3. c	9. c	14. d	19. a	24. d
4. a	10. d	15. b	20. b	25. a
5. b	11. b	16. c	21. d	26. c
6. d				

Completion

1. hydrophobic	5. rough ER	9. permeability	13. phagocytosis
2. glycocalyx	6. exocytosis	10. diffusion	14. mitosis
3. microvilli	7. endocytosis	11. isotonic	15. cytokinesis
4. cilia	8. peroxisomes	12. autolysis	

Short Essay

1. a. Circulatory—RBC
   b. Muscular—muscle cell
   c. Reproductive—sperm cell
   d. Skeletal—bone cell
   e. Nervous—neuron (Other systems and cells could be listed for this answer.)

2. a. Physical isolation
   b. Regulation of exchange with the environment
   c. Sensitivity
   d. Structural support

3. G_0—normal cell function
   G_1—cell growth, duplication of organelles, and protein synthesis
   S—DNA replication and synthesis of histones
   G_2—protein synthesis

4. Cytoskeleton, centrioles, ribosomes, mitochondria, nucleus, nucleolus, ER, Golgi apparatus, lysosomes

5. Centrioles—organization of microtubules
   Cilia—move fluids or solids across cell surfaces
   Flagella—move cell through fluids

6. (1) distance, (2) molecule size, (3) temperature, (4) gradient size, (5) electrical forces

LEVEL 3: CRITICAL THINKING AND CLINICAL APPLICATIONS

1. Cytokinesis is the cytoplasmic movement that separates two daughter cells. It completes the process of cell division.

2. Mitochondria contain RNA, DNA, and enzymes needed to synthesize proteins. The mitochondrial DNA and associated enzymes are unlike those found in the nucleus of the cell. The synthetic capability enables mitochondrial DNA to control their own maintenance, growth, and reproduction. The muscle cells of bodybuilders have high rates of energy consumption and over time their mitochondria respond to the increased energy demand by reproducing. This results in a higher number of mitochondria in each cell.

3. The isolation of internal contents of membrane-bound organelles allows them to manufacture or store secretions, enzymes, or toxins that could adversely affect the cytoplasm. A membrane supports the increased efficiency of having specialized enzyme systems concentrated in one place (i.e., the concentration of enzymes necessary for energy production in the mitochondrion increases the efficiency of cellular respiration).

4. In saltwater drowning, the body fluid becomes hypertonic to cells. The cells dehydrate and shrink, which causes the victim to go into shock, thus allowing time for resuscitation to occur. In freshwater drowning, the body fluid is hypotonic to cells, causing the cells to swell and burst. The intracellular K^+ are released into circulation and, when contacting the heart in excessive amounts, cause cardiac arrest, thus allowing little time for the possibility of resuscitation.

5. Gatorade is a combination fruit drink that is slightly hypotonic to body fluids after strenuous exercise. The greater solute concentration inside the cells causes increased osmosis of water into the cells to replenish the fluid that has been lost during exercise—hydration.

6. Because of the increase of solute concentration in the body fluid, it becomes hypertonic to the RBCs. The red blood cells dehydrate and shrink—crenation. The crenated RBCs lose their oxygen-carrying capacity and body tissues are deprived of the oxygen necessary for cellular metabolism.

7. The vegetables contain a greater solute concentration than does the watery environment surrounding them. The vegetables are hypertonic to the surrounding fluid; therefore, the osmosis of water into the vegetables causes crispness, or turgidity.

Chapter 4:
The Tissue Level of Organization

LEVEL 1: REVIEWING FACTS AND TERMS

Multiple Choice

1. d	8. a	15. c	22. a	29. c	36. b
2. c	9. b	16. a	23. b	30. d	37. d
3. d	10. c	17. a	24. d	31. a	38. b
4. b	11. d	18. b	25. c	32. d	39. c
5. c	12. d	19. c	26. a	33. d	40. a
6. a	13. d	20. a	27. c	34. d	41. b
7. d	14. b	21. c	28. b	35. a	

Completion

1. neural
2. polarity
3. mesothelium
4. endothelium
5. exocytosis
6. connective
7. collagen
8. reticular
9. ligaments
10. chondrocytes
11. lamina propria
12. superficial fascia
13. skeletal
14. neuroglia
15. necrosis
16. abscess
17. declines; increase

Matching

1. E	5. A	9. D	13. R	17. M
2. H	6. C	10. N	14. K	18. J
3. G	7. F	11. L	15. O	
4. B	8. I	12. Q	16. P	

Drawing/Illustration Labeling

Figure 4-1   Types of Epithelial Tissue

1. simple squamous epithelium
2. transitional epithelium
3. stratified squamous epithelium
4. stratified columnar epithelium
5. simple columnar epithelium
6. pseudostratified ciliated columnar epithelium
7. simple cuboidal epithelium

Figure 4-2   Types of Connective Tissue

1. connective tissue proper
2. dense regular connective tissue
3. adipose connective tissue
4. hyaline cartilage
5. fibrocartilage
6. reticular connective tissue
7. elastic cartilage
8. osseous (bone) tissue
9. elastic connective tissue
10. vascular (fluid) connective tissue (blood)

Figure 4-3   Types of Muscle Tissue

1. smooth muscle tissue   2. cardiac muscle tissue   3. skeletal muscle tissue

Figure 4-4   Identify the Type of Tissue

1. neural (nervous) tissue

LEVEL 2: REVIEWING CONCEPTS

Chapter Overview

1. stratified squamous
2. simple columnar
3. pseudostratified columnar
4. simple squamous
5. stratified cuboidal
6. simple cuboidal
7. transitional

8. stratified columnar
9. adipose
10. areolar
11. dense regular
12. dense irregular
13. hyaline cartilage
14. fibrocartilage

15. elastic cartilage
16. bone or osseous
17. blood
18. skeletal
19. cardiac
20. smooth
21. neural

Concept Maps

I Human Body Tissues

1. epithelial   2. connective   3. muscle   4. neural

II Epithelial Types

1. stratified squamous
2. lining
3. transitional

4. stratified cuboidal
5. male urethra mucosa
6. pseudostratified columnar

7. exocrine glands
8. endocrine glands

III Connective Tissue

1. loose connective tissue
2. adipose
3. regular
4. tendons or ligaments

5. tendons or ligaments
6. fluid connective tissue
7. blood

8. hyaline
9. chondrocytes in lacunae
10. bone

IV Connective Tissue Proper

1. adipose cells
2. phagocytosis

3. mast cells
4. collagen

5. branching interwoven framework
6. ground substance

V Muscle Tissue

1. cardiac
2. involuntary

3. nonstriated
4. viscera

5. multinucleate
6. voluntary

VI Neural Tissue

1. soma   2. dendrites   3. neuroglia

VII Membranes

1. mucous
2. goblet cells
3. pericardium

4. fluid formed on membrane surface
5. skin
6. thick, waterproof, dry

7. synovial
8. no basal lamina
9. phagocytosis

Multiple Choice

1. b	6. b	11. b	16. c	21. a
2. c	7. c	12. d	17. d	22. b
3. d	8. d	13. b	18. b	23. c
4. a	9. b	14. c	19. c	24. b
5. c	10. c	15. a	20. d	25. b

Completion

1. lamina propria	6. fibroblasts	11. goblet cells
2. transudate	7. platelets	12. plasma
3. tendons	8. lacunae	13. stroma
4. subserous fascia	9. serous	14. axon
5. ground substance	10. synovial fluid	15. dense regular connective tissue

Short Essay

1. Epithelial, connective, muscle, and neural tissue

2. Provides physical protection, controls permeability, provides sensations, and provides specialized secretions

3. Microvilli are abundant on epithelial surfaces where absorption and secretion take place. A cell with microvilli has at least 20 times the surface area of a cell without them. A typical ciliated cell contains about 250 cilia that beat in coordinated fashion. Materials are moved over the epithelial surface by the synchronized beating of cilia.

4. Merocrine secretion—the product is released through exocytosis.

   Apocrine secretion—loss of cytoplasm as well as secretory product.

   Holocrine secretion—product is released, cell is destroyed.

   Merocrine and apocrine secretions leave the cell intact and are able to continue secreting; holocrine does not.

5. Serous—watery solution containing enzymes

   Mucous—viscous mucous

   Mixed glands—serous and mucous secretions

6. (a) Cellularity, (b) polarity, (c) attachment, (d) avascularity, (e) regeneration

7. (a) Specialized cells, (b) extracellular protein fibers, (c) ground substance

8. Collagen, reticular, elastic

9. Mucous membranes, serous membranes, cutaneous membranes, and synovial membranes

10. Skeletal, cardiac, smooth

11. Neurons—transmission of nerve impulses from one region of the body to another

    Neuroglia—support framework for neural tissue

LEVEL 3: CRITICAL THINKING AND CLINICAL APPLICATIONS

1. New skeletal muscle is produced by the division and fusion of satellite cells, mesenchymal cells that persist in adult skeletal muscle tissue.

2. All of these regions are subject to mechanical trauma and abrasion—by food (pharynx and esophagus), feces (anus), and intercourse or childbirth (vagina).

3. Collagen fibers add strength to connective tissue. A vitamin C deficiency might result in the production of connective tissue that is weak and prone to damage because a lack of vitamin C in the diet interferes with the ability of fibroblasts to produce collagen.

4. The leukocytes (WBC) in the blood and the reticular tissue of lymphoid organs are involved with the process of phagocytosis.

5. As in the case of skeletal muscle, cardiac muscle fibers are incapable of dividing. Because this tissue lacks the satellite cells in skeletal muscle that are responsible for dividing to repair injured or damaged muscle, cardiac muscle cannot repair itself.

6. The skin is separated from the underlying structures by a layer of loose connective tissue, which provides padding and permits a considerable amount of independent movement of the two layers.

7. Cartilage lacks a direct blood supply, which is necessary for proper healing to occur. Instead of having chondrocytes repair the injury site with new cartilage, fibroblasts migrate into the area and replace the damaged cartilage with fibrous scar tissue.

8. Histamine—dilates blood vessels, thus increasing the blood flow to the area and providing nutrients and oxygen necessary for the regenerative process. Heparin—prevents blood clotting that might oppose the increased flow of blood to the injured area.

9. Any substance that enters or leaves your body must cross an epithelium. Some epithelia are easily crossed by compounds as large as proteins; others are relatively impermeable. Many epithelia contain the molecular component needed for selective absorption or secretion. The epithelial barrier can be regulated and modified in response to stimuli.

10. Connective tissue proper: consists of a permanent cell population of fibroblasts, adipocytes, and mesenchymal cells, which function in local maintenance, repair, and energy storage. Other migratory cells, including macrophages, mast cells, lymphocytes, plasma cells, and microphages, aggregate at sites of tissue injury and defend and repair damaged tissues.

Chapter 5:
The Integumentary System

LEVEL 1: REVIEWING FACTS AND TERMS

Multiple Choice

1. d	6. b	11. c	16. d	21. c
2. a	7. c	12. d	17. c	22. a
3. a	8. d	13. c	18. c	23. d
4. c	9. d	14. a	19. d	24. c
5. d	10. d	15. b	20. a	25. b

Completion

1. epidermis
2. stratum lucidum
3. stratum corneum
4. Langerhans cells
5. MSH
6. melanin
7. cholecalciferol
8. EGF
9. connective
10. reticular layer
11. vellus
12. arrector pili
13. follicle
14. melanocytes
15. sebum
16. eccrine glands
17. apocrine
18. decrease
19. eponychium
20. keloid
21. blisters
22. granulation tissue
23. glandular

Matching

1. H	4. K	7. F	10. J
2. L	5. D	8. C	11. G
3. A	6. I	9. B	12. E

Drawing/Illustration Labeling

Figure 5-1   Components of the Integumentary System

1. epidermis
2. papillary layer
3. reticular layer
4. dermis
5. hypodermis (subcutaneous layer)
6. hair shaft
7. pore of sweat gland duct
8. tactile corpuscle
9. sebaceous gland
10. arrector pili muscle
11. sweat gland duct
12. hair follicle
13. lamellated corpuscle
14. nerve fibers
15. sweat gland
16. artery
17. vein
18. fat

Figure 5-2   Hair Follicles and Hairs

1. hair shaft
2. sebaceous gland
3. arrector pili muscle
4. hair root
5. connective tissue sheath
6. hair bulb
7. hair matrix

Figure 5-3   Nail Structure (a) Superficial View (b) Longitudinal Section

1. free edge
2. lateral nail fold
3. nail body
4. lunula
5. proximal nail fold
6. eponychium
7. eponychium
8. proximal nail fold
9. nail root
10. lunula
11. nail body
12. hyponychium
13. phalanx
14. dermis
15. epidermis

LEVEL 2: REVIEWING CONCEPTS

Chapter Overview

1. cutaneous
2. accessory
3. skin
4. hair and nails
5. keratinocytes
6. basale
7. granulosum
8. corneum
9. papillary layer
10. reticular layer
11. collagen
12. elastic
13. hypodermis
14. hair
15. vellus hairs
16. terminal hairs
17. sebaceous
18. sweat
19. mammary
20. ceruminous
21. ear wax
22. hyponychium
23. eponychium
24. lunula
25. environmental

Concept Maps
I Integumentary System

1. dermis
2. vitamin D synthesis
3. sensory reception
4. exocrine glands
5. lubrication
6. produce secretions

II Integument

1. skin
2. epidermis
3. granulosum
4. papillary layer
5. nerves
6. collagen
7. hypodermis
8. connective
9. fat

III Accessory Structures

1. terminal
2. "peach fuzz"
3. arms and legs
4. glands
5. sebaceous
6. odors
7. merocrine or "eccrine"
8. cerumen (ear wax)
9. cuticle
10. lunula
11. thickened stratum corneum

Multiple Choice

1. c
2. c
3. a
4. c
5. c
6. d
7. b
8. d
9. a
10. d
11. b
12. c

Completion

1. Merkel cells
2. carotene
3. cyanosis
4. vitiligo
5. papillary
6. dermatitis
7. hypodermis
8. seborrheic dermatitis
9. keloid
10. Langerhans cells

Short Essay

1. Palms of hand; soles of feet. These areas are covered by 30 or more layers of cornified cells. The epidermis in these locations may be six times thicker than the epidermis covering the general body surface. This is a response to friction and pressure.

2. Bacteria break down the organic secretions of the apocrine glands.

3. Stratum corneum, stratum granulosum, stratum spinosum, stratum basale, papillary layer, reticular layer

4. Long-term damage can result from chronic UV radiation exposure. Alterations in underlying connective tissue lead to premature wrinkling, and skin cancer can result from chromosomal damage or breakage.

5. Hairs are dead, keratinized structures, and no amount of oiling, shampooing, or nourishment will influence either the exposed hair or the follicles buried in the dermis.

6. Creams and oils are similar to the natural secretions of the sebaceous glands. Applying them to the skin assists the sebaceous glands in maintaining a continuous, flexible, and protective integument.

7. Bedsores are a result of circulatory restrictions. They can be prevented by frequent changes in body position that vary the pressure applied to specific blood vessels.

8. Special smooth muscles in the dermis, the arrector pili muscles, pull on the hair follicles and elevate the hair, producing "goose bumps."

LEVEL 3: CRITICAL THINKING AND CLINICAL APPLICATIONS

1. Compared to our skin, the ocean is a hypertonic solution, thus causing water to leave the body by crossing the epidermis from the underlying tissues.

2. Blood with abundant O_2 is bright red, and blood vessels in the dermis normally give the skin a reddish tint. During the frightening experience, the circulatory supply to the skin is temporarily reduced because of the constriction of the blood vessels; thus, the skin becomes relatively pale.

3. Retin-A increases blood flow to the dermis and stimulates dermal repairs, resulting in a decrease in wrinkle formation and existing wrinkles become smaller.

4. Third-degree burns destroy the epidermis and dermis, extending into subcutaneous tissues. Despite swelling, these burns are less painful than second-degree burns, because sensory neurons are destroyed along with accessory structures, blood vessels, and other dermal components. Extensive third-degree burns cannot repair themselves, because granulation tissue cannot form and epithelial cells are unable to cover the injury.

5. a. Protects the scalp from ultraviolet light
   b. Cushions a blow to the head
   c. Prevents entry of foreign particles (eyelashes, nostrils, ear canals)
   d. Sensory nerves surrounding the base of each hair provide an early warning system that can help prevent injury.

6. a. Ultraviolet radiation in sunlight converts a cholesterol-related steroid into vitamin D_3, or cholecalciferol. This compound is then converted to calcitriol, which is necessary in the small intestine for absorbing the calcium and phosphorous essential for normal bone maintenance and growth.
   b. Children can drink more milk because it is routinely fortified with "vitamin D," which is easily absorbed by the intestines.

7. Sensible perspiration produced by merocrine (eccrine) glands plus a mixture of electrolytes, metabolites, and waste products is a clear secretion that is more than 99 percent water. When all of the merocrine sweat glands are working at maximum, the rate of perspiration may exceed a gallon per hour, and dangerous fluid and electrolyte losses occur.

8. When the skin is subjected to mechanical stresses, stem cells in the basale divide more rapidly, and the depth of the epithelium increases.

9. To reach the underlying connective tissues, a bacterium must:

   a. Survive the bactericidal components of sebum.

   b. Avoid being flushed from the surface by sweat gland secretions.

   c. Penetrate the stratum corneum.

   d. Squeeze between the junctional complexes of deeper layers.

   e. Escape the Langerhans cells.

   f. Cross the basal lamina.

Chapter 6:
Osseous Tissue and Bone Structure

LEVEL 1: REVIEWING FACTS AND TERMS

Multiple Choice

1. d	7. b	12. d	17. c	22. a
2. b	8. c	13. a	18. c	23. b
3. c	9. a	14. d	19. a	24. c
4. b	10. c	15. b	20. d	25. d
5. c	11. b	16. c	21. a	26. a
6. d				

Completion

1. yellow marrow	9. osteon	17. calcitriol
2. support	10. epiphysis	18. calcium
3. bone markings	11. intramembranous	19. calcitonin
4. condyle	12. endochondral	20. comminuted
5. irregular	13. ossification	21. compound
6. Wormian	14. remodeling	22. osteoporosis
7. osteoblasts	15. osteoclasts	23. cancers
8. osteocytes	16. matrix	

Matching

1. G	5. A	9. D	13. K	17. M
2. E	6. F	10. R	14. J	18. Q
3. I	7. C	11. O	15. S	19. P
4. H	8. B	12. N	16. L	

Drawing/Illustration Labeling

Figure 6-1   Structure of Compact Bone

1. small venule
2. capillary
3. concentric lamellae
4. osteon
5. trabeculae
6. spongy bone
7. compact bone
8. lacunae
9. central canal
10. perforating fibers
11. perforating canal

Figure 6-2   Structure of a Long Bone—L.S.

1. spongy bone
2. compact bone
3. medullary cavity
4. epiphysis
5. metaphysis
6. diaphysis
7. metaphysis
8. epiphysis

Figure 6-3   Major Types of Fractures

1. spiral fracture
2. greenstick fracture
3. Pott's fracture
4. transverse fracture
5. comminuted fracture
6. Colles fracture
7. compression fracture

LEVEL 2: REVIEWING CONCEPTS

Chapter Overview

1. sesamoid bones
2. long bones
3. short bones
4. flat bones
5. irregular bones
6. sutural bones
7. diaphysis
8. epiphysis
9. metaphysis
10. marrow cavity
11. spongy bone
12. osteocytes
13. lacunae
14. canaliculi
15. periosteum
16. osteon
17. trabeculae
18. red bone marrow
19. yellow bone marrow
20. ossification
21. endochondral ossification
22. intramembranous ossification
23. remodeling
24. osteopenia
25. osteoporosis

Concept Maps

I Bone Formation

1. intramembranous ossification
2. collagen
3. osteocytes
4. hyaline cartilage
5. periosteum

II Regulation of Calcium Ion Concentration

1. calcitonin
2. ↓ Ca concentration in body fluids
3. ↓ calcium level
4. parathyroid glands
5. ↑ Ca concentration in body fluids
6. releases stored Ca from bone
7. homeostasis

Multiple Choice

1. b	5. d	9. c	13. d	17. d
2. c	6. a	10. c	14. c	18. b
3. d	7. b	11. d	15. b	19. c
4. a	8. d	12. a	16. a	20. a

Completion

1. canaliculi
2. rickets
3. osteoblasts
4. intramembranous

5. endochondral
6. epiphyseal plates
7. osteomalacia
8. osteopenia

9. Pott's
10. diaphysis

Short Essay

1. (1) *Support*; (2) *storage* of calcium salts and lipids; (3) blood cell *production*; (4) *protection* of delicate tissues and organs; (5) *leverage*

2. *Compact bone:*   basic functional unit is the osteon (Haversian system); lamella (bony matrix) fills in spaces between the osteons

   Spongy bone: no osteons; lamella form bony plates (trabeculae)

3. *Periosteum:*   covers the *outer surface* of a bone; isolates bone from surrounding tissues; consists of fibrous outer layer and cellular inner layer; provides a route for circulatory and nervous supply

   Endosteum: cellular layer *inside* bone; lines the marrow cavity; covers the trabeculae of spongy bone; lines inner surfaces of the central canals

4. *Calcification:*   refers to the deposition of calcium salts within a tissue

   Ossification: refers specifically to the formation of bone

5. *Intramembranous* ossification begins when osteoblasts differentiate within a connective tissue. *Endochondral* ossification begins with the formation of a cartilaginous model.

6. Excessive or inadequate growth hormone production has a pronounced effect on activity at epiphyseal plates. *Gigantism* results from an overproduction of growth hormone before puberty. *Pituitary dwarfism* results from inadequate growth hormone production, which leads to reduced epiphyseal activity and abnormally short bones.

7. (1) Long bones—humerus, femur, tibia, fibula
   (2) Short bones—carpals, tarsals
   (3) Flat bones—sternum, ribs, scapula
   (4) Irregular bones—spinal vertebrae
   (5) Sesamoid—patella (kneecap)
   (6) Sutural (Wormian) bones—between flat bones of skull (sutures)

8. The bones of the skeleton are attached to the muscular system, extensively connected to the cardiovascular and lymphoid systems, and largely under the physiological control of the endocrine system. The digestive and excretory systems provide the calcium and phosphate minerals needed for bone growth. The skeleton represents a reserve of calcium, phosphate, and other minerals that can compensate for changes in the dietary supply of these ions.

LEVEL 3: CRITICAL THINKING AND CLINICAL APPLICATIONS

1. In premature or precocious puberty, production of sex hormones escalates early, usually by age 8 or earlier. This results in an abbreviated growth spurt, followed by premature closure of the epiphyseal plates. Estrogen and testosterone both cause early uniting of the epiphyses of long bones; thus, the growth of long bones is completed prematurely.

2. Reduction in bone mass and excessive bone fragility are symptoms of osteoporosis, which can develop as a secondary effect of many cancers. Cancers of the bone marrow, breast, or other tissues release a chemical known as osteoclast-activating factor. This compound increases both the number and activity of osteoclasts and produces a severe osteoporosis.

3. The bones of the skeleton become thinner and relatively weaker as a normal part of the aging process. Inadequate ossification, called osteopenia, causes a reduction in bone mass between the ages of 30 and 40. Once this reduction begins, women lose about 8 percent of their skeletal mass every decade, and men lose about 3 percent per decade. The greatest areas of loss are in the epiphyses, vertebrae, and the jaws. The result is fragile limbs, reduction in height, and loss of teeth.

4. Foods or supplements containing vitamins A, C, and D are necessary to maintain the integrity of bone. Vitamin D plays an important role in calcium metabolism by stimulating the absorption and transport of calcium and phosphate ions. Vitamins A and C are essential for normal bone growth and remodeling. Any exercise that is weight-bearing or that exerts a pressure on the bones is necessary to retain the minerals in the bones, especially calcium salts. As the mineral content of a bone decreases, the bone softens and skeletal support decreases.

5. The fracture probably damaged the epiphyseal cartilage in Christopher's right leg. Even though the bone healed properly, the damaged leg did not produce as much cartilage as did the undamaged leg. The result would be a shorter bone on the side of the injury.

6. The matrix of bone will absorb traces of minerals from the diet. These minerals can be identified hundreds of years later. A diet rich in calcium and vitamin B will produce denser bones than will a diet lacking these. Cultural practices such as the binding of appendages or the wrapping of infant heads will result in misshapen bones. Heavy muscular activity will result in larger bone markings, indicating an active athletic lifestyle.

7. Each bone in the body has a distinctive shape and characteristic external and internal features. The bone markings (surface features), depressions, grooves, and tunnels can be used to determine the size, weight, sex, and general appearance of an individual.

Chapter 7:
The Axial Skeleton

LEVEL 1: REVIEWING FACTS AND TERMS

Multiple Choice

1. b	8. a	15. b	21. d	27. b
2. d	9. a	16. d	22. c	28. a
3. c	10. d	17. d	23. b	29. d
4. a	11. c	18. c	24. c	30. b
5. b	12. c	19. b	25. c	31. d
6. d	13. b	20. a	26. a	32. b
7. c	14. d			

Completion

1. axial
2. muscles
3. cranium
4. foramen magnum
5. inferior conchae
6. paranasal
7. mucus
8. fontanelles
9. "soft spot"
10. compensation
11. vertebral body
12. cervical
13. costal
14. capitulum
15. floating
16. xiphoid process

Matching

1. G	5. D	8. B	11. P	14. M
2. H	6. E	9. N	12. K	15. J
3. A	7. C	10. O	13. L	16. I
4. F				

Drawing/Illustration Labeling

Figure 7-1   Bones of the Axial Skeleton

1. cranium
2. face
3. auditory ossicles
4. hyoid
5. sternum
6. ribs
7. vertebrae
8. sacrum
9. coccyx

Figure 7-2   Anterior View of the Adult Skull

1. parietal bone
2. sphenoid
3. temporal bone
4. ethmoid
5. palatine bone
6. lacrimal bone
7. zygomatic bone
8. nasal bone
9. maxilla
10. mandible
11. mental foramen
12. coronal suture
13. frontal bone
14. vomer

Figure 7-3   Lateral View of the Adult Skull

1. parietal bone
2. squamous suture
3. lambdoid suture
4. external acoustic meatus
5. mastoid process
6. styloid process
7. zygomatic arch
8. coronal suture
9. sphenoid
10. supra-orbital foramen
11. nasal bone
12. lacrimal bone
13. ethmoid
14. maxilla
15. zygomatic bone
16. mental foramen
17. mental protuberance

Figure 7-4   Inferior View of the Adult Skull

1. frontal bone
2. zygomatic bone
3. vomer
4. sphenoid
5. foramen ovale
6. styloid process
7. external acoustic meatus
8. lambdoid suture
9. occipital bone
10. external occipital protuberance
11. maxilla
12. palatine bone
13. zygomatic arch
14. temporal bone
15. occipital condyle
16. foramen magnum

Figure 7-5   Fetal Skull—Lateral View

1. coronal suture
2. frontal bone
3. nasal bone
4. maxilla
5. sphenoid
6. mandible
7. temporal bone
8. mastoid fontanelle
9. parietal bone
10. squamous suture
11. lambdoid suture
12. occipital bone

Figure 7-6   Fetal Skull—Superior View

1. frontal bone
2. coronal suture
3. frontal suture
4. anterior fontanelle
5. sagittal suture
6. parietal bone
7. lambdoid suture
8. occipital bone
9. occipital fontanelle

Figure 7-7   The Vertebral Column

1. cervical curve
2. thoracic curve
3. lumbar curve
4. sacral curve
5. cervical

6. thoracic
7. lumbar
8. sacral
9. coccygeal

Figure 7-8   The Ribs

1. sternum
2. manubrium
3. body of sternum
4. xiphoid process

5. vertebrochondral ribs
6. floating ribs
7. vertebrosternal ribs (true ribs)

8. false ribs
9. costal cartilage

LEVEL 2: REVIEWING CONCEPTS

Chapter Overview

1. axial
2. skull
3. cranium
4. zygomatic
5. nasal or lacrimal
6. nasal or lacrimal
7. mandible
8. vomer
9. occipital

10. sphenoid
11. parietal
12. Wormian (sutural)
13. sagittal
14. hyoid
15. ribs
16. sternum
17. manubrium
18. xiphoid process

19. true
20. false
21. floating
22. vertebrae
23. cervical
24. thoracic
25. lumbar
26. sacrum
27. coccyx

Concept Map
I Skeleton—Axial Division

1. skull
2. mandible
3. lacrimal
4. occipital
5. temporal

6. Wormian bones (sutural bones)
7. lambdoid
8. hyoid
9. vertebral column
10. thoracic

11. sacral
12. floating ribs
13. sternum
14. xiphoid process

Multiple Choice

1. c
2. a
3. a
4. d

5. c
6. b
7. c
8. a

9. d
10. b
11. b

Completion

1. auditory tube
2. metopic
3. lacrimal
4. auditory ossicles

5. alveolar processes
6. mental foramina
7. compensation

8. kyphosis
9. lordosis
10. scoliosis

Short Essay

1. (1) Create a framework that supports and protects organ systems in the dorsal and ventral body cavities

(2) Provide a surface area for attachment of muscles that adjust the positions of the head, neck, and trunk

(3) Provide a surface area for attachment of muscles that participate in respiratory movements

(4) Provide a surface area for attachment of muscles that stabilize or position elements of the appendicular system

2. a. *Paired bones:* parietal, temporal

    b. *Unpaired bones:* occipital, frontal, sphenoid, ethmoid

3. a. *Paired bones:* zygomatic, maxilla, nasal, palatine, lacrimal, inferior nasal concha

    b. *Unpaired bones:* mandible, vomer

4. The auditory ossicles consist of three (3) tiny bones on each side of the skull that are enclosed by the temporal bone. The hyoid bone lies below the skull suspended by the stylohyoid ligaments that connect the lesser horns of the styloid processes of the temporal bones. These are bones that do not articulate with the bones of the skull but are associated with the skull.

5. Craniostenosis is premature closure of one or more fontanelles, which results in unusual distortions of the skull.

6. The thoracic and sacral curves are called primary curves because they appear late in fetal development. They accommodate the thoracic and abdominopelvic viscera. The lumbar and cervical curves are called secondary curves because they do not appear until several months after birth. They help position the body weight over the legs.

7. *Kyphosis:* normal thoracic curvature becomes exaggerated, producing "roundback" appearance.

    Lordosis: exaggerated lumbar curvature produces "swayback" appearance.

    Scoliosis: abnormal lateral curvature.

8. The *true* ribs reach the anterior body wall and are connected to the sternum by separate cartilaginous extensions. The *false* ribs do not attach directly to the sternum.

LEVEL 3: CRITICAL THINKING AND CLINICAL APPLICATIONS

1. Vision, hearing, balance, olfaction (smell), and gustation (taste)

2. TMJ is temporomandibular joint syndrome. It occurs in the joint where the *mandible* articulates with the *temporal* bone. The mandibular condyle fits into the mandibular fossa of the temporal bone. The condition generally involves pain around the joint and its associated muscles, a noticeable clicking within the joint, and usually a pronounced malocclusion of the lower jaw.

3. The crooked nose may be a result of a deviated septum, a condition in which the nasal septum has a bend in it, most often at the junction between the bony and cartilaginous portions of the septum. Septal deviation often blocks drainage of one or more sinuses, producing chronic bouts of infection and inflammation.

4. During whiplash, the movement of the head resembles the cracking of a whip. The head is relatively massive, and it sits on top of the cervical vertebrae. Small muscles articulate with the bones and can produce significant effects by tipping the balance one way or another. If the body suddenly changes position, the balancing muscles are not strong enough to stabilize the head. As a result, a partial or complete dislocation of the cervical vertebrae can occur, with injury to muscles and ligaments and potential injury to the spinal cord.

5. A plausible explanation would be that the child has scoliosis, a condition in which the hips are abnormally tilted sideways, making one of the lower limbs shorter than the other.

6. The large bones of a child's cranium are not yet fused; they are connected by fontanelles, areas of fibrous tissue. By examining the bones, the archaeologist could see if sutures had formed. By knowing approximately how long it takes for the various fontanelles to close and by determining their sizes, he could estimate the age of the individual at death.

Chapter 8:
The Appendicular Skeleton

LEVEL 1: REVIEWING FACTS AND TERMS
Multiple Choice

1. c	5. b	9. b	13. d	17. c
2. a	6. c	10. a	14. c	18. d
3. b	7. b	11. c	15. d	19. d
4. a	8. c	12. d	16. a	

Completion

1. acromion	6. humerus	11. patella
2. clavicle	7. coxae	12. childbearing
3. pectoral girdle	8. pubic symphysis	13. teeth
4. glenoid cavity	9. femur	14. pelvis
5. styloid	10. acetabulum	15. age

Matching

1. F	4. A	7. C	10. B
2. J	5. D	8. E	11. H
3. G	6. L	9. K	12. I

Drawing/Illustration Labeling

Figure 8-1   Appendicular Skeleton

1. clavicle	6. carpal bones	11. patella	16. phalanges
2. scapula	7. metacarpal bones	12. tibia	17. pectoral girdle
3. humerus	8. phalanges	13. fibula	18. upper limbs
4. radius	9. hip bone (coxal bone)	14. tarsal bones	19. pelvic girdle
5. ulna	10. femur	15. metatarsal bones	20. lower limbs

Figure 8-2   The Scapula

1. coracoid process	5. acromion	9. body
2. glenoid cavity	6. coracoid process	10. medial border
3. lateral border	7. superior border	
4. inferior angle	8. spine	

Figure 8-3   The Humerus

1. greater tubercle	5. capitulum	9. greater tubercle
2. lesser tubercle	6. head	10. radial groove
3. deltoid tuberosity	7. coronoid fossa	11. olecranon fossa
4. lateral epicondyle	8. medial epicondyle	12. lateral epicondyle

Figure 8-4   The Radius and Ulna

1. olecranon	5. radius	8. coronoid process
2. ulna	6. styloid process of radius	9. ulna
3. ulnar head	7. trochlear notch	10. styloid process of ulna
4. head of radius		

Figure 8-5   Bones of the Wrist and Hand

1. phalanges	7. pisiform	13. trapezium
2. carpals	8. triquetrum	14. scaphoid
3. distal	9. lunate	15. capitate
4. middle	10. ulna	16. radius
5. proximal	11. metacarpals	
6. hamate	12. trapezoid	

Figure 8-6   The Pelvis (anterior view)

1. acetabulum	4. iliac crest	7. pubis
2. pubic symphysis	5. sacroiliac joint	8. ischium
3. obturator foramen	6. ilium	

Figure 8-7   The Pelvis (lateral view)

1. greater sciatic notch
2. ischial spine
3. lesser sciatic notch
4. ischial tuberosity
5. anterior superior iliac spine
6. anterior inferior iliac spine
7. acetabulum
8. ischial ramus

Figure 8-8   The Femur (anterior and posterior views)

1. greater trochanter
2. neck
3. lateral epicondyle
4. lateral condyle
5. head
6. lesser trochanter
7. medial epicondyle
8. medial condyle
9. intertrochanteric crest
10. gluteal tuberosity
11. linea aspera
12. lateral supracondylar ridge
13. lateral epicondyle

Figure 8-9   The Tibia and Fibula

1. lateral tibial condyle
2. head of the fibula
3. fibula
4. lateral malleolus
5. medial tibial condyle
6. tibial tuberosity
7. anterior margin
8. tibia
9. medial malleolus

Figure 8-10   Bones of the Ankle and Foot

1. medial cuneiform
2. intermediate cuneiform
3. navicular
4. talus
5. distal phalanx
6. middle phalanx
7. proximal phalanx
8. lateral cuneiform
9. cuboid
10. calcaneus
11. phalanges
12. metatarsals
13. tarsals

LEVEL 2: REVIEWING CONCEPTS

Chapter Overview

1. limbs
2. girdles
3. clavicles
4. scapulae
5. pectoral
6. glenohumeral
7. glenoid
8. humerus
9. ulna
10. radius
11. carpals
12. metacarpals
13. phalanges
14. coxal bones
15. ilium
16. ischium
17. pubis
18. pubic symphysis
19. acetabulum
20. femur
21. ball and socket
22. fibula
23. tibia
24. patella
25. knee
26. tarsals
27. metatarsals
28. toes

Concept Map

I Skeleton—Appendicular Division

1. pectoral girdle
2. humerus
3. radius
4. metacarpals
5. ischium
6. femur
7. fibula
8. tarsals
9. phalanges

Multiple Choice

1. c
2. a
3. a
4. d
5. d
6. b
7. a
8. b
9. c
10. d
11. d
12. c
13. c
14. b

Completion

1. clavicle
2. scapula
3. ulna
4. metacarpals
5. coxal bone
6. pubic symphysis
7. acetabulum
8. ilium
9. pelvis
10. femur
11. fibula
12. calcaneus
13. thumb
14. hallux
15. perineum

Short Essay

1. Provides control over the immediate environment; changes your position in space; and makes you an active, mobile person

2. Bones of the arms and legs and the supporting elements that connect the limbs to the trunk (pectoral girdle, upper limb, pelvic girdle, lower limb).

3. Pectoral girdle: scapula, clavicle

   Pelvic girdle: ilium, ischium, pubis

4. Both have long shafts and styloid processes for support of the wrist. The radius has a radial tuberosity and the ulna has a prominent olecranon process for muscle attachments. Both the radius and the ulna articulate proximally with the humerus and distally with the carpal bones.

5. a. Upper arm—humerus
   b. Forearm—ulna and radius
   c. Wrist (carpal bones)—scaphoid, lunate, triquetrum, pisiform, trapezium, trapezoid, capitate, hamate
   d. Metacarpal bones (bones of the hands)
   e. Phalanges—finger bones (proximal, middle, distal)

6. The tibia is the massive, weight-bearing bone of the lower leg. The fibula is the long narrow bone lateral to the tibia that is more important for muscle attachment than for support. The tibia articulates proximally with the femur and distally with the talus. The fibula articulates only with the talus.

7. Because the bones of the pelvic girdle must withstand the stresses involved in weight bearing and locomotion

8. (a) An enlarged pelvic outlet, (b) a broad pubic angle, (c) less curvature on the sacrum and coccyx, (d) a wider, more circular pelvic inlet, (e) a relatively broad pelvis that does not extend as far superiorly, (f) ilia that project farther laterally, but do not extend as far superior to the sacrum

9. (a) upper leg (femur); (b) kneecap (patella); (c) lower leg (tibia and fibula); (d) ankle (seven tarsal bones), talus, trochlea, calcaneus, cuboid, navicular, three cuneiform bones; (e) bones of the foot—five metatarsal bones; (f) toes (phalanges, proximal, middle and distal)

LEVEL 3: CRITICAL THINKING AND CLINICAL APPLICATIONS

1. Normally, the patella glides across the patellar surface of the femur. Running on hard or slanted surfaces causes improper tracking of the patella across the patellar surface, resulting in a condition known as runner's knee or patella femoral stress syndrome. The misalignment of the patella puts lateral pressure on the knee, resulting in swelling and tenderness after exercise.

2. A Pott's fracture is an ankle injury that may be caused by forceful movement of the foot outward and backward, resulting in dislocating the ankle, breaking both the lateral malleolus of the fibula and the medial malleolus of the tibia.

3. Running, while beneficial to overall health, places the foot bones under more stress than does walking. The bones are subjected to repeated shocks and impacts, sometimes resulting in hairline fractures called stress fractures. These fractures are caused by improper placement of the foot while running or by poor arch support, and usually involve one of the metatarsal bones. Running shoes with good arch support are necessary.

4. The carpal bones articulate with one another at joints that permit limited sliding and twisting. The tendons of muscles that flex the fingers pass across the anterior surface of the wrist, sandwiched between the intercarpal ligaments and a broad, superficial transverse ligament, the flexor retinaculum. The carpal bones can compress the tendons and adjacent sensory and motor nerves, producing the symptoms Janet experienced.

5. a. We can estimate a person's muscular development and muscle mass from the appearance of various ridges and the general bone mass.
   b. The condition of the teeth or the presence of healed fractures gives an indication of an individual's medical history.
   c. Sex and age can be determined or closely estimated on the basis of taking measurements of skeletal components.
   d. The skeleton may provide clues about an individual's nutritional state, handedness, and even occupation.

Chapter 9:
Articulations

LEVEL 1: REVIEWING FACTS AND TERMS

Multiple Choice

1. d	6. d	11. d	16. d	21. b
2. a	7. b	12. d	17. c	22. c
3. c	8. c	13. d	18. a	
4. b	9. b	14. d	19. c	
5. d	10. c	15. b	20. c	

Completion

1. suture	7. supination	12. hip
2. synostosis	8. gliding	13. knee
3. synovial	9. anulus fibrosus	14. osteoarthritis
4. intrinsic ligaments	10. glenohumeral	15. continuous passive motion
5. bursae	11. elbow	16. nervous
6. flexion		

Matching

1. H	5. B	9. M	13. O	16. J
2. F	6. C	10. L	14. I	17. K
3. G	7. E	11. P	15. Q	
4. A	8. D	12. N		

Drawing/Illustration Labeling

Figure 9-1   Synovial Joint—Sagittal Section

1. medullary cavity	4. synovial membrane	7. fibrous joint capsule
2. spongy bone	5. articular cartilage	8. compact bone
3. periosteum	6. joint cavity	

Figure 9-2   The Shoulder Joint—Anterior View—Frontal Section

1. coracoclavicular ligaments	7. synovial membrane	12. scapula
2. acromioclavicular ligament	8. humerus	13. articular cartilages
3. tendon of supraspinatus muscle	9. clavicle	14. joint cavity
4. acromion	10. coraco-acromial ligament	15. glenoid labrum
5. articular capsule	11. coracoid process	16. articular capsule
6. subdeltoid bursa		

Figure 9-3   Movements of the Skeleton

1. elevation	9. opposition	17. extension
2. depression	10. pronation	18. inversion
3. flexion	11. abduction	19. eversion
4. extension	12. adduction	20. retraction
5. hyperextension	13. adduction	21. protraction
6. rotation	14. abduction	22. dorsiflexion
7. abduction	15. circumduction	23. plantar flexion
8. adduction	16. flexion	

Figure 9-4   Types of Synovial Joints

1. saddle joint
2. D
3. pivot joint
4. B
5. condylar joint
   (ellipsoid joint)

6. F
7. hinge joint
8. A
9. ball-and-socket joint

10. C
11. gliding joint
12. E

Figure 9-5   Anterior View of the Knee Joint

1. patellar surface
2. fibular collateral ligament
3. lateral condyle
4. lateral meniscus

5. cut tendon
6. tibia
7. fibula
8. posterior cruciate ligament

9. medial condyle
10. tibial collateral ligament
11. medial meniscus
12. anterior cruciate ligament

LEVEL 2: REVIEWING CONCEPTS

Chapter Overview

1. articulations
2. synarthrosis
3. amphiarthrosis
4. diarthrosis
5. sutures
6. gomphoses

7. synchondroses
8. synostosis
9. syndesmosis
10. symphysis
11. synovial
12. lubrication

13. capsule
14. menisci
15. lateral
16. seven
17. patellar
18. popliteal

19. fibula
20. cruciate
21. femoral
22. tibial collateral
23. fibular collateral
24. arthritis

Concept Map

I Articulations (Joints)

1. no movement
2. sutures
3. cartilaginous

4. amphiarthrosis
5. fibrous
6. symphysis

7. synovial
8. monaxial
9. wrist

Multiple Choice

1. c
2. c
3. b
4. d

5. c
6. a
7. c
8. c

9. c
10. a
11. d
12. b

13. c
14. d
15. d

16. a
17. d
18. b

Completion

1. bunion
2. arthritis
3. gomphosis
4. synchondrosis
5. syndesmosis

6. menisci
7. fat pads
8. articular cartilage
9. hyperextension

10. rheumatism
11. tendons
12. luxation
13. intervertebral discs

Short Essay

1. 1. Synarthrosis—immovable joint
   2. Amphiarthrosis—slightly movable joint
   3. Diarthrosis—freely movable joint

2. A *synostosis* is a totally rigid immovable joint in which two separate bones actually fuse together, forming a synarthrosis.
   A *symphysis* is an amphiarthrotic joint in which bones are separated by a wedge or pad of fibrocartilage.

3. A tendon attaches muscle to bone. Ligaments attach bones to bones.

4. a. Bursa are small, synovial-filled pockets in connective tissue that form where a tendon or ligament rubs against other tissues. Their function is to reduce friction and act as a shock absorber.
   b. Menisci are articular discs that: (1) subdivide a synovial cavity, (2) channel the flow of synovial fluid, and (3) allow variations in the shape of the articular surfaces.

5. 1. Provides lubrication
   2. Acts as a shock absorber
   3. Nourishes the chondrocytes

6. a. Monaxial—hinge joint (elbow, knee)
   b. Biaxial—ellipsoidal (radiocarpal), saddle (base of thumb)
   c. Triaxial—ball-and-socket joints (shoulder, hip)

7. 1. Gliding joints—intercarpals
   2. Hinge joints—elbow, knee
   3. Pivot joints—between atlas and axis
   4. Condylar (ellipsoidal) joints—between metacarpals and phalanges
   5. Saddle joints—base of thumb
   6. Ball-and-socket joints—shoulder, hip

8. The knee joint and elbow joint are both hinge joints.

9. The shoulder joint and hip joint are both ball-and-socket joints.

10. The articulations of the carpals and tarsals are gliding diarthroses, which permit a limited gliding motion.

11. Flexion, extension, abduction, adduction

12. The glenohumeral joint has a relatively oversized loose articular capsule that extends from the scapular neck to the humerus. The loose, oversized capsule permits an extensive range of motion, but does not afford the stability found in the ball-and-socket joints.

13. Intervertebral discs are not found between the first and second cervical vertebrae (atlas and axis), the sacrum and the coccyx. C1, the atlas, sits on top of C2, the axis; the den of the axis provides for rotation of the first cervical vertebra, which supports the head. An intervertebral disc would prohibit rotation. The sacrum and the coccyx are fused bones.

14. a. Inversion   c. Plantar flexion   e. Depression
    b. Opposition   d. Protraction   f. Elevation

15. Both the elbow and knee are extremely stable joints. Stability for the elbow joint is provided by the:
   a. interlocking of the bony surfaces of the humerus and ulna
   b. thickness of the articular capsule
   c. strong ligaments reinforcing the capsule

   Stability for the hip joint is provided by the:
   a. almost complete bony socket
   b. strong articular capsule
   c. supporting ligaments
   d. muscular padding

16. The most vulnerable structures subject to possible damage if a locked knee is struck from the side are torn lateral meniscus and serious damage to the supporting ligaments.

LEVEL 3: CRITICAL THINKING AND CLINICAL APPLICATIONS

1. In the hip joint, the femur articulates with a deep, complete bony socket, the acetabulum. The articular capsule is unusually strong; there are numerous supporting ligaments, and there is an abundance of muscular padding. The joint is extremely stable and mobility is sacrificed for stability.

   In the shoulder joint, the head of the humerus articulates with a shallow glenoid cavity, and there is an oversized loose articular capsule that contributes to the weakness of the joint. The joint contains bursae that serve to reduce friction, and ligaments and muscles contribute to minimal strength and stability of the area. In this joint, stability is sacrificed for mobility.

2. The knee joint is a diarthrosis, or synovial joint, which contains synovial fluid. One of the primary functions of the synovial fluid within a joint is shock absorption. Synovial fluid cushions shocks in joints that are subjected to compression, which probably occurred during the "pile on." When the pressure across a joint suddenly increases, the synovial fluid lessens the shock by distributing it evenly across the articular surfaces and outward to the articular capsule.

3. While using a stairmaster, a lot of weight is placed on the knees in addition to tremendous stresses on the knees. Ordinarily, the medial and lateral menisci move as the position of the femur changes. Excess weight on the knee while it is partially flexed can trap a meniscus between the tibia and femur, resulting in a break or tear in the cartilage. In addition to being quite painful, the torn cartilage may restrict movement at the joint. Sometimes the meniscus can be heard and felt popping in and out of position when the knee is extended—thus a "trick knee" that feels unstable.

4. His pain is probably caused by bursitis, an inflammation of the bursae. Bursitis can result from repetitive motion, infection, trauma, chemical irritation, and pressure over the joint. Given the location of the pain, his case probably results from the repetitive motion of practicing pitches.

5. It is easier to dislocate a shoulder than a hip because the shoulder is a more mobile joint. The shoulder joint is not bound tightly by ligaments or other elements. The hip joint, although mobile, is less easily dislocated because it is stabilized by four heavy ligaments, and the bones fit tight in the joint. Additionally, the synovial capsule of the hip joint is larger than that of the shoulder, and its range of motion is smaller.

6. With age, bone mass decreases and bones become weaker; therefore, the risk of a hip fracture increases. If osteoporosis develops, the bones may weaken to the point at which fractures occur in response to stresses that could be tolerated by normal bones. With or without osteoporosis, hip fractures are the most dangerous fractures in elderly people. Hip fractures may be accompanied by hip dislocation or by pelvic fractures.

7. Bunions are the most common form of pressure-related bursitis; they form over the base of the big toe as a result of friction and distortion of the joint. A frequent cause of this friction and distortion is tight-fitting shoes with pointed toes. This style is certainly more common in women's shoes than in men's, so one would expect that women probably have more trouble with bunions.

Chapter 10:
Muscle Tissue

LEVEL 1: REVIEWING FACTS AND TERMS

Multiple Choice

1. d	7. d	13. c	19. d	24. b	29. d
2. d	8. a	14. a	20. c	25. a	30. b
3. d	9. b	15. d	21. a	26. b	31. c
4. c	10. d	16. d	22. b	27. b	32. d
5. b	11. c	17. a	23. d	28. b	33. a
6. c	12. a	18. c			

Completion

1. contraction
2. epimysium
3. fascicles
4. tendon
5. sarcolemma
6. T tubules
7. sarcomeres
8. cross-bridges
9. Z lines
10. action potential

11. recruitment
12. troponin
13. tension
14. complete tetanus
15. incomplete tetanus
16. motor unit
17. skeletal muscle
18. treppe
19. twitch
20. ATP

21. lactic acid
22. glycolysis
23. white muscles
24. red muscles
25. oxygen debt
26. anaerobic endurance
27. pacemaker
28. plasticity
29. smooth muscle
30. sphincter

Matching

1. J	6. A	11. B	15. G
2. E	7. C	12. M	16. D
3. R	8. K	13. N	17. I
4. Q	9. H	14. F	18. P
5. L	10. O		

Drawing/Illustration Labeling

Figure 10-1   Organization of Skeletal Muscles

1. epimysium	9. epimysium	17. sarcolemma
2. muscle fascicle	10. blood vessels	18. myofibril
3. endomysium	11. endomysium	19. axon
4. perimysium	12. perimysium	20. sarcoplasm
5. nerve	13. perimysium	21. capillary
6. muscle fibers	14. muscle fiber	22. endomysium
7. blood vessels	15. endomysium	23. myosatellite cell
8. tendon	16. mitochondria	24. nucleus

Figure 10-2   The Structure of a Skeletal Muscle Fiber

1. myofibril	7. myofibril	12. T tubules
2. sarcolemma	8. thin filament	13. terminal cisterna
3. sarcoplasm	9. thick filament	14. sarcolemma
4. nuclei	10. triad	15. sarcoplasm
5. mitochondria	11. sarcoplasmic reticulum	16. myofibrils
6. sarcolemma		

Figure 10-3   Types of Muscle Tissue

1. smooth	2. cardiac	3. skeletal

Figure 10-4   Structure of a Sarcomere

1. I band	5. titin	8. thin filament
2. A band	6. zone of overlap	9. thick filament
3. H band	7. M line	10. sarcomere
4. Z line		

Figure 10-5   Levels of Functional Organization in a Skeletal Muscle

1. skeletal muscle	3. muscle fiber	5. sarcomere
2. muscle fascicle	4. myofibril	

LEVEL 2: REVIEWING CONCEPTS

Chapter Overview

1. contraction	8. myosatellite	15. actin
2. skeletal	9. sarcolemma	16. myosin
3. epimysium	10. T tubules	17. Z line
4. perimysium	11. nuclei	18. A band
5. fascicle	12. myofibrils	19. thick
6. endomysium	13. myofilaments	20. sliding filament
7. muscle fibers	14. sarcomeres	21. muscle contraction

Concept Maps

I Muscle Tissue

1. heart
2. striated
3. smooth
4. involuntary
5. nonstriated
6. bones
7. multinucleate

II Muscle

1. muscle bundles (fascicles)
2. myofibrils
3. sarcomeres
4. actin
5. Z lines
6. thick filaments
7. H band

III Muscle Contraction

1. release of Ca^{2+} from sacs of sarcoplasmic reticulum
2. cross-bridging (heads of myosin attach to turned-on thin filaments)
3. shortening, i.e., contraction of myofibrils and muscle fibers they comprise
4. energy + ADP + phosphate

Multiple Choice

1. a	6. c	11. c	16. a
2. b	7. b	12. a	17. a
3. b	8. b	13. b	18. a
4. c	9. d	14. c	19. c
5. d	10. d	15. c	20. c

Completion

1. myosatellite
2. myoblasts
3. muscle tone
4. plasticity
5. recovery period
6. fatigue
7. A bands
8. motor unit
9. complete tetanus
10. concentric isotonic

Short Essay

1. a. Produce skeletal movement
   b. Maintain posture and body position
   c. Support soft tissues
   d. Guard entrances and exists
   e. Maintain body temperature

2. a. An outer *epimysium*
   b. A central *perimysium*
   c. An inner *endomysium*

3.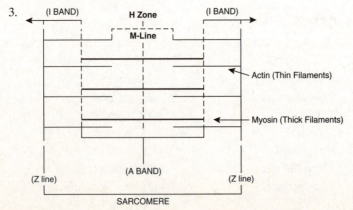

4. a. Active site exposure
   b. Cross-bridge attachment
   c. Pivoting
   d. Cross-bridging detachment
   e. Myosin activation

5. a. Release of acetylcholine
   b. Depolarization of the motor end plate
   c. Generation of an action potential
   d. Conduction of an action potential
   e. Release of calcium ions

6. A—twitch; B—incomplete tetanus; C—complete tetanus

7. *Isometric* contraction—tension rises to a maximum but the length of the muscle remains constant.

   Isotonic contraction—tension in the muscle builds until it exceeds the amount of resistance and the muscle shortens. As the muscle shortens, the tension in the muscle remains constant, at a value that just exceeds the applied resistance.

8. When fatigue occurs and the oxygen supply to muscles is depleted, aerobic respiration ceases due to the decreased oxygen supply. Anaerobic glycolysis supplies the needed energy for a short period of time. The amount of oxygen needed to restore normal pre-exertion conditions is the oxygen debt.

9. Fast fiber muscles produce powerful contractions, which use ATP in massive amounts. Prolonged activity is primarily supported by anaerobic glycolysis, and fast fibers fatigue rapidly. Slow fibers are specialized to enable them to continue contracting for extended periods. The specializations include an extensive network of capillaries so supplies of O_2 are available, and the presence of myoglobin, which binds O_2 molecules and results in the buildup of O_2 reserves. These factors improve mitochondrial performance.

LEVEL 3: CRITICAL THINKING AND CLINICAL APPLICATIONS

1. During a warm-up period when a muscle first begins to contract, its contraction may be only half as strong as those that occur shortly thereafter in response to stimuli of the same strength. These contractions show a staircase pattern called treppe. In addition to reflecting increased availability of calcium ions, the muscle begins to work and liberates heat, and the enzyme systems become more efficient. Together, these factors produce a slightly strange contraction with each successive stimulus during the beginning phase of muscle activity, and the result is greater muscle response and efficiency.

2. Training to improve *aerobic* endurance usually involves sustained low levels of muscular activity. Jogging, distance swimming, biking, and other cardiovascular activities that do not require peak tension production are appropriate for developing aerobic endurance. Training to develop *anaerobic* endurance involves frequent, brief, intensive workouts. Activities might include running sprints; fast, short-distance swimming; pole vaulting; or other exercises requiring peak tension production in a short period of time.

3. Relaxed skeletal (voluntary) muscles are almost always in a relaxed state. While muscle tone does not produce active movements, it keeps the muscles firm, healthy, and ready to respond to stimulation. Skeletal muscle tone also helps stabilize joints and maintain posture.

4. Skeletal muscles are very responsive to use and disuse. If regular exercise stops, capillary networks shrink and the number of mitochondria within the muscle fiber decreases. Actin and myosin filaments diminish, and the entire muscle atrophies. Accidents or diseases that interfere with motor nerve impulses and neurotransmitter activity at neuromuscular junctions may also cause muscle atrophy.

5. The deceased individual's body is in a state of rigor mortis (death rigor). Muscles begin to stiffen 3 to 4 hours after death, with peak rigidity occurring at approximately 12 hours and then gradually dissipating over the next 48 to 60 hours. This process illustrates the fact that cross-bridge detachment is ATP driven. Dying cells cannot exclude the higher concentration of calcium ions in the extracellular fluid, causing a calcium influx into muscle cells that promotes myosin cross-bridge binding. When breathing stops, ATP synthesis ceases, and cross-bridge detachment is impossible. This produces the stiffening of dead muscle. With the breakdown of biological molecules, rigor mortis gradually disappears.

Chapter 11:
The Muscular System

LEVEL 1: REVIEWING FACTS AND TERMS

Multiple Choice

1. c	6. d	11. a	16. c	21. d	26. b
2. b	7. b	12. c	17. a	22. d	27. d
3. d	8. b	13. b	18. a	23. c	28. a
4. c	9. c	14. c	19. d	24. c	29. d
5. b	10. d	15. a	20. b	25. a	

Completion

1. biceps brachii
2. sphincters
3. third-class
4. second-class
5. origin
6. synergist
7. popliteus
8. cervicis
9. diaphragm
10. deltoid
11. rectus femoris
12. rotator cuff
13. fibrosis
14. nervous

Matching

1. D	5. H	9. K	12. I	15. L	18. S
2. G	6. E	10. O	13. Q	16. M	19. R
3. F	7. B	11. N	14. J	17. P	
4. A	8. C				

Drawing/Illustration Labeling

Figure 11-1   Muscle Types Based on Patterns of Fascicle Organization

I (Type)		II (Name of Muscle)	
1. d	4. a	1. orbicularis oris	4. biceps brachii
2. f	5. c	2. pectoralis major	5. extensor digitorum
3. b	6. e	3. deltoid	6. rectus femoris

Figure 11-2   Major Superficial Skeletal Muscles (anterior view)

1. sternocleidomastoid
2. deltoid
3. biceps brachii
4. external obliques
5. pronator teres
6. brachioradialis
7. flexor carpi radialis
8. rectus femoris
9. vastus lateralis
10. vastus medialis
11. tibialis anterior
12. extensor digitorium longus
13. pectoralis major
14. rectus abdominis
15. ilopsoas
16. rectus femoris
17. gracilis
18. sartorius
19. gastrocnemius
20. soleus

Figure 11-3   Major Superficial Skeletal Muscles (posterior view)

1. sternocleidomastoid
2. teres major
3. latissimus dorsi
4. anconeus
5. external oblique
6. semitendinosus
7. biceps femoris
8. plantaris
9. soleus
10. trapezius
11. deltoid
12. triceps brachii
13. gluteus medius
14. gluteus maximus
15. semimembranosus
16. gracilis
17. sartorius
18. gastrocnemius
19. calcaneal tendon
20. calcaneus

Figure 11-4   Superficial View of Facial Muscles (lateral view)

1. occipital belly of occipitofrontalis
2. masseter
3. temporalis
4. frontal belly of occipitofrontalis
5. orbicularis oculi
6. zygomaticus minor
7. zygomaticus major
8. buccinator
9. orbicularis oris
10. risorius
11. platysma
12. sternocleidomastoid

Figure 11-5   Muscles of the Neck (anterior view)

1. mylohyoid
2. digastric
3. omohyoid
4. geniohyoid
5. stylohyoid
6. thyrohyoid
7. sternothyroid
8. sternohyoid
9. sternocleidomastoid

Figure 11-6   Superficial Muscles of the Forearm (anterior view)

1. biceps brachii
2. pronator teres
3. brachioradialis
4. flexor carpi radialis
5. palmaris longus
6. flexor carpi ulnaris
7. pronator quadratus

Figure 11-7   Muscles of the Hand (palmar group)

1. abductor pollicis brevis
2. flexor pollicis brevis
3. tendon of flexor pollicis longus
4. adductor pollicis
5. tendon of flexor digitorum profundus
6. flexor retinaculum
7. palmaris brevis
8. lumbricales

Figure 11-8   Superficial Muscles of the Thigh (anterior view)

1. sartorius
2. rectus femoris
3. vastus lateralis
4. quadriceps tendon
5. gracilis
6. vastus medialis

Figure 11-9   Superficial Muscles of the Thigh (posterior view)

1. adductor magnus
2. semitendinosus
3. biceps femoris
4. semimembranosus

Figure 11-10   Superficial Muscles of the Lower Leg (lateral view)

1. gastrocnemius
2. soleus
3. extensor digitorum longus
4. tibialis anterior
5. fibularis longus
6. fibularis brevis

Figure 11-11   Superficial Muscles of the Foot (plantar view)

1. lumbricales
2. flexor digiti minimi brevis
3. abductor digiti minimi
4. flexor hallucis brevis
5. abductor hallucis
6. flexor digitorum brevis
7. calcaneus

LEVEL 2: REVIEWING CONCEPTS

Chapter Overview

1. skin of eyebrow, bridge of nose
2. medial margin of orbit
3. lips
4. manubrium and clavicle
5. clavicle, scapula
6. ribs (8–12), thoracic-lumbar vertebrae
7. deltoid tuberosity of humerus
8. scapula
9. ulna-olecranon process
10. humerus-lateral epicondyle
11. humerus greater tubercle
12. ribs
13. symphysis pubis
14. lower eight ribs
15. iliotibial tract and femur
16. ischium, femur, ilium
17. tibia
18. femoral condyles
19. first metatarsal

Concept Maps
I Axial Musculature

1. buccinator
2. oculomotor muscles
3. masseter
4. hypoglossus
5. stylohyoid
6. splenius
7. capitis
8. thoracis
9. scalenes
10. external-internal intercostals
11. external obliques
12. superior constrictor
13. diaphragm
14. rectus abdominis
15. bulbospongiosus
16. urethral sphincter
17. external anal sphincter

II Appendicular Musculature

1. trapezius
2. deltoid
3. biceps brachii
4. triceps brachii
5. flexor carpi radialis
6. extensor carpi ulnaris
7. extensor digitorum
8. gluteus maximus
9. gracilis
10. obturators
11. biceps femoris
12. rectus femoris
13. gastrocnemius
14. flexor digitorum longus

Multiple Choice

1. b
2. a
3. d
4. c
5. d
6. a
7. c
8. d
9. b
10. c
11. a
12. d
13. b
14. c
15. a

Completion

1. sphincters
2. cervicis
3. antagonist
4. sartorius
5. risorius
6. perineum
7. hamstrings
8. quadriceps
9. hernia
10. fixator
11. masseter
12. iliopsoas

Short Essay

1. a. Parallel muscle—biceps brachii
   b. Convergent muscle—pectoralis major
   c. Pennate muscle—deltoid
   d. Circular muscle—orbicularis oris

2. The origin remains stationary while the insertion moves.

3. Prime mover or agonist, synergist, antagonists

4. a. Muscles of the head and neck
   b. Muscles of the vertebral column
   c. Oblique and rectus muscles
   d. Muscles of the pelvic floor

5. a. Muscles of the shoulders and arms
   b. Muscles of the pelvic girdle and legs

6. First-class lever, see-saw, muscles that extend the neck
   Second-class lever, wheelbarrow, person standing on one's toes
   Third-class lever, shovel, flexing the forearm

7. Biceps femoris, semimembranosus, semitendinosus

8. Rectus femoris, vastus intermedius, vastus lateralis, vastus medialis

9. A pennate muscle contains more muscle fibers than a parallel muscle of the same size. A muscle that has more muscle fibers also has more myofibrils and sarcomeres, resulting in contractions that generate more tension.

LEVEL 3: CRITICAL THINKING AND CLINICAL APPLICATIONS

1. A lever is a rigid structure that moves on a fixed point, called the *fulcrum*. In the human body, each bone is a lever and each joint is a fulcrum. Levers can change (a) the direction of an applied force, (b) the distance and speed of movement produced by a force, and (c) the strength of a force.

2. The rotator cuff muscles are located in the shoulder region and are responsible for acting on the arm. They include the infraspinatus, subscapularis, supraspinatus, and teres minor.

3. The "hamstrings" are found in the posterior region of the upper leg, or thigh. The muscles include the biceps femoris, semimembranosus, and semitendinosus.

4. When a crushing injury, severe contusion, or a strain occurs, the blood vessels within one or more compartments may be damaged. The compartments become swollen with blood and fluid leaked from damaged vessels. Because the connective tissue partitions are very strong, the accumulated fluid cannot escape, and pressures rise within the affected compartments. Eventually compartment pressures becomes so high that they compress the regional blood vessels and eliminate the circulatory supply to the muscles and nerves of the compartment.

5. a. Injecting into tissues rather than circulation allows a large amount of a drug to be introduced at one time because it will enter the circulation gradually. The uptake is usually faster and produces less irritation than when drugs are administered through intradermal or subcutaneous routes; also, multiple injections are possible.
   b. Bulky muscles containing few large blood vessels or nerves are ideal targets. The muscles that may serve as targets are the posterior, lateral, and superior portion of the gluteus maximus; the deltoid of the arm; and the vastus lateralis of the thigh. The vastus lateralis of the thigh is the preferred site in infants and young children whose gluteal and deltoid muscles are relatively small.

Chapter 12:
Neural Tissue

LEVEL 1: REVIEWING FACTS AND TERMS

Multiple Choice

1. a	8. a	15. d	21. d	27. b
2. c	9. c	16. a	22. a	28. a
3. d	10. d	17. c	23. b	29. d
4. c	11. a	18. d	24. b	30. b
5. c	12. b	19. c	25. b	31. d
6. d	13. d	20. c	26. d	32. a
7. b	14. c			

Completion

1. autonomic nervous system
2. collaterals
3. afferent
4. proprioceptors
5. microglia
6. electrochemical gradient
7. threshold
8. saltatory
9. electrical
10. skeletal muscle fiber
11. cholinergic
12. adrenergic
13. neuromodulators
14. temporal summation
15. spatial summation
16. facilitated
17. IPSP

Matching

1. H	5. I	9. B	13. J	17. K
2. F	6. G	10. Q	14. S	18. N
3. D	7. A	11. L	15. R	19. P
4. C	8. E	12. O	16. M	

Drawing/Illustration Labeling

Figure 12-1   Anatomy of a Multipolar Neuron

(a)

1. dendrites
2. cell body (soma)
3. axon
4. telodendria

(b)

1. dendritic branches
2. nissl bodies (RER)
3. mitochondrion
4. axon hillock
5. initial segment of axon
6. Golgi apparatus
7. neurofilament
8. nucleus
9. nucleolus
10. dendrite
11. axolemma
12. axon
13. telodendria
14. synaptic terminals

Figure 12-2   Neuron Classification (Based on Structure)

1. multipolar neuron
2. unipolar neuron
3. anaxonic
4. bipolar neuron

Figure 12-3   The Structure of a Typical Synapse

1. telodendrion
2. synaptic terminal
3. mitochondrion
4. synaptic vesicles
5. presynaptic membrane
6. postsynaptic membrane
7. synaptic cleft
8. endoplasmic reticulum

LEVEL 2: REVIEWING CONCEPTS

Chapter Overview

1. neurons
2. cell body
3. axon
4. action potential
5. dendrites
6. synapse
7. presynaptic
8. postsynaptic
9. neurotransmitters
10. neuromuscular
11. neuroglandular
12. synaptic terminal
13. axoplasmic transport
14. anterograde
15. kinesin
16. retrograde
17. dynein
18. neuroglia
19. ependymal
20. astrocytes
21. oligodendrocytes
22. microglia
23. satellite
24. Schwann

Concept Maps

I Neural Tissue

1. Schwann cells
2. astrocytes or microglia
3. astrocytes or microglia
4. transmit nerve impulses
5. surround peripheral ganglia
6. central nervous system

II Levels of Organization of the Nervous System

1. central nervous system
2. spinal cord
3. cranial nerves
4. motor (efferent) division
5. autonomic nervous system
6. voluntary nervous system
7. parasympathetic division

Multiple Choice

1. c	5. a	9. a	13. a	16. c
2. d	6. c	10. b	14. b	17. c
3. a	7. d	11. a	15. b	18. a
4. b	8. d	12. c		

Completion

1. perikaryon	6. hyperpolarization	11. interneurons
2. ganglia	7. nerve impulse	12. neurotransmitter
3. current	8. microglia	13. temporal summation
4. voltage	9. preganglionic fibers	14. nuclei
5. voltage-gated	10. postganglionic fibers	15. tracts

Short Essay

1. Central nervous system (CNS) and peripheral nervous system (PNS)

2. CNS consists of the brain and the spinal cord.
   PNS consists of all the neural tissue outside the CNS.

3. a. Astrocytes, oligodendrocytes, microglia, ependymal cells
   b. Satellite cells (amphicytes); Schwann cells (neurilemmocytes)

4. *Neurons* are responsible for information transfer and processing in the nervous system.
   Neuroglia are specialized cells that provide support throughout the nervous system.

5. a. Activation of sodium channels and membrane depolarization
   b. Sodium channel inactivation
   c. Potassium channel activation
   d. Return to normal permeability

6. In continuous propagation, an action potential spreads across the entire excitable membrane surface in a series of small steps. In saltatory propagation, an action potential appears to leap from node to node, skipping the intervening membrane surface. Saltatory propagation carries nerve impulses many times more rapidly than does continuous propagation.

7. Action potentials in muscle tissues have (1) larger resting potentials, (2) longer-lasting action potentials, and (3) slower propagation of action potentials.

8. An EPSP is a depolarization produced by the arrival of a neurotransmitter at the postsynaptic membrane. A typical EPSP produces a depolarization of around 0.5 mV, much less than the 15- to 20-mV depolarization needed to bring the axon hillock to threshold. An IPSP is a transient hyperpolarization of the postsynaptic membrane. A hyperpolarized membrane is inhibited because a larger-than-usual depolarization stimulus must be provided to bring the membrane potential to threshold.

9. a. Sensory neurons—carry nerve impulses to the CNS
   b. Motor neurons—carry nerve impulses from CNS to PNS
   c. Interneurons—situated between sensory and motor neurons within the brain and spinal cord

10. At an electrical synapse, the presynaptic and postsynaptic cell membranes are bound by interlocking membrane proteins at a gap junction. Pores within these proteins permit the passage of local currents, and the two neurons act as if they share a common cell membrane. At chemical synapses, excitatory neurotransmitters cause depolarization and promote the generation of action potentials. Inhibitory neurotransmitters cause hyperpolarization and suppress the generation of action potentials.

LEVEL 3: CRITICAL THINKING AND CLINICAL APPLICATIONS

1. Facilitation results from the summation of EPSPs, or from the exposure to certain drugs in the extracellular fluid. The larger the degree of facilitation, the smaller the additional stimulus needed to trigger an action potential. The nicotine in cigarettes stimulates postsynaptic ACh receptors, producing prolonged EPSPs that facilitate CNS neurons. Nicotine also increases the release of the neurotransmitter dopamine, producing feelings of pleasure and reward, and may lead to addiction.

2. In myelinated fibers, saltatory propagation transmits nerve impulses at rates over 300 mph, allowing the impulses to reach the neuromuscular junctions fast enough to initiate muscle contraction and promote normal movements. In a demyelinating disease, the nerve impulses can't propagate along the demyelinated axon; therefore, the muscle is not stimulated, leading to paralysis, or lack of movement. Eventually the muscles atrophy because of a lack of adequate activity involving contraction.

3. Microglia do not originate in neural tissue; they are phagocytic defense cells that have migrated across capillary walls in the neural tissue of the CNS. They engulf cellular debris, waste products, and pathogens. In times of infection or injury their numbers increase dramatically, as other phagocytic cells are attracted to the damaged area.

4. Based on their effects on postsynaptic membranes, neurotransmitters are often classified as excitatory or inhibitory. This may be useful, but not always precise. For example, acetylcholine is released at synapses in the CNS and PNS, and typically produces a depolarization in the postsynaptic membrane. The acetylcholine released at neuromuscular junctions in the heart produces a transient hyperpolarization of the membrane, moving the transmembrane potential farther from threshold.

5. A subthreshold membrane potential is referred to as an excitatory postsynaptic potential (EPSP), and the membrane is said to be *facilitated*. EPSPs may combine to reach threshold and initiate an action potential in two ways: (a) by spatial summation, during which several presynaptic neurons simultaneously release neurotransmitter to a single postsynaptic neuron; and (b) by temporal summation, during which the EPSPs result from the rapid and successive discharges of neurotransmitter from the same presynaptic terminal.

Chapter 13:
The Spinal Cord, Spinal Nerves, and Spinal Reflexes

LEVEL 1: REVIEWING FACTS AND TERMS

Multiple Choice

1. d	6. c	11. c	16. c	20. d	24. d
2. d	7. a	12. c	17. b	21. c	25. a
3. a	8. c	13. b	18. c	22. a	26. b
4. a	9. b	14. b	19. b	23. b	
5. a	10. a	15. d			

Completion

1. ganglia
2. conus medullaris
3. filum terminale
4. nuclei
5. columns
6. epineurium
7. dorsal ramus
8. dermatome
9. convergence
10. somatic reflexes
11. receptor
12. innate reflexes
13. acquired reflexes
14. cranial reflexes
15. polysynaptic reflex
16. crossed extensor reflex
17. flexor reflex
18. reinforcement
19. Babinski sign

Matching

1. H	4. A	7. C	10. L	13. I	16. P
2. G	5. E	8. D	11. K	14. M	17. N
3. F	6. B	9. Q	12. J	15. O	

Drawing/Illustration Labeling

Figure 13-1   Anterior (sectional) View of the Adult Spinal Cord

1. posterior white column
2. lateral gray horn
3. dorsal root ganglion
4. posterior median sulcus
5. posterior gray commissure
6. sensory nuclei
7. motor nuclei
8. ventral root
9. anterior gray commissure
10. anterior median fissure

Figure 13-2   Organization of the Reflex Arc

1. receptor
2. sensory
3. interneuron
4. CNS
5. motor
6. effector

Figure 13-3   The Organization of Neuronal Pools

1. divergence
2. convergence
3. serial processing

4. parallel processing
5. reverberation

LEVEL 2: REVIEWING CONCEPTS

Chapter Overview

1. receptors
2. afferent
3. sensory

4. interneurons
5. visceral motor neurons
6. motor

7. postganglionic efferent fiber
8. smooth muscles
9. effectors

Concept Maps

I Spinal Cord

1. gray matter
2. glial cells
3. posterior gray horns
4. sensory information

5. viscera
6. somatic motor neurons
7. skeletal muscles
8. anterior white columns

9. ascending tracts
10. brain
11. motor commands

II Spinal Cord—Spinal Meninges

1. dura mater
2. stability, support
3. simple squamous epithelium

4. subarachnoid space
5. shock absorber
6. diffusion medium

7. pia mater
8. blood vessels

Multiple Choice

1. c	6. b	11. b	15. a	19. d
2. b	7. a	12. d	16. c	20. c
3. a	8. c	13. b	17. a	21. a
4. d	9. a	14. d	18. d	22. c
5. d	10. c			

Completion

1. perineurium
2. descending tracts
3. dura mater
4. spinal meninges
5. pia mater
6. cauda equina

7. sensory nuclei
8. motor nuclei
9. dermatome
10. gray commissures
11. gray ramus

12. dorsal ramus
13. nerve plexus
14. brachial plexus
15. reflexes
16. gamma efferents

Short Essay

1. (a) Dura mater, (b) arachnoid, (c) pia mater

2. Mixed nerves contain both afferent (sensory) and efferent (motor) fibers.

3. The white matter contains large numbers of myelinated and unmyelinated axons. The gray matter is dominated by bodies of neurons and glial cells.

4. A dermatome is a specific region of the body surface monitored by a pair of spinal nerves. Damage to a spinal nerve or dorsal root ganglion will produce characteristic loss of sensation in the skin.

5. a. Arrival of a stimulus and activation of a receptor
   b. Activation of a sensory neuron
   c. Information processing
   d. Activation of a motor neuron
   e. Response of a peripheral effector

6.　a.　Their development
　　b.　The site where information processing occurs
　　c.　The nature of the resulting motor response
　　d.　The complexity of the neural circuit involved

7.　a.　The interneurons can control several different muscle groups.
　　b.　The interneurons may produce either excitatory or inhibitory postsynaptic potentials at CNS motor nuclei; thus the response can involve the stimulation of some muscles and the inhibition of others.

8.　When one set of motor neurons is stimulated, those controlling antagonistic muscles are inhibited. The term *reciprocal* refers to the fact that the system works both ways. When the flexors contract, the extensors relax; when the extensors contract, the flexors relax.

9.　In a contralateral reflex arc, the motor response occurs on the side opposite the stimulus. In an ipsilateral reflex arc, the sensory stimulus and the motor response occur on the same side of the body.

10.　The Babinski reflex in an adult is a result of an injury in the CNS. Usually the higher centers or the descending tracts are damaged.

LEVEL 3: CRITICAL THINKING AND CLINICAL APPLICATIONS

1.　Because the spinal cord extends to the L_2 level of the vertebral column and the meninges extend to the end of the vertebral column, a needle can be inserted through the meninges inferior to the medullary cone into the subarachnoid space with minimal risk to the cauda equina.

2.　Meningitis is an inflammation of the meninges. It is caused by viral or bacterial infection and produces symptoms that include headache and fever. Severe cases result in paralysis, coma, and sometimes death.

3.　An epidural block is a temporary sensory loss or a sensory and motor paralysis, depending on the anesthetic selected. Injecting the anesthetic into the epidural space in the lower lumbar region inferior to the conus medullaris has at least two advantages:

　　1.　It affects only the spinal nerves in the immediate area of the injection, and

　　2.　It provides mainly sensory anesthesia. In caudal anesthesia, the injection site is into the epidural space of the sacrum, making the paralysis and anesthetizing more widespread in the abdominal and perineal areas.

4.　The positive patellar reflex will confirm that the spinal nerves and spinal segments L_2–L_4 are not damaged.

5.　a.　Eliminates sensation and motor control of the arms and legs. Usually results in extensive paralysis—quadriplegia.
　　b.　Motor paralysis and major respiratory muscles such as the diaphragm—patient needs mechanical assistance to breathe.
　　c.　The loss of motor control of the legs—paraplegia.
　　d.　Damage to the elements of the cauda equina causes problems with peripheral nerve function.

Chapter 14:
The Brain and Cranial Nerves

LEVEL 1: REVIEWING FACTS AND TERMS

Multiple Choice

1. d	8. d	15. c	22. a	29. b	36. d
2. b	9. d	16. b	23. c	30. a	37. a
3. a	10. b	17. c	24. a	31. b	38. a
4. d	11. d	18. d	25. b	32. c	39. c
5. a	12. b	19. d	26. d	33. d	40. b
6. d	13. a	20. b	27. c	34. b	41. d
7. a	14. b	21. d	28. c	35. a	

Completion

1. cerebrum
2. cerebellum
3. brain stem
4. pons
5. myelencephalon
6. cerebrospinal fluid
7. cerebral aqueduct
8. falx cerebri
9. tentorium cerebelli
10. choroid plexus
11. edendymal
12. spinal cord
13. cranial nerves
14. thalamus
15. respiratory centers
16. white matter
17. ataxia
18. red nuclei
19. diencephalon
20. third ventricle
21. amygdaloid body
22. hypothalamus
23. basal nuclei
24. lentiform nucleus
25. theta waves
26. seizures
27. direct light reflex
28. corneal reflex
29. vestibulo-ocular

Matching

1. E
2. G
3. A
4. H
5. B
6. D
7. C
8. F
9. M
10. P
11. I
12. O
13. K
14. J
15. L
16. N

Drawing/Illustration Labeling

Figure 14-1   Lateral View of the Human Brain

1. central sulcus
2. precentral gyrus
3. frontal lobe
4. lateral sulcus
5. temporal lobe
6. pons
7. medulla oblongata
8. cerebellum
9. occipital lobe
10. parietal lobe
11. postcentral gyrus

Figure 14-2   An Introduction to Brain Structures and Functions

1. cerebrum
2. diencephalon
3. thalamus
4. hypothalamus
5. midbrain
6. pons
7. medulla oblongata
8. cerebellum

LEVEL 2: REVIEWING CONCEPTS

Chapter Overview

Part I: Review of Cranial Nerves

1. olfactory
2. oculomotor
3. trigeminal
4. facial
5. glossopharyngeal
6. accessory
7. S
8. M
9. M
10. S
11. B
12. M

Part II: Origins of Cranial Nerves—Inferior View of the Human Brain—Partial

1. olfactory bulb
2. N III oculomotor
3. N V trigeminal
4. N VII facial
5. N IX glossopharyngeal
6. N X vagus
7. N II optic
8. N IV trochlear
9. N VI abducens
10. N VIII vestibulocochlear
11. N XII hypoglossal
12. N XI accessory

Concept Maps

I Major Regions of the Brain

1. diencephalon
2. hypothalamus
3. corpora quadrigemina
4. 2 cerebellar hemispheres
5. pons
6. medulla oblongata

II Region I: Cerebrum

1. temporal
2. somatic sensory cortex
3. occipital
4. cerebral cortex
5. projections
6. commissural
7. basal nuclei

III Region II: Diencephalon

1. thalamus
2. anterior
3. ventral
4. geniculates
5. preoptic area
6. autonomic centers
7. pineal gland
8. cerebrospinal fluid

IV Region III: Midbrain

1. cerebral peduncles
2. gray matter
3. corpora quadrigemina
4. inferior colliculi
5. red nucleus
6. substantia nigra

V Region IV: Cerebellum

1. vermis
2. arbor vitae
3. gray matter
4. Purkinje cells
5. cerebellar nuclei

Region V: Pons

1. respiratory centers
2. apneustic
3. white matter
4. superior
5. inferior
6. transverse fibers

VII Region VI: Medulla Oblongata

1. white matter
2. brain and spinal cord
3. cuneatus
4. olivary
5. reflex centers
6. cardiac
7. distribution of blood flow
8. respiratory rhythmicity centers

Multiple Choice

1. d	6. c	11. b	16. a	21. c	26. c
2. a	7. a	12. a	17. c	22. d	27. c
3. d	8. d	13. d	18. d	23. b	28. c
4. d	9. c	14. c	19. b	24. b	29. c
5. b	10. b	15. b	20. a	25. c	

Completion

1. thalamus
2. pituitary gland
3. arcuate
4. commissural
5. corpus striatum
6. hippocampus
7. fornix
8. drives
9. third ventricle
10. cerebral aqueduct
11. spinal cord
12. sulci
13. fissures
14. Wernicke
15. hypothalamus
16. global aphasia
17. aphasia
18. dyslexia
19. arousal
20. dopamine

Short Essay

1. The brain's versatility results from (a) the tremendous number of neurons and neuronal pools in the brain, and (b) the complexity of the interconnections between the neurons and neuronal pools.

2. 1. Cerebrum
   2. Diencephalon
   3. Midbrain
   4. Cerebellum
   5. Pons
   6. Medulla oblongata

3. The limbic system includes nuclei and tracts along the border between the cerebrum and diencephalon. It is involved in the processing of memories, creation of emotional states, drives, and associated behaviors.

4. Most endocrine organs are under direct or indirect hypothalamic control. Releasing hormones and inhibiting hormones secreted by nuclei in the tuberal area of the hypothalamus promote or inhibit secretion of hormones by the anterior pituitary gland. The hypothalamus also secretes the hormones ADH (antidiuretic hormone) and oxytocin.

5. (a) The cerebellum oversees the postural muscles of the body, making rapid adjustments to maintain balance and equilibrium. (b) The cerebellum programs and fine tunes movements controlled at the conscious and subconscious levels.

6. (a) Sensory and motor nuclei for four of the cranial nerves; (b) nuclei concerned with the involuntary control of respiration; (c) tracts that link the cerebellum with the brain stem, cerebrum, and spinal cord; (d) ascending, descending and transverse tracts.

7. It isolates neural tissue in the CNS from the general circulation.

8. (a) It provides cushioning, (b) it provides support, and (c) it transports nutrients, chemical messengers, and waste products.

9. N I—olfactory, N II—optic, N VIII—vestibulocochlear

10. Cranial reflexes provide a quick and easy method for checking the condition of cranial nerves and specific nuclei and tracts in the brain.

LEVEL 3: CRITICAL THINKING AND CLINICAL APPLICATIONS

1. Even though cerebrospinal fluid exits in the brain are blocked, CSF continues to be produced. The fluid continues to build inside the brain, causing pressure that compresses the nerve tissue and causes the ventricles to dilate. The compression of the nervous tissue causes irreversible brain damage.

2. Smelling salts (i.e., ammonia) stimulate trigeminal nerve endings in the nose, sending impulses to the reticular formation in the brain stem and the cerebral cortex. The axons in the reticular formation cause arousal and maintenance of consciousness. The reticular formation and its connections form the reticular activating system that is involved with the sleep-wake cycle.

3. Activity in the thirst center produces the conscious urge to take a drink. Hypothalamic neurons in this center detect changes in the osmotic concentration of the blood. When the concentration rises, the thirst center is stimulated. The thirst center is also stimulated by ADH (antidiuretic hormone). Stimulation of the thirst center in the hypothalamus triggers a behavior response (drinking) that complements the physiological response to ADH (water conservation by the kidneys).

4. Many drugs, including alcohol, have profound effects on the function of the cerebellum. A person who is under the influence cannot properly anticipate the range and speed of limb movement, resulting in a disturbance of muscular coordination called ataxia.

5. The objects serve as stimuli initiating tactile sensations that travel via sensory neurons to the spinal cord. From the ascending tracts in the spinal cord the impulses are transmitted to the somesthetic association cortex of the left cerebral hemisphere, where the objects are "recognized." The impulses are then transmitted to the speech comprehension area (Wernicke's area), where the objects are give names. From there the impulses travel to Broca's area for formulation of the spoken words, and finally the impulses arrive at the premotor and motor cortex for programming and for producing the movements to form and say the words that identify the objects.

Chapter 15:
Neural Integration I: Sensory Pathways and the Somatic Nervous System

LEVEL 1: REVIEWING FACTS AND TERMS

Multiple Choice

1. c	6. c	11. d	16. a	20. d	24. c
2. b	7. d	12. b	17. d	21. d	25. d
3. a	8. a	13. c	18. b	22. b	26. b
4. d	9. b	14. c	19. b	23. a	27. b
5. b	10. c	15. a			

Completion

1. sensory receptors
2. diencephalon
3. transduction
4. generator potential
5. adaptation
6. interoceptors
7. substance P
8. proprioceptors
9. Merkel's discs
10. lamellated corpuscles
11. muscle spindles
12. first order
13. thalamus
14. spinothalamic
15. Purkinje cells
16. solitary nucleus
17. sensory homunculus
18. corticobulbar
19. corticospinal
20. cerebral peduncles
21. lateral
22. hypothalamus
23. spinal cord
24. cerebral hemispheres
25. cerebellum

Matching

1. I	5. H	9. E	13. O	17. L
2. F	6. J	10. C	14. K	18. S
3. G	7. D	11. Q	15. R	19. N
4. B	8. A	12. T	16. M	20. P

Drawing/Illustration Labeling

Figure 15-1   Ascending and Descending Tracts of the Spinal Cord

(*Decending*)
1. lateral corticospinal
2. rubrospinal
3. reticulospinal
4. vestibulospinal

5. tectospinal
6. anterior corticospinal

(*Ascending*)
7. fasciculus gracilis
8. fasciculus cuneatus

9. posterior spinocerebellar
10. lateral spinothalamic
11. anterior spinocerebellar
12. anterior spinothalamic

Figure 15-2   Tactile Receptors in the Skin

1. free nerve endings
2. root hair plexus
3. Merkel cells and tactile discs

4. tactile corpuscle
5. lamellated corpuscle
6. Ruffini corpuscle

LEVEL 2: REVIEWING CONCEPTS

Chapter Overview

1. left cerebral hemisphere
2. midbrain
3. medulla oblongata
4. spinal cord
5. motor nuclei of cranial nerves
6. decussation of pyramids
7. to skeletal muscles

Concept Maps

I Afferent Division of Nervous System—Sensory Tracts (Ascending)

1. posterior column
2. medulla (nucleus gracilis)
3. fasciculus cuneatus
4. posterior tract
5. cerebellum (cortex)
6. proprioceptors
7. spinothalamic
8. thalamus (ventral nuclei)
9. anterior tract
10. interneurons

II Efferent Division of Nervous System—Motor Tracts (Descending)

1. basal nuclei
2. anterior corticospinal tracts
3. pyramids
4. cranial nerves
5. control of skeletal muscles
6. vestibular nuclei
7. control of muscle tone and balance
8. tectospinal tracts
9. regulation of eye, head, neck, upper limb position
10. reticulospinal tracts
11. uncrossed
12. regulation of reflex activity
13. lateral pathway
14. brain stem
15. control of muscle tone and precise distal limb movements

Multiple Choice

1. d
2. a
3. c
4. c
5. b
6. a
7. b
8. d
9. c
10. a
11. b
12. c
13. a
14. d
15. d
16. c
17. a
18. b
19. a
20. c
21. d
22. a

Completion

1. descending
2. ascending
3. GABA
4. anencephaly
5. basal nuclei
6. Meissner's
7. deep pressure
8. nociceptors
9. medial lemniscus
10. interoceptors
11. posterior
12. cerebellum
13. thalamus
14. homunculus
15. primary sensory cortex

Short Essay

1. Posterior column, spinothalamic, and spinocerebellar pathways

2. Corticobulbar tracts, lateral corticospinal tract, and anterior corticospinal tract

3. Lateral pathway: rubrospinal tract

   Medial pathway: reticulospinal tract, vestibulospinal tract, and tectospinal tract.

4. (a) Overseeing the postural muscles of the body, and (b) adjusting voluntary and involuntary motor patterns.

5. Tonic receptors are always active and convey relative stimulus activity by changing the rate of action potential generation. Phasic receptors are normally inactive and only become active for a short time whenever stimulated by an appropriate change in conditions. Phasic receptors are likely to be fast-adapting receptors; tonic receptors are likely to be slow-adapting receptors.

6. The CNS determines information about a stimulus in the following ways: (a) the labeled line identifies the type of stimulus; (b) the area of the sensory cortex stimulated by the sensory information determines the perceived location of the stimulus; and (c) the frequency and pattern of action potential generation (sensory coding) conveys information about stimulus strength, duration, and/or variation.

7. General senses: temperature, touch, pain, pressure, vibration, proprioception
   Special senses: olfaction (smell), vision, gustation (taste), equilibrium (balance), hearing

LEVEL 3: CRITICAL THINKING AND CLINICAL APPLICATIONS

1. This is an example of referred pain, where strong visceral pain from the heart has stimulated spinal cord interneurons of the spinothalamic pathway and created pain sensations in a specific part of the body surface—in this case, radiating down the left arm.

2. An individual can experience painful sensations that are not real. It is called a phantom limb pain, caused by activity in the sensory neurons or interneurons along the spinothalamic pathway. The neurons involved were once part of the labeled line that monitored conditions in the intact limb. These labeled lines and pathways are developmentally programmed.

3. a. One group of axons synapses with thalamic neurons, which then send their axons to the primary motor cortex. A feedback loop is created that changes the sensitivity of the pyramidal cells and alters the pattern of instructions carried by the corticospinal tracts.
   b. A second group of axons synapses on the reticular formation, altering the excitatory or inhibitory output of the reticulospinal tracts.

4. Injuries to the motor cortex eliminate the ability to exert fine control over motor units, but gross movements may still be produced by the basal nuclei using the tracts of the still-functioning medial and lateral pathways.

5. The suspicious cause of death could be anencephaly, a rare condition in which the brain fails to develop at levels above the midbrain or lower diencephalon. The brain stem, which controls complex involuntary motor patterns, is still intact; therefore, all of the normal behavior patterns expected of a newborn can still occur.

Chapter 16:
Neural Integration II: The Autonomic Nervous System and Higher Order Functions

LEVEL 1: REVIEWING FACTS AND TERMS

Multiple Choice

1. c	11. d	21. a	31. c
2. d	12. b	22. a	32. a
3. b	13. d	23. d	33. b
4. b	14. d	24. c	34. d
5. a	15. c	25. b	35. c
6. d	16. d	26. d	36. d
7. d	17. b	27. a	37. b
8. d	18. b	28. d	
9. c	19. d	29. d	
10. b	20. c	30. c	

Completion

1. involuntary
2. "fight or flight"
3. collateral ganglia
4. sympathetic activation
5. acetylcholine
6. epinephrine
7. excitatory
8. norepinephrine
9. "rest and digest"
10. vagus nerve
11. anabolic
12. cholinergic
13. acetylcholinesterase
14. receptor
15. muscarinic
16. opposing
17. autonomic tone
18. limbic
19. hypothalamus
20. visceral reflexes
21. memory consolidation
22. CNS
23. rapid eye movement
24. schizophrenia
25. plaques

Matching

1. H
2. E
3. F
4. K
5. A
6. D
7. J
8. B
9. C
10. G
11. I
12. L
13. M
14. N
15. X
16. Y
17. S
18. Q
19. P
20. W
21. O
22. V
23. T
24. U
25. R

Drawing/Illustration Labeling

Figure 16-1   Organization of the Sympathetic Division of the ANS

1. preganglionic neurons
2. paired sympathetic chain ganglia
3. unpaired collateral ganglia
4. visceral effectors in abdominopelvic cavity
5. paired adrenal medullae
6. organs and systems throughout body

Figure 16-2   Organization of the Parasympathetic Division of the ANS

1. nuclei in brain stem
2. N VII (facial)
3. N X (vagus)
4. ciliary ganglion
5. intrinsic eye muscles (pupil and lens shape)
6. nasal, tear, and salivary glands
7. otic ganglion
8. parotid salivary gland
9. intramural ganglia

LEVEL 2: REVIEWING CONCEPTS

Chapter Overview

1. P
2. P
3. S
4. P
5. S
6. S
7. P
8. S
9. S
10. S
11. P
12. S
13. P
14. P

Concept Maps

I Autonomic Nervous System—Sympathetic Division

1. thoracolumbar
2. spinal segments T_1–L_2
3. postganglionic
4. sympathetic chain of ganglia (paired)
5. visceral effectors
6. adrenal medulla (paired)
7. general circulation

II Autonomic Nervous System—Parasympathetic Division

1. craniosacral
2. brain stem
3. ciliary ganglion
4. N VII (facial)
5. nasal, tear, salivary glands
6. otic ganglia
7. N X (vagus)
8. segments S_2–S_4
9. intramural ganglia
10. lower abdominopelvic cavity

III Levels of Autonomic Control

1. hypothalamus
2. parasympathetic
3. medulla oblongata
4. cardiovascular
5. respiratory
6. spinal cord
7. sympathetic
8. parasympathetic

Multiple Choice

1. a
2. d
3. b
4. c
5. a
6. d
7. a
8. c
9. b
10. a
11. c
12. d
13. d
14. b
15. d
16. d

Completion

1. postganglionic
2. norepinephrine
3. adrenal medulla
4. collateral
5. gray ramus
6. white ramus
7. splanchnic
8. hypothalamus
9. acetylcholine
10. blockers

Short Essay

1. The sympathetic division stimulates tissue metabolism, increases alertness, and generally prepares the body to deal with emergencies.

2. The parasympathetic division conserves energy and promotes sedentary activities, such as digestion.

3. a. All preganglionic fibers are cholinergic; they release acetylcholine (ACh) at their synaptic terminals. The effects are always excitatory.
   b. Postganglionic parasympathetic fibers are also cholinergic, but the effects may be excitatory or inhibitory, depending on the nature of the receptor.
   c. Most postganglionic sympathetic terminals are adrenergic; they primarily release norepinephrine (NE). The effects are usually excitatory.

4. a. Preganglionic neurons located within segments T1 and L2 of the spinal cord
   b. Ganglionic neurons located in ganglia near the vertebral column (sympathetic chain ganglia, collateral ganglia)
   c. Specialized ganglionic neurons in the medulla of the adrenal gland

5. a. The reduction of blood flow and energy use by visceral organs such as the digestive tract that are not important to short-term survival
   b. The release of stored energy reserves through the action of the adrenal medulla

6. a. Preganglionic neurons in the brain stem and in sacral segments of the spinal cord
   b. Ganglionic neurons in peripheral ganglia located within or adjacent to the target organs

7. III (oculomotor), VII (facial), IX (glossopharyngeal) and X (vagus)

8. Stimulation of the parasympathetic system leads to a general increase in the nutrient content of the blood. Cells throughout the body respond to this increase by absorbing nutrients and using them to support growth and other anabolic activities.

9. alpha-1, alpha-2; beta-1, beta-2, beta-3

10. Nicotinic, muscarinic

11. When dual innervation exists, the two divisions of the ANS often but not always have opposing effects. Sympathetic-parasympathetic opposition can be seen along the digestive tract, at the heart, in the lungs, and elsewhere throughout the body.

12. If a nerve maintains a background level of activity, it may either increase or decrease its activity, thus providing a range of control options.

13. a. Defecation and urination reflexes
    b. Papillary and vasomotor reflexes

LEVEL 3: CRITICAL THINKING AND CLINICAL APPLICATIONS

1. Sympathetic postganglionic fibers that enter the thoracic cavity in autonomic nerves cause the heart rate to accelerate, increase the force of cardiac contractions, and dilate the respiratory passageways. Because of these functions, the heart works harder, moving blood faster. The muscles receive more blood, and their utilization of stored and absorbed nutrients accelerates. Lipids, a potential energy source, are released. The lungs deliver more oxygen and prepare to eliminate the carbon dioxide produced by contracting muscles. Sweat glands become active and the pupils dilate to provide more light.

2. When the sympathetic division of the ANS is fully activated, it produces what is known as the "fight or flight" response, which readies the body for a crisis that may require sudden, intense physical activity. It generally stimulates tissue metabolism and increases alertness. Amanda's responses may have included an increased metabolic rate to as much as twice the resting level, temporary suspension of her digestive and urinary activities, and increased blood flow to her skeletal muscles. Other responses may include: breathing more quickly and more deeply, increased heart rate and blood pressure, and a feeling of warmth that may induce perspiring.

3. Even though most sympathetic postganglionic fibers are adrenergic, releasing norepinephrine, a few are cholinergic, releasing acetylcholine. The distribution of the cholinergic fibers via the sympathetic division provides a method of regulating sweat gland secretion and selectively controlling blood flow to skeletal muscles while reducing the flow to other tissues in a body wall.

4. Blocking the beta receptors on cells decreases or prevents sympathetic stimulation of the tissues that contain those cells. The heart rate, the force of contraction of cardiac muscle, and the contraction of smooth muscle in the walls of blood vessels would decrease. These changes would contribute to a lowering of the blood pressure.

Chapter 17:
The Special Senses

LEVEL 1: REVIEWING FACTS AND TERMS

Multiple Choice

1. c	7. d	13. c	19. a	25. b
2. a	8. a	14. b	20. b	26. c
3. d	9. c	15. c	21. d	27. b
4. b	10. a	16. d	22. b	28. d
5. c	11. d	17. c	23. b	29. c
6. b	12. a	18. c	24. c	30. b

Completion

1. cerebral cortex
2. olfactory
3. olfactory discrimination
4. taste buds
5. gustducins
6. sclera
7. pupil
8. rods
9. amacrine
10. accommodation
11. astigmatism
12. cones
13. occipital
14. photoreception
15. convergence
16. endolymph
17. round window
18. saccule
19. midbrain

Matching

1. G
2. I
3. E
4. B
5. H
6. C
7. F
8. J
9. D
10. A
11. O
12. R
13. T
14. K
15. U
16. S
17. M
18. L
19. V
20. P
21. N
22. Q

Drawing/Illustration Labeling

Figure 17-1   Sagittal Sectional Anatomy of the Left Eye

1. posterior cavity
2. choroid
3. fovea
4. optic nerve
5. optic disc
6. retina
7. sclera
8. fornix
9. palpebral conjunctiva
10. ocular conjunctiva
11. ciliary body
12. iris
13. lens
14. cornea
15. limbus
16. suspensory ligaments
17. vitreous body

Figure 17-2   Anatomy of the Ear: External, Middle, and Internal Ear

1. external ear
2. middle ear
3. internal ear
4. auricle
5. external acoustic meatus
6. cartilage
7. tympanic membrane
8. auditory ossicles
9. tympanic cavity
10. petrous portion of temporal bone
11. vestibular cochlear nerve (VIII)
12. cochlea
13. auditory tube

Figure 17-3   Anatomy of the Ear (Bony Labyrinth)

1. semicircular canals
2. vestibular duct
3. cochlear duct
4. tympanic duct
5. maculae
6. cochlea
7. spiral organ (organ of Corti)

Figure 17-4   Anatomy of the Cochlea (Details Visible in Section)

1. vestibular duct
2. body cochlear wall
3. cochlear duct
4. tectorial membrane
5. spiral organ (organ of Corti)
6. basilar membrane
7. tympanic duct
8. spiral ganglion
9. cochlear nerve

Figure 17-5   Three-Dimensional Structure of Tectorial Membrane and Hair Complex of the Spiral Organ

1. tectorial membrane
2. outer hair cell
3. basilar membrane
4. inner hair cell
5. nerve fibers

Figure 17-6   Gustatory Receptors—Landmarks and Receptors on the Tongue

1. sour
2. bitter
3. salty
4. sweet
5. circumvallate papilla
6. fungiform papilla
7. filiform papilla

LEVEL 2: REVIEWING CONCEPTS

Chapter Overview

1. olfactory epithelium
2. olfactory cilia
3. second messenger
4. depolarize
5. modified neurons
6. taste receptors
7. papillae
8. salty
9. sour
10. sweet
11. bitter
12. umami
13. water
14. rods
15. cones
16. green
17. continually
18. open
19. transducin
20. close
21. bipolar
22. bleaching
23. internal ear
24. semicircular ducts
25. utricle and saccule
26. vestibule
27. stimulates
28. inhibits
29. accessory structures
30. rotational planes
31. floats
32. distorts
33. gravity
34. spiral organ
35. tympanic membrane
36. oval window
37. perilymph
38. cochlear duct
39. endolymph
40. sterocilia
41. tectorial
42. depolarize
43. neurotransmitter

Concept Maps
I Special Senses

1. olfaction
2. smell
3. tongue
4. taste buds
5. ears
6. balance and hearing
7. audition
8. retina
9. rods and cones

II Sound Perception

1. external acoustic canal
2. middle ear
3. incus
4. oval window
5. endolymph in cochlear duct
6. round window
7. hair cells of organ of Corti
8. vestibulocochlear nerve VIII

Multiple Choice

1. c
2. a
3. b
4. b
5. d
6. a
7. b
8. b
9. d
10. b
11. a
12. d
13. a
14. c
15. b
16. a
17. c
18. d
19. c
20. b
21. a
22. d
23. b
24. c
25. d

Completion

1. sensory receptor
2. sensation
3. receptive field
4. transduction
5. receptor potential
6. afferent fiber
7. adaptation
8. phasic
9. ampulla
10. nystagmus
11. hertz
12. odorants
13. olfactory stimulation
14. taste buds
15. pupil
16. aqueous humor
17. retina
18. cataract
19. accommodation
20. myopia
21. hyperopia

Short Essay

1. Sensations of olfaction, gustation, vision, equilibrium and hearing

2. a. The stimulus alters the permeability of the receptor membrane.
   b. The receptor potential produces a generator potential.
   c. The action potential travels to the CNS over an afferent fiber.

3. Sensations leaving the olfactory bulb travel along the olfactory tract to reach the olfactory cortex, the hypothalamus, and portions of the limbic system.

4. Sweet, salty, sour, bitter

5. Receptors in the saccule and utricle provide sensations of gravity and linear acceleration.

6. a. Provides mechanical support and physical protection
   b. Serves as an attachment site for the extrinsic eye muscles
   c. Assists in the focusing process
   d. Permits light to enter the eye

7. a. Provides a route for blood vessels and lymphatics to the eye
   b. Secretes and reabsorbs aqueous humor
   c. Controls the shape of the lens; important in focusing process
   d. Controls the amount of light entering the posterior chamber

8. The pigment layer: (a) absorbs light after it passes through the retina, and (b) biochemically interacts with the photoreceptor layer of the retina.

   The retina contains: (a) photoreceptors that respond to light, (b) supporting cells and neurons that perform preliminary processing and integration of visual information, and (c) blood vessels supplying tissues lining the vitreous chamber.

9. During accommodation, the lens becomes rounder to focus the image of a nearby object on the retina.

10. As aging proceeds, the lens becomes less elastic, takes on a yellowish hue, and eventually begins to lose its transparency. Visual clarity begins to fade, and when the lens turns completely opaque, the person becomes functionally blind despite the fact that the retinal receptors are alive and well.

LEVEL 3: CRITICAL THINKING AND CLINICAL APPLICATIONS

1. The chemical stimuli of smell are eventually converted into neural events that reach the olfactory cortex, the hypothalamus, and portions of the limbic system. The extensive limbic and hypothalamic connections help to explain the profound emotional and behavioral responses that can be produced by certain scents. The perfume industry understands the practical implications of these connections.

2. The total number of olfactory receptors declines with age, and elderly individuals have difficulty detecting odors unless large quantities are used. Younger individuals can detect the odor in lower concentrations because the olfactory receptor population is constantly producing new receptor cells by division and differentiation of basal cells in the epithelium.

3. The middle ear begins at the tympanic membrane, which forms a boundary between the middle ear and the external ear. The tympanic membrane is the site of the first transduction that takes place in the ear—external sound waves (pressure waves) are transduced into mechanical movements by the auditory ossicles of the middle ear. The auditory ossicles include the *malleus*, or "hammer"; the *incus*, or "anvil"; and the *stapes*, or "stirrups." The second transduction that occurs takes place where the stapes meets the oval window (a membranous boundary between the air-filled middle ear and the fluid-filled internal ear). At the oval window, the mechanical movements of the auditory ossicles (which correspond to the sound waves from the external environment) are transduced back into pressure waves that move through the perilymph-filled chambers of the cochlea.

4. Contrary to popular belief, carrots do not contain vitamin A. Carrots contain carotene, an orange pigment, which is converted to retinol (vitamin A) in the body. Retinal, the pigment in the rhodopsin molecule, is synthesized from vitamin A. A deficiency of vitamin A can cause night blindness.

5. When light falls on the eye, it passes through the cornea and strikes the retina, bleaching many molecules of the pigment rhodopsin that lie within them. Retinal, a part of the rhodopsin molecule, is broken off when bleaching occurs. After an intense exposure to light, a photoreceptor cannot respond to further stimulation until its rhodopsin molecules have been regenerated by recombining with a retinal molecule.

Chapter 18:
The Endocrine System

LEVEL 1: REVIEWING FACTS AND TERMS

Multiple Choice

1. a	10. b	19. d	28. b	37. b	46. d	55. d
2. c	11. b	20. a	29. c	38. a	47. d	56. d
3. a	12. d	21. b	30. d	39. d	48. a	57. b
4. d	13. d	22. c	31. c	40. b	49. b	58. a
5. c	14. c	23. d	32. a	41. c	50. c	59. d
6. d	15. a	24. a	33. b	42. d	51. c	60. c
7. b	16. b	25. c	34. a	43. c	52. d	61. b
8. a	17. a	26. b	35. c	44. a	53. c	62. d
9. d	18. b	27. a	36. d	45. c	54. b	

Completion

Part I

1. long-term
2. paracrine
3. negative feedback
4. neurotransmitters
5. adrenal gland
6. catecholamines
7. plasma membrane
8. hypothalamus
9. sella turcica
10. infundibulum
11. ADH; oxytocin
12. tropic hormones
13. diabetes insipidus
14. gigantism
15. isthmus
16. trachea
17. calcitonin
18. Graves' disease
19. thyroid gland
20. parathyroid hormone (PTH)
21. bones
22. peritoneal cavity
23. cortex
24. aldosterone
25. Na$^+$ ions and water

Part II

26. glucocorticoids
27. Addison's disease
28. aldosteronism
29. epithalamus
30. antioxidant
31. islets of Langerhans
32. abdominopelvic cavity
33. insulin
34. glucagon
35. diabetes mellitus
36. thymosins
37. androgens
38. estrogens
39. leptin
40. synergism
41. integrative
42. GH
43. thyroid hormone
44. PTH and calcitriol
45. general adaptation syndrome
46. exhaustion
47. stress
48. precocious puberty
49. kidneys

Matching

1. H	6. L	11. G	16. X	21. V
2. M	7. C	12. I	17. N	22. S
3. E	8. D	13. J	18. W	23. P
4. K	9. F	14. U	19. Y	24. T
5. A	10. B	15. R	20. Q	25. O

Drawing/Illustration Labeling

Figure 18-1   The Endocrine System—Organs and Tissues

1. hypothalamus
2. pituitary gland
3. thyroid gland
4. thymus
5. adrenal glands
6. pineal gland
7. parathyroid glands
8. heart
9. kidney
10. adipose tissue
11. digestive tract
12. pancreatic islets
13. gonads

Figure 18-2   Classification of Hormones by Chemical Structure

1. amino acid derivatives
2. derivatives of tyrosine
3. derivatives of tryptophan
4. thyroid hormones
5. catecholamines
6. peptide hormones
7. glycoproteins
8. short polypeptides and small proteins
9. lipid derivatives
10. eicosanoids
11. steroid hormones

LEVEL 2: REVIEWING CONCEPTS

Chapter Overview

1. pineal gland
2. melatonin
3. hypothalamus
4. pituitary
5. sella turcica
6. thyroid
7. parathyroid
8. thymus
9. heart
10. pancreas
11. insulin
12. adrenal
13. kidneys
14. erythropoietin
15. red blood cells
16. ovaries
17. estrogen
18. progesterone
19. testes
20. testosterone

Concept Maps

I Endocrine Glands

1. hormones
2. epinephrine
3. peptide hormones
4. testosterone
5. pituitary
6. parathyroids
7. heart
8. male/female gonads
9. pineal
10. bloodstream

II Pituitary Gland

1. sella turcica
2. adenohypophysis
3. melanocytes
4. pars distalis
5. thyroid
6. adrenal cortex
7. gonadotropic hormones
8. mammary glands
9. posterior pituitary
10. pars nervosa
11. oxytocin
12. ADH (vasopressin)

III Endocrine System Functions

1. cellular communication
2. homeostasis
3. target cells
4. contraction
5. ion channel opening
6. hormones

Multiple Choice

1. b
2. a
3. a
4. c
5. c
6. a
7. c
8. d
9. d
10. d
11. c
12. b
13. d
14. b
15. d
16. c
17. b
18. a

Completion

1. hypothalamus
2. target cells
3. nucleus
4. PTH
5. tyrosine
6. leptin
7. cyclic-AMP
8. calmodulin
9. somatomedins
10. fenestrated
11. portal
12. aldosterone
13. reticularis
14. epinephrine
15. glucocorticoids
16. G protein
17. cytoplasm

Short Essay

1. (a) Amino acid derivatives; (b) peptide hormones; (c) lipid derivatives

2. (a) Activation of adenylate cyclase; (b) release or entry of calcium ions; (c) activation of phosphodiesterase; (b) activation of other intracellular enzymes (such as kinases)

3. The hypothalamus:(a) contains autonomic centers that exert direct neural control over the endocrine cells of the adrenal medulla; sympathetic activation causes the adrenal medulla to release hormones into the bloodstream; (b) acts as an endocrine organ to release hormones into the circulation at the posterior pituitary; and (c) secretes regulatory hormones that control activities of endocrine cells in the pituitary glands.

4. (a) Control by releasing hormones; (b) control by inhibiting hormones; (c) regulation by releasing and inhibiting hormones

5. Thyroid hormones elevate oxygen consumption and rate of energy consumption in peripheral tissues, causing an increase in the metabolic rate. As a result, more heat is generated, replacing the heat lost to the environment.

6. The secretion of melatonin by the pineal gland is lowest during daylight hours and highest in the darkness of night. The cyclic nature of this activity parallels daily changes in physiological processes that follow a regular pattern.

7. (a) The two hormones may have opposing, or *antagonistic*, effects. (b) The two hormones may have an additive, or *synergistic*, effect. (c) One can have a *permissive* effect on another. In such cases the first hormone is needed for the second to produce its effect. (d) The hormones may have *integrative* effects (i.e., the hormones may produce different but complementary results in specific tissues and organs).

LEVEL 3: CRITICAL THINKING AND CLINICAL APPLICATIONS

1. Androgens are known to produce several complications, including
   a. premature closure of epiphyseal cartilages
   b. liver dysfunctions
   c. prostate gland enlargement and urinary tract obstruction
   d. testicular atrophy and infertility

   Additional side effects in women include:
   a. hirsutism
   b. enlargement of the laryngeal cartilages

   Links to heart attacks, impaired cardiac function, and strokes have also been suggested.

2. a. *Patient A:* Decreased insulin production; diabetes mellitus
      Patient B: Underproduction of thyroxine (T_4); myxedema
      Patient C: Hypersecretion glucocorticoids; Cushing's disease

   b. *Patient A:* Pancreatic disorder
      Patient B: Thyroid disorder
      Patient C: Adrenal disorder

3. Structurally, the hypothalamus is a part of the diencephalon; however, this portion of the brain secretes ADH and oxytocin that target peripheral effectors; the hypothalamus controls secretory output of the adrenal medulla, and it releases hormones that control the pituitary gland.

4. The anti-inflammatory activities of these steroids result from their effects on white blood cells and other components of the immune system. The glycocorticoids (steroid hormones) slow the migration of phagocytic cells into an injury site, and phagocytic cells already in the area become less active. In addition, mast cells exposed to these steroids are less likely to release the chemicals that promote inflammation, thereby slowing the wound-healing process. Because the region of the open wound's defenses are weakened, the area becomes an easy target for infecting organisms.

5. Iodine must be available for thyroxine to be synthesized. Thyroxine deficiencies result in decreased rates of metabolism, decreased body temperature, poor response to physiological stress, and an increase in the size of the thyroid gland (goiter). People with goiter usually suffer from sluggishness and weight gain.

6. Anxiety, anticipation, and excitement may stimulate the sympathetic division of the ANS or release of epinephrine from the adrenal medulla. The result is increased cellular energy utilization and mobilization of energy reserves. Catecholamine secretion triggers a mobilization of glycogen reserves in skeletal muscles and accelerates the breakdown of glucose to provide ATP. This combination increases muscular power and endurance.

Chapter 19:
Blood

LEVEL 1: REVIEWING FACTS AND TERMS

Multiple Choice

1. d	8. c	15. d	22. a	29. a
2. d	9. c	16. d	23. d	30. b
3. d	10. c	17. a	24. c	31. c
4. a	11. b	18. c	25. a	32. d
5. b	12. c	19. b	26. a	33. d
6. d	13. b	20. d	27. c	34. c
7. d	14. d	21. a	28. d	35. b

Completion

1. blood
2. leukocytes
3. plasma
4. formed elements
5. venipuncture
6. arterial puncture
7. globulins
8. serum
9. rouleaux
10. hemoglobin
11. transferrin
12. vitamin B_{12}
13. erythroblasts
14. agglutinogens
15. agglutinins
16. lymphocytes
17. nonspecific
18. diapedesis
19. lymphopoiesis
20. platelets
21. megakaryocytes
22. vascular spasm
23. plasmin
24. anticoagulants

Matching

1. G	6. H	11. L	16. V	21. W
2. A	7. I	12. D	17. P	22. X
3. J	8. E	13. S	18. R	23. N
4. C	9. B	14. O	19. M	24. Q
5. K	10. F	15. T	20. U	

Drawing/Illustration Labeling

Figure 19-1   Formed Elements of the Blood

1. red blood cell (erythrocyte)
2. white blood cells (leukocytes)
3. platelets
4. neutrophil
5. eosinophil
6. basophil
7. monocyte
8. lymphocyte

LEVEL 2: REVIEWING CONCEPTS

Chapter Overview

1. myeloid stem cells
2. progenitor cells
3. blast cells
4. formed elements of blood
5. hemocytoblasts
6. proerythroblast
7. reticulocyte
8. erythrocyte
9. megakaryocyte
10. platelets
11. myelocytes
12. basophil
13. eosinophil
14. neutrophil
15. monoblast
16. monocyte
17. lymphoid stem cells
18. lymphocyte

Concept Maps

I Whole Blood

1. plasma
2. solutes
3. albumins
4. metalloproteins
5. leukocytes
6. neutrophils
7. monocytes

II Hemostasis

1. vascular spasm
2. platelet phase
3. plug
4. blood clot
5. clot retraction
6. clot dissolves
7. plasminogen
8. plasmin

III Coagulation

1. extrinsic pathway
2. factor VII and Ca^{2+}
3. intrinsic pathway
4. PF-3 (platelet factor)
5. factor X
6. thrombin
7. fibrinogen
8. fibrin

Multiple Choice

1. a	6. a	11. b	16. a	21. d
2. b	7. c	12. d	17. c	22. a
3. c	8. a	13. b	18. c	23. a
4. a	9. d	14. c	19. d	24. c
5. d	10. b	15. b	20. b	25. d

Completion

1. fractionated
2. hypovolemic
3. fibrin
4. metalloproteins
5. lipoproteins
6. hematocrit
7. hemoglobinuria
8. hemolysis
9. bilirubin
10. leukopenia
11. leukocytosis
12. thrombocytopenia
13. differential
14. hematopoiesis
15. hematocytoblasts
16. hemoglobin

Short Essay

1. a. Blood *transports* dissolved gases.
   b. Blood *distributes* nutrients.
   c. Blood *transports* metabolic wastes.
   d. Blood *delivers* enzymes and hormones.
   e. Blood *regulates* the pH and electrolyte composition of interstitial fluid.
   f. Blood *restricts* fluid losses through damaged vessels.
   g. Blood *defends* the body against toxins and pathogens.
   h. Blood helps *regulate* body temperature by absorbing and redistributing heat.

2. (a) Water; (b) electrolytes; (c) nutrients; (d) organic wastes; (e) proteins

3. (a) Red blood cells (erythrocytes); (b) white blood cells (leukocytes); (c) platelets

4. (a) Albumins; (b) globulins; (c) fibrinogen

5. Mature red blood cells do not have mitochondria. They also lack ribosomes and a nucleus.

6. Granular WBCs: neutrophils, eosinophils, basophils

   Agranular WBCs: monocytes, lymphocytes

7. a. Release of chemicals important to the clotting process
   b. Formation of a temporary patch in the walls of damaged blood vessels
   c. Reducing the size of the break in the vessel wall.

8. a. Vascular phase: spasm in damaged smooth muscle
   b. Platelet phase: platelet adhesion and platelet aggregation
   c. Coagulation phase: activation of clotting system and clot formation
   d. Clot retraction: contraction of blood clot
   e. Fibrinolysis: enzymatic destruction of clot

9. Erythropoietin: (a) stimulates increased rates of mitotic divisions in erythroblasts and in stem cells that produce erythroblasts; and (b) speeds up the maturation of RBCs by accelerating the rate of hemoglobin synthesis.

LEVEL 3: CRITICAL THINKING AND CLINICAL APPLICATIONS

1. a. 154 lb ÷ 2.2 kg/lb = 70 kg
        1 kg blood = approximately 1 liter
        therefore, 70 kg × 0.07 = 4.9 kg or 4.9 l

   b. Hematocrit: 45%
      therefore, total cell volume = 0.45 × 4.9 l = 2.2 l
   c. plasma volume = 4.9 l – 2.2 l = 2.7 l
   d. plasma volume in % = 2.7 l ÷ 4.9 l = 55%
   e. % formed elements = 2.2 l ÷ 4.9 l = 45%

2. a. increasing hematocrit
   b. increasing hematocrit
   c. decreasing hematocrit
   d. decreasing hematocrit

3. Plasma expanders provide a temporary solution to low blood volume. Those used for clinical purposes often contain large carbohydrate molecules, rather than proteins, to maintain proper osmotic concentrations. The plasma expanders are easily stored, and their sterile preparation avoids viral or bacterial contamination, which can be a problem with donated plasma.

4. The person with type B blood has surface antigen-B only, and his or her plasma contains anti-A antibodies, which will attack type A surface antigens, causing the red blood cells to clump or agglutinate, potentially blocking blood flow to various organs and tissues.

5. RhoGAM is given to prevent the mother from producing anti-Rh antibodies. The RhoGAM contains anti-Rh antibodies that remove fetal Rh-postive antigens from the mother's circulation before the mother's immune system recognizes their presence and begins to produce anti-Rh antibodies.

6. During differentiation, the RBCs of humans and other mammals lose most of their organelles, including nuclei. Because they lack nuclei and ribosomes, circulating mammalian RBCs cannot divide or synthesize structural proteins or enzymes. The RBCs cannot perform repairs, and, in the absence of mitochondria, they obtain their energy demands through the anaerobic metabolism of glucose absorbed from the surrounding plasma.

7. (a) Vitamin K deficiencies are rare because vitamin K is produced by bacterial synthesis in the digestive tract; (b) dietary sources include liver, green leafy vegetables, cabbage-type vegetables, and milk; (c) adequate amounts of vitamin K must be present for the liver to be able to synthesize four of the clotting factors, including prothrombin. A deficiency of vitamin K leads to the breakdown of the common pathway, inactivating the clotting system.

Chapter 20:
The Heart

LEVEL 1: REVIEWING FACTS AND TERMS

Multiple Choice

1. d	9. a	16. d	23. d	30. c	37. c
2. c	10. a	17. b	24. a	31. c	38. b
3. a	11. d	18. d	25. c	32. c	39. a
4. d	12. c	19. c	26. b	33. d	40. b
5. c	13. a	20. d	27. d	34. b	41. c
6. a	14. a	21. c	28. b	35. a	42. c
7. b	15. b	22. c	29. b	36. c	43. a
8. a					

Completion

1. exchange vessels
2. veins
3. right ventricle
4. atria
5. cardiac skeleton
6. intercalated discs
7. epicardium
8. pericardium
9. deoxygenated
10. left ventricle
11. pulmonary veins
12. pulmonary
13. myocardium
14. endocardium
15. coronary sinus
16. repolarization
17. nodal cells
18. automaticity
19. electrocardiogram
20. auscultation
21. atrial reflex
22. chemoreceptors
23. medulla oblongata
24. SA node
25. ESV
26. preload
27. stroke volume

Matching

1. K
2. E
3. H
4. I
5. F
6. L
7. J
8. A
9. C
10. G
11. D
12. B
13. X
14. N
15. U
16. R
17. T
18. M
19. O
20. P
21. W
22. S
23. V
24. Q

Drawing/Illustration Labeling

Figure 20-1   Anatomy of the Heart (Frontal Section through the Heart)

1. aortic arch
2. brachiocephalic trunk
3. superior vena cava
4. right pulmonary arteries
5. ascending aorta
6. fossa ovalis
7. opening of coronary sinus
8. right atrium
9. pectinate muscles
10. conus arteriosus
11. cusp of tricuspid valve
12. chordae tendineae
13. papillary muscle
14. right ventricle
15. inferior vena cava
16. left common carotid artery
17. left subclavian artery
18. ligamentum arteriosum
19. pulmonary trunk
20. pulmonary valve
21. left pulmonary arteries
22. left pulmonary veins
23. interatrial septum
24. aortic valve
25. cusp of bicuspid valve
26. left ventricle
27. interventricular septum
28. trabeculae carneae
29. moderator band
30. descending aorta

LEVEL 2: REVIEWING CONCEPTS

Chapter Overview

1. superior vena cava
2. right atrium
3. tricuspid valve
4. right ventricle
5. pulmonary semilunar valve
6. pulmonary arteries
7. pulmonary veins
8. left atrium
9. bicuspid valve
10. left ventricle
11. aortic semilunar valve
12. L. common carotid artery
13. aorta
14. systemic arteries
15. systemic veins
16. inferior vena cava

Concept Maps

I The Heart

1. two atria
2. blood from atria
3. endocardium
4. tricuspid
5. oxygenated blood
6. two semilunar
7. aortic
8. deoxygenated blood
9. epicardium
10. pacemaker cells

II Factors Affecting Cardiac Output

1. autonomic innervation
2. heart rate
3. filling time
4. hormones
5. preload
6. end-systolic volume
7. afterload
8. cardiac output

III Heart Rate Controls

1. increasing blood pressure, decreasing blood pressure
2. chemoreceptors
3. decreasing CO_2
4. increasing CO_2
5. increasing acetylcholine
6. increasing epinephrine, norepinephrine
7. decreasing heart rate
8. increasing heart rate

Multiple Choice

1. a	5. b	9. a	13. d	17. d
2. b	6. c	10. d	14. c	18. a
3. b	7. d	11. d	15. b	19. c
4. a	8. b	12. b	16. c	20. b

Completion

1. anastomoses
2. systemic
3. pectinate muscles
4. trabeculae carneae
5. endocardium
6. myocardium
7. bradycardia
8. tachycardia
9. pacemaker
10. filling time
11. cardiac reserve
12. afterload
13. auricle
14. angina pectoris
15. infarct

Short Essay

1. $CO = SV \times HR$

    $CO = 75$ m$l \times 80$ beats/min $= 6000$ ml/min

    6000 ml $= 6.0$ l/min

2. $SV = EDV - ESV$

    $SV = 140$ m$l - 60$ m$l = 80$ ml

3. $CO = SV \times HR$

    Therefore, $\dfrac{CO}{HR} = \dfrac{SV \times \cancel{HR}}{\cancel{HR}}$

    Therefore, $SV = \dfrac{CO}{HR}$

    $SV = \dfrac{5l/\cancel{min}}{100 \text{ B}/\cancel{min}} = 0.05 l$/beat

4. The visceral pericardium, or epicardium, covers the outer surface of the heart. The parietal pericardium lines the inner surface of the pericardial sac that surrounds the heart.

5. The chordae tendineae and papillary muscles are located in both ventricles, the trabeculae carneae are found on the interior walls of both ventricles, and the pectinate muscles are found on the interior of both atria and a part of the right atrial wall. The chordae tendineae are tendinous fibers that brace each cusp of the AV valves and are connected to the papillary muscles.

    The trabeculae carneae contain a series of deep grooves and folds. The pectinate muscles are prominent muscular ridges that run along the surfaces of the atria and across the anterior atrial wall.

6. (a) Epicardium; (b) myocardium; (c) endocardium

7. a. Stabilizes positions of muscle fibers and valves in heart
   b. Provides support for cardiac muscle fibers and blood vessels and nerves in the myocardium
   c. Helps distribute the forces of contraction
   d. Prevents overexpansion of the heart
   e. Helps to maintain the shape of the heart
   f. Provides elasticity that helps the heart return to its original shape after each contraction
   g. Physically isolates the muscle fibers of the atria from those in the ventricles

8. Cardiac muscle fibers are connected by gap junctions at intercalated discs, which allow ions and small molecules to move from one cell to another. This creates a direct electrical connection between the two muscle fibers, and an action potential can travel across an intercalated disc, moving quickly from one cardiac muscle fiber to another. Because the cardiac muscle fibers are mechanically, chemically, and electrically connected to one another, the entire tissue resembles a single, enormous muscle fiber. For this reason cardiac muscle has been called a functional syncytium.

9. SA node → AV node → AV bundle (bundle of His) → bundle branches → Purkinje cells → contractile cells of ventricular myocardium

10. Bradycardia: heart rate slower than normal

Tachycardia: heart rate faster than normal

11. Under normal circumstances, (1) autonomic activity and (2) circulating hormones are responsible for making delicate adjustments to the heart rate as circulatory demands change. These factors act by modifying the natural rhythm of the heart.

LEVEL 3: CRITICAL THINKING AND CLINICAL APPLICATIONS

1. During tachycardia, the heart beats at an abnormally fast rate. The faster the heart beats, the less time there is in between contractions for it to fill with blood again. As a result, over a period of time the heart fills with less and less blood and thus pumps less blood out. The stroke volume and cardiac output decrease. When the cardiac output decreases to the point where not enough blood reaches the central nervous system, loss of consciousness occurs.

2. A drug that blocks the calcium channels in cardiac muscle cells will decrease the force of cardiac contraction, therefore causing a lowering of the stroke volume.

3. There are four heart sounds, designated as S_1 through S_4. The first heart sound, known as the "lubb" (S_1), lasts a little longer than the second, called "dupp" (S_2). S_1, which marks the start of ventricular contraction, is produced as the AV valves close; S_2 occurs at the beginning of ventricular filling, when the semilunar valves close. The third and fourth heart sounds are usually very faint and seldom audible in healthy adults. These sounds are associated with blood flowing into the ventricles (S_3) and atrial contraction (S_4) rather than valve action. If the valves are malformed or there are problems with the papillary muscles or chordae tendineae, the heart valves may not close properly and regurgitation occurs during ventricular systole. The surges, swirls, and eddies that accompany regurgitation create a rushing, gurgling sound known as a heart murmur.

4. One of the first symptoms of coronary artery disease (CAD) is commonly angina pectoris. In its most common form, a temporary ischemia develops when the workload of the heart increases. The symptoms that Joe experienced after exercising are typical for this condition. Angina can be controlled by a combination of drug treatment and changes in lifestyle, including (1) limiting activities such as strenuous exercise, and avoiding stressful situations; (2) stopping smoking; and (3) lowering fat consumption. Medications include drugs that block sympathetic stimulation (propanolol), vasodilators such as nitroglycerin, and drugs that block calcium movement into the cardiac and vascular smooth muscle cells (calcium channel blockers). Angina may also be treated surgically.

5. The small P wave accompanies the depolarization of the atria. The QRS complex appears as the ventricles depolarize. The small T wave indicates ventricular depolarization. Of particular diagnostic importance is the amount of depolarization occurring during the P wave and the QRS complex. An excessively large QRS complex often indicates that the heart has become enlarged. A smaller-than-normal electrical signal may mean that the mass of the heart muscle has decreased. The size and shape of the T wave may be affected by a condition that slows ventricular depolarization, starvation, low cardiac reserves, coronary ischemia, or abnormal ion concentrations, which will reduce the size of the T wave.

6. The amount of preload, and hence the degree of myocardial stretching, varies with the demands on the heart. When standing at rest, the EDV is low. When you begin exercising, venous return increases and more blood flows into the heart. The EDV increases, and the myocardium stretches further. As the sarcomeres approach optimal lengths, the ventricular muscle cells contract more efficiently and produce more forceful contractions and shortening, forcing more blood to be pumped out of the heart.

7. Anastomoses are the interconnections between arteries in coronary circulation. Because of the interconnected arteries, the blood supply to the cardiac muscle remains relatively constant, regardless of pressure fluctuations within the left and right coronary arteries.

8. Both of the atria are responsible for pumping blood into the ventricles. The right ventricle normally does not need to push very hard to propel blood through the pulmonary circuit because of the close proximity of the lungs to the heart, and the pulmonary arteries and veins are relatively short and wide. The wall of the right ventricle is relatively thin. When it contracts it squeezes the blood against the mass of the left ventricle, moving blood efficiently with minimal effort, developing relatively low pressures. The left ventricle has an extremely thick muscular wall, generating powerful contractions and causing the ejection of blood into the ascending aorta and producing a bulge into the right ventricular cavity. The left ventricle must exert six to seven times as much force as the right ventricle to push blood around the systemic circuit.

Chapter 21:
Blood Vessels and Circulation

LEVEL 1: REVIEWING FACTS AND TERMS

Multiple Choice

1. a	8. d	15. b	22. d	29. c	35. a	41. b
2. d	9. c	16. a	23. c	30. a	36. a	42. b
3. b	10. a	17. d	24. c	31. d	37. d	43. d
4. c	11. d	18. b	25. c	32. b	38. c	44. b
5. a	12. c	19. a	26. c	33. c	39. a	45. c
6. d	13. d	20. b	27. b	34. d	40. c	46. c
7. b	14. b	21. a	28. c			

Completion

1. arterioles
2. venules
3. fenestrated
4. precapillary sphincter
5. viscosity
6. pulse pressure
7. sphygmomanometer
8. circulatory pressure
9. metarteriole
10. vasomotion
11. angiogenesis
12. turbulence
13. autoregulation
14. vasodilators
15. vasoconstriction
16. respiratory pump
17. shock
18. stroke
19. anastomoses
20. alveoli
21. hepatic portal vein
22. femoral artery
23. right atrium
24. foramen ovale
25. atrioventricular septal
26. arteriosclerosis
27. integumentary

Matching

1. I	6. J	11. A	16. X	21. Q	26. V
2. G	7. D	12. K	17. AA	22. BB	27. Z
3. E	8. F	13. C	18. N	23. U	28. S
4. B	9. H	14. CC	19. W	24. Y	29. R
5. M	10. L	15. T	20. O	25. P	

Drawing/Illustration Labeling

Figure 21-1   An Overview of the Major Systemic Arteries

1. vertebral
2. right subclavian
3. brachiocephalic trunk
4. aortic arch
5. ascending aorta
6. celiac trunk
7. brachial
8. radial
9. ulnar
10. palmar arches
11. external iliac
12. popliteal
13. posterior tibial
14. anterior tibial
15. fibular
16. plantar arch
17. right common carotid
18. left common carotid
19. left subclavian
20. axillary
21. pulmonary trunk
22. descending aorta
23. diaphragm
24. renal
25. superior mesenteric
26. gonadal
27. inferior mesenteric
28. common iliac
29. internal iliac
30. deep femoral
31. femoral
32. descending genicular
33. dorsalis pedis

Figure 21-2   An Overview of the Major Systemic Veins

1. vertebral
2. external jugular
3. subclavian
4. axillary
5. cephalic
6. brachial
7. basilic
8. hepatic veins
9. median cubital
10. radial
11. median antebrachial
12. ulnar
13. palmar venous arches
14. digital veins
15. great saphenous
16. popliteal
17. small saphenous
18. fibular
19. plantar venous arch
20. dorsal venous arch
21. internal jugular
22. brachiocephalic
23. superior vena cava
24. intercostal veins
25. inferior vena cava
26. renal
27. gonadal
28. lumbar veins
29. L. and R. common iliac
30. external iliac
31. internal iliac
32. deep femoral
33. femoral
34. posterior tibial
35. anterior tibial

Figure 21-3   Major Arteries of the Head and Neck

1. superficial temporal artery
2. cerebral arterial circle
3. posterior cerebral arteries
4. basilar artery
5. internal carotid artery
6. vertebral artery
7. thyrocervical trunk
8. subclavian artery
9. internal thoracic artery
10. anterior cerebral arteries
11. maxillary artery
12. facial artery
13. external carotid artery
14. carotid sinus
15. common carotid artery
16. brachiocephalic artery

Figure 21-4   Major Veins of the Head and Neck

1. superior sagittal sinus
2. straight sinus
3. right transverse sinus
4. vertebral vein
5. external jugular vein
6. temporal vein
7. maxillary vein
8. facial vein
9. internal jugular vein
10. right brachiocephalic vein
11. left brachiocephalic vein

LEVEL 2: REVIEWING CONCEPTS

Chapter Overview

1. aortic valve
2. aortic arch
3. descending aorta
4. brachiocephalic
5. L. common carotid
6. L. subclavian
7. thoracic aorta
8. mediastinum
9. intercostal
10. superior phrenic
11. abdominal aorta
12. inferior phrenic
13. celiac
14. adrenal
15. renal
16. superior mesenteric
17. gonadal
18. inferior mesenteric
19. lumbar
20. common iliacs

Concept Maps

I The Cardiovascular System

1. pulmonary veins
2. arteries and arterioles
3. veins and venules
4. pulmonary arteries
5. systemic circuit

II Endocrine System and Cardiovascular Regulation

1. epinephrine, norepinephrine
2. adrenal cortex
3. increasing blood pressure
4. ADH (vasopressin)
5. increasing plasma volume
6. kidneys
7. increasing fluid
8. erythropoietin
9. atrial natriuretic peptide (ANP)

III ANP Effects on Blood Volume and Blood Pressure

1. increasing H_2O loss by kidneys
2. decreasing H_2O intake
3. decreasing blood pressure
4. increasing blood flow (l/min)

IV Major Branches of the Aorta

1. ascending aorta
2. brachiocephalic artery
3. L. subclavian artery
4. thoracic artery
5. celiac trunk
6. superior mesenteric artery
7. R. gonadal artery
8. L. common iliac artery

V Major Veins Draining into the Superior and Inferior Venae Cavae

1. superior vena cava
2. azygos vein
3. L. hepatic veins
4. R. adrenal vein
5. L. renal vein
6. L. common iliac vein

Multiple Choice

1. b
2. d
3. d
4. c
5. d
6. b
7. c
8. a
9. c
10. d
11. a
12. c
13. d
14. a
15. c
16. d
17. b
18. d
19. b
20. a
21. d
22. c
23. b
24. a
25. b

Completion

1. cerebral arterial circle
2. elastic rebound
3. recall of fluids
4. edema
5. precapillary sphincters
6. thrombus
7. endothelium
8. embolus
9. veins
10. venous return
11. aorta
12. brachial
13. radial
14. great saphenous
15. lumen

Short Essay

1. Tunica intima, tunica media, tunica externa

2. Heart → arteries → arterioles → capillaries (gas exchange area) → venules → veins → heart

3. a. Sinusoids are specialized fenestrated capillaries.
   b. They are found in the liver, spleen bone marrow, and the adrenal glands.
   c. They form flattened, irregular passageways, so blood flows through the tissues slowly, maximizing time for absorption and secretion and molecular exchange.

4. In the pulmonary circuit, oxygen stores are replenished, carbon dioxide is excreted, and the "reoxygenated" blood is returned to the heart for distribution in the systemic circuit. The systemic circuit supplies the capillary beds in all parts of the body with oxygenated blood, and returns deoxygenated blood to the heart of the pulmonary circuit for removal of carbon dioxide and replenishment of oxygen.

5. a. Vascular resistance, viscosity, turbulence
   b. Only vascular resistance can be adjusted by the nervous and endocrine systems.

6. $F \propto \dfrac{BP}{PR}$

   Flow is directly proportional to the blood pressure and inversely proportional to peripheral resistance (i.e., increasing pressure, increasing flow; decreasing pressure, decreasing flow; increasing PR, decreasing flow; decreasing PR, increasing flow).

7. $\dfrac{120 \text{ mm Hg}}{80 \text{ mm Hg}}$ is a "normal" blood pressure reading.

   The top number, 120 mm Hg, is the *systolic* pressure (i.e., the peak blood pressure measured during ventricular systole).

   The bottom number, 80 mm Hg, is the *diastolic* (i.e., the minimum blood pressure at the end of ventricular diastole).

8. $\text{MAP} = \dfrac{\text{pulse pressure}}{3} + \text{diastolic pressure}$

Therefore, pulse pressure = 110 mm Hg – 80 mm Hg = 30 mm Hg

$\text{MAP} = \dfrac{30}{3} + 80 = 10 + 80 = 90$ mm Hg

MAP = 90 mm Hg

9. a. Distributes nutrients, hormones, and dissolved gases throughout tissues
   b. Transports insoluble lipids and tissue proteins that cannot enter circulation by crossing capillary linings
   c. Speeds removal of hormones and carries bacterial toxins and other chemical stimuli to cells of the immune system
   d. Ensures that plasma and interstitial fluid are in constant communication

10. Cardiac output, blood volume, peripheral resistance

11. Aortic baroreceptors, carotid sinus baroreceptors, atrial baroreceptors

12. Epinephrine and norepinephrine, ADH, angiotensin II, erythropoietin, and atrial and brain natriuretic peptides

13. Decreasing hematocrit, thrombus formation, pulmonary embolism, and pooling of blood in the veins

14. Arteries lose their elasticity, increasing the risk of aneurism; calcium deposits increase, causing risk of stroke and myocardial infarction.

LEVEL 3: CRITICAL THINKING AND CLINICAL APPLICATIONS

1. L. ventricle → aortic arch → L. subclavian artery → L. axillary artery → L. brachial artery → L. radial and ulnar arteries → L. palm and wrist arterial anastomoses → L. digital arteries → L. palm and wrist venous anastomoses → L. cephalic and basilic veins → L. radius and ulnar veins → L. brachial vein → L. axillary vein → L. subclavian vein → L. brachiocephalic vein → superior vena cava → R. atrium.

2. When a person rises rapidly from a lying position, a drop in blood pressure in the neck and thoracic regions occurs because of the pull of gravity on the blood. Owing to the sudden decrease in blood pressure, the blood flow to the brain is reduced enough to cause dizziness or loss of consciousness.

3. By applying pressure on the carotid artery at frequent intervals during exercise, the pressure to the region of the carotid sinus may be sufficient to stimulate the baroreceptors. The increased action potentials from the baroreceptors initiate reflexes in parasympathetic impulses to the heart, causing a decrease in the heart rate.

4. During exercise the following changes occur:
   a. increased blood flow to tissues, supplying O_2 and nutrients, removing wastes
   b. increased blood flow to skin—thermoregulatory—gets rid of excess heat in the body
   c. increased venous return due to skeletal muscle movement and the respiratory pump
   d. decreased oxygen tension resulting from increased muscular activity
   e. increased blood pressure
   f. increased skeletal blood vessel dilation
   g. increased sympathetic stimulation to the heart—increased cardiac output
   h. increased sympathetic activation—this causes vasoconstriction in blood vessels of skin and viscera, shunting blood to skeletal muscles
   i. increased vasodilation of capillaries because of presence of CO_2, K^+, and lactic acid

Chapter 22:
The Lymphatic System and Immunity

LEVEL 1: REVIEWING FACTS AND TERMS

Multiple Choice

1. d	7. a	13. d	19. c	25. b	31. a	37. d
2. c	8. a	14. c	20. d	26. b	32. c	38. d
3. d	9. c	15. b	21. d	27. c	33. b	39. d
4. c	10. c	16. a	22. a	28. a	34. d	40. b
5. d	11. b	17. c	23. b	29. c	35. b	41. c
6. b	12. d	18. a	24. a	30. c	36. d	

Completion

1. immunity
2. lymphocytes
3. lacteals
4. cytotoxic T
5. antibodies
6. lymphatic capillaries
7. phagocytes
8. diapedesis
9. interferons
10. passive
11. cell-mediated
12. innate
13. active
14. suppressor T
15. helper T
16. costimulation
17. memory T cells
18. sensitization
19. memory B cells
20. plasma cells
21. neutralization
22. precipitation
23. haptens
24. immunological competence
25. IgG
26. immunodeficiency disease
27. autoimmune disorder
28. inflammation
29. lymphokines
30. monokines
31. antigens
32. thymic hormones
33. endocrine

Matching

1. G	6. F	11. J	16. BB	21. X	26. M
2. I	7. D	12. B	17. W	22. N	27. P
3. A	8. C	13. Q	18. CC	23. O	28. Z
4. H	9. E	14. DD	19. T	24. AA	29. V
5. K	10. L	15. Y	20. R	25. U	30. S

Drawing/Illustration Labeling

Figure 22-1   The Lymphoid Organs and Lymphatic Vessels

1. tonsil
2. cervical lymph nodes
3. R. lymphatic duct
4. thymus
5. thoracic duct
6. cisterna chyli
7. lumbar lymph nodes
8. appendix
9. lymphatics of lower limb
10. lymphatics of upper limb
11. axillary lymph nodes
12. thoracic duct
13. lymphatics of mammary gland
14. spleen
15. MALT in intestinal tract
16. pelvic lymph nodes
17. inguinal lymph nodes

Figure 22-2   Innate (Nonspecific) Defenses

1. physical barriers
2. phagocytes
3. immunological surveillance
4. interferons
5. complement system
6. inflammatory response
7. fever

LEVEL 2: REVIEWING CONCEPTS

Chapter Overview

1. viruses
2. macrophages
3. natural killer cells
4. helper T cells
5. B cells
6. antibodies
7. killer T cells (cytotoxic T cells)
8. suppressor T cells
9. memory T and B cells

Concept Maps

I Immune System

1. innate immunity
2. phagocytic cells
3. inflammation
4. adaptive immunity
5. active
6. active immunization
7. transfer of antibodies via placenta
8. passive immunization

II Inflammation Response

1. tissue damage
2. increasing vascular permeability
3. phagocytosis of bacteria
4. tissue repaired
5. bacteria not destroyed

Multiple Choice

1. b
2. d
3. c
4. b
5. a
6. c
7. a
8. d
9. d
10. d
11. c
12. b
13. a
14. c
15. d
16. c
17. b
18. c
19. b
20. d

Completion

1. T cells
2. microglia
3. Kupffer cells
4. Langerhans cells
5. costimulation
6. cytokines
7. properdin
8. mast
9. pyrogens
10. opsonization
11. NK cells
12. interferon
13. IgG
14. IgM
15. helper T

Short Essay

1. a. Lymphatic vessels
   b. Lymph
   c. Lymphoid organs

2. a. Production, maintenance, and distribution of lymphocytes
   b. Maintenance of normal blood volume
   c. Elimination of local variations in the composition of the interstitial fluid

3. a. T cells (thymus)
   b. B cells (bone marrow)
   c. NK cells—natural killer (bone marrow)

4. a. Cytotoxic T cells—cell-mediated immunity
   b. Helper T cells—release lymphokines that coordinate specific and nonspecific defenses
   c. Suppressor T cells—depress responses of other T cells and B cells

5. Activated B cells differentiate into plasma cells that are responsible for the production and secretion of antibodies. B cells are said to be responsible for humoral immunity.

6. a. Lymph nodes
   b. Thymus
   c. Spleen

7. a. Physical barriers
   b. Phagocytes
   c. Immunological surveillance
   d. Complement system
   e. Inflammatory response
   f. Fever

8. NK (natural killer) cells are sensitive to the presence of abnormal cell membranes and respond immediately. When the NK cell makes contact with an abnormal cell, it releases secretory vesicles that contain proteins called perforins. The perforins create a network of pores in the target cell membrane, releasing free passage of intracellular materials necessary for homeostasis, thus causing the cell to disintegrate.

   T cells and B cells provide defenses against specific threats, and their activation requires a relatively complex and time-consuming sequence of events required to build a response to a specific antigen.

9. a. Destruction of target cell membranes
   b. Stimulation of inflammation
   c. Attraction of phagocytes
   d. Enhancement of phagocytosis

10. a. Specificity
    b. Versatility
    c. Memory
    d. Tolerance

11. Active immunity appears following exposure to an antigen, as a consequence of the immune response. Passive immunity is produced by transfer of antibodies from another individual.

12. a. Direct attack by activated T cells (cellular immunity)
    b. Attack by circulating antibodies released by plasma cells derived from activated B cells (humoral immunity)

13. a. Neutralization
    b. Agglutination and precipitation
    c. Activation of complement
    d. Attraction of phagocytes
    e. Opsonization
    f. Stimulation of inflammation
    g. Prevention of bacterial and viral adhesion

14. a. IgG
    b. IgE
    c. IgD
    d. IgM
    e. IgA

15. a. Interleukins
    b. Interferons
    c. Tumor necrosis factor
    d. Phagocyte-activating chemicals
    e. Colony-stimulating factors

16. Autoimmune disorders develop when the immune response mistakenly targets normal body cells and tissues. When the immune recognition system malfunctions, activated B cells begin to manufacture antibodies against other cells and tissues.

LEVEL 3: CRITICAL THINKING AND CLINICAL APPLICATIONS

1. Tissue transplants normally contain protein molecules called human lymphocyte antigen (HLA) genes that are foreign to the recipient. These antigens trigger the recipient's immune responses, activating the cellular and humoral mediated responses that may act to destroy the donated tissue.

2.

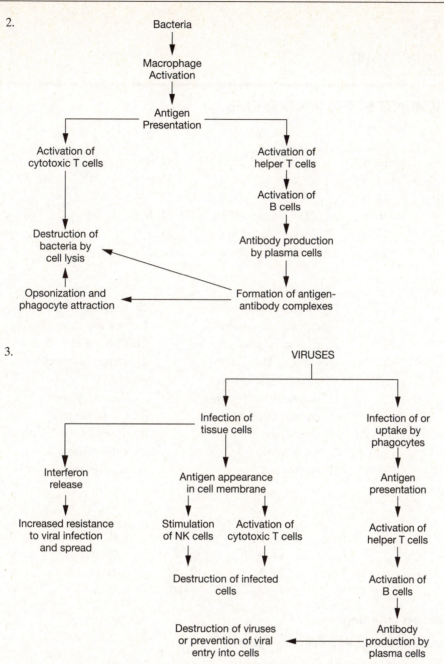

Bacteria
↓
Macrophage Activation
↓
Antigen Presentation

Activation of cytotoxic T cells

Activation of helper T cells
↓
Activation of B cells
↓
Antibody production by plasma cells
↓
Formation of antigen-antibody complexes

Destruction of bacteria by cell lysis

Opsonization and phagocyte attraction

3.

VIRUSES

Infection of tissue cells

Infection of or uptake by phagocytes

Interferon release
↓
Increased resistance to viral infection and spread

Antigen appearance in cell membrane

Stimulation of NK cells

Activation of cytotoxic T cells
↓
Destruction of infected cells

Antigen presentation
↓
Activation of helper T cells
↓
Activation of B cells
↓
Antibody production by plasma cells

Destruction of viruses or prevention of viral entry into cells

4. Viruses reproduce inside living cells, beyond the reach of antibodies. Infected cells incorporate viral antigens into their cell membranes. NK cells recognize these infected cells as abnormal. By destroying virally infected cells, NK cells slow or prevent the spread of viral infection.

5. High body temperatures may inhibit some viruses and bacteria, or may speed up their reproductive rates so that the disease runs its course more rapidly. The body's metabolic processes are accelerated, which may help mobilize tissue defenses and speed the repair process.

6. During times of stress the body responds by increasing the activity of the adrenal glands, which may cause a decrease in the inflammatory response, reduce the activities and numbers of phagocytes in peripheral tissues, and inhibit interleukin secretion. "Stress reduction" may cause a decrease in adrenal gland secretions and maintain homeostatic control of the immune response.

Chapter 23:
The Respiratory System

LEVEL 1: REVIEWING FACTS AND TERMS

Multiple Choice

1. d	7. b	13. c	19. b	25. b	31. d	36. c	41. b
2. b	8. c	14. a	20. d	26. c	32. c	37. d	42. a
3. c	9. b	15. a	21. a	27. d	33. c	38. d	43. b
4. c	10. b	16. b	22. c	28. b	34. b	39. c	44. c
5. d	11. a	17. d	23. c	29. c	35. a	40. c	45. b
6. c	12. d	18. a	24. b	30. a			

Completion

1. expiration
2. inspiration
3. surfactant
4. mucus escalator
5. external nares
6. glottis
7. larynx
8. sound waves
9. bronchioles
10. sympathetic
11. respiratory bronchioles
12. simple squamous cells
13. external respiration
14. internal respiration
15. gas diffusion
16. alveolar ventilation
17. intrapulmonary pressure
18. skeletal muscles
19. diaphragm
20. temperature
21. equilibrium
22. residual volume
23. iron ion
24. Bohr effect
25. carbaminohemoglobin
26. apnea
27. baroreceptors
28. chemoreceptor
29. apneustic
30. emphysema
31. cardiovascular

Matching

1. E	5. M	9. F	13. J	17. P	21. N	25. U
2. B	6. H	10. L	14. X	18. V	22. AA	26. S
3. A	7. D	11. I	15. T	19. O	23. Q	27. R
4. C	8. G	12. K	16. Z	20. BB	24. Y	28. W

Drawing/Illustration Labeling

Figure 23-1   Structures of the Upper Respiratory System

1. nasal cavity
2. internal nares
3. nasopharynx
4. pharyngeal tonsil
5. entrance to auditory tube
6. soft palate
7. palatine tonsil
8. oropharynx
9. epiglottis
10. laryngopharynx
11. glottis
12. vocal fold
13. esophagus
14. frontal sinus
15. nasal conchae
16. nasal vestibule
17. external nares
18. hard palate
19. oral cavity
20. tongue
21. mandible
22. lingual tonsil
23. hyoid bone
24. thyroid cartilage
25. cricoid cartilage
26. trachea
27. thyroid gland

Figure 23-2   Bronchi and Lobules of the Lungs

1. left primary bronchus
2. visceral pleura
3. cartilage plate
4. tertiary bronchi
5. smaller bronchi
6. alveoli in pulmonary lobule
7. trachea
8. bronchioles
9. terminal bronchiole
10. respiratory bronchiole

LEVEL 2: REVIEWING CONCEPTS

Chapter Overview

1. nose
2. external nares
3. nasal cavity
4. nasal vestibule
5. nasal conchae
6. hard palate
7. soft palate
8. nasopharynx
9. pharynx
10. larynx
11. glottis
12. epiglottis
13. vocal folds
14. vocal cords
15. trachea
16. cilia
17. pharynx
18. primary bronchi
19. smooth muscle
20. bronchioles
21. alveoli
22. oxygen
23. carbon dioxide
24. pulmonary capillaries

Concept Maps

I External–Internal Respiration

1. pharynx
2. trachea
3. secondary bronchi
4. respiratory bronchioles
5. alveoli
6. pulmonary veins
7. left ventricle
8. systemic capillaries
9. right atrium
10. pulmonary arteries

II Respiratory System

1. nose
2. nasal sinuses
3. laryngopharynx
4. larynx
5. cricoid
6. vocal folds
7. trachea
8. left primary bronchus
9. tertiary bronchi
10. respiratory bronchioles
11. alveoli

Multiple Choice

1. c
2. b
3. d
4. b
5. b
6. a
7. d
8. a
9. b
10. c
11. d
12. c
13. b
14. b
15. d
16. c
17. c
18. d
19. b
20. c
21. b
22. c
23. a
24. c
25. a

Completion

1. lamina propria
2. vestibule
3. oropharynx
4. laryngopharynx
5. epiglottis
6. thyroid cartilage
7. phonation
8. hilum
9. eupnea
10. hypoxia
11. anoxia
12. atelectasis
13. pneumothorax
14. intrapleural
15. hypercapnia
16. carina
17. parenchyma
18. anatomic dead space
19. alveolar ventilation
20. Henry's law
21. Dalton's law

Short Essay

1. a. Provides an area for gas exchange between air and blood
   b. To move air to and from exchange surfaces of the lungs along the respiratory passageways
   c. Protects respiratory surfaces from pathogens, dehydration, temperature changes, or other environmental variations
   d. Permits communication via production of sound
   e. Facilitating detection of olfactory stimuli by olfactory receptors

2. The air is warmed to within 1 degree of body temperature before it enters the pharynx. This results from the warmth of rapidly flowing blood in an extensive vasculature of the nasal mucosa. For humidifying the air, the nasal mucosa is supplied with small mucus glands that secrete a mucoid fluid inside the nose. The warm and humid air prevents drying of the pharynx, trachea, and lungs; facilitates

the flow of air through the respiratory passageways; and helps to minimize the impact of external environmental conditions.

3. Surfactant cells are scattered among the simple squamous epithelial cells of the respiratory membranes. They produce a secretion containing a mixture of phospholipids that coats the alveolar epithelium, lowers surface tension, and keeps the alveoli from collapsing.

4. Hairs, cilia, and mucus traps in the nasal cavity; mucus escalator in the larynx, trachea, and bronchi—these areas have a mucus lining and contain cilia that beat toward the pharynx; alveolar macrophages.

5. External respiration includes the diffusion of gases between the alveoli and the circulating blood. Internal respiration is the exchange of dissolved gases between the blood and the interstitial fluids in peripheral tissues.

   Cellular respiration is the absorption and utilization of oxygen by living cells via biochemical pathways that generate carbon dioxide.

6. Increasing volume; decreasing pressure and decreasing volume; increasing pressure

7. A *pneumothorax* is an injury to the chest wall that penetrates the parietal pleura or damages the alveoli and the visceral pleura, allowing air into the pleural cavity.

   An *atelectasis* is a collapsed lung.

   A pneumothorax breaks the fluid bond between the pleurae and allows the elastic fibers to contract; the result is a collapsed lung.

8. The vital capacity is the maximum amount of air that can be moved into and out of the respiratory system in a single respiratory cycle.

   Vital capacity = Inspiratory reserve + Expiratory reserve + Tidal volume.

9. Alveolar ventilation, or $\dot{V}_a$ is the amount of air reaching the alveoli each minute. It is calculated by subtracting the anatomic dead space, V_d, from the tidal volume (V_t) using the formula:

$$\dot{V}_a = f_x (V_t - V_d)$$

10. a. CO_2 may be dissolved in the plasma (7 percent).
    b. CO_2 may be bound to the hemoglobin of RBCs (23 percent).
    c. CO_2 may be converted to a molecule of carbonic acid (70 percent).

11. a. chemoreceptors sensitive to the P_{CO_2}, pH, or P_{O_2} of the blood or cerebrospinal fluid
    b. baroreceptors in the aortic carotid sinuses sensitive to changes in blood pressure
    c. stretch receptors that respond to changes in the volume of the lungs
    d. protective reflexes

12. A rise in arterial P_{CO_2} stimulates peripheral and central chemoreceptors. These receptors stimulate the respiratory center, causing an increase in the rate and depth of respiration, or hyperventilation.

LEVEL 3: CRITICAL THINKING AND CLINICAL APPLICATIONS

1. Running usually requires breathing rapidly through the mouth, which eliminates much of the preliminary filtration, heating, and humidifying of the inspired air. When these conditions are eliminated by breathing through the mouth, the delicate respiratory surfaces are subject to chilling, drying out, and possible damage.

2. If a pulmonary embolus is in place for several hours the alveoli will permanently collapse. If the blockage occurs in a major pulmonary vessel rather than a minor tributary, pulmonary resistance increases. The resistance places extra strain on the right ventricle, which may be unable to maintain cardiac output, and congestive heart failure can result.

3. The esophagus lies immediately posterior to the cartilage-free posterior wall of the trachea. Distention of the esophagus due to the passage of a bolus of food exerts pressure on surrounding structures. If the tracheal cartilages are complete rings, the tracheal region of the neck and upper thorax cannot accommodate the necessary compression to allow the food bolus to pass freely through the esophagus.

4. In emphysema, respiratory bronchioles and alveoli are functionally eliminated. The alveoli expand and their walls become infiltrated with fibrous tissue without alveolar capillary networks. As connective tissues are eliminated, compliance increases; air moves into and out of the lungs more easily than

before. However, the loss of respiratory surface area restricts oxygen absorption, so the individual becomes short of breath.

5. The passengers experienced decompression sickness, a painful condition that develops when a person is exposed to a sudden drop in atmospheric pressure. Nitrogen is responsible for the problems experienced, owing to its high partial pressure in air. When the pressure drops, nitrogen comes out of solution, forming bubbles. The bubbles may form in joint cavities, in the bloodstream, and in the cerebrospinal fluid. It is typical to curl up from the pain in the affected joints, which accounts for the condition's common name: the bends.

Chapter 24:
The Digestive System

LEVEL 1: REVIEWING FACTS AND TERMS

Multiple Choice

1. c	8. c	15. c	22. a	29. c	36. b	42. a
2. c	9. b	16. b	23. d	30. b	37. a	43. b
3. b	10. d	17. a	24. b	31. d	38. d	44. c
4. d	11. b	18. d	25. c	32. d	39. b	45. c
5. a	12. c	19. b	26. d	33. a	40. c	46. c
6. d	13. d	20. c	27. d	34. b	41. c	47. b
7. b	14. b	21. c	28. d	35. c		

Completion

1. esophagus
2. digestion
3. enteroendocrine
4. lamina propria
5. plicae
6. pacesetter
7. peristalsis
8. mucins
9. cuspids
10. carbohydrates
11. uvula
12. tongue
13. nasopharynx
14. cardiac sphincter
15. gastrin
16. chyme
17. duodenum
18. cholecystokinin
19. Kupffer cells
20. acini
21. haustra
22. colon
23. kidneys
24. vitamin K
25. lipase
26. pepsin
27. vitamin B_{12}
28. exocytosis
29. heartburn
30. olfactory
31. integumentary

Matching

1. D	7. J	12. F	17. Y	22. T
2. H	8. A	13. S	18. M	23. Q
3. K	9. L	14. N	19. X	24. R
4. B	10. I	15. W	20. V	25. U
5. G	11. C	16. P	21. Z	26. O
6. E				

Drawing/Illustration Labeling

Figure 24-1   The Components of the Digestive System

1. oral cavity, teeth, tongue
2. liver
3. gallbladder
4. pancreas
5. large intestine
6. salivary glands
7. pharynx
8. esophagus
9. stomach
10. small intestine

Figure 24-2   The Structure of the Digestive Tract

1. mesenteric artery and vein
2. mesentery
3. plica circularis
4. mucosa
5. submucosa
6. lumen
7. muscularis externa
8. submucosal gland
9. visceral peritoneum (serosa)

Figure 24-3   Structure of a Typical Tooth—Sectional View

1. crown
2. neck
3. root
4. enamel
5. dentin
6. pulp cavity
7. gingiva
8. cementum
9. root canal
10. bone of alveolus

LEVEL 2: REVIEWING CONCEPTS

Chapter Overview

1. oral cavity
2. lubrication
3. mechanical processing
4. gingivae (gums)
5. teeth
6. mastication
7. palates
8. tongue
9. salivary glands
10. salivary amylase
11. complex carbohydrates
12. pharynx
13. uvula
14. swallowing
15. esophagus
16. peristalsis
17. lower esophageal sphincter
18. stomach
19. gastric juices
20. chyme
21. rugae
22. mucus
23. hydrochloric
24. pepsins
25. proteins
26. pyloric canal
27. duodenum
28. pancreas
29. absorption
30. feces
31. defecation
32. rectum
33. anus

Concept Maps

I Digestive System

1. saliva
2. pancreas
3. bile
4. hormones
5. digestive tract movements
6. stomach
7. large intestine
8. hydrochloric acid
9. intestinal mucosa

II Chemical Events in Digestion

1. esophagus
2. small intestine
3. polypeptides
4. amino acids
5. complex sugars and starches
6. disaccharides, trisaccharides
7. simple sugars
8. monoglycerides, fatty acids in micelles
9. triglycerides
10. lacteal

III Action of Major Digestive Tract Hormones

1. material in jejunum
2. GIP
3. VIP
4. acid production
5. insulin
6. bile in gall bladder
7. nutrient utilization by tissues

Multiple Choice

1. a	5. c	9. d	13. b	17. a
2. d	6. a	10. b	14. c	18. c
3. b	7. d	11. c	15. b	19. b
4. c	8. d	12. a	16. d	20. d

Completion

1. submucosal plexus
2. parietal peritoneum
3. mesenteries
4. adventitia
5. segmentation
6. ankyloglossia
7. incisors
8. bicuspids
9. alkaline tide
10. duodenum
11. ileum
12. lacteals
13. Brunner's glands
14. Peyer's patches
15. enterocrinin

Short Essay

1. a. Ingestion
   b. Mechanical processing
   c. Digestion
   d. Secretion
   e. Absorption
   f. Excretion (defecation)

2. a. Epithelium
   b. Lamina propria
   c. Muscularis mucosa
   d. Submucosa
   e. Muscularis externa
   f. Serosa (in most portions of the digestive tract inside the peritoneal cavity) and adventitia (in the oral cavity, pharynx, esophagus, and rectum)

3. During a peristaltic movement, the circular muscles first contract behind the digestive contents. Longitudinal muscles contract next, shortening adjacent segments. A wave of contraction in the circular muscles then forces the materials in the desired direction. During segmentation, a section of intestine resembles a chain of sausages owing to the contractions of circular muscles in the muscularis externa. A given pattern exists for a brief moment before those circular muscles relax and others contact, subdividing the "sausages."

4. a. Parotid glands
   b. Sublingual glands
   c. Submandibular glands

5. a. Incisors—clipping or cutting
   b. Cuspids (canines)—tearing or slashing
   c. Bicuspids (premolars)—crushing, mashing, and grinding
   d. Molars—crushing and grinding

6. a. Upper esophageal sphincter
   b. Lower esophageal sphincter
   c. Pyloric sphincter
   d. Ileocecal valve
   e. Internal anal sphincter
   f. External anal sphincter

7. Parietal cells and chief cells are types of secretory cells found in the wall of the stomach. Parietal cells secrete intrinsic factors and hydrochloric acid. Chief cells secrete an inactive proenzyme, pepsinogen.

8. a. Cephalic, gastric, intestinal
   b. CNS regulation; short neural reflexes coordinated by the submucosal and myenteric plexuses, and release of gastrin into circulation; enterogastric reflexes, secretion of cholecystokinin (CCK) and secretin

9. Secretin, cholecystokinin (CCK), gastric inhibitory peptide (GIP), vasoactive intestinal peptide (VIP), gastrin, enterocrinin

10. a. Resorption of water and compaction of feces
    b. The absorption of important vitamins liberated by bacterial action
    c. The storing of fecal material prior to defecation

11. a. Metabolic regulation
    b. Hematological regulation
    c. Bile production

12. a. Endocrine function—pancreatic islets secrete insulin and glucagon into the bloodstream.
    b. Exocrine function—secrete a mixture of water, ions, and digestive enzymes into the small intestine.

LEVEL 3: CRITICAL THINKING AND CLINICAL APPLICATIONS

1. The largest category of digestive disorders includes those resulting from inflammation or infection of the digestive tract. The relative delicacy of the epithelial lining of the digestive tract makes it susceptible to damage from chemical attack and abrasion. Peptic ulcers develop if acids and enzymes contact and erode the gastric mucosa, causing inflammation that results in pain in the affected areas.

2. General motility decreases, and peristaltic contractions are weaker as a result of a decrease in smooth muscle tone. The changes slow the rate of fecal movement and promote constipation. Sagging and inflammation of the haustra in the colon can occur. Straining to eliminate compacted feces can stress the less resilient wall of blood vessels, producing hemorrhoids.

3. Heartburn, or acid indigestion, is not related to the heart, but is caused by a backflow, or reflux, of stomach acids into the esophagus, a condition called gastroesophageal reflux. Caffeine and alcohol stimulate the secretion of HCl in the stomach. When the HCl is refluxed into the esophagus, it causes the burning sensation.

4. Any condition that severely damages the liver represents a serious threat to life. It is one of the most versatile organs in the body, performing many metabolic and synthetic functions. The liver has a limited ability to regenerate itself after tissue damage, and will not fully recover unless the normal vascular pattern is restored. Cirrhosis is characterized by the replacement of lobules by fibrous tissue, affecting liver function and normal vascular patterns. It is the largest blood reservoir in the body, receiving approximately 25 percent of the cardiac output.

5. This young adolescent is lactose intolerant. This is a common disorder caused by a lack of lactase, and enzyme secreted in the walls of the small intestine that is needed to break down lactose, the sugar in cow's milk. The intestinal mucosa often stops producing lactase by adolescence, causing lactose intolerance. The condition is treated by avoiding cow's milk and other products that contain lactose. Babies may be given formula based on soy or another milk substitute. There are also milk products in which the lactose is predigested. Yogurt and certain cheeses usually can be tolerated because the lactose already has been broken down.

Chapter 25:
Metabolism and Energetics

LEVEL 1: REVIEWING FACTS AND TERMS

Multiple Choice

1. c	7. b	13. b	19. d	25. d
2. a	8. a	14. d	20. c	26. a
3. b	9. c	15. b	21. a	27. a
4. c	10. d	16. b	22. b	28. b
5. d	11. b	17. a	23. c	29. d
6. a	12. b	18. b	24. b	

Completion

1. anabolism	9. transamination	16. malnutrition
2. glycolysis	10. glycogen	17. minerals
3. oxidative phosphorylation	11. liver	18. avitaminosis
4. triglycerides	12. adipose	19. hypervitaminosis
5. beta oxidation	13. insulin	20. calorie
6. free fatty acids	14. glucagon	21. thermoregulation
7. deamination	15. nutrition	22. pyrexia
8. urea		

Matching

1. F	5. H	9. A	13. Q	17. M
2. E	6. G	10. C	14. S	18. T
3. I	7. D	11. J	15. L	19. P
4. B	8. K	12. O	16. N	20. R

LEVEL 2: REVIEWING CONCEPTS

Chapter Overview

1. metabolism
2. cellular metabolism
3. homeostasis
4. catabolism
5. anabolism
6. nutrient pool
7. carbohydrates
8. cytosol
9. anaerobic
10. aerobic
11. glycolysis
12. pyruvic acid
13. two
14. NAD
15. eight
16. three
17. twenty-four
18. $FADH_2$
19. four
20. GTP
21. ETS (electron transport system)
22. thirty-six

Concept Maps

I Food Intake

1. vegetables
2. meat
3. carbohydrates
4. 9 cal/gram
5. proteins
6. tissue growth and repair
7. vitamins
8. metabolic regulators

II Anabolism-Catabolism of Carbohydrates, Lipids, and Proteins

1. lipolysis
2. lipogenesis
3. glycolysis
4. beta oxidation
5. gluconeogenesis
6. amino acids

III Fate of the Lipoproteins

1. chylomicrons
2. skeletal muscle
3. cardiac muscle
4. adipose cells
5. liver cells
6. VLDL
7. IDL
8. LDL
9. LDL
10. HDL
11. HDL

IV Control of Blood Glucose

1. pancreas
2. liver
3. fat cells
4. glucagon
5. epinephrine
6. glycogen → glucose
7. protein

Multiple Choice

1. c	5. c	9. c	13. d	17. d
2. b	6. c	10. a	14. c	18. b
3. a	7. b	11. d	15. b	19. c
4. d	8. c	12. b	16. d	20. d

Completion

1. nutrient pool
2. lipoproteins
3. chylomicrons
4. LDLs
5. HDLs
6. transamination
7. deamination
8. evaporation
9. calorie
10. metabolic rate
11. BMR
12. convection
13. insensible water loss
14. sensible perspiration
15. acclimatization

Short Essay

1. Linoleic acid, linolenic acid

2. a. Chylomicrons—95 percent triglycerides
   b. Very-low-density lipoproteins (VDLs)—triglycerides + small amounts of phospholipids and cholesterol
   c. Intermediate-density lipoproteins (IDLs)—small amounts of triglycerides, more phospholipids and cholesterol.
   d. Low-density lipoproteins (LDLs)—cholesterol, lesser amounts of phospholipids, very few triglycerides
   e. High-density lipoproteins (HDLs)—equal amounts of lipid and protein

3. Transamination—attaches the amino group of an amino acid to keto acid.

   Deamination—the removal of an amino group in a reaction that generates an ammonia molecule.

4. Essential amino acids are necessary in the diet because they cannot be synthesized by the body. Nonessential amino acids can be synthesized by the body on demand.

5. a. Proteins are difficult to break apart.
   b. Their energy yield is less than that of lipids.
   c. A by-product, ammonium ions, is a toxin that can damage cells.
   d. Proteins form the most important structural and functional components of any cell. Extensive protein catabolism threatens homeostasis at the cellular and system levels.

6. a. Liver
   b. Adipose tissues
   c. Skeletal muscle
   d. Neural tissues
   e. Other peripheral tissues

7. During the absorptive state, the intestinal mucosa is busily absorbing the nutrients from the food you've eaten. Attention in the postabsorptive state is focused on the mobilization of energy reserves and the maintenance of normal blood glucose levels.

8. a. Milk group
   b. Meat and beans group
   c. Vegetable group
   d. Fruit group
   e. Grains group
   f. Oils

9. The amount of nitrogen absorbed from the diet balances the amount lost in the urine and feces. N-compound synthesis and breakdown are equivalent.

10. Minerals (i.e., inorganic ions) are important because:
    a. They determine the osmolarities of fluids.
    b. They play major roles in important physiological processes.
    c. They are essential co-factors in a variety of enzymatic reactions.

11. a. Fat-soluble—A, D, E, K
    b. Water-soluble—B complex and C (ascorbic acid)

12. a. Radiation
    b. Conduction
    c. Convection
    d. Evaporation

13. a. Physiological mechanisms
    b. Behavioral modifications

LEVEL 3: CRITICAL THINKING AND CLINICAL APPLICATIONS

1. A high total cholesterol value linked to a high LDL level spells trouble. In effect, an excessive amount of cholesterol is being exported to peripheral tissues. Greg's problem is a high total cholesterol and a low HDL level. In his case, excess cholesterol delivered to the tissues cannot easily be returned to the liver for excretion. The amount of cholesterol in peripheral tissues, especially arterial walls, is likely to increase, creating a valid predictor for the development of atherosclerosis, causing problems for the heart and general circulation.

2. Diets too low in calories, especially carbohydrates, will bring about physiological responses that are similar to fasting. During brief periods of fasting or low calorie intake, the increased production of

ketone bodies resulting from lipid catabolism results in ketosis, a high concentration of ketone bodies in body fluids. When keto acids dissociate in solution, they release a hydrogen ion. The appearance of ketone bodies in the circulation lowers the plasma pH. During prolonged starvation or low-calorie dieting, a dangerous drop in pH occurs. This acidification of the blood is called ketoacidosis. Charlene's symptoms represent warning signals to what could develop into more disruptive tissue activities, ultimately causing coma, cardiac arrhythmias, and death, if left unchecked.

3. Fats = 9 cal/g; 1 tbsp = 13 grams of fat
   3 tbsp × 13 grams of fat = 39 grams of fat
   39g × 9 cal/g = 351 calories

4. a. 40 grams protein: 40 g × 4 cal/g = 160 cal
      50 grams fat: 50 g × 9 cal/g = 450 cal
      60 grams carbohydrate: 69 g × 4 cal/g = <u>276 cal</u>
      Total   = 886 cal

   b. 160 cal ÷ 886 cal = 18 percent protein
      450 cal ÷ 886 cal = 50 percent fat
      276 cal ÷ 886 cal = 31 percent carbohydrate

5. Daily energy expenditures for a given individual vary widely with activity. If your daily energy intake exceeds your total energy demands, you will store the excess energy primarily as triglycerides in adipose tissue. If your daily caloric expenditure exceeds your dietary supply, the result is a net reduction in your body's energy reserves and a corresponding loss in weight. This relationship accounts for the significance of both calorie counting and exercise in a weight-control program.

Chapter 26:
The Urinary System

LEVEL 1: REVIEWING FACTS AND TERMS

Multiple Choice

1. a	9. c	17. a	24. a	31. a
2. b	10. d	18. d	25. b	32. b
3. b	11. d	19. a	26. c	33. a
4. d	12. c	20. a	27. b	34. c
5. c	13. b	21. b	28. a	35. b
6. a	14. d	22. c	29. c	36. c
7. c	15. a	23. c	30. b	37. a
8. b	16. d			

Completion

1. kidneys
2. ureters
3. urethra
4. glomerulus
5. nephrons
6. interlobar veins
7. filtrate
8. nephron loop
9. secretion
10. renal threshold
11. glomerular hydrostatic
12. glomerular filtration
13. composition
14. concentration
15. parathyroid hormone
16. countertransport
17. countercurrent multiplication
18. antidiuretic hormone
19. internal urethral sphincter
20. rugae
21. neck
22. micturition reflex
23. cerebral cortex
24. glomerular filtration rate (GFR)
25. endocrine

Matching

1. I	5. A	9. E	13. T	17. P
2. F	6. H	10. C	14. R	18. S
3. G	7. J	11. D	15. Q	19. L
4. K	8. B	12. N	16. O	

20. M
21. V
22. U

Drawing/Illustration Labeling

Figure 26-1   Anterior View of the Urinary System

1. kidney   2. ureter   3. urinary bladder   4. urethra

Figure 26-2   Sectional View of the Left Kidney

1. inner layer of fibrous capsule
2. renal sinus
3. adipose tissue in renal sinus
4. hilum
5. renal pelvis
6. renal papilla
7. ureter
8. cortex
9. medulla
10. renal pyramid
11. connection to minor calyx
12. minor calyx
13. major calyx
14. renal lobe
15. renal columns
16. outer layer of fibrous capsule

Figure 26-3   Functional Anatomy of a Representative Nephron

1. nephron
2. proximal convoluted tubule
3. distal convoluted tubule
4. renal corpuscle
5. nephron loop
6. collecting system
7. connecting collecting duct
8. papillary duct

LEVEL 2: REVIEWING CONCEPTS

Chapter Overview

1. glomerular capsule
2. glomerulus
3. filtrate
4. proximal
5. ions
6. organic
7. out
8. isotonic
9. nephron loop
10. peritubular
11. tubular
12. water
13. distal
14. aldosterone
15. ADH
16. concentrating
17. collecting
18. urine
19. ureters
20. urinary bladder
21. urethra

Concept Maps

I Urinary System

1. urinary bladder
2. nephrons
3. renal tubules
4. glomerulus
5. proximal convoluted tubule
6. papillary duct
7. renal (inner) medulla
8. renal sinus
9. major calyces
10. ureter

II Kidney Circulation

1. renal artery
2. arcuate artery
3. afferent arteriole
4. efferent arteriole
5. interlobular vein
6. interlobar vein

III Renin–Angiotensin–Aldosterone System

1. $\downarrow$ plasma volume
2. renin
3. liver
4. angiotensin I
5. adrenal cortex
6. Na^+ reabsorption
7. $\downarrow$ Na^+ excretion
8. $\downarrow$ H_2O excretion

Multiple Choice

1. d	5. c	9. c	13. b	17. c
2. c	6. d	10. b	14. d	18. d
3. b	7. b	11. d	15. c	19. a
4. d	8. d	12. a	16. a	20. c

Completion

1. retroperitoneal
2. glomerular filtration rate
3. osmotic gradient
4. transport maximum
5. macula densa
6. cortical
7. vasa recta
8. filtration
9. reabsorption
10. secretion
11. glomerular filtration
12. aldosterone
13. diabetes insipidus
14. angiotensin II
15. renin

Short Essay

1. a. Regulates plasma concentrations of ions
   b. Regulates blood volume and blood pressure
   c. Contributes to stabilization of blood pH
   d. Conserves valuable nutrients
   e. Eliminates organic wastes
   f. Assists liver in detoxification and deamination

2. Kidney → ureters → urinary bladder → urethra

3. a. Fibrous capsule
   b. Perinephritic fat capsule
   c. Renal fascia

4. Glomerulus → proximal convoluted tubule → descending limb of nephron loop → ascending limb of nephron loop → distal convoluted tubule

5. a. Reabsorption of organic substrates
   b. Reabsorption of water and ions
   c. Secreting waste products

6.
    (a)
    capillary endothelium
    dense layer
    visceral epithelium

    (b)
    fenestrated capillaries
    dense and thick
    podocytes with pedicels separated by slit pores

7. Renin and erythropoietin

8. a. Produces a powerful vasoconstriction of the afferent arteriole, thereby decreasing the GFR and slowing the production of filtrate
   b. Stimulation of renin release
   c. Alters the GFR by changing the regional pattern of blood circulation

9. Filtration, reabsorption, secretion

10. $FP = GHP - (CsHP + BCOP)$

    $$\begin{matrix} \text{filtration} \\ \text{pressure} \end{matrix} = \begin{matrix} \text{glomerular} \\ \text{hydrostatic} \\ \text{pressure} \end{matrix} - \left\{ \begin{matrix} \text{capsular} \\ \text{hydrostatic} \\ \text{pressure} \end{matrix} + \begin{matrix} \text{blood colloid} \\ \text{osmotic} \\ \text{pressure} \end{matrix} \right\}$$

11. Muscle fibers breaking down glycogen reserves release lactic acid.

    The number of circulating ketoacids increases.

12. a. Sodium and chloride are pumped out of the filtrate in the ascending limb and into the peritubular fluid.
    b. The pumping elevates the osmotic concentration in the peritubular fluid around the descending limb.
    c. The result is an osmotic flow of water out of the filtrate held in the descending limb and into the peritubular fluid.

13. a. Autoregulation
    b. Hormonal regulation
    c. Autonomic regulation

14. a. ADH—decreased urine volume
    b. Renin—causes angiotensin II production; stimulates aldosterone production
    c. Aldosterone—increased sodium ion reabsorption; decreased urine concentration and volume
    d. Atrial natriuretic peptide (ANP)—inhibits ADH production; results in increased urine volume

LEVEL 3: CRITICAL THINKING AND CLINICAL APPLICATIONS

1. The alcohol acts as a diuretic that inhibits ADH secretion from the posterior pituitary, causing the distal convoluted tubule to be relatively impermeable to water. Inhibiting the osmosis of water from the tubule with the increased fluid intake results in an increase in urine production, and increased urination becomes necessary.

2. a. If plasma proteins and numerous WBCs are appearing in the urine, there is obviously increased permeability of the filtration membrane. This condition usually results from inflammation of the filtration membrane within the renal corpuscle. If the condition is temporary, it is probably an acute glomerular nephritis usually associated with a bacterial infection such as streptococcal sore throat. If the condition is long term, resulting in a nonfunctional kidney, it is referred to as chronic glomerular nephritis.
    b. The plasma proteins in the filtrate increase the osmolarity of the filtrate, causing the urine volume to be greater than normal.

3. a. Filtration
    b. Primary site of nutrient reabsorption
    c. Primary site for secretion of substances into the filtrate
    d. The nephron loop and collecting system interact to regulate the amount of water and the number of sodium and potassium ions lost in the urine.

4. a.
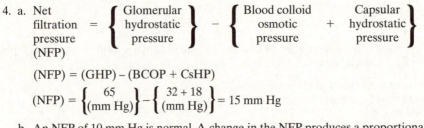

$$\text{Net filtration pressure (NFP)} = \left\{ \begin{array}{c} \text{Glomerular} \\ \text{hydrostatic} \\ \text{pressure} \end{array} \right\} - \left\{ \begin{array}{c} \text{Blood colloid} \\ \text{osmotic} \\ \text{pressure} \end{array} + \begin{array}{c} \text{Capsular} \\ \text{hydrostatic} \\ \text{pressure} \end{array} \right\}$$

$$(\text{NFP}) = (\text{GHP}) - (\text{BCOP} + \text{CsHP})$$

$$(\text{NFP}) = \left\{ \begin{array}{c} 65 \\ (\text{mm Hg}) \end{array} \right\} - \left\{ \begin{array}{c} 32 + 18 \\ (\text{mm Hg}) \end{array} \right\} = 15 \text{ mm Hg}$$

    b. An NFP of 10 mm Hg is normal. A change in the NFP produces a proportional change in the GFR.
    c. An NFP of 15 mm Hg indicates some type of kidney disease. Problems that affect filtration pressure disrupt kidney function and cause a variety of clinical signs and symptoms.

5. Strenuous exercise causes sympathetic activation to produce powerful vasoconstriction of the afferent arteriole, which delivers blood to the renal capsule. This causes a decrease in the GFR and alters the GFR by changing the required pattern of blood circulation. Dilation of peripheral blood vessels during exercise shunts blood away from the kidney and the GFR declines. A decreased GFR slows the production of filtrate. As the blood flow increases to the skin and skeletal muscles, kidney perfusion gradually declines and potentially dangerous conditions develop as the circulating concentration of metabolic wastes increases and peripheral water losses mount.

Chapter 27:
Fluid, Electrolyte, and Acid–Base Balance

LEVEL 1: REVIEWING FACTS AND TERMS

Multiple Choice

1. b	7. d	13. c	19. a	24. c	29. c
2. d	8. b	14. b	20. b	25. b	30. d
3. b	9. d	15. c	21. d	26. d	31. a
4. b	10. d	16. a	22. c	27. b	32. a
5. d	11. c	17. c	23. b	28. a	33. b
6. a	12. a	18. d			

Completion

1. electrolyte
2. fluid
3. osmoreceptors
4. aldosterone
5. antidiuretic hormone
6. fluid shift
7. hypertonic
8. hypotonic
9. edema
10. net hydrostatic pressure
11. colloid osmotic pressure
12. kidneys
13. calcium
14. buffers
15. hemoglobin
16. respiratory compensation
17. renal compensation
18. acidosis
19. alkalosis
20. hypercapnia
21. skeletal mass

Matching

1. I
2. K
3. E
4. D
5. B
6. H
7. C
8. J
9. F
10. A
11. G
12. P
13. Q
14. L
15. N
16. U
17. R
18. T
19. M
20. O
21. S

Drawing/Illustration Labeling

Figure 27-1   The pH Scale

1. pH 6.80
2. acidosis
3. pH 7.35
4. pH 7.45
5. alkalosis
6. pH 7.80

LEVEL 2: REVIEWING CONCEPTS

Chapter Overview

(A)

1. increased P_{co_2}
2. acidosis
3. H^+; HCO_3^-
4. increased
5. decreased

(B)

1. decreased P_{co_2}
2. alkalosis
3. reabsorption; secretion
4. respiratory rate
5. P_{co_2}

Concept Maps

I Homeostasis of Total Volume of Body Water

1. ↓ volume of body H_2O
2. ↑ H_2O retention at kidneys
3. aldosterone secretion by adrenal cortex

II Fluid and Electrolyte Imbalance

1. ↓ pH
2. ECF hypotonic to ICF
3. ↓ ECF volume
4. ↑ ICF volume

III Respiratory Mechanisms for Control of pH

1. ↑ blood CO_2
2. ↑ depth of breathing
3. hyperventilation
4. ↑ blood pH
5. normal blood pH

IV Urinary Mechanisms for Maintaining Homeostasis of Blood pH

1. ↓ blood pH
2. HCO_3^-
3. ↑ blood pH

V Homeostasis Fluid Volume Regulation–Sodium Ion Concentrations

1. ↓ B.P. at kidneys
2. ↑ aldosterone release
3. ↓ H_2O loss
4. ↑ plasma volume
5. ↑ ANP release
6. ↓ ADH release
7. ↓ aldosterone release
8. ↑ H_2O loss

Multiple Choice

1. b	5. d	9. a	13. b	17. c
2. d	6. d	10. b	14. c	18. b
3. c	7. b	11. c	15. c	19. c
4. a	8. d	12. d	16. a	20. d

Completion

1. kidneys
2. angiotensin II
3. volatile acid
4. fixed acids

5. organic acids
6. buffer system
7. respiratory acidosis
8. hypoventilation

9. hyperventilation
10. lactic acidosis
11. ketoacidosis
12. alkaline tide

Short Essay

1. a. Fluid balance
   b. Electrolyte balance
   c. Acid–base balance

2. a. Antidiuretic hormone (ADH)
   b. Aldosterone
   c. Atrial natriuretic peptide (ANP)

3. a. It stimulates water conservation at the kidney, reducing urinary water losses.
   b. It stimulates the thirst center to promote the drinking of fluids. The combination of decreased water loss and increased water intake gradually restores normal plasma osmolarity.

4. a. Alterations in the potassium ion concentration in the ECF
   b. Changes in pH
   c. Aldosterone levels

5. a. An initial fluid shift into or out of the ICF
   b. "Fine tuning" via changes in circulating levels of ADH

6. $CO_2 + H_2O \rightarrow H_2CO_3 \rightarrow H^+ + HCO_3^-$

7. a. Protein buffer system, phosphate buffer system, and carbonic acid–bicarbonate buffer system
   b. Respiratory mechanisms, renal mechanisms

8. a. Secrete or absorb hydrogen ions
   b. Control excretion of acids and bases
   c. Generate additional buffers

9. Hypercapnia— $\uparrow$ plasma P_{CO_2}; $\downarrow$ plasma pH—respiratory acidosis

   Hypocapnia— $\downarrow$ plasma P_{CO_2}; $\uparrow$ plasma pH—respiratory alkalosis

10. a. Impaired ability to excrete H^+ at the kidneys
    b. Production of a large quantity of fixed and/or organic acids
    c. Severe bicarbonate loss

LEVEL 3: CRITICAL THINKING AND CLINICAL APPLICATIONS

1. The comatose teenager's ABG studies reveal a severe respiratory acidosis. A pH of 7.17 (low) and a P_{CO_2} of 73 mm Hg (high) cause his respiratory centers to be depressed, resulting in hypoventilation, CO_2 retention, and consequent acidosis. His normal HCO_3^- value indicates that his kidneys haven't had time to retain significant amounts of HCO_3^- to compensate for the respiratory condition.

2. The 62-year-old woman's ABG studies reveal that she had metabolic alkalosis. A pH of 7.65 (high) and an HCO_3^- of 55 mEq/liter (high) are abnormal values. Her P_{CO_2} of 52 mm Hg indicates that her lungs are attempting to compensate for the alkalosis by retaining CO_2 in an effort to balance the HCO_3^- value. Her vomiting caused a large acid loss from her body via HCl, which means a loss of H^+, the acid ion. Predictable effects include slow respirations, an overexcitable CNS, leading to irritability, and, if untreated, possible tentany and convulsions.

3. a. Metabolic acidosis
   b. Respiratory alkalosis
   c. Respiratory acidosis
   d. Metabolic alkalosis

4. Total body water content gradually decreases with age. The body's ability to concentrate urine declines, so more water is lost in urine. In addition, the rate of insensible perspiration increases as the skin becomes thinner and more delicate. A reduction in ADH and aldosterone sensitivity makes older people less able than younger people to conserve body water when losses exceed gains.

Chapter 28:
The Reproductive System

LEVEL 1: REVIEWING FACTS AND TERMS

Multiple Choice

1. d	8. d	15. c	22. a	28. a
2. b	9. c	16. b	23. c	29. b
3. d	10. d	17. a	24. d	30. c
4. c	11. d	18. c	25. d	31. b
5. a	12. a	19. b	26. b	32. c
6. a	13. d	20. b	27. d	33. c
7. b	14. b	21. c		

Completion

1. fertilization
2. gonads
3. seminiferous tubules
4. testes
5. spermiogenesis
6. spermatids
7. ductus deferens
8. raphe
9. prostate gland
10. fructose
11. LH
12. ovaries
13. graafian
14. ovulation
15. infundibulum
16. uterus
17. oogenesis
18. greater vestibular
19. inhibin
20. placenta
21. orgasm
22. menopause
23. muscular

Matching

1. J	6. I	11. H	16. R	20. V
2. F	7. A	12. N	17. M	21. S
3. C	8. B	13. P	18. O	22. U
4. G	9. K	14. T	19. Q	23. W
5. E	10. D	15. L		

Drawing/Illustration Labeling

Figure 28-1   Male Reproductive Organs (Sagittal Section)

1. pubic symphysis
2. prostatic urethra
3. corpus cavernosum
4. corpus spongiosum
5. spongy urethra
6. ductus deferens
7. penis
8. epididymis
9. testis
10. external urethral orifice
11. scrotum
12. ureter
13. urinary bladder
14. seminal gland
15. rectum
16. prostate gland
17. ejaculatory duct
18. bulbo-urethral gland
19. anus

Figure 28-2 The Testes and Epididymis

1. efferent ductules
2. spermatic cord
3. head of epididymis
4. ductus deferens
5. rete testis
6. seminiferous tubule
7. tunica albuginea
8. body of epididymis
9. tail of epididymis

Figure 28-3 Female External Genitalia

1. mons pubis
2. glans of clitoris
3. labia minora
4. vestibule
5. hymen (torn)
6. prepuce of clitoris
7. urethral opening
8. labia majora
9. vaginal entrance
10. anus

Figure 28-4 Female Reproductive System (Sagittal Section)

1. ovarian follicle
2. ovary
3. uterine tube
4. uterus
5. urinary bladder
6. pubic symphysis
7. urethra
8. clitoris
9. labium minus
10. labium majus
11. sigmoid colon
12. fornix
13. cervix
14. vagina
15. rectum
16. anus

Figure 28-5 Female Reproductive Organs (Posterior View)

1. ampulla
2. isthmus
3. fundus of uterus
4. uterine tube
5. Suspensory ligament of the ovary
6. ovarian ligament
7. round ligament of uterus
8. cervix
9. infundibulum
10. fimbriae
11. perimetrium
12. myometrium
13. endometrium
14. internal os of uterus
15. isthmus of uterus
16. cervical canal
17. external os of uterus
18. vaginal rugae

Figure 28-6 Spermatogenesis

1. spermatogonia
2. primary spermatocyte
3. secondary spermatocyte
4. spermatids
5. spermatozoa

Figure 28-7 Oogenesis

1. oogorium
2. primary oocyte
3. secondary oocyte
4. mature ovum

LEVEL 2: REVIEWING CONCEPTS

Chapter Overview

1. seminiferous tubules and rete testis
2. body of epididymis
3. ductus deferens
4. ejaculatory duct
5. urethra
6. spongy urethra
7. external urethral orifice

Concept Maps

I Male Reproductive Tract

1. ductus deferens
2. penis
3. seminiferous tubules
4. produce testosterone
5. FSH
6. seminal vesicles
7. bulbo-urethral glands
8. urethra

II Penis

1. crus
2. body
3. corpus spongiosum
4. prepuce
5. external urethral orifice
6. frenulum

III Male Hormone Regulation

1. anterior pituitary
2. FSH
3. testes
4. interstitial cells
5. inhibin
6. CNS
7. male secondary sex characteristics

IV Female Reproductive Tract

1. uterine tubes
2. follicles
3. granulosa and thecal cells
4. endometrium
5. supports fetal development
6. vagina
7. vulva
8. labia majora and minora
9. clitoris
10. nutrients

V Female Hormone Regulation

1. GnRH
2. LH
3. follicles
4. progesterone
5. bone and muscle growth
6. accessory glands and organ

Multiple Choice

1. a
2. c
3. a
4. c
5. d
6. b
7. b
8. c
9. b
10. d
11. d
12. c
13. b
14. d
15. a
16. d
17. b
18. b
19. d

Completion

1. androgens
2. inguinal canals
3. raphe
4. cremaster
5. ejaculation
6. rete testis
7. acrosome
8. prepuce
9. smegma
10. detumescence
11. menopause
12. zona pellucida
13. corona radiata
14. corpus luteum
15. corpus albicans
16. mesovarium
17. tunica albuginea
18. fimbriae
19. cervical os
20. fornix
21. hymen
22. clitoris
23. menses
24. ampulla

Short Essay

1. a. Maintenance of the blood–testis barrier
   b. Support of spermiogenesis
   c. Secretion of inhibin
   d. Secretion of androgen-binding protein

2. a. It monitors and adjusts the composition of tubular fluid.
   b. It acts as a recycling center for damaged spermatozoa.
   c. It is the site of physical maturation of spermatozoa.

3. a. Seminal vesicles, prostate gland, bulbo-urethral glands
   b. Activating the sperm, providing nutrients for sperm motility, propelling sperm and fluids, producing buffers to counteract acid conditions

4. Seminal fluid is the fluid component of semen. Semen consists of seminal fluid, sperm, and enzymes.

5. *Emission* involves peristaltic contractions of the ampulla, pushing fluid and spermatozoa into the prostatic urethra. Contractions of the seminal vesicles and prostate gland move the seminal mixture into the membranous and penile walls of the prostate gland.

   Ejaculation occurs as powerful, rhythmic contractions of the ischiocavernosus and bulbocavernosus muscles push semen toward the external urethral orifice.

6. a. Promotes the functional maturation of spermatozoa
   b. Maintains accessory organs of male reproductive tract
   c. Responsible for male secondary sexual characteristics
   d. Stimulates bone and muscle growth
   e. Stimulates sexual behaviors and sexual drive

7. a. Serves as a passageway for the elimination of menstrual fluids
    b. Receives penis during coitus; holds sperm prior to passage into uterus
    c. In childbirth, forms the lower portion of the birth canal

8. a. *Arousal*—parasympathetic activation leads to an engorgement of the erectile tissues of the clitoris and increased secretion of the greater vestibular glands.
    b. *Coitus*—rhythmic contact with the clitoris and vaginal walls provides stimulation that eventually leads to orgasm.
    c. *Orgasm*—accompanied by peristaltic contractions of the uterine and vaginal walls and rhythmic contractions of the bulbocavernosus and ischiocavernosus muscles giving rise to pleasurable sensations.

9. 1. Formation of primary follicles
    2. Formation of secondary follicle
    3. Formation of a tertiary follicle
    4. Ovulation
    5. Formation of the corpus luteum
    6. Formation of the corpus albicans

10. Estrogens:
    a. stimulate bone and muscle growth;
    b. maintain female secondary sex characteristics;
    c. stimulate sex-related behaviors and drives;
    d. maintain functional accessory reproductive glands and organs;
    e. initiate repair and growth of the endometrium.

11. a. Menses
    b. Proliferative phase
    c. Secretory phase

LEVEL 3: CRITICAL THINKING AND CLINICAL APPLICATIONS

1. a. The normal temperature of the testes in the scrotum is 1–2 degrees lower than the internal body temperature—the ideal temperature for developing sperm. Mr. Hurt's infertility is caused by the inability of sperm to tolerate the higher temperature in the abdominopelvic cavity.
    b. Three major factors are necessary for fertility in the male:
       • adequate motility of sperm—30%–35% motility necessary
       • 20–100 million/m*l* of semen—adequate numbers of sperm
       • 60 percent of sperm must be normal in appearance

2. This 19-year-old female likely has a very different mix of sex hormones than her peers. Females, like males, secrete estrogens and androgens; however, in females, estrogen secretion usually "masks" the amount of androgen secreted in the body. In females, an excess of testosterone secretion may cause a number of conditions, such as sterility, fat distribution like a male, facial hair, low-pitched voice, skeletal muscle enlargement, clitoral enlargement, and a diminished breast size.

3. The contraceptive pill decreases the stimulation of FSH and prevents ovulation. It contains large quantities of progesterone and a small quantity of estrogen. It is usually taken for 20 days beginning on day 5 of a 28-day cycle. The increased level of progesterone and decreased levels of estrogen prepare the uterus for egg implantation. On day 26 the progesterone level decreases. If taken as directed, the pill will allow for a normal menstrual cycle.

4. In males, the disease-causing organism can move up the urethra to the bladder or into the ejaculatory duct to the ductus deferens. There is no direct connection into the pelvic cavity in the male. In females, the pathogen travels from the vagina to the uterus, to the uterine tubes, and into the pelvic cavity, where it can infect the peritoneal lining, resulting in peritonitis.

Chapter 29:
Development and Inheritance

LEVEL 1: REVIEWING FACTS AND TERMS

Multiple Choice

1. b	7. d	13. c	19. d	24. a	29. a
2. d	8. c	14. b	20. c	25. c	30. b
3. c	9. b	15. d	21. c	26. c	31. c
4. a	10. b	16. b	22. a	27. d	32. c
5. b	11. a	17. d	23. d	28. c	33. a
6. a	12. c	18. a			

Completion

1. development
2. embryological development
3. fertilization
4. capacitation
5. polyspermy
6. induction
7. second trimester
8. first trimester
9. yolk sac
10. chorion
11. human chorionic gonadotropin
12. placenta
13. true labor
14. parturition
15. expulsion
16. childhood
17. infancy
18. meiosis
19. gametogenesis
20. autosomal
21. homozygous
22. heterozygous

Matching

1. G	6. A	11. M	16. S	21. N
2. B	7. K	12. D	17. U	22. V
3. H	8. C	13. J	18. T	23. P
4. E	9. F	14. X	19. R	24. Q
5. L	10. I	15. W	20. O	

Drawing/Illustration Labeling

Figure 29-1   Major Forms of Inheritance

1. autosomal
2. sex
3. simple
4. polygenic
5. strict dominance
6. codominance
7. incomplete dominance

LEVEL 2: REVIEWING CONCEPTS

Chapter Overview

1. secondary oocyte
2. fertilization
3. zygote
4. 2-cell stage
5. 8-cell stage
6. morula
7. early blastocyst
8. implantation

Concept Maps
I Initiation of Labor and Delivery

1. placenta
2. fetal oxytocin release
3. estrogen
4. distortion of myometrium
5. maternal oxytocin
6. positive feedback
7. parturition

II Hormones During Pregnancy

1. anterior pituitary
2. ovary
3. progesterone
4. placenta
5. relaxin
6. mammary gland development

Multiple Choice

1. b	5. a	9. a	13. b	17. d
2. c	6. b	10. d	14. d	18. b
3. d	7. d	11. c	15. c	19. d
4. c	8. b	12. c	16. a	20. b

Completion

1. inheritance
2. genetics
3. chromatids
4. synapsis
5. tetrad
6. genetic recombination
7. spontaneous mutations
8. amniocentesis
9. alleles
10. X-linked
11. simple inheritance
12. polygenic inheritance
13. corona radiata
14. hyaluronidase
15. cleavage
16. chorion
17. differentiation
18. activation

Short Essay

1. In simple inheritance, phenotypic characters are determined by interactions between a single pair of alleles.

   Polygenic inheritance involves interactions between alleles on several genes.

2. Capacitation is the activation progress that must occur before a spermatozoon can successfully fertilize an egg. It occurs in the vagina following ejaculation.

3. (a) Cleavage, (b) implantation, (c) placentation, (d) embryogenesis

4. (a) Ectoderm, (b) mesoderm, (c) endoderm

5.
(a)	(b)
• yolk sac	• endoderm and mesoderm
• amnion	• ectoderm and mesoderm
• allantois	• endoderm and mesoderm
• chorion	• mesoderm and trophoblast

6. a. The respiratory rate goes up and the tidal volume increases.
   b. The maternal blood volume increases.
   c. The maternal requirements for nutrients increase.
   d. The glomerular filtration rate increases.
   e. The uterus increases in size.

7. (a) Estrogens, (b) oxytocin, (c) prostaglandins

8. a. Secretion of relaxin by the placenta—softens symphysis pubis
   b. Weight of the fetus—deforms cervical orifice
   c. Rising estrogen levels
   d. Both b and c promote release of oxytocin.

9. (a) Dilation stage, (b) expulsion stage, (c) placental stage

10. Neonatal, infancy, childhood, adolescence, maturity, senescence

11. a. Hypothalamus—increasing production of GnRH
    b. Increasing circulatory levels of FSH and LH by the anterior pituitary
    c. FSH and LH initiate gametogenesis and the production of male or female sex hormones that stimulate the appearance of secondary sexual characteristics and behaviors.

12. a. Some cell populations grow smaller throughout life.
    b. The ability to replace other cell populations decreases.
    c. Genetic activity changes over time.
    d. Mutations occur and accumulate.

LEVEL 3: CRITICAL THINKING AND CLINICAL APPLICATIONS

1. Color blindness is an X-linked trait. The Punnett square shows that sons produced by a normal father and a heterozygous mother will have a 50 percent chance of being color blind, whereas the daughters will all have normal color vision.

Maternal alleles

		X^C	X^c
Paternal alleles	X^C	$X^C X^C$	$X^C X^c$
	Y	$X^C Y$	$X^c Y$ (color blind)

2. The Punnett square reveals that 50 percent of their offspring have the possibility of inheriting albinism.

Maternal alleles

		a	a
Paternal alleles	A	Aa	Aa
	a	aa (albino)	aa (albino)

3. Both the mother and father are heterozygous-dominant.

   T—tongue roller; t—non-tongue roller

   The Punnett square reveals that there is a 25 percent chance of having children who are not tongue rollers and a 75 percent chance of having children with the ability to roll the tongue.

Maternal alleles

		T	t
Paternal alleles	T	TT (yes)	Tt (yes)
	t	Tt (yes)	tt (no)

4. Risks to the fetus are higher in breech births than in normal deliveries, because the umbilical cord can become constricted, cutting off placental blood flow. The head is normally the widest part of the fetus; the mother's cervix may dilate enough to pass the baby's legs and body, but not the head. This entrapment compresses the umbilical cord, prolongs delivery, and subjects the fetus to severe distress and potential injury. If attempts to reposition the fetus or promote further dilation are unsuccessful over the short term, delivery by Caesarean section may be required.